Green Polymer Chemistry:
Biocatalysis and Biomaterials

ACS SYMPOSIUM SERIES **1043**

Green Polymer Chemistry: Biocatalysis and Biomaterials

H. N. Cheng, Editor
Southern Regional Research Center
USDA - Agricultural Reseach Service

Richard A. Gross, Editor
Polytechnic Institute of New York University (NYU-POLY)

Sponsored by the
ACS Division of Polymer Chemistry

American Chemical Society, Washington, DC

Library of Congress Cataloging-in-Publication Data

100699 3324
Green polymer chemistry : biocatalysis and biomaterials / H. N. Cheng, Richard A. Gross, editors.
     p. cm. -- (ACS symposium series ; 1043)
  Includes bibliographical references and index.
  ISBN 978-0-8412-2581-7 (alk. paper)
  1. Biodegradable plastics--Congresses. 2. Environmental chemistry--Industrial applications--Congresses. 3. Biopolymers--Congresses. I. Cheng, H. N. II. Gross, Richard A., 1957-
  TP1180.B55G74 2010
  547'.7--dc22
                                        2010023453

The paper used in this publication meets the minimum requirements of American National Standard for Information Sciences—Permanence of Paper for Printed Library Materials, ANSI Z39.48n1984.

Foreword

The ACS Symposium Series was first published in 1974 to provide a mechanism for publishing symposia quickly in book form. The purpose of the series is to publish timely, comprehensive books developed from the ACS sponsored symposia based on current scientific research. Occasionally, books are developed from symposia sponsored by other organizations when the topic is of keen interest to the chemistry audience.

Before agreeing to publish a book, the proposed table of contents is reviewed for appropriate and comprehensive coverage and for interest to the audience. Some papers may be excluded to better focus the book; others may be added to provide comprehensiveness. When appropriate, overview or introductory chapters are added. Drafts of chapters are peer-reviewed prior to final acceptance or rejection, and manuscripts are prepared in camera-ready format.

As a rule, only original research papers and original review papers are included in the volumes. Verbatim reproductions of previous published papers are not accepted.

ACS Books Department

Contents

Novel Biobased Materials

New or Improved Biocatalysts

Syntheses of Polyesters and Polycarbonates

Syntheses of Polyamides and Polypeptides

Syntheses and Modifications of Polysaccharides

Biocatalytic Redox Polymerizations

Enzymatic Hydrolyses and Degradations

Grafting and Functionalization Reactions

Indexes

Preface

Green Polymer Chemistry is a crucial area of research and product development that continues to grow in its influence over industrial practices. Developments in these areas are driven by environmental concerns, interest in sustainability, desire to decrease our dependence on petroleum, and commercial opportunities to develop "green" products. Publications and patents in these fields are increasing as more academic, industrial, and government scientists become involved in research and commercial activities.

The purpose of this book is to publish new work from a cutting-edge group of leading international researchers from academia, government, and industrial institutions. Because of the multidisciplinary nature of Green Polymer Chemistry, corresponding publications tend to be spread out over numerous journals. This book brings these papers together so that the reader can gain a better appreciation of the breadth and depth of activities in Green Polymer Chemistry.

This book is based on contributions by oral and poster presenters at the international symposium, Biocatalysis in Polymer Science, held at the ACS National Meeting in Washington D.C. on August 17-20, 2009. Whereas many aspects of Green Polymer Chemistry were covered during the symposium, a particular emphasis was placed on biocatalysis and biobased materials. Many exciting new findings in basic research and applications were reported. In addition, several leaders in these areas who were unable to attend the symposium contributed important reviews of their ongoing work. As a result this book provides a good representation of activities at the forefront of research in Green Polymer Chemistry emphasizing activities in biocatalysis and biobased chemistry.

This book will be useful to scientists and engineers (chemists, biochemists, chemical engineers, biochemical engineers, material scientists, microbiologists, molecular biologists, and enzymologists) as well as graduate students who are engaged in research and developments in polymer biocatalysis and biomaterials. It can also be a useful reference book for those interested in these topics.

We thank the authors for their timely contributions and their cooperation while the manuscripts were being reviewed and revised. In addition we also thank the ACS Division of Polymer Chemistry, Inc. for sponsoring the 2009 symposium and providing generous funding for the symposium.

H. N. Cheng

Southern Regional Research Center
USDA – Agricultural Research Service
1100 Robert E. Lee Blvd.
New Orleans, LA 70124

Richard A. Gross

Herman F. Mark Professor
Director: NSF I/UCRC for Biocatalysis and Bioprocessing of Macromolecules
Polytechnic Institute of NYU (NYU-POLY)
Six Metrotech Center
Brooklyn, NY 11201

Chapter 1

Green Polymer Chemistry: Biocatalysis and Biomaterials[‡]

H. N. Cheng[1,*] and Richard A. Gross[2]

[1]Southern Regional Research Center, USDA/Agriculture Research Service,
1100 Robert E. Lee Blvd., New Orleans, LA 70124
[2]NSF I/UCRC for Biocatalysis and Bioprocessing of Macromolecules,
Polytechnic Institute of NYU (NYU-POLY), Six Metrotech Center, Brooklyn,
NY 11201, http://www.poly.edu/grossbiocat/
*hn.cheng@ars.usda.gov
[‡]Names of products are necessary to report factually on available data;
however, the USDA neither guarantees nor warrants the standards of the
products, and the use of the name USDA implies no approval of the products
to the exclusion of others that may also be suitable.

This overview briefly surveys the practice of green chemistry in
polymer science. Eight related themes can be discerned from the
current research activities: 1) biocatalysis, 2) bio-based building
blocks and agricultural products, 3) degradable polymers,
4) recycling of polymer products and catalysts, 5) energy
generation or minimization during use, 6) optimal molecular
design and activity, 7) benign solvents, and 8) improved
synthesis to achieve atom economy, reaction efficiency, and
reduced toxicity. All of these areas are experiencing an increase
in research activity with the development of new tools and
technologies. Examples are given of recent developments in
green chemistry with a focus on biocatalysis and biobased
materials.

Introduction

Green chemistry is the design of chemical products and processes that reduce or eliminate the use or generation of hazardous substances (*1*). Sustainability refers to the development that meets the needs of the present without compromising the ability of future generations to meet their own needs (*2*). In the past few years these concepts have caught on and have become popular topics for research. Several books and review articles have appeared in the past few years (*3–6*).

In the polymer area, there is also increasing interest in green chemistry. This is evident by many recent symposia organized on this topic at national ACS meetings. In our view, developments in green polymer chemistry can be roughly grouped into the following eight related themes. These eight themes also agree well with most of the themes described in recent articles and books on green chemistry (*3–6*).

1) Greener catalysts (e.g., biocatalysts such as enzymes and whole cells)
2) Diverse feedstock base (especially agricultural products and biobased building blocks)
3) Degradable polymers and waste minimization
4) Recycling of polymer products and catalysts (e.g., biological recycling)
5) Energy generation or minimization of use
6) Optimal molecular design and activity
7) Benign solvents (e.g., water, ionic liquids, or reactions without solvents)
8) Improved syntheses and processes (e.g., atom economy, reaction efficiency, toxicity reduction)

In this article, we provide an overview of green polymer chemistry, with a particular emphasis on biocatalysis (*7, 8*) and biobased materials (*9, 10*). Examples are taken from the recent literature, especially articles in this symposium volume (*11–39*) and the preprints (*40–62*) from the international symposium on "Biocatalysis in Polymer Science" at the ACS national meeting in Washington, DC in August 2009.

Green Polymer Chemistry - The Eightfold Path

Biocatalysts

Biocatalysis is an up-and-coming field that has attracted the attention and participation of many researchers. Several reviews (*7*) and books (*8*) are available on biocatalysis. This current symposium volume documents important new research that uses biocatalysis and biobased materials as tools to describe practical and developing strategies to implement green chemistry practices. A total of 22 articles (and 17 symposium preprints) describe biocatalysis and biotransformations. Among these papers, 31 articles focus on cell-free enzyme catalysts and 8 utilize whole-cell catalysts to accomplish biotransformations.

Biobased Materials

Interest in biobased materials (*9, 10*) appears to be increasing proportionally with increases or increased volatility of crude oil prices. There is also general recognition that the resources of the world are limited, and *sustainability* has become a rallying point for many organizations and industries participating in chemical product development. Thus, there is growing interest in using readily renewable materials as ingredients for commercial products or raw materials for synthesis and polymerization. In this book, 14 articles deal with biobased raw materials or products. In addition, 11 symposium preprints focus on this topic.

Degradable Polymers and Waste Minimization

One advantage of agricultural raw materials and bio-based building blocks is that they are potentially biodegradable and have less negative environmental impact. In addition to the potential economic benefits, the use of agricultural by-products minimizes waste and mitigates disposal problems. Biocatalysis is helpful in this effort because enzyme-catalysts often catalyzed reactions of natural substrates at high rates. Many biobased products are biodegradable, and hydrolytic enzymes are critically important for the break down of biomass to usable building blocks for fermentation processes. Four of the articles in this book deal specifically with polymer degradation and hydrolysis (*20, 34–36*). In addition, most of the polymers described in this book (polyesters, polyamides, polypeptides, polysaccharides) are biodegradable or potentially biodegradable.

Recycling

Another advantage of agricultural raw materials and bio-based building blocks is that they can often be recycled. Some resulting polymers that are biodegradable can undergo biological recycling by which they are converted to biomass, CO_2, CH_4 (anaerobic conditions) and water. Recycling is also important for biocatalysts in order to decrease process cost; this is one of the reasons for the use of immobilized enzymes. Several examples of immobilized enzymes appear in this book (*vide infra*). A popular enzyme used thus far is Novozym® 435 lipase from Novozymes A/S, which is an immobilized lipase from *Candida antarctica*.

Energy Generation and Minimization of Use

An active area of research is biofuels, and many review articles are available (*63, 64*). First generation products have largely been based on biotransformation of sugars and starch. The second-generation products, based on lignocelluloses conversion to sugars, are still under development. Biocatalysis is compatible with energy savings because their use often involves lower reaction temperatures and, therefore, lower energy input (e.g., refs. (*27, 48*)). The reactive extrusion technique is another process methodology that can decrease energy use (*38, 39*).

Molecular Design and Activity

In polymer science, structure-property and structure-activity correlations are often employed as part of synthetic design, and many articles on synthesis in this book inherently incorporated this feature. In biochemistry, a good example of molecular design is the development via protein engineering of protein variants that are optimized for a particular activity or characteristic (e.g. thermal stability). For example, Kiick (40) used *in vivo* methods to produce resilin, and McChalicher and Srienc (50) used site-specific mutagenesis for the synthase that produces poly(hydroxyalkanoate)s. In a different way, Ito et al (18) used molecular recognition to optimize biological activity of aptamers. Li et al (29) used biopathway engineering to produce lipopolysaccharides and their analogs.

Benign Solvents

A highly desirable goal of green chemistry is to replace organic solvents in chemical reactions with water. Biocatalytic reactions are highly suited for this. In fact, all whole-cell biotransformations and many enzymatic reactions in this book are performed in aqueous media. An alternative is to carry out the reaction without any solvents, as exemplified by several articles in this book.

Improved Syntheses and Processes

Optimization of experimental parameters in synthesis and process improvement during scale-up and commercialization are part of the work that synthetic scientists and engineers do. Biocatalysis certainly brings a new dimension to reactions and processes. Biocatalytic reactions often involve fewer by-products and less (or no) toxic chemical reagents. Several new or improved synthetic and process methodologies are described in the following sections. In addition, it is noteworthy that Matos et al (52) used microwave energy to assist in lipase-catalyzed polymerization, and Fishman et al (15) used microwave for extraction. Wang et al (38, 39) used reactive extrusion to facilitate polymer modification reactions.

From the foregoing discussion, it is clear that *biocatalysis* and *biobased materials* are major contributors to current research and development activities in green polymer chemistry. Active researchers in these fields have been working with different polymers, different biocatalysts, and different strategies. For convenience, the rest of this review is divided into eight sections: 1) Novel Biobased Materials, 2) New or Improved Biocatalysts, 3) Synthesis of Polyesters and Polycarbonates, 4) Synthesis of Polyamides and Polypeptides, 5) Synthesis and Modification of Polysaccharides, 6) Biocatalytic Redox Polymerizations, 7) Enzymatic Hydrolysis and Degradation, and 8) Grafting and Functionalization Reactions.

Novel Biobased Materials

As noted earlier, biobased materials constitute one of the most active research areas today. These include polypeptides/proteins, carbohydrates, lipids/triglycerides, microbial polyesters, plant fibers, and many others.

Polypeptides/Proteins

An active area of research is to use polypeptides and proteins for various applications. Kiick (*40*) worked with resilin, the insect energy storage protein that shows useful mechanical properties. This work involved incorporation of unnatural amino acids to produce biomaterials for possible use in engineering the vocal folds (more commonly known as vocal cords). Liu et al (*12*) carried out biofabrication based on enzyme-catalyzed coupling and crosslinking of pre-formed biopolymers for potential use as medical adhesives. Renggli and Bruns (*11*) reviewed polymer-protein hybrid materials and their use as biomaterials and biocatalytic polymers. Zhang and Chen (*13*) made novel blends of soy proteins and biodegradable thermoplastics, which exhibit excellent mechanical properties. Jong (*57*) made composites from rubber and soy protein modified with phthalic anhydride and found they provide a significant reinforcement effect. Venkateshan and Sun (*14*) made urea-soy protein composites and characterized their thermodynamic behavior and structural changes.

Polysaccharides

DeAngelis (*30*) and Schwach-Abdellaoui et al (*31*) both worked with glycoaminoglycans, which are useful in drug delivery, implantable gels, and cell scaffolds. Li et al (*29*) carried out extensive work in *in vitro* biosynthesis of O-polysaccharides and *in vivo* production of liposaccharides. Bulone et al (*48*) described a low-energy biosynthetic approach for the production of high-strength nanopaper from compartmentalized bacterial cellulose fibers. Fishman et al (*15*) extracted polysaccharides from sugar beet pulp and extensively characterized the resulting fractions.

Lipids and Triglycerides

In their article, Lu and Larock (*16*) provided a good overview of their work on converting agricultural oils into plastics, rubbers, composites, coatings and adhesives. Zini, et al (*41*) reviewed their work on poly(sophorolipid) and its potential as a biomaterial. Lu, et al (*44*) produced new ω-hydroxy and ω-carboxy fatty acids as building blocks for functional polyesters.

Specialty Polymeric Materials

Dinu et al (*17*) made smart coatings by immobilizing enzymes on carbon nanotubes and incorporating them into latex paints. The resulting materials can detect and eliminate hazardous agents to combat chemical and biological agents.

Xue, at al (*61*) made poly(ester-urethanes) based on poly(ε-caprolactone) that exhibit shape-memory effect at body temperature. Rovira-Truitt and White (*43*) prepared poly(D,L-lactide)/tin-supported mesoporous nanocomposites by in-situ polymerization.

Biomaterials

Most of the above aforementioned materials can be used in medical and dental applications as *biomaterials*, e.g., tissue engineering, implants, molecular imprinting, stimuli responsive systems for drug delivery and biosensing. Moreover, the nucleic acid-based aptamer described by Ito et al (*18*) has potential use for biosensing, diagnostic, and therapeutics. In addition, most of the polyesters described in this book are biodegradable and also have potential use in medical applications. For example, polylactides and polyglycolides are well known bioresorbable polyesters used as sutures, stents, dialysis media, drug delivery devices and others.

New and Improved Biocatalysts

Not surprisingly, one of the active research areas of biocatalysis and biotransformation is the development of new and improved biocatalysts.

New or Improved Enzymes

Methods to improve protein activity, specificity, stability and other characteristics are rapidly developing both through high-throughput as well as information-rich small library strategies. An example was given by McChalicher and Srienc (*50*) who modified the synthase to facilitate the synthesis of poly(hydroxyalkanoate) (PHA). Ito et al (*18*) described a different class of enzymes ("aptazymes") based on oligonucleotides, which bind to hemin (iron-containing porphyrin) and also show peroxidase activity.

Ganesh and Gross (*34*) embedded enzymes within a bioresorbable polymer matrix, thereby demonstrating a new concept by which the lifetime of existing bioresorbable materials can be "fine tuned." Gitsov et al (*45*) made enzyme-polymer complexes that form "nanosponges." Renggli and Bruns (*11*) also made enzyme-polymer hybrid materials. Schoffelen et al (*19, 46*) developed a method to introduce an azide group onto an enzyme, which allowed subsequent coupling via click chemistry to other structures such as a polymer or enzyme(s) to facilitate reactions that require multiple enzymes. Immobilized enzymes were also used by a large number of authors in this book.

Whole Cell Approaches

Whole cell approaches were used by Yu (*21*) and by Smith (*47*) to produce PHA. Li et al (*29*) conducted *in vitro* biosynthesis of O-polysaccharides and *in vivo* production of liposaccharides. Schwach-Abdellaoui et al (*31*) used a transferred gene in *Bacillus subtilis* to produce hyaluronic acid through an advanced fermentation process. Bulone et al (*48*) produced cellulose nanofibrils via *Gluconacetobacter xylinus* in the presence of hydroethylcellulose. Lu et al (*44*) produced ω-hydroxy and ω-carboxy fatty acids by engineering a *Candida tropicalis* strain and the corresponding fermentation processes. Uses of ω-hydroxy and ω-carboxy fatty as biobased monomers for next-generation poly(hydroxyanoates) was discussed. In their review on Baeyer-Villiger biooxidation Lau et al (*33*) included whole cell approaches. In her article, Kawai (*36*) summarized microorganisms capable of degrading polylactic acid.

Syntheses of Polyesters and Polycarbonates

Many examples of biocatalytic routes to polyesters and polycarbonates are discussed in this book and corresponding symposium preprints. In order to facilitate accessing these contributions to the book, the specific polymers, biocatalysts and authors for each polymer system are summarized in the following Table 1.

Syntheses of Polyamides and Polypeptides

Resilin-like polypeptides were made via whole cell biocatalysis described by Kiick (*40*). Co-oligopeptides consisting of glutamate and leucine residues were prepared via protease catalysis by Li et al (*42*). Polyamides were synthesized via lipase catalysis by Gu et al (*49*), by Cheng and Gu (*27*), and by Loos et al (*53*). Palmans et al (*53*) used dynamic kinetic resolution method to form chiral esters and amides, which can potentially lead to chiral polyamides.

Syntheses and Modifications of Polysaccharides

As noted earlier, DeAngelis (*30*) and Schwach-Abdellaoui et al (*31*) both produced glycoaminoglycans through cell-free enzyme and whole-cell approaches, respectively, and Bulone et al (*48*) produced bacterial cellulose through a whole-cell approach. Li et al (*29*) produced O-polysaccharides and liposaccharides through *in vitro* and *in vivo* biosynthesis. Fishman et al (*15*) made carboxymethylcellulose with materials obtained from sugar beet pulp. In addition, Biswas et al (*56*) grafted polyacrylamide onto starch using horseradish peroxidase as a catalyst.

Biocatalytic Redox Polymerizations

In their chapter, Bouldin et al (*32*) provided a good review of the use of oxidoreductase as a catalyst for the synthesis of electrically conducting polymers based on aniline, pyrrole, and thiophene. In a preprint (*58*), they reported a low-temperature, template-assisted polymerization of pyrrole using soybean oxidase in an aqueous solvent system. In another study,

Table 1. Examples of polyester and polycarbonate synthesis via biocatalysis

Polymer[a]	Biocatalyst[b]	Authors	Ref.
PHA	Whole cell	Yu	(21)
PHA	Whole cell with mutant enzyme	McChalicher, Srienc	(50)
PHA (Mirel™)	Whole cell	Smith	(47)
Functional polycarbonates	Lipase (N-435)	Bisht, Al-Azemi	(22)
Copolymers of PDL, caprolactone, valerolactone, dioxanone, trimethylenecarbonate	Lipase (N-435)	Scandola et al	(23)
Poly(PDL-co-glycolate)	Lipase (N-435)	Jiang, Liu	(24)
Polycaprolactone	Embedded N-435	Ganesh, Gross	(34)
Polyol polyesters	Lipase (N-435)	Gross, Sharma	(51)
Polycaprolactone	Lipase (N-435)	Matos et al	(52)
Chiral polyesters	Lipase (N-435)	Palmans et al	(53)
Poly(carbonate-co-ester), terpolymer	Lipase (N-435)	Jiang et al	(54)
Poly(carbonate-co-ester), diblock	Lipase (N-435)	Dai et al	(55)
Poly(PDL-co-butylene-co-succinate)	Lipase (N-435)	Mazzocchetti et al	(60)
Polycaprolactone-based poly(ester-urethanes)	Lipase (N-435)	Xue et al	(61)
Polycaprolactone diol	Immobilized lipase from Y. lipolytica	Barrera-Rivera et al	(25)
Polyester elastomer from 12-hydroxystearate, itaconate, and 1,4-butanediol	Immobilized lipase from B. cepacia	Yasuda et al	(26)
Polyesters	(Immobilized) Cutinase	Baker, Montclare	(20)

[a] PHA = poly(hydroxyalkanoate), PDL = ω-pentadecalactone, [b] N-435 = Novozym® 435

Cruz-Silva et al (59) polymerized pyrrole using horseradish peroxidase/H_2O_2 and 2,2'-azino-bis(3-ethylbenzthiazoline-6-sulfonic acid) as mediator.

Lau et al (33) provided a useful review on Baeyer-Villiger biooxidative transformations, covering both cell-free enzyme and whole-cell approaches. Liu et al (12) used a tyranosinase to conjugate pre-formed biopolymers.

Enzymatic Hydrolyses and Degradation

Ganesh and Gross (*34*) demonstrated the concept of controlled biomaterial lifetime by embedded Novozym® 435 lipase into poly(ε-caprolactone). By using different quantities of embedded enzyme in films, they controlled the degradation rate and tuned the lifetime of these biomaterials. Ronkvist et al (*35*) discovered surprisingly rapid enzymatic hydrolysis of poly(ethylene terephthalate) using cutinases. The ability of cutinases to carry out polymer hydrolysis and degradation was also noted by Baker and Montclare (*20*) in their review on cutinase. A good review was provided by Kawai (*36*) on poly(lactic acid)-degrading microorganisms and depolymerases. Some proteases were found to be specific to poly(L-lactic acid), but lipases active for poly(lactic acid) hydrolysis preferred degrading poly(D-lactic acid).

Grafting and Functionalization Reactions

Puskas and Sen (*37*) used the immobilized lipase-catalyst system Novozym 435 to catalyze methacrylation of hydroxyl functionalized polyisobutylene and polydimethylsiloxane as well as conjugation of thymine onto poly(ethylene glycol). Wang and Schertz (*39*) grafted poly(lactic acid) onto poly(hydroxyalkanoate) using a reactive extrusion process. Wang and He (*38*) modified poly(lactic acid) and poly(butylene succinate) with a diol or a functionalized alcohol via a catalyst, also using a reactive extrusion process. Moreover, as noted earlier, polyacrylamide was grafted onto starch using horseradish peroxidase by Biswas et al (*56*).

References

1. http://en.wikipedia.org/wiki/Green_chemistry.
2. http://www.epa.gov/sustainability/basicinfo.htm.
3. Horvath, I. T.; Anastas, P. T. *Chem. Rev.* **2007**, *107*, 2169–2173.
4. Stevens, F. S. *Green Plastics: An Introduction to the New Science of Biodegradable Plastics*; Princeton University Press: Princeton, NJ, 2002.
5. Lancaster, M. *Green Chemistry: An Introductory Text*; Royal Society of Chemistry: Cambridge, U.K., 2002.
6. Matlack, A. S. *Introduction to Green Chemistry*; Marcel Dekker: New York, NY, 2001.
7. Reviews on biocatalysis include (a) Gross, R. A.; Kumar, A.; Kalra, B. *Chem. Rev.* **2001**, *101*, 2097−2124. (b) Kobayashi, S.; Uyama, H.; Kimura, S. *Chem. Rev.* **2001**, *101*, 3793−3818. (c) Kobayashi, S.; Makino, A. *Chem. Rev.* **2009**, *109*, 5288−5353.
8. Books on polymer biocatalysis include (a) *Polymer Biocatalysis and Biomaterials II*; ACS Symposium Series 999; Cheng, H. N.; Gross, R. A., Eds.; American Chemical Society: Washington, DC, 2008. (b) *Polymer Biocatalysis and Biomaterials*; ACS Symposium Series 900; Cheng, H. N.; Gross, R .A., Eds.; American Chemical Society: Washington, DC, 2005.

(c) *Biocatalysis in Polymer Science*; Gross, R. A.; Cheng, H. N., Eds.; American Chemical Society: Washington, DC, 2003.

9. Reviews on biobased materials include (a) Roach, P.; Eglin, D.; Rohde, K.; Perry, C. C. *J. Mater. Sci.: Mater. Med.* **2007**, *18*, 1263. (b) Meier, M. A. R.; Metzgerb, J. O.; Schubert, U. S. *Chem. Soc. Rev.* **2007**, *36*, 1788–1802. (c) Bhardwaj, R.; Mohanty, A. K. *J. Biobased Mater. Bioenergy* **2007**, *1* (2), 191.

10. Books on biobased materials include (a) *Biomaterials*, 2nd ed.; Bhat, S. V.; Alpha Science: 2005. (b) *Biorelated Polymers: Sustainable Polymer Science and Technology*; Chiellini, E.; Springer: 2001.

11. Renggli, K.; Bruns, N. Solid or Swollen Polymer-Protein Hybrid Materials. *Green Polymer Chemistry: Biocatalysis and Biomaterials*; ACS Symposium Series 1043 (this volume); American Chemical Society: Washington, DC, 2010; Chapter 2.

12. Liu, Y.; Yang, X.; Shi, X.; Bentley, W. E.; Payne, G. F. Biofabrication Based on the Enzyme-Catalyzed Coupling and Crosslinking of Pre-Formed Biopolymers. *Green Polymer Chemistry: Biocatalysis and Biomaterials*; ACS Symposium Series 1043 (this volume); American Chemical Society: Washington, DC, 2010; Chapter 3.

13. Zhang, J.; Chen, F. Development of Novel Soy Protein-Based Polymer Blends. *Green Polymer Chemistry: Biocatalysis and Biomaterials*; ACS Symposium Series 1043 (this volume); American Chemical Society: Washington, DC, 2010; Chapter 4.

14. Venkateshan, K.; Sun, X. S. Thermodynamic and Microscopy Studies of Urea-Soy Protein Composites. *Green Polymer Chemistry: Biocatalysis and Biomaterials*; ACS Symposium Series 1043 (this volume); American Chemical Society: Washington, DC, 2010; Chapter 5.

15. Fishman, M. L.; Cooke, P. H.; Hotchkiss, A. T., Jr. Extraction and Characterization of Sugar Beet Polysaccharides. *Green Polymer Chemistry: Biocatalysis and Biomaterials*; ACS Symposium Series 1043 (this volume); American Chemical Society: Washington, DC, 2010; Chapter 6.

16. Lu, Y.; Larock, R. C. Novel Biobased Plastics, Rubbers, Composites, Coatings and Adhesives from Agricultural Oils and By-Products. *Green Polymer Chemistry: Biocatalysis and Biomaterials*; ACS Symposium Series 1043 (this volume); American Chemical Society: Washington, DC, 2010; Chapter 7.

17. Dinu, C. Z.; Borkar, I. V.; Bale, S. S.; Zhu, G.; Sanford, K.; Whited, G.; Kane, R. S.; Dordick, J. S. Enzyme-Nanotube-Based Composites Used for Antifouling, Chemical and Biological Decontamination. *Green Polymer Chemistry: Biocatalysis and Biomaterials*; ACS Symposium Series 1043 (this volume); American Chemical Society: Washington, DC, 2010; Chapter 8.

18. Liu, M.; Abe, H.; Ito, Y. Hemin-Binding Aptamers and Aptazymes. *Green Polymer Chemistry: Biocatalysis and Biomaterials*; ACS Symposium Series 1043 (this volume); American Chemical Society: Washington, DC, 2010; Chapter 9.

19. Schoffelen, S.; Schobers, L.; Venselaar, H.; Vriend, G.; van Hest, C. C. M. Synthesis of Covalently Linked Enzyme Dimers. *Green Polymer Chemistry: Biocatalysis and Biomaterials*; ACS Symposium Series 1043 (this volume); American Chemical Society: Washington, DC, 2010; Chapter 10.

20. Baker, P. J.; Montclare, J. K. Biotransformations Using Cutinase. *Green Polymer Chemistry: Biocatalysis and Biomaterials*; ACS Symposium Series 1043 (this volume); American Chemical Society: Washington, DC, 2010; Chapter 11.

21. Yu, J. Biosynthesis of Polyhydroxyalkanoates from 4-Ketovaleric Acid in Bacterial Cells. *Green Polymer Chemistry: Biocatalysis and Biomaterials*; ACS Symposium Series 1043 (this volume); American Chemical Society: Washington, DC, 2010; Chapter 12.

22. Bisht, K. S.; Al-Azemi, T. F. Synthesis of Functional Polycarbonates from Renewable Resources. *Green Polymer Chemistry: Biocatalysis and Biomaterials*; ACS Symposium Series 1043 (this volume); American Chemical Society: Washington, DC, 2010; Chapter 13.

23. Scandola, M.; Focarete, M. L.; Gross, R. A. Polymers from Biocatalysis: Materials with a Broad Spectrum of Physical Properties. *Green Polymer Chemistry: Biocatalysis and Biomaterials*; ACS Symposium Series 1043 (this volume); American Chemical Society: Washington, DC, 2010; Chapter 14.

24. Jiang, Z.; Liu, J. Lipase-Catalyzed Copolymerization of ω-Pentadecalactone (PDL) and Alkyl Glycolate: Synthesis of Poly(PDL-co-GA). *Green Polymer Chemistry: Biocatalysis and Biomaterials*; ACS Symposium Series 1043 (this volume); American Chemical Society: Washington, DC, 2010; Chapter 15.

25. Barrera-Rivera, K. A.; Marcos-Fernández, A.; Martínez-Richa, A. Chemo-Enzymatic Syntheses of Polyester-Erethanes. *Green Polymer Chemistry: Biocatalysis and Biomaterials*; ACS Symposium Series 1043 (this volume); American Chemical Society: Washington, DC, 2010; Chapter 16.

26. Yasuda, M.; Ebata, H.; Matsumura, S. Enzymatic Synthesis and Properties of Novel Biobased Elastomers Consisting of 12-Hydroxystearate, Itaconate and Butane-1,4-diol. *Green Polymer Chemistry: Biocatalysis and Biomaterials*; ACS Symposium Series 1043 (this volume); American Chemical Society: Washington, DC, 2010; Chapter 17.

27. Cheng, H. N.; Gu, Q. Synthesis of Poly(aminoamides) via Enzymatic Means. *Green Polymer Chemistry: Biocatalysis and Biomaterials*; ACS Symposium Series 1043 (this volume); American Chemical Society: Washington, DC, 2010; Chapter 18.

28. Schwab, L. W.; Baum, I.; Fels, G.; Loos, K. Mechanistic Insight in the Enzymatic Ring-Opening Polymerization of β-Propiolactam. *Green Polymer Chemistry: Biocatalysis and Biomaterials*; ACS Symposium Series 1043 (this volume); American Chemical Society: Washington, DC, 2010; Chapter 19.

29. Li, L.; Yi, W.; Chen, W.; Woodward, R.; Liu, X.; Wang, P. G. Production of Natural Polysaccharides and Their Analogues via Biopathway Engineering. *Green Polymer Chemistry: Biocatalysis and Biomaterials*; ACS Symposium

Series 1043 (this volume); American Chemical Society: Washington, DC, 2010; Chapter 20.

30. DeAngelis, P. L. Glycosaminoglycan Synthases: Catalysts for Customizing Sugar Polymer Size and Chemistry. *Green Polymer Chemistry: Biocatalysis and Biomaterials*; ACS Symposium Series 1043 (this volume); American Chemical Society: Washington, DC, 2010; Chapter 21.

31. Schwach-Abdellaoui, K.; Fuhlendorff, B. L.; Longin, F.; Lichtenberg, J. Development and Applications of a Novel, First-in-Class Hyaluronic Acid from *Bacillus*. *Green Polymer Chemistry: Biocatalysis and Biomaterials*; ACS Symposium Series 1043 (this volume); American Chemical Society: Washington, DC, 2010; Chapter 22.

32. Bouldin, R.; Kokil, A.; Ravichandran, S.; Nagarajan, S.; Kumar, J.; Samuelson, L. A.; Bruno, F. F.; Nagarajan, R. Enzymatic Synthesis of Electrically Conducting Polymers. *Green Polymer Chemistry: Biocatalysis and Biomaterials*; ACS Symposium Series 1043 (this volume); American Chemical Society: Washington, DC, 2010; Chapter 23.

33. Lau, P. C. K.; Leisch, H.; Yachnin, B. J.; Mirza, I. A.; Berghuis, A. M.; Iwaki, H.; Hasegawa, Y. Sustained Development in Baeyer-Villiger Biooxidation Technology. *Green Polymer Chemistry: Biocatalysis and Biomaterials*; ACS Symposium Series 1043 (this volume); American Chemical Society: Washington, DC, 2010; Chapter 24.

34. Ganesh, M.; Gross, R. A. Embedding Enzymes to Control Biomaterial Lifetime. *Green Polymer Chemistry: Biocatalysis and Biomaterials*; ACS Symposium Series 1043 (this volume); American Chemical Society: Washington, DC, 2010; Chapter 25.

35. Ronkvist, Å. M.; Xie, W.; Lu, W.; Gross, R. A. Surprisingly Rapid Enzymatic Hydrolysis of Poly(ethylene terephthalate). *Green Polymer Chemistry: Biocatalysis and Biomaterials*; ACS Symposium Series 1043 (this volume); American Chemical Society: Washington, DC, 2010; Chapter 26.

36. Kawai, F. Polylactic Acid (PLA)-Degrading Microorganisms and PLA Depolymerases. *Green Polymer Chemistry: Biocatalysis and Biomaterials*; ACS Symposium Series 1043 (this volume); American Chemical Society: Washington, DC, 2010; Chapter 27.

37. Puskas, J. E.; Sen, M. Y. Green Polymer Chemistry: Enzymatic Functionalization of Liquid Polymers in Bulk. *Green Polymer Chemistry: Biocatalysis and Biomaterials*; ACS Symposium Series 1043 (this volume); American Chemical Society: Washington, DC, 2010; Chapter 28.

38. Wang, J. H.; He, A. Bio-Based and Biodegradable Aliphatic Polyesters Modified by a Continuous Alcoholysis Reaction. *Green Polymer Chemistry: Biocatalysis and Biomaterials*; ACS Symposium Series 1043 (this volume); American Chemical Society: Washington, DC, 2010; Chapter 29.

39. Wang, J. H.; Schertz, D. M. Synthesis of Grafted Polylactic Acid and Polyhydroxyalkanoate by a Green Reactive Extrusion Process. *Green Polymer Chemistry: Biocatalysis and Biomaterials*; ACS Symposium Series 1043 (this volume); American Chemical Society: Washington, DC, 2010; Chapter 30.

40. Kiick, K. Modular biomolecular materials for engineering mechanically active tissues. *ACS Polym. Prepr.* **2009**, *50* (2), 53.
41. Zini, E.; Gazzano, M.; Scandola, M.; Gross, R. A. Glycolipid biomaterials: Synthesis and solid-state properties of a poly(sophorolipid). *ACS Polym. Prepr.* **2009**, *50* (2), 31.
42. Li, G.; Viswanathan, K.; Xie, W.; Gross, R.A. Protease-catalyzed synthesis of co-oligopeptides consisting of glutamate and leucine residues. *ACS Polym. Prepr.* **2009**, *50* (2), 60.
43. Rovira-Truitt, R.; White, J. L. Growing organic-inorganic biopolymer nanocomposites from the inside out. *ACS Polym. Prepr.* **2009**, *50* (2), 36.
44. Lu, W.; Yang, Y.; Zhang, X.; Xie, W.; Cai, M.; Gross, R. A. Fatty acid biotransformations: omega-hydroxy- and omega-carboxy fatty acid building blocks using a engineered yeast biocatalyst. *ACS Polym. Prepr.* **2009**, *50* (2), 29.
45. Gitsov, I.; Simonyan, A.; Tanenbaum, S. Hybrid enzymatic catalysts for environmentally benign biotransformations and polymerizations. *ACS Polym. Prepr.* **2009**, *50* (2), 40.
46. Schoffelen, S.; van Dongen, S. F. M.; Teeuwen, R.; van Hest, J. C. M. Azide-functionalized *Candida Antarctica* lipase B for conjugation to polymer-like materials. *ACS Polym. Prepr.* **2009**, *50* (2), 3.
47. Smith, P. B. Renewable chemistry at ADM: Materials for the 21st century. *ACS Polym. Prepr.* **2009**, *50* (2), 62.
48. Zhou, Q.; Malm, E.; Nilsson, H.; Larsson, P. T.; Iversen, T.; Berglund, L. A.; Bulone, V. Biomimetic design of cellulose-based nanostructured composites using bacterial cultures. *ACS Polym. Prepr.* **2009**, *50* (2), 7.
49. Gu, Q.; Michel, A.; Maslanka, W. W.; Staib, R.; Cheng, H. N. New polyamide structures based on methyl acrylate and diamine. *ACS Polym. Prepr.* **2009**, *50* (2), 54.
50. McChalicher, C.; Srienc, F. Synthesis of mixed class polyhydroxyalkanoates using a mutated synthase enzyme. *ACS Polym. Prepr.* **2009**, *50* (2), 67.
51. Gross, R.A.; Sharma, B. Lipase-catalyzed routes to polyol-polyesters. *ACS Polym. Prepr.* **2009**, *50* (2), 48.
52. Matos, T. D.; King, N.; Simmons, L.; Walker, C.; McClain, A.; Mahapatro, A.; Rispoli, F.; McDonnell, K.; Shah, V. Mixture design to optimize microwave assisted lipase catalyzed polymerizations. *ACS Polym. Prepr.* **2009**, *50* (2), 52.
53. Palmans, A. R. A.; Veld, M.; Deshpande, S.; Meijer, E. W. *Candida antarctica* lipase B for the synthesis of (chiral) polyesters and polyamides. *ACS Polym. Prepr.* **2009**, *50* (2), 11.
54. Jiang, Z.; Liu, C.; Gross, R. A. Lipase-catalysis provides an attractive route for poly(carbonate-co-esters) synthesis. *ACS Polym. Prepr.* **2009**, *50* (2), 46.
55. Dai, S.; Xue, L.; Li, Z. Enzymatic preparation and characterization of di-block co-polyester-carbonates consisting of poly[(R)-3-hydroxybutyrate] and poly(trimethylene carbonate) blocks via ring-opening polymerization. *ACS Polym. Prepr.* **2009**, *50* (2), 21.
56. Shogren, R. L.; Willett, J. L.; Biswas, A. HRP-mediated synthesis of starch-polyacrylamide graft copolymers. *ACS Polym. Prepr.* **2009**, *50* (2), 38.

57. Jong, L. Effect of phthalic anhydride modified soy protein on viscoelastic properties of polymer composites. *ACS Polym. Prepr.* **2009**, *50* (2), 15.

58. Bouldin, R.; Ravichandran, S.; Garhwal, R.; Nagarajan, S.; Kumar, J.; Bruno, F.; Samuelson, L.; Nagarajan, R. Enzymatically synthesized water-soluble polypyrrole. *ACS Polym. Prepr.* **2009**, *50* (2), 23.

59. Cruz-Silva, R.; Roman, P.; Escamilla, A.; Romero-Garcia, J. Enzymatic and biocatalytic synthesis of polyaniline and polypyrrole colloids. *ACS Polym. Prepr.* **2009**, *50* (2), 475.

60. Mazzocchetti, L.; Scandola, M.; Jiang, Z. Enzymatic synthesis and thermal properties of poly(omega-pentadecalactone-co-butylene-co-succinate). *ACS Polym. Prepr.* **2009**, *50* (2), 477.

61. Xue, L.; Dai, S.; Li, Z. Synthesis and characterization of three-arm poly(epsilon-caprolactone)-based poly(ester–urethanes) with shape-memory effect at body temperature. *ACS Polym. Prepr.* **2009**, *50* (2), 579.

62. *ACS Polymer Preprints* can be accessed at http://www.polyacs.org/11.html?sm=87279.

63. Books on biofuels include (a) *Handbook on Bioethanol: Production and Utilization*; Wyman, C. E., Ed.; Taylor and Francis: Washington, DC, 1996. (b) Mousdale, D. M. *Biofuels: Biotechnology, Chemistry and Sustainable Development*; CRC Press, New York, 2008. (c) Demirbas, A. *Biofuels: Securing the Planet's Future Energy Needs*; Springer-Verlag, London, 2009.

64. Recent reviews include (a) Ragauskas, A. J.; et al. *Science* **2006**, *311*, 484. (b) von Blottnitz, H.; Curran, M. A. *J. Cleaner Prod.* **2007**, *15* (7), 607. (c) Gomez, L. D.; Steele-King, C. G.; McQueen-Mason, S. J. *New Phytol.* **2008**, *178*, 473. (d) Nia, M.; Leung, D. Y. C.; Leung, M. K. H. *Int. J. Hydrogen Energy* **2007**, *32* (15), 3238. (e) Rajagopal, D.; Zilberman, D. *Review of Environmental, Economic and Policy Aspect of Biofuels*; The World Bank Development Research Group: September, 2007.

Novel Biobased Materials

Chapter 2

Solid or Swollen Polymer-Protein Hybrid Materials

Kasper Renggli and Nico Bruns*

Department of Chemistry, University of Basel, Klingelbergstr. 80, CH-4056 Basel, Switzerland
***Fax: +41 61 2673855; e-mail: nico.bruns@unibas.ch**

Hybrid materials comprising synthetic polymers and proteins or active enzymes combine the best of two worlds: The structural properties, the processibility and the moldability of man-made plastics or gels and the highly evolved functionality and responsiveness of nature's polypeptides. In this chapter we review the body of literature on these smart hybrid materials and classify them according to their function. Biocatalytic plastics and polymers, stimuli-responsive hybrid hydrogels, self-assembled hydrogels with protein crosslinks, hybrid materials for selective binding of heavy metal ions, materials for tissue engineering, materials for controlled drug release, biodegradable materials, smart hydrogels with improved mechanical properties, and self-reporting materials are covered.

Introduction

Over the last century, we have witnessed an amazing rise of man-made polymeric materials, which by now have made their way into nearly every realm of our life. The reasons why polymeric materials found so widespread applications (e.g., as structural components in every-day applications, as construction materials, and as biomedical materials) and in many cases replaced natural materials are that polymers are cheap and easy to manufacture, they are easy to process and to mold into any desired shape. Furthermore, their properties can be superior to their natural counterparts and the properties can be tailored to fulfill specific needs by, e.g., the design of the chemical structure of the polymer. More recently, polymers that respond to an external stimulus with a change in their properties, so called smart materials, have attracted much attention (*1–4*). Their potential application ranges from drug-delivery devices, to actuators, sensors and microfluidic valves. Although a lot of polymer systems have been labeled as smart, their responsiveness is most often limited to a single, quite simple stimulus such as a change in temperature, a change in pH, or an increase in ion concentration.

When it comes to multifunctional and smart materials, nature leads the way with a number of responsive and adaptive materials (*5*). Form a few building blocks, e.g., lipids, proteins, DNA, and carbohydrates, a variety of mesoscopic materials are formed, such as cell membranes and cell walls, scaffold structures, skin, plant leaves, etc. These materials are time-dependant, adaptive, responsive and multifunctional. They can provide structural and mechanical stability, the ability to integrate tissue, to regenerate and to grow. Moreover, they can process information and sense optical, chemical and magnetic stimuli.

Hybrid materials that comprise biomolecules and man-made polymers can be a means to combine the best of two worlds in one single material: The highly evolved functionality and responsiveness of nature's building blocks and the structural properties, the processibility and moldability of synthetic polymers. Although these hybrid materials do not reach the degree of functionality of some natural materials yet, they offer a significant increase in functionality of smart polymers.

This review will summarize the emerging field of polymer-protein hybrid materials in the solid or the swollen state with a special focus on materials that contain dispersed proteins in the polymer matrix. The various hybrid materials reported in literature will be classified according to their function: Biocatalytic plastics and polymers, stimuli-responsive hybrid hydrogels, self-assembled hydrogels with protein crosslinks, hybrid materials for selective removal of heavy metal ions from water, materials for tissue engineering, materials for controlled drug release, biodegradable materials, smart hydrogels with improved mechanical properties, and self-reporting materials.

Polymer-protein hybrid materials that are soluble in water (e.g., protein-polymer conjugates) (*6, 7*), or form colloidal nanostructures such as protein-containing block copolymer vesicles (*8, 9*) have been extensively reviewed elsewhere and are beyond the scope of this book chapter. The same applies for peptide-polymer block copolymers and other peptide-polymer hybrid structures (*7, 10–13*).

Polymer-Protein Hybrid Materials

Biocatalytic Plastics and Polymers

The immobilization of enzymes in or onto polymeric supports has long been used as a means to improve the properties of the biocatalysts and might be the earliest form of preparing polymer-protein hybrid materials (*14*). The methods available to incorporate enzymes into polymeric supports include entrapment, covalent attachment as well as adsorption and have been reviewed extensively (*15–17*). Most often in the realm of biocatalysis, immobilized enzymes on or in polymeric supports were not regarded as hybrid materials because the focus laid on the biocatalyst itself and the enhancement of its properties due to the immobilization, e.g., by improving its catalytic properties in water and non-aqueous media, by improving its stability and by allowing for an easy recovery of the biocatalyst at the end of the reaction. On the other hand, solid, powdered enzymes were blended into epoxy and other resins to produce protein-aggregate-filled composites, e.g., for biosensor applications (15). However, starting in the mid 1990s, biocatalytic materials began to emerge that incorporated dispersed enzymes, i.e., starting with enzymes that were soluble in the polymerization mixture (15, 18–20). Dordick and coworkers drew the attention to the materials side of these systems and coined the term biocatalytic plastics (20). Prior, Russell and coworkers had modified subtilisin Carlsberg and thermolysin with pendant poly(ethyleneglycol) (PEG) acrylates to yield organo-soluble, copolymerizable enzymes. They were copolymerized by free radical polymerization with methyl methacrylate in the presence of the crosslinker trimethylol propane trimethacrylate to yield hybrid materials that retained high activities in aqueous, aqueous-organic and organic media with improved long-term operational stabilites (18, 19). In their original work on biocatalytic plastics, Dordick and coworkers modified α-chymotrypsin and subtilisin Carlsberg with acryloyl chloride in order to introduce copolymerizable groups. The enzymes were solubilized in organic solvents by the formation of non-covalent ion pairs of enzymes and surfactants. The solubilized enzymes were copolymerized in the organic phase via free radical polymerization with hydrophobic monomers such as methyl methacrylate, styrene, vinyl acetate, and ethyl vinyl ether, using trimethylol propane trimethacrylate or divinyl benzene as crosslinker. The hybrid materials were used to synthesize peptides, sugars and nucleotide esters in organic solvents such as THF and ethyl acetate. In later work, the researchers extended the concept to more hydrophilic biocatalytic plastics based on 2-hydroxyethyl methacrylate and examined the influence of the chemical nature of the polymer on the enzymatic activity of α-chymotrypsin in n-hexane (21). The activity correlates with the hydrophilicity of the polymer and was found to be lowest for poly(methyl methacrylate)-based hybrid materials and highest for poly(2-hydroxyethyl methacrylate)-based materials. In an extension of the work on biocatalytic plastics, Kim, Dordick and Clark reported the synthesis of biocatalytic films, coats, membranes and paints (22). α-Chymotrypsin and pronase were incorporated into poly(dimethylsiloxane) (PDMS)-based materials by dispersing enzymes entrapped in sol-gel particles in solutions of silanol-terminated PDMS (without prior solvation of the enzyme in organic solvents). Alternatively, enzymes were used that were modified with 3-aminopropyl triethoxy silane to provide chemical functionality that can react with silanol-terminated PDMS. The PDMS/enzyme mixtures were subsequently cured at room temperature by condensation of the PDMS's silanol groups and with the curing agent tetramethyl orthosilicate and/or with the ethoxy silane groups on the enzyme. The materials showed proteolytic activity and as a result of this a reduced protein adsorption compared to materials that did not contain any enzyme. Therefore, these coatings and paints could be used to reduce fouling by proteins on surfaces. Biocatalytic silicone elastomers were also reported more recently by Ragheb et al. (*23–25*). These rubbers contained entrapped lipase from *Candida rugosa* and were highly active in the esterification of lauric acid with octanol in isooctane.

As mentioned above, epoxy resins can be doped with enzymes. The amino groups on proteins and enzymes can react directly with the epoxy prepolymers. However, protein-aggregate filled resins were obtained because the biomolecules are not soluble in the commonly used, hydrophobic prepolymers (*15*). With polymer precursors partially miscible with water, such as epoxy compounds based on glycidyl-terminated poly(methacrylate-*co*-2-hydroxyethyl acrylate) or bisphenol-A-glycidylether-terminated glycerol ethoxylates, enzymes could be incorporated from aqueous solution into epoxy materials under retention of activity (*15*). Amongst others, solid monoliths containing β-glucosidase or cytochrome C were reported.

Biocatalytic materials that beautifully combine enzymatic activity with flexibility of shape and tunable mechanical properties of polymeric materials are foams, sponges, sheets and coatings that contain organophosphorous hydrolases (OPH, also referred to as phosphotriesterases) (*26, 27*). These biocatalysts can degrade highly toxic organophosphate nerve agents found in chemical weapons and in pesticides. Thus, several approaches have been pursued to incorporate these enzymes into polymeric materials to generate detoxification materials. Successful applications that also have been commerzialized (*28*), rely on the reaction between polyurethane-prepolymers containing isocyanate end-groups with water, the amino-, thiol-, and hydroxyl-groups of proteins and optionally polyether polyol or polyether polyamines as crosslinkers (*15*). Carbon-dioxide blown, solid polyurethane foams were obtained. Their mechanical properties could be tuned by variations in the ratio of hard segments (polyurethane prepolymers) and soft segments (polyether polyol or polyether polyamine). Also, solid polyurethane-enzyme hybrids could be synthesized, e.g., as water-borne coatings, if an aqueous polyester-based polyol dispersion was reacted with a water-dispersible aliphatic isocyanate in the presence of enzymes (*29*). The proteins were covalently bound to the polyurethane network at multiple points. As they were added to the polymerization mixture in form of an aqueous solution, they were dispersed in the polymer matrix. This kind of chemistry was successfully applied to the immobilization of OPH, but also to the immobilization of diisopropyl fluorophosphatase, parathion hydrolase, butyrylcholine esterase, acetylcholine esterase, and amyloglucosidase (*26, 30–36*). OPH-containing elastomeric polyurethane sponges were reported that could be used to clean and degrade toxic organophosphate spills (*30*). Rigid foams were synthesized for the decontamination of gases and liquids and for use in topical creams (*31, 33–36*). Furthermore, the polyurethane foams could potentially be used in the large-scale destruction of nerve agent stockpiles accumulated for chemical warfare (*26*). Polyurethane based coatings of fibers could further be used in protective clothing, to enhance the sate-of-the-art of garments containing activated carbon. Another approach towards detoxification materials was followed by Gill and Balesteros (*36, 37*). They adsorbed OPH and other enzymes onto poly(hydroxymethyl siloxane), fumed silica, or trimethyl siloxy-silica and encapsulated these primary immobilizates into conventional curing silicones to produce, amongst others, bioactive granulates, monolithic samples, thick-film coatings and macroporous foams. The materials were effective in the liquid and gas phase detoxification of organophosphate nerve agents.

More recently, another class of polymeric materials was shown to easily incorporate active enzymes while allowing for various shapes, forms and applications. Amphiphilic conetworks are polymer networks that consist of hydrophilic and hydrophobic chain segments (38). Because of the chain's immiscibility, they separate into two phases. However, the covalent crosslinks present in the network prevent macroscopic phase separation and the phase separation occurs only on the nanoscale, resulting in nanostructured materials. Over a wide range of compositions, bicontinuous morphologies were observed in which both phases interpenetrate each other and were continuous from the materials surface throughout its bulk (39). Bruns and Tiller were able to show that these materials could be loaded with enzymes by simply incubating them in aqueous solutions of the biomolecules (Figure 1) (40, 41). In water, the hydrophilic phase swelled and allowed proteins to diffuse into and throughout the network. Upon drying, the phase shrinked and encapsulated the enzymes into nanoscopic hydrophilic compartments that were surrounded by the hydrophobic phase. Thus, as long as the network was not exposed to water, the enzymes were entrapped in the amphiphilic conetwork.

Nanophase-separated amphiphilic conetworks based on poly(2-hydroxyethylacrylate) (PHEA) as the hydrophilic component and PDMS as the hydrophobic component, termed PHEA-l-PDMS, were synthesized as coatings (39), free-standing membranes (39), and as micro particles (42). Networks that comprise a PHEA-phase and a perfluorinated hydrophobic phase based perfluoropolyether (PHEA-l-PFPE) were also synthesized (41). Amphiphilic conetworks were loaded with a variety of enzymes and proteins (horseradish peroxidase (HRP) (40, 43), chloroperoxidase (40), lipases (41, 44), α-chymotrypsin (42), myoglobin (45), and haemoglobin (45)) showing the generality of the immobilization approach. The enzyme-loaded networks were applied to catalyze reactions in organic solvents (39, 42) and in supercritical CO_2 (44), as these non-aqueous solvents swell the PDMS or the PFPE phase, respectively. Substrates were able to diffuse into and throughout this phase and access the enzyme through the large interface between the two phases. The enzymes showed a strong increase in apparent catalytic activity in non-aqueous solvents and an increased stability when incorporated into the networks, compared to free enzymes. As an example, HRP incorporated into PHEA-l-PDMS was up to 100-fold more active in catalyzing an oxidative coupling reaction in n-heptane and possessed a significant longer operational stability (40). The materials point of view came into focus, when disposable sensor chips were constructed from enzyme-loaded amphiphilic conetworks to determine peroxides in non-polar organic solvents (43, 46). To this end, conetworks were synthesized as surface-attached, several micrometer thick films on modified glass slides. As these coatings are clear and transparent, they could be used as optical biochemical sensor matrixes when placed into the beam of a UV/Vis spectrometer. HRP was co-immobilized with its colorimetric substrate 2,2'-azino-bis(3-ethylbenzothiazolin-6-sulfonic acid) diammonium salt (ABTS) into the hydrophilic phase of the films (*43*). Upon exposure to the hydrophobic analyte *tert*-butyl hydroperoxide dissolved in *n*-heptane, the sensor chips reported the presence of the analyte by a change in color due to the peroxidase catalyzed reaction of the analyte with ABTS. Sensitivity towards the peroxide between approximately 1 and at least 50 mmol L^{-1} and a response time between 1.7 to 5.0 min was reported. The sensitivity is in the same range as sensitivities of optical sensors containing immobilized enzymes in a sol-gel, hydrogel or spongiform matrix (*43*).

Stimuli-Responsive Hydrogels

Hydrogels that respond to external stimuli such as temperature, pH and antigen-antibody recognition with a change in volume are the prototype of smart materials (*1*, *2*). Their potential applications range from drug delivery systems, to bioactive surfaces, microfluidics, diagnostics, and bioreactors. As proteins are well-known to undergo conformational changes upon these and other stimuli, it is not surprising that they are exquisite candidates to render synthetic hydrogels stimuli-responsive. Two-component hybrid hydrogels consisting of synthetic polymers crosslinked by recombinant proteins were pioneered by Kopecek and coworkers (*47*). They prepared a linear copolymer of *N*-(2-hydroxypropyl)-methacrylamide (HPMA), and *N*-(*N'*,*N'*-dicarboxymethyl aminopropyl)-methacrylamide (DAMA) by radical copolymerization. Ni^{2+} was complexed by the metal-chelating pendant group of the DAMA and by histidine tags of a coiled-coil, a well-defined protein-folding motif. Thus the coiled-coil crosslinked the poly(HPMA-*co*-DAMA) chains to yield hydrogels (Figure 2). These hydrogels shrank to 10% of their volume at room temperature with a mid-point transition temperature at 39 °C, an effect which was attributed to a temperature-induced conformational change in the coiled-coil: Upon heating, the rod-like helical protein unfolded and collapsed. In further work, recombinant block proteins with varying number of coiled-coil blocks were used to crosslink the poly(HPMA-co-DAMA) in order to investigate the influence of higher-order structure and stability of coiled-coil cross-links on the properties of hybrid hydrogels (*48*). It was shown, that the temperature-induced deswelling can be tuned over a wide temperature-range by using coiled-coils with varying melting temperatures.

Using a similar crosslinking strategy, the authors also crosslinked acrylamide copolymers with the I28 immunoglobin-like module of human cardiac titin, an elastic muscle protein, through metal coordination bonding between terminal His tags of the I28 module and metal-chelating nitrilotriacetic acid-containing side chains on the copolymer (*49*). At temperatures above the melting temperature of the protein, the hydrogels swelled to 3 times their initial volume.

The competitive binding of a free and hydrogel-bound antigen to the corresponding polymer-bound antibody was exploited as trigger mechanism for a hydrogel which undergoes reversible volume changes in response to the presence of a specific antigen in solution (Figure 3) (*50*). To generate these hybrid hydrogels, goat anti-rabbit IgG antibody and rabbit IgG antigen were modified with *N*-succinimidyl acrylate to introduce copolymerizable acrylamide groups to

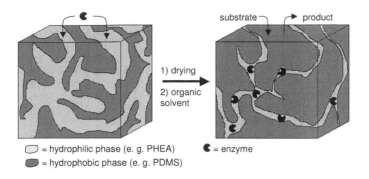

substrate → product

1) drying
2) organic solvent

◯ = hydrophilic phase (e. g. PHEA)   ◖ = enzyme
◖ = hydrophobic phase (e. g. PDMS)

Figure 1. Loading of an amphiphilic conetwork with enzymes in aqueous solution and use as biocatalytic material in organic solvents. (Reproduced from reference (40).)

the proteins. Then, they were copolymerized in two steps with acrylamide and the crosslinker N,N'-methylene bis(acrylamide) (MBAA) to yield antigen-antibody semiinterpenetrating networks. In addition to the chemical crosslinks, the antigen-antibody binding interaction introduces further crosslinks into the hydrogel. However, these non-covalent crosslinks could be broken by immersing the hydrogel into a solution containing free antigen. Swelling of the gel was observed. Upon removal of the free antigen, the intra-network antigen-antibody complexes reformed and the gel deswelled back to its original volume. Using the swelling/deswelling mechanism, the permeability of hemoglobin through membranes made out of the network could be controlled.

Antibody-antigen interactions are not the only biological binding events that have been used to form non-permanent crosslinks in a polymeric hydrogel in order to control the swelling state of hybrid hydrogels. Another example for the concept of using competitive binding between a free and a network-bound biological ligand to a network-bound protein is based on the simultaneous binding of Ca^{2+} and an anti-psychotic drug from the class of phenothiazines to the calcium-binding protein calmodulin (CaM) (*51*). Acrylamide, the crosslinker MBAA, allylamine-moiety modified CaM, and an acrylamide-functionalized phenothiazine were copolymerized to form a poly(acrylamide) hydrogel (PAAm) that contained immobilized protein and immobilized ligand. In the presence of Ca^{2+}, the phenothiazine was bound to the CaM, giving rise to physical cross-links (Figure 4). On removal of the metal ion by complexation with a chelating agent, the hydrogel swelled mainly due to the release of the phenothiazine from the protein binding site. A change in conformation of the protein might as well contribute to the swelling. The process was reversible and the original, less swollen sate was regained by placing the hybrid network in a Ca^{2+} solution. Furthermore, swelling could be triggered in the presence of Ca^{2+} by the addition of free chlorpromazine, which competes with the immobilized ligand for the binding to the protein, resulting in the cleavage of physical crosslinks within the polymer network. The hydrogels were successfully tested as a Ca^{2+}-responsive valve in a microfluidic device and as a membrane with controlled permeability for vitamin B_{12} (*51*). In a more recent report, such CaM-based hydrogels were

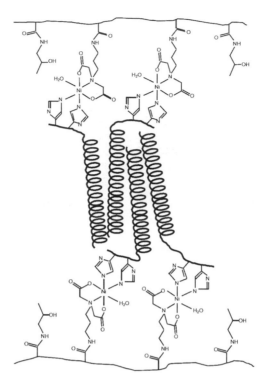

Figure 2. Temperature-responsive hybrid hydrogel: Poly(HPMA-co-DAMA) crosslinked by coiled-coil protein domains. (Reproduced with permission from reference (47). Copyright 1999 Macmillan Publishers Ltd.)

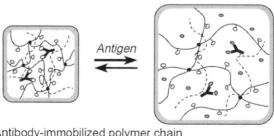

`⌁` : Antibody-immobilized polymer chain
`⌁` : Antigen-immobilized polymer chain
`◦` : Free antigen

Figure 3. Antigen-responsive hybrid hydrogel acts by competitive binding of free and polymer-bound antigen to polymer-bound antibody. (Reproduced with permission from reference (50). Copyright 1999 Macmillan Publishers Ltd.)

used to synthesize dynamic microlens arrays. The optical properties of the lenses were tunable using soluble chlorpromazine (52).

The changes in volume in the CaM containing hydrogel reviewed above were mostly due to changes in the network's crosslinking density. However, CaM has two distinct conformational states and undergoes substantial change in shape upon

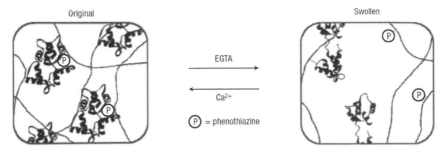

Figure 4. Ion-responsive hybrid hydrogel: Upon removal of Ca²⁺, non-covalent crosslinks are broken due to dissociation of a ligand from calmodulin. EGTA = ethylene glycol-bis(β-aminoethyl ether)-N,N,N',N'-tetraacetic acid. (Reproduced with permission from reference (51). Copyright 1999 Macmillan Publishers Ltd.)

binding of ligands. In the presence of calcium, CaM is a dumbbell-shaped protein, i.e., in an extended conformation. Upon binding of ligands, CaM undergoes a pronounced hinge motion to a collapsed conformation. This molecular motion was exploited to induce macroscopic volume changes of a hybrid hydrogel in the presence of the ligand trifluoperazine (*53–55*). CaM was engineered to expose cysteine residues at both ends of the dumbbell. In one example, these residues were reacted in an Michael-type addition with acrylate-end-groups of a four-arm, star-shaped PEG to yield hybrid hydrogels (*53*). In a second example, both cysteine residues of the CaM were modified with linear PEG diacrylate in such a way, that two acrylate-terminated PEG chains were conjugated to the protein (*54, 55*). Next, the protein-polymer conjugates were polymerized by photoinitiation. Varying amounts of PEG diacrylate were added to the polymerization mixture in order to vary the total protein content of the resulting hydrogel. Upon incubation in a solution of the ligand, the volume of the hydrogels decreased up to 65%. This effect could be attributed to the change in conformation of the CaM. The conformational changes in response to ligand binding were reversible and the hydrogels could be cycled between high and low volume states. Further to triggered volume changes, some of the hydrogels also showed tunable and reversible changes in optical transparency upon exposure to a CaM-ligand. This property was exploited to construct a label-free optical biosensor for the drug trifluoperazine. The authors attribute the change in optical transparency to varying degrees of light scattering due to changes in crosslinking density and crosslink homogeneity of the hybrid network (*55*).

Quite a few proteins other than CaM are also known to undergo substantial conformational changes upon binding of ligands or substrates. A hybrid hydrogel crosslinked by an active enzyme that undergoes these changes upon substrate binding was generated by reacting linear poly(HPMA-*co*-N-(3-aminopropyl)-methacrylamide) (APMA) copolymers bearing pendant maleimide side groups with mutants of the enzyme adenylate kinase (Figure 5) (*56*). Some hydrogels were additionally crosslinked with dithiothreitol in order to control the overall crosslinking density. The enzyme catalyzes the phosphoryl transfer reaction between ATP and AMP. When ATP binds to the enzyme, a bulky lid domain closes over the active site. Two cysteine groups at the edge of this lid were engineered into the enzyme as attachment points for the polymer. Thus, upon transfer of the hydrogels from ATP-free buffer to 4 mM ATP solution, the gels shrank between ~5-17% in proportion to their enzyme content. Upon washing with ATP-free buffer, the gels expanded again. The deswelling upon addition of ATP could be repeated several times. Similar ATP-responsive gels were prepared with maleimide terminated 4-arm PEG, crosslinked by the adenylate kinase (56).

Protein-crosslinked hydrogels can not only report an external stimulus by changes in their volume, but also by changes in the fluorescence of embedded fluorescent proteins, as reported by Francis and coworkers (57). A copolymer based on poly(HPMA-co-APMA) was crosslinked with enhanced green fluorescent protein (eGFP) in order to obtain hybrid hydrogels. The authors developed orthogonal reactions to activate the N and C termini of the protein, in order to crosslink the polymer chains with the termini of the protein. The fluorescence of eGFP and thus of the hydrogel decreased when the pH of the incubation media was lowered from pH 6.5 to 5.5 and recovered when the pH was readjusted to 6.5. Furthermore, eGFP denatures from 60 to 80 °C, which manifests in a loss of fluorescence. The hydrogels showed the same sensitivity towards a rise in temperature and lost their fluorescence between 70 and 75°C. Furthermore, the gels shrank due to the denaturing of the protein.

Self-Assembled Hybrid Hydrogels with Protein Crosslinks

Some proteins form well-defined dimerization motifs such as coiled-coils. Polymers containing such protein grafts can self-assemble to hybrid hydrogels. Non-covalent crosslinks are formed by the protein dimers. This concept was proven by Kopecek and coworkers in a series of papers (Figure 6) (58–61). Two HPMA graft-copolymers containing complementary coiled-coil forming domains were prepared. Maleimide-group bearing polymer precursors were synthesized by radical copolymerization of HPMA and APMA, followed by functionalization with a maleimde containing linker. Protein domains were linked to the precursor polymer via a genetically introduced cysteine residue. Upon mixing of equimolar ratios of HPMA copolymers containing complementary coiled-coil domains at defined conditions (pH, temperature, concentration), formation of antiparallel heterodimeric coiled-coils occurred, resulting in gelation of the mixture. The formation of the obtained hydrogels was reversible and degelation could be achieved by denaturation of the coiled-coil with high concentrations of guanidinium hydrochloride. On the other hand, removal of the denaturing agent by dialysis caused hydrogel reassembly due to coiled-coil refolding. Recently, the concept of self-assembly hybrid hydrogels was extended to HPMA copolymers grafted with a beta-sheet peptide (62).

Selective Removal of Heavy Metal Ions from Water

In an extension of their previous work (57), Francis and coworkers crosslinked polymer chains by a 75 residue segment of the metal binding protein metallothionein (63). Their coupling strategy of preformed polymer chains to the C and N-terminus of proteins tolerated the presence of cysteine groups in the

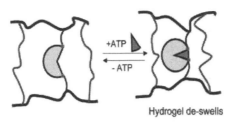

Figure 5. Substrate-responsive hybrid hydrogel. (Reproduced from reference (56).)

protein. Metallothioneins posses a series of cysteine-lined pockets that tightly bind heavy metal atoms such as copper, zinc, cadmium, and mercury. This property was exploited to create hydrogels that can remove toxic metal from solutions and that report the binding through changes in their swelling state. Cu^{2+}, Zn^{2+}, Cd^{2+}, and Hg^{2+} ions at a concentration of ~50 ppb were effectively removed from environmental water samples in the presence of other ions that do not bind to the protein, such as Ca^{2+}. The binding event caused folding of the protein, which in turn resulted in a decrease of the hydrogel's volume.

Support for Tissue Engineering

For therapeutic tissue regeneration, major efforts are being directed towards the design of synthetic matrixes that mimic the extracellular matrix. An approach pursued by Hubbell and coworkers was to design and produce genetically engineered proteins that carry specific key features of the extracellular matrix and to incorporate them into PEG-based hydrogels (64–66). Different strategies were followed. PEG-grafted recombinant proteins were crosslinked by photopolymerization of acrylate end-groups on the PEG chains (64). In a more recent approach recombinant proteins were crosslinked with PEG via Michael-type conjugation addition of vinylsulfone groups at the chain ends of PEG with the protein's cysteine residues (65). The proteins of these networks were designed to contain a cell-adhesion motif (cell-binding site for ligation of cell-surface integrin receptors) and specific protease cleavable sites. They were shown to indeed promote specific cellular adhesion in vitro and to degrade in the presence of proteases (66). Three-dimensional cell migration inside the gels was dependant on the proteolytic sensitivity and on suitable mechanical properties. In vivo experiments showed that these hydrogels can be applied as matrix in order to heal critical-sized defects in rat calvaria (66). To this end, the hydrogels had to be sufficient degradable and carry bone morphogenetic protein. Non-degradable hydrogels or the absence of the bone morphogenetic protein prevented replacement of the artificial matrix with osteoblasts and calcified bone.

Controlled Drug Release

Bovine serum albumin (BSA)-crosslinked PAAm was envisaged by Tada et al. as biodegradable device for sustained drug-delivery (67, 68). To this end, amino groups of lysine residues on BSA were modified to yield acrylamide groups (*67*). The functionalized protein was copolymerized with acrylamide in order to synthesize BSA-crosslinked hydrogels. The hydrogels were loaded with salicylic acid, a drug having affinity for albumin, by swelling the dried gel in a drug solution. In subsequent drug-release studies in vitro, the hydrogels released the salicylic acid for up to 50 h. In contrast, release of a drug that does not bind to BSA, sodium benzoate, was finished within 5 h. In a subsequent study, other, structurally related benzoic acid derivatives were tested as well. It was found that the amount of compounds loaded into a hydrogel of high BSA content and the duration of their release were dependent on the compound's affinity for BSA (*68*). The concept was further extended to albumin-crosslinked alginate hydrogels (*69*).

A more sophisticated drug release system was achieved with hybrid hydrogels that combined two properties, the sensing of drugs and triggered release of a therapeutic protein (*70*). Bacterial gyrase subunit B was grafted via a His-tag to PAAm functionalized with nitrilotriacetic acid in the presence of Ni^{2+} ions. Subsequently, the protein was dimerized by addition of coumermycin, an aminocoumarin antibiotic, resulting in the formation of crosslinks and gelation. Gels were loaded with human vascular endothelial growth factor that carried a His-tag at the N-terminus by adding the payload to the gel-forming mixture. When the antibiotic novobiocin was added, the dimerized protein subunits dissociated and as a consequence, the hydrogel dissolved and the therapeutic payload was released. This resulted in a significantly increased proliferation of human umbilical vein endothelial cells.

Protein conformational changes due to ligand binding were recently used to modulate the release of a biotherapeutic from a protein-containing hydrogel (*71*). To this end, PEG-calmodulin hydrogels were prepared as reviewed above (*54, 55*). The hydrogels decreased in volume by 10% to 80% when exposed to the ligand trifluoroperazine. Calmodulin undergoes a hinge motion upon ligand binding and this motion is transduced into a contraction of the gel. Hydrogels loaded with vascular endothelial growth factor released the therapeutic protein more efficiently in the presence of the ligand than in its absence.

Biodegradability

Proteins are susceptible to hydrolytic cleavage and to enzymatic degradation by proteases. Therefore, biodegradability is an intrinsic property of polymer networks crosslinked by proteins. However, only a few reports investigated this property explicitly. Tada et al. reported BSA-crosslinked hydrogels as biodegradable drug-delivery device (*67*). Upon treatment with a 0.25 wt% trypsin solution, the hydrogels markedly swelled after 10 days and dissolved within 20 days. PAAm networks crosslinked by a conventional crosslinker did not show this behavior. The hydrogels crosslinked by eGFP reported by Francis and coworkers were exposed to a 3.3 wt% trypsin solution and disintegrated completely within 3 h, while gels immersed only in buffer did not (*57*).

Site-specific enzymatic degradation of polymer-protein hybrid hydrogels was reported by Hubbell and coworkers for their protein-PEG hybrid hydrogels designed as cell-adhesive and proteolytically degradable hydrogel matrixes for therapeutic tissue regeneration (*65*). The proteins of the network were designed to contain protease-cleavable sites, namely substrate sequences for plasmin and matrix metalloproteinase. Proteolytic degradation upon incubation in solutions of metalloproteinase 1 or plasmin was followed by monitoring the protein's degree of swelling (*66*). Gels formed by PEG-crosslinked protein chains containing designed protease substrates showed a rapid increase of swelling due to degradation and finally dissolved within 250 h. Controls with non-substrate protein chains did not swell dramatically and where stable against dissolution.

Smart Hydrogels with Improved Mechanical Properties

Some hydrogels have rather poor mechanical properties. This is especially true for hydrogels that undergo strong volume changes due to changes in their swelling state as a reaction to external stimuli. Therefore, Zhang et al. aimed at improving the mechanical properties of temperature sensitive poly(*N*-isopropyl acrylamide) (PNIPAAm) gels by grafting PNIPAAm onto crosslinked BSA (*72*). BSA-based hydrogels were formed by the coupling of the protein's carboxyl groups with its amine groups via carbodiimide chemistry. In the presence of monocarboxyl-functionalized PNIPAAm, the polymer got incorporated into the hydrogel by the formation of amide bonds from PNIPAAm-COOH with BSA-NH$_2$. The resulting hybrid hydrogels show significant morphological changes in response to an increase in temperature from 24 to 37 °C without structural damage. (However, a stability study while cycling several times through the transition temperature was not performed.) As evaluated from SEM images, the gel changes from an expanded hydrated state below the transition temperature to a more compact structure at elevated temperatures. This effect was exploited in order to produce temperature responsive membranes, which were obtained by crosslinking the BSA/PNIPAAm within pores of sintered glass filter discs. The permeability of lysozyme or riboflavin through the membranes was found to be tunable by temperature. Diffusion of the model proteins through the membrane significantly increased above the transition temperature of the hybrid hydrogel.

Self-Reporting Materials

In all of the above examples of polymer/protein hybrid materials, the proteins are used to alter and control properties of the polymer matrix. The reverse case, in which changes of the polymer matrix trigger a response from the embedded protein, was for the first time reported by Bruns et al. (*73*). A protein was added in small amounts to the polymer matrix in order to act as a reporter for structural deformation and damage of the protein matrix, thus creating a hybrid material that autonomously reports structural damage (Figure 7). Such "self-reporting" materials could prevent catastrophic failure of load-bearing materials (e.g., fiber reinforced composites in aerospace and automotive applications) and biomedical applications (e.g., tubings) by making defects visible before they enlarge to cause structural collapse or leakage.

In order to engineer the protein reporter, a pair of fluorescent proteins was encapsulated by means of point directed mutagenesis and chemical linkers into

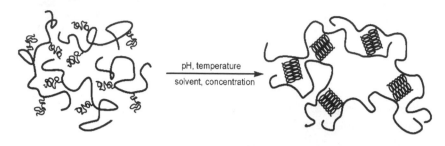

Figure 6. Self-assembled hybrid hydrogel: Formation through antiparallel heterodimeric coiled-coil association. (Reproduced from reference (58).)

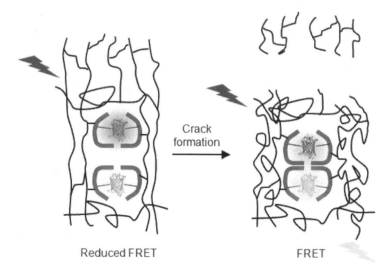

Reduced FRET   FRET

Figure 7. Self-reporting hybrid material that reports the formation of cracks by a change in FRET efficiency. (Reproduced with permission from reference (73).Copyright 2009 Wiley-VCH Verlag GmbH & Co. KGaA)

a third, cage-like protein, the thermosome from *Thermoplasma acidophilum*. It is a group II chaperonin, that is composed of 16 protein subunits forming two stacked rings. Each ring encloses a central cavity with a diameter of approx. 5 nm. They are accessible for macromolecular guests through a huge pore. The pores of the thermosome are gated by an ATP-driven, build-in lid. It is formed by helical protrusions at the tip of the apical domains of each subunit. In the absence of ATP, the thermosome rests in an open conformation, so that guests can enter the cavities. However, they would diffuse out of the cavity readily. Therefore selective attachment points, i.e., cysteine residues, were introduced into the wall of the cavities by point directed mutagenesis. They were modified with chemical linkers that form stable bisaryl hydrazone bonds to their counterparts on the surface of guest proteins, covalently entrapping the guests in the cavities. By using a two fluorescent proteins, enhanced cyan fluorescent protein and enhanced yellow fluorescent protein, a pair of fluorophores capable of fluorescence resonance energy transfer (FRET) was incorporated into the thermosome. The structure of the thermosome placed the fluorescent proteins in a distance of 5.2 nm, which is close to the Förster distance of these fluorophores and allows for energy transfer from one to the other. The possibility to use this assembly as a sensor for mechanical deformation relies on the fact that the thermosome has a mechanically weak plane right between the two cavities and therefore between the fluorescent proteins. Thus, if stress is transferred from a polymer matrix onto the protein, the distance between the fluorescent proteins changes, which manifests in a change in FRET efficiency.

In order to synthesize polymer-protein hybrid materials, the protein complex was modified with an excess of *N*-succinimidyl acrylate which introduced acrylamide groups. Then, 0.2 wt% of the protein were copolymerized with acrylamide and the cross-linker MBAA to form PAAm hydrogels, which were subsequently dried to give transparent, fluorescent, and glassy materials. These polymers were used as model systems to investigate the effect of mechanical deformation of the polymer on the biomechanical sensor.

It turned out that FRET in the dried polymer was surprisingly low and increased around microcracks upon strain fracture of the materials. During the solidification process, some mechanical stress built up within the material that was transferred onto the protein complex. The two halves of the thermosome drifted apart, the distance between the fluorescent proteins increased and thus the FRET decreased. However, cracks formed when the plastic was damaged. The formation of a crack is accompanied by a plastic deformation of the surroundings. Around a crack the polymer chains and the embedded protein had the chance to relax. Thus, FRET efficiency was recovered in the vicinity of the damage. With this change in fluorescence it was possible to detect microscopic cracks by two methods, intensity based fluorescence microscopy and fluorescence lifetime imaging (FLIM).

Conclusion

By combining proteins and synthetic polymers in a single material, hybrid gels or plastics are obtained that offer much more diverse functionality than conventional smart polymeric materials. Although the beginnings of this field of research can be traced back to the early days of immobilized enzymes, the tremendous potential of polymer-protein hybrid materials has only been realized over the last couple of years. Possible applications for the hybrids reflect the diversity of nature's protein toolbox, and range from biocatalysts and detoxification materials, to responsive microfluidic actuators, to biomedical applications and to materials that report structural damage, so called self-reporting materials. The field is still in its infancy and offers a lot of room for innovation, as only few of the known functional proteins or enzymes have been married with a limited set of synthetic polymers.

Acknowledgments

Generous financial support for N. B. by the Marie Curie Intra European Fellowship "ThermosomeNanoReact" is gratefully acknowledged.

References

1. Chaterji, S.; Kwon, I. K.; Park, K. *Prog. Polym. Sci.* **2007**, *32*, 1083–1122.
2. Kopecek, J.; Yang, J. Y. *Polym. Int.* **2007**, *56*, 1078–1098.
3. Kumar, A.; Srivastava, A.; Galaev, I. Y.; Mattiasson, B. *Prog. Polym. Sci.* **2007**, *32*, 1205–1237.
4. Roy, D.; Cambre, J. N.; Sumerlin, B. S. *Prog. Polym. Sci.* **2009**, in press, accepted manuscript.
5. Mann, S. *Angew. Chem., Int. Ed.* **2008**, *47*, 5306–5320.
6. Heredia, K. L.; Maynard, H. D. *Org. Biomol. Chem.* **2007**, *5*, 45–53.
7. Löwik, D. W. P. M.; Ayres, L.; Smeenk, J. M.; Van Hest, J. C. M. *Adv. Polym. Sci.* **2006**, *202*, 19–52.
8. Kita-Tokarczyk, K.; Grumelard, J.; Haefele, T.; Meier, W. *Polymer* **2005**, *46*, 3540–3563.
9. Mecke, A.; Dittrich, C.; Meier, W. *Soft Matter* **2006**, *2*, 751–759.
10. van Dongen, S. F. M.; de Hoog, H.-P. M.; Peters, R. J. R. W.; Nallani, M.; Nolte, R. J. M.; van Hest, J. C. M. *Chem. Rev. (Washington, DC, U. S.)* **2009**, *109*, 6212–6274.
11. Vandermeulen, G. W. M.; Klok, H.-A. *Macromol. Biosci.* **2004**, *4*, 383–398.
12. Klok, H. A. *J. Polym. Sci., Part A: Polym. Chem.* **2005**, *43*, 1–17.
13. Klok, H. A. *Macromolecules* **2009**, *42*, 7990–8000.
14. Faber, K. *Biotransformations in Organic Chemistry: A Textbook*, 5th ed.; Springer: Berlin, 2004.
15. Gill, I.; Ballesteros, A. *Trends Biotechnol.* **2000**, *18*, 469–479.
16. Sheldon, R. A. *Adv. Synth. Catal.* **2007**, *349*, 1289–1307.
17. Brady, D.; Jordaan, J. *Biotechnol. Lett.* **2009**, *31*, 1639–1650.
18. Yang, Z.; Williams, D.; Russell, A. J. *Biotechnol. Bioeng.* **1995**, *45*, 10–17.
19. Yang, Z.; Mesiano, A. J.; Venkatasubramanian, S.; Gross, S. H.; Harris, J. M.; Russell, A. J. *J. Am. Chem. Soc.* **1995**, *117*, 4843–50.
20. Wang, P.; Sergeeva, M. V.; Lim, L.; Dordick, J. S. *Nat. Biotechnol.* **1997**, *15*, 789–793.
21. Novick, S. J.; Dordick, J. S. *Biotechnol. Bioeng.* **2000**, *68*, 665–671.
22. Kim, Y. D.; Dordick, J. S.; Clark, D. S. *Biotechnol. Bioeng.* **2001**, *72*, 475–482.
23. Ragheb, A. M.; Brook, M. A.; Hrynyk, M. *Biomaterials* **2005**, *26*, 1653–1664.
24. Ragheb, A. M.; Hileman, O. E.; Brook, M. *Biomaterials* **2005**, *26*, 6973–6983.
25. Ragheb, A.; Brook, M. A.; Hrynyk, M. *Chem. Commun. (Cambridge, U. K.)* **2003**, 2314–2315.
26. Russell, A. J.; Berberich, J. A.; Drevon, G. F.; Koepsel, R. R. *Annu. Rev. Biomed. Eng.* **2003**, *5*, 1–27.

27. Yair, S.; Ofer, B.; Arik, E.; Shai, S.; Yossi, R.; Tzvika, D.; Amir, K. *Crit. Rev. Biotechnol.* **2008**, *28*, 265–275.

28. Agentase, a company of icx technologies. www.icxt.com.

29. Drevon, G. F.; Danielmeier, K.; Federspiel, W.; Stolz, D. B.; Wicks, D. A.; Yu, P. C.; Russell, A. J. *Biotechnol. Bioeng.* **2002**, *79*, 785–794.

30. Havens, P. L.; Rase, H. F. *Ind. Eng. Chem. Res.* **1993**, *32*, 2254–8.

31. Yang, F.; Wild, J. R.; Russell, A. J. *Biotechnol. Prog.* **1995**, *11*, 471–4.

32. LeJeune, K. E.; Frazier, D. S.; Caranto, G. R.; Maxwell, D. M.; Amitai, G.; Russel, A. J.; Doctor, B. P. *Med. Def. Biosci. Rev., Proc.* **1996**, *1*, 223–230.

33. LeJeune, K. E.; Russell, A. J. *Biotechnol. Bioeng.* **1996**, *51*, 450–457.

34. Lejeune, K. E.; Mesiano, A. J.; Bower, S. B.; Grimsley, J. K.; Wild, J. R.; Russell, A. J. *Biotechnol. Bioeng.* **1997**, *54*, 105–114.

35. LeJeune, K. E.; Swers, J. S.; Hetro, A. D.; Donahey, G. P.; Russell, A. J. *Biotechnol. Bioeng.* **1999**, *64*, 250–254.

36. Gill, I.; Ballesteros, A. *Biotechnol. Bioeng.* **2000**, *70*, 400–410.

37. Gill, I.; Pastor, E.; Ballesteros, A. *J. Am. Chem. Soc.* **1999**, *121*, 9487–9496.

38. Erdodi, G.; Kennedy, J. P. *Prog. Polym. Sci.* **2006**, *31*, 1–18.

39. Bruns, N.; Scherble, J.; Hartmann, L.; Thomann, R.; Ivan, B.; Mülhaupt, R.; Tiller, J. C. *Macromolecules* **2005**, *38*, 2431–2438.

40. Bruns, N.; Tiller, J. C. *Nano Lett.* **2005**, *5*, 45–48.

41. Bruns, N.; Tiller, J. C. *Macromolecules* **2006**, *39*, 4386–4394.

42. Savin, G.; Bruns, N.; Thomann, Y.; Tiller, J. C. *Macromolecules* **2005**, *38*, 7536–7539.

43. Hanko, M.; Bruns, N.; Tiller, J. C.; Heinze, J. *Anal. Bioanal. Chem.* **2006**, *386*, 1273–1283.

44. Bruns, N.; Bannwarth, W.; Tiller, J. C. *Biotechnol. Bioeng.* **2008**, *101*, 19–26.

45. Bruns, N. Ph.D. Thesis, Albert-Ludwigs University of Freiburg, Freiburg, 2006.

46. Hanko, M.; Bruns, N.; Rentmeister, S.; Tiller, J. C.; Heinze, J. *Anal. Chem.* **2006**, *78*, 6376–6383.

47. Wang, C.; Stewart, R. J.; Kopecek, J. *Nature (London)* **1999**, *397*, 417–420.

48. Wang, C.; Kopecek, J.; Stewart, R. J. *Biomacromolecules* **2001**, *2*, 912–920.

49. Chen, L.; Kopecek, J.; Stewart, R. J. *Bioconjugate Chem.* **2000**, *11*, 734–740.

50. Miyata, T.; Asami, N.; Uragami, T. *Nature (London)* **1999**, *399*, 766–769.

51. Ehrick, J. D.; Deo, S. K.; Browning, T. W.; Bachas, L. G.; Madou, M. J.; Daunert, S. *Nat. Mater.* **2005**, *4*, 298–302.

52. Ehrick, J. D.; Stokes, S.; Bachas-Daunert, S.; Moschou, E. A.; Deo, S. K.; Bachas, L. G.; Daunert, S. *Adv. Mater. (Weinheim, Ger.)* **2007**, *19*, 4024–4027.

53. Murphy, W. L.; Dillmore, W. S.; Modica, J.; Mrksich, M. *Angew. Chem., Int. Ed.* **2007**, *46*, 3066–3069.

54. Sui, Z. J.; King, W. J.; Murphy, W. L. *Adv. Mater. (Weinheim, Ger.)* **2007**, *19*, 3377–3380.

55. Sui, Z. J.; King, W. J.; Murphy, W. L. *Adv. Funct. Mater.* **2008**, *18*, 1824–1831.

56. Yuan, W.; Yang, J.; Kopeckova, P.; Kopecek, J. *J. Am. Chem. Soc.* **2008**, *130*, 15760–15761.

57. Esser-Kahn, A. P.; Francis, M. B. *Angew. Chem., Int. Ed.* **2008**, *47*, 3751–3754.

58. Yang, J. Y.; Xu, C. Y.; Wang, C.; Kopecek, J. *Biomacromolecules* **2006**, *7*, 1187–1195.

59. Yang, J. Y.; Xu, C. Y.; Kopeckova, P.; Kopecek, J. *Macromol. Biosci.* **2006**, *6*, 201–209.

60. Yang, J.; Wu, K.; Konak, C.; Kopecek, J. *Biomacromolecules* **2008**, *9*, 510–517.

61. Dusek, K.; Duskova-Smrckova, M.; Yang, J.; Kopecek, J. *Macromolecules* **2009**, *42*, 2265–2274.

62. Radu-Wu, L. C.; Yang, J.; Wu, K.; Kopecek, J. *Biomacromolecules* **2009**, *10*, 2319–2327.

63. Esser-Kahn, A. P.; Iavarone, A. T.; Francis, M. B. *J. Am. Chem. Soc.* **2008**, *130*, 15820–15822.

64. Halstenberg, S.; Panitch, A.; Rizzi, S.; Hall, H.; Hubbell, J. A. *Biomacromolecules* **2002**, *3*, 710–723.

65. Rizzi, S. C.; Hubbell, J. A. *Biomacromolecules* **2005**, *6*, 1226–1238.

66. Rizzi, S. C.; Ehrbar, M.; Halstenberg, S.; Raeber, G. P.; Schmoekel, H. G.; Hagenmueller, H.; Mueller, R.; Weber, F. E.; Hubbell, J. A. *Biomacromolecules* **2006**, *7*, 3019–3029.

67. Tada, D.; Tanabe, T.; Tachibana, A.; Yamauchi, K. *J. Biosci. Bioeng.* **2005**, *100*, 551–555.

68. Tada, D.; Tanabe, T.; Tachibana, A.; Yamauchi, K. *Mater. Sci. Eng., C* **2007**, *27*, 895–897.

69. Tada, D.; Tanabe, T.; Tachibana, A.; Yamauchi, K. *Mater. Sci. Eng., C* **2007**, *27*, 870–874.

70. Ehrbar, M.; Schoenmakers, R.; Christen, E. H.; Fussenegger, M.; Weber, W. *Nat. Mater.* **2008**, *7*, 800–804.

71. King, W. J.; Mohammed, J. S.; Murphy, W. L. *Soft Matter* **2009**, *5*, 2399–2406.

72. Zhang, R.; Bowyer, A.; Eisenthal, R.; Hubble, J. *Adv. Polym. Technol.* **2008**, *27*, 27–34.

73. Bruns, N.; Pustelny, K.; Bergeron, L. M.; Whitehead, T. A.; Clark, D. S. *Angew. Chem., Int. Ed.* **2009**, *48*, 5666–5669.

Chapter 3

Biofabrication Based on the Enzyme-Catalyzed Coupling and Crosslinking of Pre-Formed Biopolymers

Yi Liu,[†] Xiaohua Yang,[†] Xiao-Wen Shi,[†] William E. Bentley,[†,‡] and Gregory F. Payne[*,†]

[†]Center for Biosystems Research, University of Maryland Biotechnology Institute, College Park, Maryland 20742, USA
[‡]Fischell Department of Bioengineering, University of Maryland, College Park, Maryland 20742, USA
[*]payne@umbi.umd.edu

Biology provides abundant materials and inspiring mechanisms for biofabrication. Our work forcuses on the use of pre-formed biological polymers and enzymes to fabricate functionalized macromolecular structures. Biological polymers often possess many unique properties and they are renewable and biodegradable (i.e. environmentally friendly). Enzymes provide catalytic mechanism for introducing covalent bonds for the site-selective crosslinking and conjugation of these pre-formed biopolymers. In one study, we employed microbial transglutaminase (mTG)-catalyzed crosslinking of gelatin to confer mechanical function for medical adhesive applications. This biomimetic adhesive should be less expensive and simpler to use than the commercial fibrin-based medical sealants. In another study, we combined the stimuli-responsive properties of chitosan and the residue specificity of two enzymes, tyrosinase and mTG, to assemble proteins at specific device addresses. This enzymatic assembly process is reagentless and the target proteins can be controllably assembled onto electrode surfaces. In sum, we believe these two studies illustrate that biofabrication has broad range of potential applications for tissue engineering and for interfacing biology to microelectronics.

Introduction

We define biofabrication as the use of biological materials and mechanisms for the construction of macromolecular structures. Biological materials such as polysaccharides and proteins are readily available from natural resources; they are renewable, biodegradable and environmentally friendly. Importantly, these materials possess unique properties (e.g. stimuli-responsive) that can be exploited in biofabrication. The biological mechanisms (i.e. enzymatic) to macromolecular construction are considered a "milder" alternative to conventional polymer synthesis and modification chemistries (1). In addition, enzymes allow the precise coupling of pre-formed biopolymers (2–4) or the hierarchical assembly of macromolecules to surfaces (5).

Enzymes Known To Couple Pre-Formed Biopolymers

We build the macromolecular structures using pre-formed biopolymers as starting materials. Biopolymers may serve as reactants toward enzymatic reactions but many reactions are hydrolytic, leading to the reduction in molecular weight. Few enzymes are known for the *in vitro* construction with biopolymers and we focus on two enzymes, *transglutaminases* and *tyrosinases*.

Transglutaminases catalyze the transamidation of glutamine (Gln) and lysine (Lys) residues to crosslink proteins (6) (Figure 1). One of the well-known tissue transglutaminases is factor XIIIa that is responsible for crosslinking fibrin monomers during the late stages of blood coagulation (7). Tissue transglutaminases have been studied for various potential applications and fibrin-based medical sealants are commercially available (8–11). However, applications of tissue transglutaminases beyond medicine have been limited by the cost of these blood-derived proteins and the needs for Ca^{2+} ions and thrombin for activation. In our study, we use Ca^{2+}-independent microbial transglutaminase (mTG) to crosslink and couple proteins. The mTG is less expensive and simpler to use than tissue transglutaminase (12, 13). In addition, mTG is expected to be safe because it has been developed for food applications (13–17). Importantly, mTG-catalyzed crosslinking involves an enzyme-bound intermediate with coupling occurring at the active site which offers considerable specificity to the coupling reactions (18, 19).

Another enzyme that is known to couple pre-formed biopolymers is *tyrosinase* (20). Tyrosinases are copper-containing phenol oxidases that use O_2 to selectively convert phenolic residues such as tyrosine (Tyr) into reactive *o*-quinones that can diffuse away from the enzyme's active site and undergo subsequent non-enzymatic reactions (Figure 2). Tyrosinase-catalyzed oxidation reactions are involved in many biological processes such as food browning (21, 22), wound sealing (23, 24), and most interestingly mussel glue curing (25–28). There are several examples of tyrosinase-mediated coupling of proteins to amine-containing polymers (29–32). This protein coupling method is simple, safe and selective toward Tyr residues. A possible disadvantage is that the "activated" quinone intermediates are reactive and can potentially undergo undesired side-reactions.

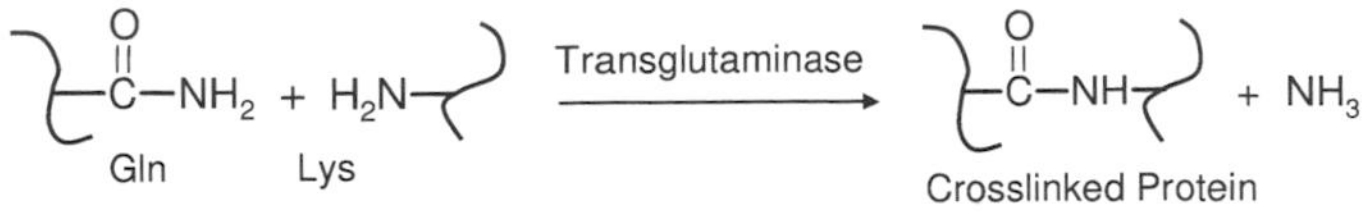

$$\text{Gln: } \underset{\text{Gln}}{\overset{\overset{\displaystyle O}{\|}}{-C-NH_2}} + \underset{\text{Lys}}{H_2N-} \quad \xrightarrow{\text{Transglutaminase}} \quad \underset{\text{Crosslinked Protein}}{\overset{\overset{\displaystyle O}{\|}}{-C-NH-}} + NH_3$$

Figure 1. Transglutaminase-catalyzed crosslinking of proteins.

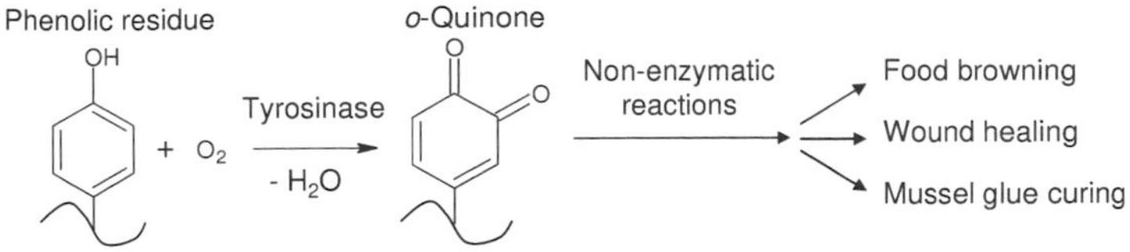

Phenolic residue   o-Quinone

OH   O

$+ O_2 \quad \xrightarrow[\text{- } H_2O]{\text{Tyrosinase}}$

Non-enzymatic reactions $\longrightarrow$ Food browning / Wound healing / Mussel glue curing

Figure 2. Tyrosinase-catalyzed oxidation of phenolic residues and subsequent non-enzymatic reactions.

Fusion Tags for Enzymatic Reactions of Globular Proteins

One advantage of enzymatic crosslinking/coupling is the amino acid residue specifity, however, the residues must be accessible to the enzyme. The residues on open chain proteins (e.g. gelatin) are readily accessible to enzymes (*33, 34*), whereas those on globular proteins may not be accessible (*35–37*). To facilitate the enzymatic coupling of globular proteins, one can genetically engineer the gene so that the protein is translated to have a short unstructured amino acids sequence (i.e. fusion tag) that is "exposed" for the enzyme. For example, Tyr residues have been fused to enhance the tyrosinase-catalyzed oxidation reaction (*38, 39*); Gln or Lys residues have been fused to augment the number of accessible residues toward mTG (*40, 41*).

Applications of Enzymatic Crosslinking and Coupling of Biopolymers

We envision that enzymatic crosslinking and coupling of pre-formed biopolymers provides new opportunities for biofabrication. As will be discussed in the following, our biofabrication approaches exploit the unique capabilities of enzymes, fusion tag proteins, and biopolymers.

Enzyme-Catalyzed Crosslinking of Gelatin as Medical Adhesive

When microbial transglutaminase (mTG) is added to gelatin-containing solutions, a crosslinked hydrogel network is formed as a result of the reaction in Figure 1 (*6*). Previous studies in our group have shown that during gelation, the mTG-catalyzed gel bonds with moist tissue and the adhesive strength is comparable to, or better than, fibrin-based sealants (*42, 43*). Recently, we evaluated the potential of the mTG-gelatin adhesive as a hemostatic sealant (*44*). To serve as a hemostatic sealant, the adhesive needs to adhere on wet tissue surface and offer the adhesive and cohesive strength to restrain bleeding. We performed *in vitro* tests to evaluate the gel-forming process of the mTG-gelatin

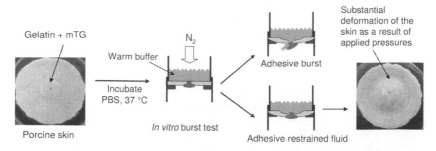

Figure 3. In vitro burst pressure test of the mTG-gelatin adhesive.

adhesive's capability to restrain fluid at physiological pressure (i.e. blood pressure). As illustrated in Figure 3, a porcine skin sample was cut into a circular shape and a small hole was punched at the center. The sample was put in a petri dish with PBS buffer so that the skin surface remained wet. Then a warm mTG-gelatin mixture was applied over the hole and the sample was cured at 37 °C. The prepared sample was then assembled in a custom-built apparatus where a warm buffer was poured over the skin and N_2 gas was introduced to gradually raise the pressure. The burst pressure is the maximum pressure observed prior to bursting (i.e. failure) of the sealant. We obtained an average burst pressure of 320 ± 50 mmHg, indicating that the strength of the gel formed by mixing mTG and gelatin are appropriate to restrain fluid at blood pressure.

Subsequent animal studies indicate that this mTG-gelatin adhesive cures within 5 minutes, adheres to tissue in the presence of a modest amounts of blood, and can stop arterial bleeding (*44*). These *in vitro* and *in vivo* results are promising and the next step would require long-term tests of biocompatibility and biodegradability of the adhesive to fully assess the potential of the mTG-gelatin adhesive as a surgical sealant.

Enzyme-Mediated Protein Assembly on Electrode Surface

There is considerable interest in interfacing biology to devices and we believe that stimuli-responsive aminopolysaccharide chitosan offers a number of unique properties to integrate proteins and microelectronics (*45, 46*). As indicated in Figure 4, chitosan is soluble under acidic condition and becomes insoluble as the pH increases (i.e. chitosan is pH-responsive). The pKa of chitosan is ~6.5, indicating that the soluble/insoluble transition occurs under conditions that are sufficiently mild to accommodate the labile nature of biological components. When pH is raised, chitosan can form a stable hydrogel film. In addition, chitosan has nucleophilic amines which allows biopolymers to be covalently attached through a range of mechanisms (*38, 47–50*).

Electrodeposited Chitosan Film as Bio-Device Interface

A unique feature of chitosan's pH-responsive film-forming properties is that it "recognizes" electric stimuli and responds by depositing as a thin film on an

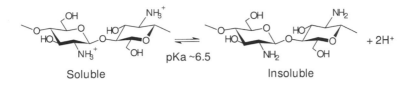

Soluble   pKa ~6.5   Insoluble   + 2H⁺

Figure 4. The primary amines in chitosan's repeating glucosamine residues confer pH-responsive properties.

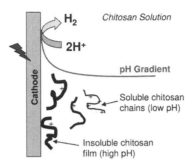

Figure 5. The working mechanism of chitosan electrodeposition process.

electrode surface. As shown in Figure 5, when a voltage is applied, a localized pH gradient is generated near the cathode as a result of electrochemical consumption of protons at the electrode surface. The localized high pH induces chitosan to undergo a sol-gel transition to form a thin film at the cathode surface (*51–54*). The electrodeposited chitosan film is stable in the absence of an applied voltage as long as the pH is kept above 6.5, however, the chitosan film can redissolve at lower pHs. We have demonstrated that the thickness of the electrodeposited chitosan film can be controlled by the deposition conditions (*51*) and that deposition is spatially-selective in the lateral dimensions (*52*). Importantly, the electrodeposited chitosan film can be (bio)functionalized to incorporate various components (*55–59*) and the biofunctionalized chitosan may assist in converting biological events into device-compatible (e.g. electric or optical) signals (*60–62*).

Tyrosinase-Mediated Protein Coupling on Chitosan-Coated Electrodes

The interfacing of proteins to electronic devices may be achieved by functionalizing electrodeposited chitosan films using enzyme-mediated protein coupling reactions. As shown in Figure 6, chitosan is first deposited on an electrode surface and the film is washed with water. The chitosan-coated electrode is immersed in a buffer solution containing Tyr tagged proteins, and then a small amount of tyrosinase is added to catalyze the oxidation of Tyr residues to generate active intermediates that react with chitosan to form protein-chitosan conjugates.

The tyrosinase-mediated protein assembly approach is simple, safe and robust, and a number of protein-chitosan conjugates have been successfully prepared in our lab and our collaborators' (*38, 50, 57, 63*). In addition, this

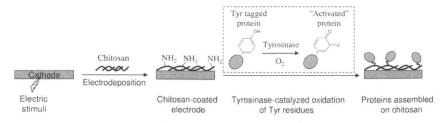

Figure 6. Tyrosinase-mediated coupling of Tyr-tagged protein on chitosan.

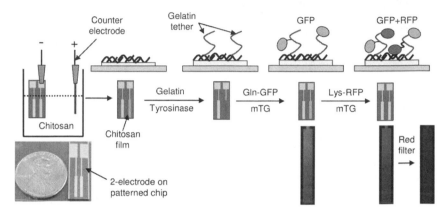

Figure 7. General procedure of orthogonal enzyme-mediated assembly of target proteins on electrode addresses.

tyrosinase-mediated protein coupling is being extended to Lab-on-a-Chip applications (*64, 65*).

Orthogonal Enzymatic Reactions To Assemble Proteins on Electrodes

In some cases, tethers have been found to be useful for the surface assembly of proteins. Recently, we developed a surface assembly approach that uses orthogonal enzymatic reactions to assemble target proteins on specific electrode addresses (*41*). The general procedure of this surface protein assembly is illustrated in Figure 7. First, chitosan is electrodeposited onto one of the gold electrodes on a patterned chip. Second, gelatin molecules are grafted to chitosan by tyrosinase-mediated coupling reaction. Gelatin has very few Tyr residues which are located at the end of the polypeptide chain (*33, 34*). As discussed above, tyrosinase catalyzes the oxidation of the terminal Tyr residues and the "activated" gelatin undergoes uncatalyzed coupling reactions with chitosan (following similar reaction scheme shown in Figure 6). Since gelatin has an open chain structure, it can serve as a flexible tether on which further protein assembly may be carried out.

The final steps shown in Figure 7 are the sequential assembly of two model target proteins to the gelatin tether using mTG-catalyzed coupling. The first target protein, Gln-tagged green fluorescent protein (Gln-GFP), is assembled by coupling with Lys residues on the gelatin tethers. The grafting of the Gln-GFP is confirmed by the green fluorescence image observed on the electrode. Subsequently, the second protein, Lys-tagged red fluorescent protein (Lys-RFP), is assembled on the same electrode address by coupling with Gln residues on the gelatin tethers. The yellow color of the image is the result of a combination of green and red fluorescence from Gln-GFP and Lys-RFP, respectively. When only the red filter is used, the fluoscence of the electrode surface appears red, demonstrating that the second target protein, Lys-RFP, has been successfully assembled on the same electrode address. We should note the possibility that mTG could crosslink the gelatin tethers, although the results in Figure 7 indicate that the tethers retain sufficient Lys and Gln residues to allow the mTG-catalyzed assembly of the Gln-GFP and Lys-RFP target proteins.

The orthogonal enzyme-mediated protein assembly approach provides an alternative to existing chemical coupling methods. The most obvious potential advantage of this enzymatic assembly approach is that it does not require reactive reagents or harsh conditions. In addition, the biopolymers (chitosan and gelatin) and enzymes (tyrosinase and mTG) used for protein assembly are either derived from foods or approved for food use and thus would be expected to be appropriate for devices that directly contact food, pharmaceutical, or cosmetic ingredients.

Conclusion

In this paper, we demonstrate the use of enzymatic reactions of biological polymers to fabricate gelatin-based medical adhesive and to assemble proteins on electrode surfaces. We believe these two studies illustrate a broad range of possibilities. First, the mTG-catalyzed crosslinking of gelatin suggests that enzymes and pre-formed biopolymers are capable of producing biocompatible macromolecular networks. Potentially, enzymatic construction with biopolymers may have broader impact in creating soft materials, such as tissue engineering scaffolds, which would be useful in regenerative medicine. Second, the hierarchical assembly of proteins on the pH-responsive chitosan suggests that biopolymers have unique properties to integrate biology and electronics. We envision significant opportunities for the fabrication of bio-electronics (e.g. biosensors, lab-on-a-chip and point-of-care devices), in which biological and electric signals "communicate" with each other. In general, we believe biofabrication based on enzymes and pre-formed biopolymers provide exciting opportunities to build functionalized macromolecular structures.

Acknowledgments

The authors gratefully acknowledge financial support from the R. W. Deutsch Foundation and the National Science Foundation (CBET-0650650 and EFRI-0735987).

References

1. Cheng, H. N.; Gross, R. A. *Polymer Biocatalysis and Biomaterials*; American Chemical Society: Washington, DC, 2005.
2. Mao, H.; Hart, S. A.; Schink, A.; Pollok, B. A. *J. Am. Chem. Soc.* **2004**, *126*, 2670–2671.
3. Popp, M. W.; Antos, J. M.; Grotenbreg, G. M.; Spooner, E.; Ploegh, H. L. *Nat. Chem. Biol.* **2007**, *3*, 707–708.
4. Pritz, S.; Wolf, Y.; Kraetke, O.; Klose, J.; Bienert, M.; Beyermann, M. *J. Org. Chem.* **2007**, *72*, 3909–3912.
5. Parthasarathy, R.; Subramanian, S.; Boder, E. T. *Bioconjugate Chem.* **2007**, *18*, 469–476.
6. Babin, H.; Dickinson, E. *Food Hydrocolloids* **2001**, *15*, 271–276.
7. Adany, R.; Bardos, H. *Cell Mol. Life Sci.* **2003**, *60*, 1049–1060.
8. Jackson, M. R. *Am. J. Surg.* **2001**, *182*, 1S–7S.
9. Spotnitz, W. D. *Am. J. Surg.* **2001**, *182*, 8S–14S.
10. Albala, D. M. *Cardiovasc. Surg.* **2003**, *11* (1), 5–11.
11. Buchta, C.; Hedrich, H. C.; Macher, M.; Hocker, P.; Redl, H. *Biomaterials* **2005**, *26*, 6233–6241.
12. Shimba, N.; Yokoyama, Y.; Suzuki, E. *J. Agric. Food Chem.* **2002**, *50*, 1330–1334.
13. Yokoyama, K.; Nio, N.; Kikuchi, Y. *Appl. Microbiol. Biotechnol.* **2004**, *64*, 447–454.
14. Broderick, E. P.; O'Halloran, D. M.; Rochev, Y. A.; Griffin, M.; Collighan, R. J.; Pandit, A. S. *J. Biomed. Mater. Res., Part B* **2005**, *72B*, 37–42.
15. Tang, C. H.; Wu, H.; Yu, H. P.; Li, L.; Chen, Z.; Yang, X. Q. *J. Food Biochem.* **2006**, *30*, 35–55.
16. Dube, M.; Schafer, C.; Neidhart, S.; Carle, R. *Eur. Food Res. Technol.* **2007**, *225*, 287–299.
17. Ozer, B.; Kirmaci, H. A.; Oztekin, S.; Hayaloglu, A.; Atamer, M. *Int. Dairy J.* **2007**, *17*, 199–207.
18. Tominaga, J.; Kamiya, N.; Doi, S.; Ichinose, H.; Maruyama, T.; Goto, M. *Biomacromolecules* **2005**, *6*, 2299–2304.
19. Kamiya, N.; Doi, S.; Tominaga, J.; Ichinose, H.; Goto, M. *Biomacromolecules* **2005**, *6*, 35–38.
20. Mayer, A. M. *Phytochemistry* **2006**, *67*, 2318–2331.
21. Whitaker, J. R.; Lee, C. Y. *Enzymatic Browning and Its Prevention*; American Chemical Society: Washington, DC, 1995; Vol. 600.
22. Yoruk, R.; Marshall, M. R. *J. Food Biochem.* **2003**, *27*, 361–422.
23. Sugumaran, H. *Pigment Cell Res.* **2002**, *15*, 2–9.
24. Sugumaran, M.; Hennigan, B.; O'Brien, J. *Arch. Insect Biochem. Physiol.* **2005**, *6*, 9–25.
25. Burzio, L. A.; Waite, J. H. *Biochemistry* **2000**, *39*, 11147–11153.
26. Burzio, L. A.; Waite, J. H. *Protein Sci.* **2001**, *10*, 735–740.
27. McDowell, L. M.; Burzio, L. A.; Waite, J. H.; Schaefer, J. *J. Biol. Chem.* **1999**, *274*, 20293–20295.
28. Waite, J. H. *Int. J. Biol. Macromol.* **1990**, *12*, 139–144.

29. Ahmed, S. R.; Lutes, A. T.; Barbari, T. A. *J. Membr. Sci.* **2006**, *282*, 311–321.
30. Freddi, G.; Anghileri, A.; Sampaio, S.; Buchert, J.; Monti, P.; Taddei, P. *J. Biotechnol.* **2006**, *125*, 281–294.
31. Kang, G. D.; Lee, K. H.; Ki, C. S.; Nahm, J. H.; Park, Y. H. *Macromol. Res.* **2004**, *12*, 534–539.
32. Selinheimo, E.; Lampila, P.; Mattinen, M. L.; Buchert, J. *J. Agric. Food Chem.* **2008**, *56*, 3118–3128.
33. King, G.; Brown, E. M.; Chen, J. M. *Protein Eng.* **1996**, *9*, 43–49.
34. Mayo, K. H. *Biopolymers* **1996**, *40*, 359–370.
35. Matsumura, Y.; Chanyongvorakul, Y.; Kumazawa, Y.; Ohtsuka, T.; Mori, T. *Biochim. Biophys. Acta, Protein Struct. Mol. Enzymol.* **1996**, *1292*, 69–76.
36. Nieuwenhuizen, W. F.; Dekker, H. L.; De Koning, L. J.; Groneveld, T.; De Koster, C. G.; De Jong, G. A. H. *J. Agric. Food Chem.* **2003**, *51*, 7132–7139.
37. Nieuwenhuizen, W. F.; Dekker, H. L.; Groneveld, T.; de Koster, C. G.; de Jong, G. A. H. *Biotechnol. Bioeng.* **2004**, *85*, 248–258.
38. Lewandowski, A. T.; Small, D. A.; Chen, T. H.; Payne, G. F.; Bentley, W. E. *Biotechnol. Bioeng.* **2006**, *93*, 1207–1215.
39. Stayner, R. S.; Min, D. J.; Kiser, P. F.; Stewart, R. J. *Bioconjugate Chem.* **2005**, *16*, 1617–1623.
40. Tanaka, T.; Kamiya, N.; Nagamune, T. *Bioconjugate Chem.* **2004**, *15*, 491–497.
41. Yang, X. H.; Shi, X. W.; Liu, Y.; Bentley, W. E.; Payne, G. F. *Langmuir* **2009**, *25*, 338–344.
42. McDermott, M. K.; Chen, T. H.; Williams, C. M.; Markley, K. M.; Payne, G. F. *Biomacromolecules* **2004**, *5*, 1270–1279.
43. Chen, T.; Janjua, R.; McDermott, M. K.; Bernstein, S. L.; Steidl, S. M.; Payne, G. F. *J. Biomed. Mater. Res., Part B* **2006**, *77*, 416–422.
44. Liu, Y.; Kopelman, D.; Wu, L.-Q.; Hijji, K.; Attar, I.; Preiss-Bloom, O.; Payne, G. F. *J. Biomed. Mater. Res., Part B* **2009**, *91B*, 5–16.
45. Yi, H. M.; Wu, L. Q.; Bentley, W. E.; Ghodssi, R.; Rubloff, G. W.; Culver, J. N.; Payne, G. F. *Biomacromolecules* **2005**, *6*, 2881–2894.
46. Payne, G. F.; Raghavan, S. R. *Soft Matter* **2007**, *3*, 521–527.
47. Shi, X. W.; Liu, Y.; Lewandowski, A. T.; Wu, L. Q.; Wu, H. C.; Ghodssi, R.; Rubloff, G. W.; Bentley, W. E.; Payne, G. F. *Macromol. Biosci.* **2008**, *8*, 451–457.
48. Vazquez-Duhalt, R.; Tinoco, R.; D'Antonio, P.; Topoleski, L. D. T.; Payne, G. F. *Bioconjugate Chem.* **2001**, *12*, 301–306.
49. Shi, X. W.; Yang, X. H.; Gaskell, K. J.; Liu, Y.; Kobatake, E.; Bentley, W. E.; Payne, G. F. *Adv. Mater.* **2009**, *21*, 984–988.
50. Chen, T. H.; Small, D. A.; Wu, L. Q.; Rubloff, G. W.; Ghodssi, R.; Vazquez-Duhalt, R.; Bentley, W. E.; Payne, G. F. *Langmuir* **2003**, *19*, 9382–9386.
51. Wu, L. Q.; Gadre, A. P.; Yi, H. M.; Kastantin, M. J.; Rubloff, G. W.; Bentley, W. E.; Payne, G. F.; Ghodssi, R. *Langmuir* **2002**, *18*, 8620–8625.
52. Wu, L. Q.; Yi, H. M.; Li, S.; Rubloff, G. W.; Bentley, W. E.; Ghodssi, R.; Payne, G. F. *Langmuir* **2003**, *19*, 519–524.
53. Fernandes, R.; Wu, L. Q.; Chen, T. H.; Yi, H. M.; Rubloff, G. W.; Ghodssi, R.; Bentley, W. E.; Payne, G. F. *Langmuir* **2003**, *19*, 4058–4062.

54. Pang, X.; Zhitomirsky, I. *Mater. Chem. Phys.* **2005**, *94*, 245–251.
55. Pang, X.; Zhitomirsky, I. *Mater. Charact.* **2007**, *58*, 339–348.
56. Wu, L. Q.; Lee, K.; Wang, X.; English, D. S.; Losert, W.; Payne, G. F. *Langmuir* **2005**, *21*, 3641–3646.
57. Fernandes, R.; Tsao, C. Y.; Hashimoto, Y.; Wang, L.; Wood, T. K.; Payne, G. F.; Bentley, W. E. *Metab. Eng.* **2007**, *9*, 228–239.
58. Yi, H. M.; Nisar, S.; Lee, S. Y.; Powers, M. A.; Bentley, W. E.; Payne, G. F.; Ghodssi, R.; Rubloff, G. W.; Harris, M. T.; Culver, J. N. *Nano Lett.* **2005**, *5*, 1931–1936.
59. Luo, X. L.; Xu, J. J.; Zhang, Q.; Yang, G. J.; Chen, H. Y. *Biosens. Bioelectron.* **2005**, *21*, 190–196.
60. Luo, X. L.; Xu, J. J.; Du, Y.; Chen, H. Y. *Anal. Biochem.* **2004**, *334*, 284–289.
61. Meyer, W. L.; Liu, Y.; Shi, X. W.; Yang, X. H.; Bentley, W. E.; Payne, G. F. *Biomacromolecules* **2009**, *10*, 858–864.
62. Liu, Y.; Gaskell, K. J.; Cheng, Z. H.; Yu, L. L.; Payne, G. F. *Langmuir* **2008**, *24*, 7223–7231.
63. Wu, H. C.; Shi, X. W.; Tsao, C. Y.; Lewandowski, A. T.; Fernandes, R.; Hung, C. W.; DeShong, P.; Kobatake, E.; Valdes, J. J.; Payne, G. F.; Bentley, W. E. *Biotechnol. Bioeng.* **2009**, *103*, 231–240.
64. Lewandowski, A. T.; Bentley, W. E.; Yi, H.; Rubloff, G. W.; Payne, G. F.; Ghodssi, R. *Biotechnol. Prog.* **2008**, *24*, 1042–1051.
65. Luo, X. L.; Lewandowski, A. T.; Yi, H. M.; Payne, G. F.; Ghodssi, R.; Bentley, W. E.; Rubloff, G. W. *Lab Chip* **2008**, *8*, 420–430.

Chapter 4

Development of Novel Soy Protein-Based Polymer Blends

Jinwen Zhang* and Feng Chen

Composite Materials and Engineering Center, Washington State University,
Pullman, WA 99163
*jwzhang@wsu.edu

A novel approach was used in the preparation of soy protein concentrate (SPC) and biodegradable thermoplastic blends. In contrast to many other soy protein (SP) blends where SP functioned merely as a filler, in this study SPC was processed as a plastic in blending with poly(lactic acid) (PLA) and poly(butylene adipate-co-terephthalate) (PBAT), respectively. The plastication of SP and blending of the resulting SP plastic with PLA or PBAT were performed in the same extrusion process. The plastication involved gelation of SPC in the presence of water and subsequent plasticization of the gelated SPC by water and/or other plasticizer. The resulting PLA/SPC blends displayed a co-continuous phase structure and PBAT/SPC blends a percolated SPC network structure. Consequently, the blends demonstrated superior mechanical properties.

Introduction

As a renewable polymer, SP has received great interest in non-food industrial applications. Bioplastics development is one of the high interest areas where SP may play an important role in the near future. Current plastic materials are almost entirely made from petroleum feedstock and consume a significant portion of limited petroleum resources. The demand for plastics is still growing steadily, which will further complicate the situation in view of an unstable oil supply and other associated problems. In addition, the disposal of traditional plastics such as polyethylene (PE) and polypropylene (PP) after use has posed a serious threat to the environment due to their non-biodegradability. SP is an abundantly available and inexpensive plant polymer; as a result, SP-based plastics have received considerable attention in recent years.

The use of SP as an ingredient in plastics can be traced back to the 1930s and 1940s when soy flour was incorporated into phenolic resins (1). In that case SP was used as a reactive component in phenolic resins and was hardened by formaldehyde, so the resulting resins displayed very low water absorption. Since then little progress has been made on SP plastics because of the advent of low-cost petroleum-based plastics; not until the 1990s did SP-based plastic regain research interest due to its environmentally benign advantages. In this chapter, the current status of SP plastics will be briefly reviewed, and the results from our recent new investigation of SP blends will be discussed in detail.

Current Status of SP Plastics Research

In general, the role of SP in current bioplastics research can be classified into two types: as a thermoplastic for neat SP plastics or matrix polymer, and as filler for thermoplastics or thermosetting resins. In addition, chemical modification of SP has also studied to improve the physical and mechanical properties of SP plastics.

SP as a Thermoplastic

When heated, SP can undergo gelation in the presence of adequate water. The gelated SP can be further plasticized by water and other plasticizers and behave like a thermoplastic under shear stress. Compression molding is a simple method for processing SP plastics. By adding a small amount of extra water and/or plasticizers, native SP can be compression-molded (2–4). However, the strong intra- and intermolecular interactions of SP usually result in very high melt viscosity that makes melt processing such as extrusion and injection molding difficult unless a large amount of plasticizer is present. We previously demonstrated that in the presence of sufficient water and/or other plasticizers, soy protein isolate (SPI) could be extruded into clear sheets (5). Water is not only necessary for the gelation of SP, but is also the most efficient plasticizer for SP. However, water evaporates easily during storage and the products can become very brittle. Therefore, other plasticizers are needed for SP plastics. In 1939, Brother and McKinney (1) tested 70 commercially available plasticizers on formaldehyde-hardened SP and found that ethylene glycol performed better than the others. They also found that oleanolic acid and aluminum stearate in combination with ethylene glycol worked very well in reducing water absorption. Currently, glycerol, ethylene glycol, propylene glycol are usually employed in SP plastics (3, 5, 6).

Table 1 shows the tensile property data of neat SPI sheets with varying concentrations of glycerol as plasticizer. The sheets containing 10% glycerol demonstrated fairly high strength but very low elongation. Large decreases in strength and Young's modulus were noted when glycerol concentration increased from 20 to 30 parts. Further increase sin glycerol concentration resulted in

Table 1. Effect of glycerol concentration on tensile properties of SPI sheets[a]

Glycerol (parts)[b]	Strength (MPa)	Modulus (MPa)	Elongation (%)
10	40.6±1.0	1226±85	3±1
20	34.0±0.8	1119±70	74±8
30	15.6±0.4	374±70	133±10
40	9.1±0.2	176±12	159±12
50	7.1±0.4	144±16	185±15

[a] Modified from Table 1 in the reference (5).   [b] On the basis of 100-part dry SPC by weight.

seriously detrimental effects on the overall mechanical properties. Hence, it seems that the formulation with 20% glycerol demonstrated optimal overall properties. Partial substitution of glycerol with methyl glucoside resulted in a higher T_g and therefore more rigid sheets, but reduced the flowability of the SP plastic melt (5). Other processing aids may be employed, such as sodium tripolyphosphate for interrupting soy protein ionic interactions (3, 7) or sodium sulfite as a reducing agent to break the disulfide bonds (8). These processing aids help to improve the denaturation of protein and slightly increase the moisture resistance of SP-based plastics (9).

SP can also act as a matrix material in composites. For example, natural fibers such as kenaf (10), chitin whisker (11), ramie (12) and konjack (13) have been used to reinforce SP plastics. However, the already low flowability of SP plastic is further reduced by the incorporation of reinforcing fillers. Lubricant which facilitates the processibility of high molecular polymers in both molding and extrusion is also an important additive in soy plastic processing. Nevertheless, neat SP plastics and blends or composites with SP as matrix polymer still retain some serious problems including water/moisture sensitivity, narrow processing window, low impact strength and brittleness.

SP as a Filler

Blending is often the most efficient and economic means to address many of the problems SP-based bioplastics face. Blending SP with biodegradable thermoplastic polymers is particularly interested because the resulting blends retain complete biodegradability. These polymers include poly (butylenes succinate-co-adipate) (14), polycaprolactone (3, 15), poly (hydroxyl ester ether) (16) and poly (butylenes adipate-co-terephthalate) (17). Blending SP with non-biodegradable polymers including poly(ethylene-co-ethyl acrylate-co-maleic anhydride) (18), polyurethane (19) and styrene-butadiene latex (20) have also been studied. SP is a pressure-sensitive thermoset polymer; it will unfold/melt (paste-like) and cure under both elevated temperature and pressure. Most functional groups in SP are polar, while most thermoplastics are nonpolar (21). Because of the large disparity in molecular structures between SP and these thermoplastics, the interfacial adhesion in many blends is not strong enough to yield satisfactory mechanical properties. Improving interfacial adhesion by

47

adding appropriate compatibilizer or coupling agent in the blends has proved to be an efficient means to obtain fine phase structures and improved mechanical properties. However, SP was merely used as organic particulate filler dispersed in the matrix polymer in these blends. Although the blends generally displayed greater water resistance, processibility and toughness than neat SP plastics, and greater stiffness than the soft matrix polymers, they often exhibited less strength than both the neat SP plastics and polymers.

Because the free hydroxyl and amino groups in SP can participate in the cure reactions of thermosetting resin, SP can be used as a reactive additive in thermosetting resins. These functional groups can also be utilized for chemical modification to improve physical and mechanical properties. For example, Wu *et al.* blended SP with polyurethane prepolymer based on polycaprolactone polyol/hexamethylene diisocyanate (*22*); and the formation of urea-urethane linkages was detected in the cured resins, suggesting that the amino groups of SP were involved in the curing. The polyurethane prepolymer modified soy protein plastic demonstrated superior toughness and water resistance.

A Novel Method for SP Blends- Processing SP as a Plastic in Blending

Since SP can be processed like a thermoplastic in the presence of sufficient water and/or other plasticizers, it is important to understand how this thermoplastic behavior will influence the morphology and properties of the resulting blends. Numerous studies have shown that the properties of a polymer blend are largely determined by its phase morphology. The advantage of processing SP as plastic rather than as a filler is that it allows a flexible design of the morphological structure of the blends, and hence the ability to manipulate their properties. In fact, processing SP as a thermoplastic involves the very same two fundamental requirements as does processing starch as a thermoplastic: gelation in the presence of sufficient water and subsequent plasticization of the gelated polymer by water and/or other plasticizers. Blends based on thermoplastic starch have demonstrated diverse morphological structures and hence varying performance. Thermoplastic starch blends have found some niche commercial applications. It is reasonable to consider that SP could also be similarly used as a plastic in blends with other thermoplastic polymers. In this study, two biodegradable thermoplastic polymers, PLA and PBAT, were selected to blend with soy protein concentrate (SPC). PLA exhibits high strength and modulus but low elongation, while PBAT displays low strength but high flexibility and high ductility. A one-step method was adopted for the preparation of blends, meaning that the plasticization of SPC in the presence of water and/or other plasticizers and blending between the resulting SPC plastic and polymer were accomplished in one extrusion process. One benefit of using a one-step method is that the blending can be carried out at a lower temperature than it usually takes to process the neat polymer because of the water and/or other plasticizers present in the pre-compounding SPC; another benefit is that there is no need for a large amount of water and other plasticizers in the plasticization of SPC because the thermoplastic polymer introduces good flowability.

General Preparation Method for SP as a Plastic in Polymer Blends

SPC (Arcon F) used in this study was obtained from Archer Daniels Midland Company, containing ca. 69% protein (on dry basis), 20% carbohydrate and 7.5% moisture as received. PLA (4032D) was a commercial product from NatureWorks. PBAT (Ecoflex F) was purchased from BASF. Poly(2-ethyl-2-oxazoline) (PEOX) with a weight average molecular weight of ca. 500 KDa was obtained from Aldrich, and was used as the compatibilizer for PLA/SPC blends. Maleic anhydride grafted PBAT (MA-g-PBAT) was prepared in our laboratory and used as the compatiblizer for PBAT/SPC blends. The reaction was conducted in melt reaction, and dicumyl peroxide was used as the initiator by forming free radicals. The unreacted anhydride was removed by sublimation at 80 °C in a vacuum oven. The degree of grafting was measured by the titration method and was found to be ca. 1.4%.

Prior to blending, SPC was first mixed with a predetermined amount of water, plasticizer, Na_2SO_3 and other additives using a kitchen mixer. The formulated SP was sealed in a plastic bag and kept overnight at room temperature to equilibrate before compounding. Compounding extrusion was carried out using a co-rotating twin-screw extruder (Leistriz ZSE-18) equipped with 17.8 mm screws having an L/D ratio of 40. The extruder has eight controlled heating zones ranging from the zone next to the feeding segment to the die adaptor. The temperature profile of these heating zones ranged from 90 to 160 °C for blending SPC with PLA and from 90 to 145 °C for blending SPC with PBAT. The mixture of SP, polymer and compatibilizer was fed by a volumetric feeder, and the extrudate was cooled in a water bath and subsequently granulated by a strand pelletizer. Test specimens were prepared using a Sumitomo injection molding machine (SE 50D) with four barrel temperature zones from the feeding section to the nozzle. The pellets obtained were dried in a convection oven at 90 °C for at least 8 h before injection molding.

Blending SPC with a Rigid Thermoplastic – *PLA/SPC Blends*

In this part of study, the formulation of pre-compounding SP was: SPC (100 parts, dry weight), Na_2SO_3 (0.5 parts), sodium tripolyphosphate (1 part), lubricant (0.8 parts), glycerol (2 parts) and water (10 parts).

Phase Structure

Figure 1 shows the phase morphology of PLA/SPC blends with different compositions. The samples for SEM analysis were prepared by etching the cryo-fractured surfaces of the injection-molded specimens with a buffer solution to remove the SP phase or with chloroform to remove the PLA phase. The SEM micrographs (Figure 1) show that blends with the SPC/PLA (w/w) ratio ranging from 30/70 to 70/30 all exhibited a fine co-continuous phase structure. This result demonstrates that SPC functioned like a plastic component in PLA/SPC blends, in contrast to most other polymer/SP blends in which SP performs merely as a particulate filler. The water added in pre-compounding SPC played a critical role

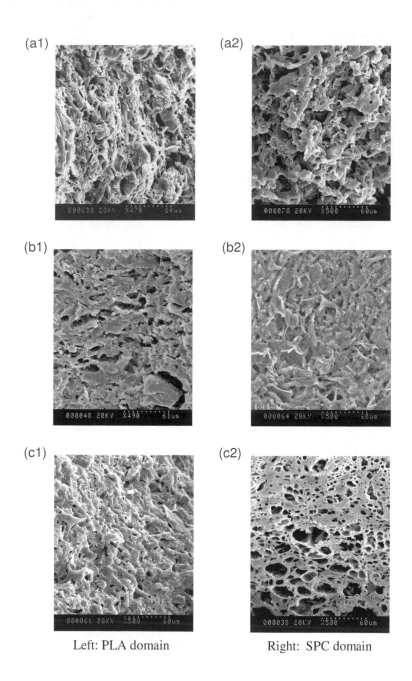

Left: PLA domain   Right: SPC domain

Figure 1. SEM micrographs of SPC and PLA phases in blends containing 3 phr PEOX. (a1) & (a2): SPC/PLA = 70/30 (w/w); (b1) & (b2): SPC/PLA = 50/50 (w/w); (c1) & (c2): SPC/PLA = 30/70 (w/w). Samples were fractured in the transverse direction; surfaces were etched with a buffer solution to remove SP or with CHCl₃ to remove PLA.

in converting SPC into plastic. In the compounding extrusion process, gelation of SP occurred under heating and in the presence of sufficient water; subsequently the gelated SP was plasticized by water and other plasticizer(s). At certain water and plasticizer concentrations, the gelated SP possessed appropriate flowability (or rheological properties) as in the case where neat SPI was extruded and was able to flow under the processing conditions, therefore achieving a good blend with PLA.

Dynamic Rheological Properties

In the terminal (low frequency) region, the PLA melt exhibited a typical liquid-like behavior like most other thermoplastic polymers. Logarithmic storage (G') or loss (G'') modulus versus logarithmic angular frequency (ω) showed a smooth linear relationship by G' $\propto$ $\omega^{1.64}$ and G'' $\propto$ ω (Figures 2a and b) which was close to the theoretical prediction (G' $\propto$ ω^2 and G'' $\propto$ ω) for a typical linear polymer with narrow molecular weight distribution. PLA/SPC blends displayed not only much higher η^*, G' and G'' in the terminal region than PLA but also a drastically different terminal behavior. In the terminal region, log(G') and log(G'')) vs. log(ω) of the blends tended to approach a (secondary) plateau, while log (η^*) vs. log (ω) changed from a Newtonian (primary) plateau (Figure 2c) for the neat PLA to a continuous decrease with increasing ω for the blends. The deviation of the blends from the neat PLA at the terminal region increased with increasing SPC content in the blends. These changes in rheological properties suggest that a characteristic solid- or gel-like network structure existed in the melt (*23*). The rheological behavior of the blends corresponded to the evidence of SPC morphological structure in the blends. The rheological behavior of the blends indicates that the presence of SP phase greatly hindered the movement of polymer molecular chains in the melt state. Considering that both the residual moisture and glycerol concentrations were very low in the final blend products and crosslinking of SP also occurred to a certain degree during the compounding and molding processes, it is understandable that the SPC phase would possess very low flowability in the testing environment and behaved more like a solid domain.

Mechanical Properties and Water Absorption

Figure 3 shows the tensile strength and modulus of SPC/PLA blends. Tensile strength increased steadily with increasing PLA concentration in the blends. The blends containing 30 and 50% SPC-H_2O retained 81 and 65% strength of neat PLA (63 MPa), respectively. On the other hand, modulus of the blends decreased slightly with increasing PLA concentration but to a lesser degree in comparison with the influence of PLA content on strength. Nevertheless, all blends displayed higher modulus than that of neat PLA (3.43 GPa), suggesting the SPC phase in the blend was more rigid than PLA. However, elongation at break of the blends was approximately half the already low elongation of neat PLA (3.8%), which leaves

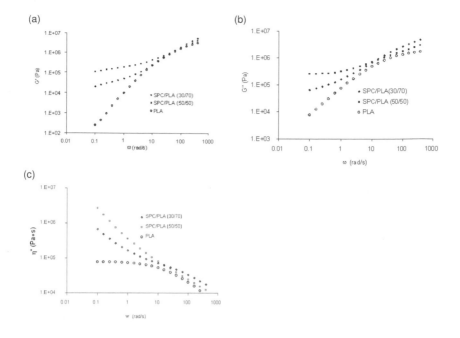

Figure 2. Dynamic frequency sweep of PLA and SP/PLA blends. strain = 5%, temperature = 175 °C.

space for toughening and plasticization in the future investigation. The results of tensile property testing suggest that when SPC is used as a plastic, the resulting blends can still demonstrate high performance even at high SPC levels.

Blending SPC with PLA not only improved the processibility of SP plastics because of reduced viscosity compared to that of neat SP plastic, it also greatly increased water resistance (Figure 4). Although the blend containing 70% SPC exhibited relatively high water absorption (~ 15%) in 2-h immersion in water compared to other blends, it still demonstrated impressively higher water resistance than neat SP plastics, which could uptake as much as 92% water in 2-h immersion (5). Increasing PLA content in the blends resulted in a progressive increase in water resistance. For example, the blend with 30% SPC displayed excellent water resistance, absorbing only 1.7 and 4.55% water in 24 and 72 h immersion, respectively.

Blending SPC with a Flexible Thermoplastic - PBAT/SPC Blends

In this part of study, the formulation of pre-compounding SP was: SPC (100 parts, on the basis of dry weight), sodium sulfite (0.5 parts), glycerol (10 parts) and/or water. Two levels of water content in the SPC powder were selected: less than 1.0 wt% (vacuum dried at 70°C for 12 h) and 22.5wt% (by adding 15% extra water to native SPC), respectively. Hereinafter, the blends are coded as PBAT/dry-SPC and PBAT/SPC-H$_2$O.

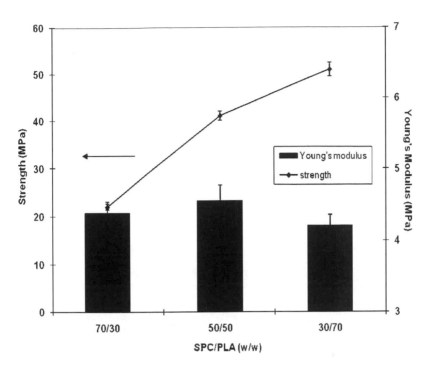

Figure 3. Tensile strength and modulus of different SPC/PLA blends

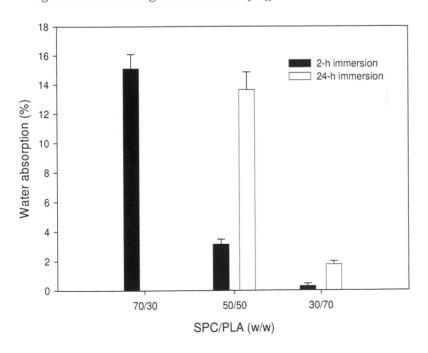

Figure 4. Water absorption of SPC/PLA blends in immersion tests

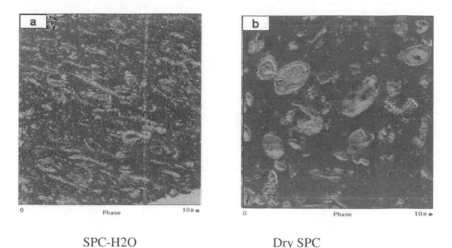

SPC-H2O   Dry SPC

Figure 5. Tapping mode AFM micrographs of the PBAT/SPC blends containing 3% MA-g-PBAT. (a): PBAT/SPC-H₂O 70/30 (w/w) (b): PBAT/Dry SPC 70/30 (w/w); bright yellow area is the SPC phase and the dark red area is the PBAT phase

Phase Structure

Figure 5 shows the high resolution tapping mode AFM micrographs of the blends phase structure. It was evident that SPC presented as fine threads in the PBAT/SPC-H_2O blends, and the connection of SPC threads was clearly noted. On the other hand, SPC appeared as homogenously dispersed particulates in the PBAT/dry-SPC blends. These micrographs indicate that a percolated SPC network structure was formed in some blends. It is known that percolation threshold (volume fraction) depends on several characteristics of the filler component in the system, including shape and aspect ratio. If the SPC domains are approximately assumed to be ellipsoid in each blend, according to Garboczi *et al.* (*24*), who predicted the percolation threshold as a function of the aspect ratio of the ellipsoids for randomly oriented ellipsoids of revolution, dry SPC particles (aspect ratio ~ 1) would reach the percolation threshold at 28.5 vol%. Therefore, 30% dry SPC in the blend was probably in the vicinity of percolation threshold, while 50% dry SPC in the blend had already formed percolated structure. Because the formed SPC threads in the PLA/SPC-H_2O blends had a much higher aspect ratio than the dry particles, the SPC concentration was well above the percolation threshold in both blends containing 30% SPC-H_2O.

Dynamic Rheological Properties

Dynamic rheological properties are correlated to phase structure and interactions within the polymer melt. Because the solid structure could be

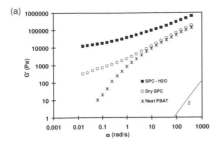 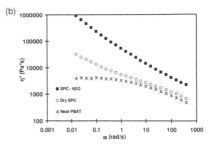

Figure 6. Effect of water content in pre-compounding SPC on dynamic rheological properties of PBAT/SPC (70/30) blends (containing 3 wt% MA-g-PBAT). Strain = 3%, temperature = 160°C.

Table 2. Tensile mechanical properties of PBAT/SPC blends[a]

PBAT/ SPC (parts)	H_2O in SPC (%)[b]	Dry test			Wet test[d]		
		Strength (MPa)	Mod- ulus (MPa)	Elon- gation (%)	Strength (MPa)	Mod- ulus (MPa)	Elon- gation (%)
50/50	22.5	24.5±0.7	937±77	11.6±1.6	--	--	--
	<1	14.2±0.4	659±18	8.0±1.4	--	--	--
70/30	22.5	19.3±0.6	411±48	40.1±5.2	7.6±0.4	59.9±4.4	200±27
	<1	11.0±0.1	274±17	80.5±6.5	7.0±0.0	46.9±0.2	271±11
85/15	22.5	14.0±0.1	156±4	408±25.9	14.4±0.3	131±5.2	524±34
	<1	13.4±0.2	158±10	451±24.1	14.0±0.2	143±5.6	510±14
Neat PBAT	0	ND[c]	100	ND	ND	ND	ND

[a] All the blends contain 3% MA-g-PBAT.   [b] Total water content including the 7.5% water of the native SPC as received.   [c] ND: not determined because sample was not broken.   [d] 24 h immersion in water.

maintained under small-strain test conditions, the dynamic rheology test is a very useful tool to evaluate morphological structures of polymer blends and composites (*25, 26*). Similar to PLA/SPC blends, PBAT/SPC blends not only exhibited much higher G' and η* values than neat PBAT, but also demonstrated drastically different pseudo solid-like terminal behaviors with varying water content in pre-compounding SPC. In the terminal region in Figure 6, log(G') vs. log(ω) of the PBAT/SPC blends tended to approach a (secondary) plateau; the terminal slope was much lower than that of neat PBAT, following a decreasing trend with increasing water content in SPC; log(G'') vs. log(ω) also exhibited the same trend (curve not shown); while log (η*) vs. log (ω) changed from a Newtonian (primary) plateau for neat PBAT to a continuous decrease with increasing ω for the blends (Figure 6). The rheological results of the blends confirm the formation of an interconnected network structure of anisometric filler - a characteristic solid or gel-like structure formed in the system (*23*). At higher water content levels in the pre-compounding SPC, a higher order of pseudo solid-like structure tends to form.

Mechanical Properties

Since the elastic modulus of the SPC plastic phase exceeds 4 GPa as shown in PLA/SPC blends, almost 40 times that of the PBAT matrix, it is reasonable to assume the SPC phase to be a non-deformable domain in PBAT/SPC blends. Therefore, the blend became an in-situ formed composite. Table 2 shows when percolated SPC network structure was formed, the corresponding blends displayed substantially higher yield strength but diminished yield strain compared to that of neat PBAT. With SPC ranging from 15% to 50% in the presence of compatibilizer MA-g-PBAT, the trend that the blends having 15% extra water exhibited substantially greater strength and modulus but lower yield strain and elongation compared to blends with native and dry SPC was strictly followed. The large increases in yield strength and modulus and decrease in yield strain were direct consequences of the suppression of the global cooperative chain movement of PBAT by the percolated SPC network structure in the blend with 15% extra water. For the blends containing 30% SPC, wet strength after 24-h immersion in water retained only 39 and 64% corresponding dry strength for the SPC-H2O and dry SPC blends, respectively. In contrast, for the blends containing 15% SPC, wet strength after 24-h immersion showed almost no change with respect to dry strength. For the 30%SPC blends, water absorptions after 24-h immersion was 5.0 and 4.5% for the SPC-H2O and dry SPC blends, respectively. Similarly, for the 15% SPC blends, water absorption was 2.3 and 2.2% for the SPC-H2O and dry SPC blends, respectively.

Conclusion

When SPC was blended with PLA or PBAT, processing SP as plastic resulted in an overall higher performance of the blends compared to blends formed when SP wasmixed in as an organic particulate filler. Water content in pre-compounding SPC was critical in determining whether SPC acted as plastic or filler in the blends. PLA/SPC blends still retained large portions of the high strength and modulus of neat PLA at high levels of SPC content, while the PBAT/SPC blends displayed significantly higher strength and modulus than neat PBAT and higher ductility than neat SPC plastics. The high performance of SPC blends was attributed to the formation of co-continuous phase structure in the PLA/SPC system, and SPC percolated network structure in the PBAT/SPC system, respectively. The processibility and water resistance of SP plastics were greatly improved through blending with biodegradable thermoplastic polymers. These results indicate that when SP is utilized as a plastic in blending with other biodegradable polymers, it could be an important raw material for high performance, cost effective and biodegradable bioplastics.

References

1. Brother, G. H.; McKinney, L. L. *Ind. Eng. Chem.* **1939**, *31*, 84–87.
2. Paetau, I.; Chen, C.; Jane, J. *Ind. Eng. Chem. Res.* **1994**, *33*, 1821–1827.
3. Mungara, P.; Zhang, J.; Zhang, S.; Jane, J. In *Protein-based Films and Coatings*; Gennadios, A., Ed.; CRC Press: Boca Raton, FL, 2002; pp 621–638.
4. Mo, X; Sun, X. *J. Am. Oil Chem. Soc.* **2002**, *79*, 197–202.
5. Zhang, J.; Mungara, P.; Jane, J. *Polymer* **2001**, *42*, 2569–2578.
6. Wang, S.; Sue, H.; Jane, J. *J. Macromol. Sci., Part A: Pure Appl. Chem.* **A1996**, *33*, 557–569.
7. Otaigbe, J.; Adams, D. *J. Environ. Polym. Degrad.* **1997**, *5*, 199–208.
8. Jane, J.; Wang, S. U.S. Patent 5,523,293, 1996.
9. Zhang, J.; Jiang, L.; Zhu, L.; Jane, J.; Mungara, P. *Biomacromolecules* **2006**, *7*, 1551–1561.
10. Liu, W.; Drzal, L. T.; Mohanty, A. K.; Misra, M. *Composites, Part B* **2007**, *38*, 352–359.
11. Lu, Y.; Weng, L.; Zhang, L. *Biomacromolecules* **2004**, *5*, 1046–1051.
12. Lodha, P.; Netravali, A. N. *Compos. Sci. Technol.* **2005**, *65*, 1211–1225.
13. Chen, Y.; Zhang, L.; Lu, Y.; Ye, C.; Du, L. *J. Appl. Polym. Sci.* **2003**, *90*, 3790–3796.
14. John, J.; Bhattacharya, M. *Polym Int.* **1999**, *48*, 1165–1172.
15. Zhong, Z.; Sun, X. S. *Polymer* **2001**, *42*, 6961–6969.
16. Wang, C.; Carriere, C. J.; Willett, J. L. *J. Polym. Sci., Part B: Polym. Phys.* **2002**, *40*, 2324–2332.
17. Graiver, D.; Waikul, L. H.; Berger, C.; Narayan, R. *J. Appl. Polym. Sci.* **2004**, *92*, 3231–3239.
18. Zhong, Z.; Sun, S. *J. Appl. Polym.Sci.* **2003**, *88*, 407–413.
19. Chang, L.; Xue, Y.; Hsieh, F. *J. Appl. Polym. Sci.* **2001**, *80*, 10–19.
20. Jong, L. *J. Appl. Polym. Sci.* **2005**, *98*, 353–361.
21. Yang, X.; Tang, J. *Advances in bioprocessing engineering*; World Scientific: Singapore, 2002; Vol. I, p154.
22. Wu, Q.; Sakabe, H.; Isobe, S. *Polymer* **2003**, *44*, 3901–3908.
23. Utracki, L. A. *Rheology and processing of multiphase systems*; Carl Hanser: New York, 1987; Vol. II, pp7–59.
24. Garboczi, E. J. *Phys. Rev. B* **1988**, *37*, 318–320. Garboczi, E. J. *Phys. Rev. B* **1988**, *37*, 318–320.
25. Utracki, L. A. *Polymer Alloys & Blends*; Carl Hanser: New York, 1989; pp131–174.
26. Du, M.; Zheng, Q.; Yang, H. *Nihon Reoroji Gakkaishi* **2003**, *31*, 305–311.

Chapter 5

Thermodynamic and Microscopy Studies of Urea-Soy Protein Composites

K. Venkateshan* and X. S. Sun

Bio-Materials and Technology Lab, Department of Grain Science and Industry, Kansas State University, Manhattan, Kansas 66506, USA.
*kcvenkat@ksu.edu

Biocomposites composed of organic molecules and confined in soy protein matrices have potential for various medical and drug-delivery applications. The objective of this study was to characterize the thermodynamic behavior and corresponding structural changes of urea-soy protein composite. Large melting temperature depression was observed for urea crystals interlayered with soy protein in comparison with bulk urea. In addition to a broader melting peak, the shape and size of the urea crystals interlayered with soy protein were significantly altered from those of bulk urea crystals. Formation of the interlayered morphology of urea crystals and soy protein involved dissolution and interpenetration of urea in the soy protein solution, which resulted in dissolution and denaturation of soy protein followed by moisture dehydration and subsequent precipitation and confinement of urea crystals in soy protein layers.

Laser scanning microscopy was used to characterize the structural changes of the urea crystals and soy protein. The altered structure was composed of rod-shaped urea crystals of width 300 to 1500 nm confined between soy protein layers of width 500 to 1000 nm forming an interlayered morphology. The main conclusion drawn from the thermodynamic and microscopy studies is that the melting temperature depression and the corresponding crystal size of urea do not agree with the Gibbs-Thomson predictions. We attribute this to the configurational effects at the urea-soy protein interfaces being more dominant than size effects as considered by Gibbs-Thomson theory.

Introduction

In the past two decades, extensive studies have been performed to investigate the effect of confinement of low-molecular-weight organic materials and polymeric materials in inorganic and polymeric matrices on transition temperatures (1–3). Small sized molecules are embedded in the matrices by confined nucleation and growth or precipitation in inorganic matrix pores or polymer cross-linked network pores of well-known pore sizes (4–6). The studies have focused on both glass and melting transition temperatures. The melting temperature depression (ΔT_m) studies (4, 7, 8) have focused on determining $\Delta T_m = T_m^{bulk} - T_m^r$, where T_m^{bulk} and T_m^r are the melting temperatures of bulk and confined materials of size, r, respectively. In certain studies (4, 7, 8) in which size effects dominated, good agreement between the melting point depression and confined material size based on the Gibbs-Thomson equation (9) has been reported. Some studies (5, 6), however, showed that the Gibbs-Thomson equation does not always agree well when the bulk heat of melting in the Gibbs-Thomson equation is invalid because of interfacial inhomogeneities. Small sized molecules were also embedded in cross-linked polymer networks, such as highly cross-linked elastomer, and ΔT_m has been reported (7, 8). In these studies, changes in ΔT_m were explained by modifications in contributions to specific heats of the components due to the dominance of enthalpic and entropic effects.

Similar to structural modifications of polymer networks by addition of small molecules, studies (10–12) describing protein structural changes with addition of organic molecules, acids, bases, and salts are extensive and well known. Among these modification studies, protein denaturation or structural changes with urea have been studied in depth by using calorimetry. Thermodynamic and simulation studies have focused on dissolution and protein denaturation mechanisms combined with the study of urea-protein interactions in solution to determine the enthalpic and entropic components of the protein modification process (10–15). These studies also described basic features of the hydrophobic effect and hydrogen bonding in relation to protein stability. Primarily, two mechanisms have been proposed to account for urea-induced perturbation of the hydrophobic regions of proteins in solutions. One mechanism is based on the indirect role of urea in perturbing the structure and dynamics of water, which, in turn, leads to the perturbation of structure and dynamics of hydrophobic regions (16). The other mechanism is based on the direct role of urea in interpenetrating the hydrophobic regions and forming open structures around hydrophobic regions (17).

Despite the scope of these studies of urea-protein in solution, characterization of urea-soy protein thermal behavior and structure in the precipitated solid state has not been extensive. The modification and depression of melting temperature of urea confined between protein layers and corresponding morphological changes have not been reported and characterized. Furthermore, the mechanism leading to the open structure at the urea-protein interface is not well understood.

In this study, we investigate (1) changes in melting behavior of urea interlayered with protein by using differential scanning calorimetry (DSC) and (2) structural changes of urea-protein composite by using laser scanning microscopy (LSM).

Materials and Method

Materials

The water-based soy proteins used in this work were extracted from defatted soy flour with a protein dispersability index (PDI) of 90 according to the acidic precipitation method described by Sun et al. (*18, 19*) and labeled as soy protein isolate (SPI). The defatted soy flour was purchased from Cargill, Inc. The weight fraction of soy protein in solution is 32%. The molecular weight of the protein is approximately 100 to 400 kDa. The soy protein is predominantly composed of two globulins, 7S and 11S. It has been reported that the extracted soy protein in solution is partially denatured (*18, 19*). The semidenatured soy protein will undergo further denaturation with addition and dissolution of urea. Urea in granular form was purchased from Fisher Scientific. The molecular weight of urea is 60 Da.

Synthesis of Urea-Soy Protein Composites

The urea-soy protein composite was synthesized by the following procedure. An accurately weighed amount of SPI was added to a clean glass beaker, followed by the addition of weighed urea to the same beaker. The weight fraction ratio of urea and SPI was 33:67. Since SPI contains 68 wt% water, the concentration of urea in SPI was 7 M. The urea-SPI mixture was mixed gently for 10 minutes and then dried in an oven at 363 K for one hour. Further, the mixture was removed from the oven and stored at room temperature for 24 hours until drying was complete. During the thermal conditioning and drying process, urea and SPI precipitated and evolved from a clear transparent solution to a solid composite. The moisture content in the dried composite sample was 4.5 % by weight and was determined by using thermal gravimetric analysis (TGA) measurements (scan not shown) and heater oven weight loss measurements. Furthermore, the peak and final temperature of water vaporization obtained from TGA scan was 425 K and 438 K respectively.

Experimental Methods

Differential Scanning Calorimetry (DSC) Technique

Melting behavior was studied by using a differential scanning calorimeter (DSC 7; PerkinElmer, Norwalk, CT) calibrated with indium and zinc. About 3 mg of the sample was accurately weighed and transferred into the DSC pans, and the pans were sealed. The DSC scans were obtained between temperature range of 313 K to 415 K and for a heating rate of 10 K/minute. The DSC scan were obtained for bulk SPI, bulk urea, and urea-SPI composite. Onset, peak, and final melting transition temperatures of urea were determined, and the corresponding enthalpy of melting was calculated.

Laser Scanning Microscopy (LSM) Technique

Images were obtained by using an Axioplan 2 MOT research microscope (Carl Zeiss, Inc., Thornwood, NY, USA) equipped with a Zeiss Axiocam HR digital camera, a fully motorized stage with mark-and-find software, plan neofluor objectives (1.25_/0.035, 10_/0.3, 20_/0.5, 40_/0.75, 40_/1.3 oil), plan apochromat objectives (63_/1.4 oil, 100_/1.4 oil), an achroplan objective (4_/0.1), differential contrast interference (DIC), phase contrast (ph), dark field, bright field, and Axiovision 3.1 software with interactive measurements and D deconvolution modules.

A small amount of the solid composite sample was placed onto a 3-inch × 1-inch glass slide (Fisher Scientific) without spreading force. The sample was allowed to set at room conditions for 2 min. Differential contrast interference images were taken at various magnifications and locations in the sample. Images were obtained for bulk SPI, bulk urea, and urea-SPI composite.

Results

Melting Studies

Figures 1A and 1B show the plots of heat flow, dH/dt, where H is enthalpy and t is time against temperature, T, of bulk urea and urea-soy protein composite. The melting peak temperature of bulk urea is was 407.7 K with $\Delta T = 11$ K, the difference between the onset and end temperatures of the melting endotherm and enthalpy of melting of bulk urea, ΔH_m^{bulk} was 15.37 kJ/mol, which are in agreement with the literature values (20, 21). The DSC scan of urea-soy protein composite (Figure 1B) showed significant depression of urea melting peak temperature, observed at 373.9 K, a larger ΔT of 52 K, and a lower enthalpy of melting, ΔH_m^r (enthalpy of melting of confined urea crystal of size r in composite) value of 9.44 kJ/mol in comparison with ΔH_m^{bulk}. In calculating ΔH_m^r, it is important to note that the accurate weight fraction of urea in the urea-soy protein composite was considered and that ΔH_m^r is reported as enthalpy per mol of urea. In the scan for urea-soy protein composite, peaks corresponding to denaturing of 7S and 11S were absent, evidence that the soy protein globulins had undergone complete denaturation.

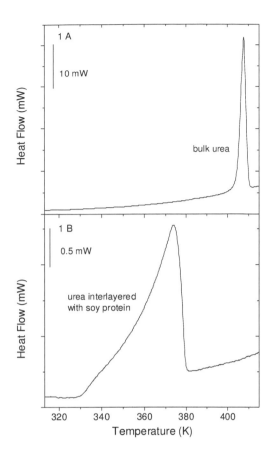

Figure 1. DSC scans of bulk urea (A) and urea-soy protein composite (B) wherein urea is interlayered with soy protein obtained by heating at a rate of 10 K/minute.

Laser Scanning Microscope

Figure 2 shows the LSM image of bulk SPI. The bright regions in the image are protein clusters, and it is evident from the image that the clusters are interconnected and the average size of the clusters is one micron. Because the clusters were nonconventional in shape, the size of the clusters corresponds to the largest dimension of the cluster. Figure 3 shows the LSM image of bulk urea. The urea particles exhibited rectangular and tetragonal faceted shapes. Based on the rectangular geometry, the size of urea crystals were determined to range in size from 50 to 300 microns.

The LSM images of urea-soy protein composite (Figure 4) show evidence of size reduction and structure modification of both protein clusters and urea particles. The urea crystals (dark rods) are randomly oriented and interlayered with the protein matrix (bright layers). The average size, or width, of the layers of urea and protein are 500 nm and 750 nm, respectively. These images depict the coexistence of urea and soy protein in an interlayered morphology. The structural changes observed with the interpenetration of urea in the protein matrix are a result of fine dispersion of the urea in protein matrix in both the particulate and emulsion states.

Discussion

Melting Temperature Depression of Confined Urea in Protein Matrix

In previous studies, the melting temperature depression has been attributed to the dominance of either (1) size effects as described by Gibbs-Thomson theory (19) or (2) configurational effects arising from higher interatomic spacing at surface and interfacial regions. In the following section, we consider these two aspects that are attributed to melting temperature depression.

1. Size Effects and Validity of Gibbs-Thomson Predictions

According to the Gibbs-Thomson equation, the equilibrium melting temperature, T_m^r, of a crystal of radius r with a free surface is written as (22):

$$\frac{T_m^r}{T_m^{bulk}} = \exp\left[-\left(\frac{2\gamma_{s,l}V_s}{\Delta H_m^r}\right)\frac{1}{r}\right] \tag{1}$$

also written alternatively as:

$$\frac{T_m^r}{T_m^{bulk}} = \left[1 - \left(\frac{2\gamma_{s,l}V_s}{\Delta H_m^{bulk}}\right)\frac{1}{r}\right] \tag{1a}$$

By rearranging Eq. 1a, we obtain:

$$r = \frac{2\gamma_{s,l}V_s}{\Delta H_m^{bulk}}\frac{T_m^{bulk}}{\left(T_m^{bulk} - T_m^r\right)} \tag{1b}$$

where T_m^{bulk} is the melting temperature of the bulk crystal, $\gamma_{s,l}$ is the interfacial tension between the solid and its melt, V_s is the molar volume of the solid, and ΔH_m^r is the molar enthalpy of melting of crystal of radius r at T_m^r, which is taken to be equal to that of the enthalpy of melting of the bulk crystal, ΔH_m^{bulk}, in Eq. 1a.

In the course of investigating the validity of Gibbs-Thomson theory with our experimental findings, assumptions that were used to derive the Gibbs-Thomson equation and its relevance to our work are briefly summarized as follows:

    i)   Inhomogeneities composed of coexisting crystalline and amorphous regions and contributions from heterogeneous nucleation are ignored (2).

ii) Surface interactions between the confined crystals and surrounding heterogeneous and amorphous phase are considered to be minimal, and the theory considers the attainment of equilibrium to be exclusively between the confined crystals and their liquid phase (22).

iii) The theory is derived for crystals that are in equilibrium with their liquid and vapor phase and not for crystals that are tightly confined within a cavity, where pressure may change with changes in both temperature and volume on melting (22).

iv) The presence of crystals in the solvated state and the resulting incomplete crystallization are considered to be negligible (12).

These assumptions lead to the consideration that the enthalpy of melting of the confined urea crystals of size r, ΔH_m^r, is equal to the enthalpy of melting of the bulk urea crystal, ΔH_m^{bulk}, in deriving the Gibbs-Thomson equation. In addressing ΔH_m^r and ΔH_m^{bulk} with our results, values of ΔH_m^r and ΔH_m^{bulk} were calculated and compared. The melting peaks in Figures 1B and 1A were integrated, and the calculated values of enthalpies of melting of urea, ΔH_m^r and ΔH_m^{bulk} were 9.44 and 15.37 kJ/mol, respectively. The different ΔH_m^r and ΔH_m^{bulk} values indicate the possibilities of (1) coexisting crystalline and amorphous regions, (2) surface interactions where urea is consumed because of chemical bonding with SPI, and (3) incomplete crystallization. Hence, we consider using $\Delta H_{m,r}$ in Eq. 1b to determine r for the observed melting point depression value of 33.8 K.

In studying the validity of Gibbs-Thomson theory, we consider Figures 1A and 1B, which show that the difference in melting temperature between the confined urea crystals in soy protein and the bulk urea crystal ($\Delta T_m = T_m^{bulk} - T_m^r$) is 33.8 K. By substituting the known values for urea of $\gamma_{s,l} = 45.3$ mJ/m² (23), $V_s = 45.45$ cm³/mol (molecular wt = 60 Da and density = 1.32 g/cc), $\Delta H_m^r = 9.44$ kJ/mol (from Figure 1B), and $\Delta T_m = 33.8$ K in Eq. 1b, the size of confined urea crystals in soy protein was calculated to be 5.26 nm. In contrast, the size of interlayered urea crystals determined from Figure 4, ranged from 300 nm to 1500 nm. Evidently, the decrease in size of urea crystals is much less than that predicted by the Gibbs-Thomson equation, thus exhibiting a clear disagreement with the Gibbs-Thomson prediction.

2. Configurational Contribution to the Melting Process

In our studies, because the calculated size of the nanocrystal (5.26 nm) from Gibbs-Thomson equation for $\Delta T = 33.8$ K is much lower than the experimentally observed crystal size (ranging from 300 nm to 1500 nm), it is very probable that

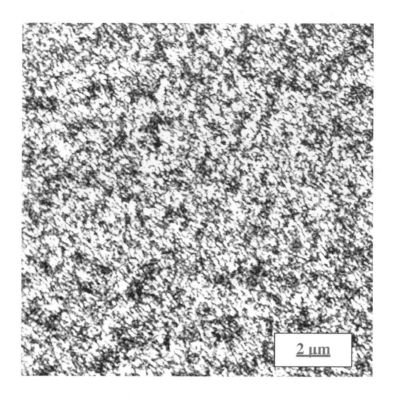

Figure 2. High-magnification LSM image of bulk SPI.

configurational effects at the solid-liquid and urea-soy protein interfaces play a more significant role than size effects in causing a high melting temperature depression of 33.8 K.

Surface Energy and Configurational Entropy

The size-dependent melting temperature, T_m^r, of metallic nanocrystals is described as (*23*):

$$\frac{T_m^r}{T_m^{bulk}} = \exp\left[\frac{-2S_{vib}}{3R}\frac{1}{\frac{r}{r_o}-1}\right] \qquad (2)$$

where r is the radius of the crystal, S_{vib} is the vibrational component of the melting entropy of bulk crystals at the bulk melting temperature, R is the ideal gas constant, and r_o is a critical radius at which all atoms of a particle are located on its surface. For most crystals, the overall melting entropy, S_m, is dominated by its vibrational component, $S_m = S_{vib}$ (*24, 25*). For semiconductors and embedded

Figure 3. LSM image of bulk urea crystals. The magnification scale is shown in red.

crystals (*26*), the electronic and configurational components of the melting entropy are not negligible, and $S_m \neq S_{vib}$.

When $r \gg r_o$, Eq. 2 can be mathematically expressed as:

$$\frac{T_m^r}{T_m^{bulk}} = \left[1 - \left(\frac{2S_m}{3R} \right) \frac{r_o}{r} \right] \tag{3}$$

By comparing Eqns. 3 and 1a, we obtain:

$$\gamma_{s,l} = \frac{r_o S_m \Delta H_m^{bulk}}{3RV_s} \tag{4}$$

From Eq. 4, it is evident that $\gamma_{s,l}$ is dependent on S_m, which, in turn, is affected by vibrational and configurational components. In scenarios in which the configurational component is significant, the assumption of $S_m = S_{vib}$ would be erroneous and thus underestimate the $\gamma_{s,l}$ value. Instead, $S_m = S_{vib} + S_{config}$ must be incorporated, where S_{config} is the configurational entropy. The significant configurational contribution to the melting processes arises from higher interatomic spacing and disorder and anharmonic forces at surface and interfacial

Figure 4. High-magnification LSM image of urea-soy protein composite. The urea crystals (dark rods) are randomly oriented and interlayered with the protein matrix (bright layers).

regions that result in lowering of vibrational frequencies (*22*). On the basis of the water amount (4.5 % by weight) and the effects of dissolution or solvation of urea in water, we consider the role of water in influencing the melting point depression of urea in the composite sample to be small.

Furthermore, the evidence of lack of secondary chemical interactions between urea and soy protein at the interface leads us to believe that the configurational effects at the interface are the primary cause of melting temperature depression. The molecular origin of the configurational component of entropy in nanocrystals and hydrated proteins and its consequence in exhibiting higher heat capacity, C_p, has been studied by using both standard and temperature-modulated calorimetry (*22, 27, 28*).

Conclusion

The observed melting temperature depression and corresponding reduction in urea crystal size do not obey the Gibbs-Thomson theory. We attribute this to configurational effects at the urea-soy protein interface being more dominant than size effects as considered by Gibbs-Thomson theory. The higher configurational entropy, S_{config}, of urea nanocrystals interlayered with soy protein can be attributed to (1) a large fraction of atoms in the interfacial region and a more open structure that lead to weaker interatomic coupling that decreases the vibrational frequencies and increases the vibrational and configurational entropy, (2) thermally induced variation of the vibrational and configurational entropy of materials due to lattice vibrations and variation of equilibrium defect concentration, especially in the interface region, and (3) strong anharmonic forces that lower the vibrational frequencies because of increased interatomic spacing. In general, the disordered interfacial atoms of interlayered urea crystals have a higher configurational entropy contribution and heat capacity (C_p) than the bulk crystal.

Acknowledgments

The authors thank the U.S. Department of Agriculture and Kansas Agricultural Experiment Station for grant sponsorship, contribution no. 10-195-B from the Kansas Agricultural Experiment Station. The authors also appreciate Dr. Dan Boyle's assistance with the LSM images.

References

1. Sanz, N.; Boudet, A.; Ibanez, A. *J. Nanopart. Res.* **2002**, *4* (1/2), 99–105.
2. Jones, B. A.; Torkelson, J. M. *J. Polym. Sci., Part B: Polym. Phys.* **2004**, *42* (18), 3470–3475.
3. Ellison, C. J.; Ruszkowski, R. L.; Fredin, N. J.; Torkelson, J. M. *Phys. Rev. Lett.* **2004**, *92* (11), 119901–119911.
4. Jackson, C. L.; McKenna, G. B. *J. Chem. Phys.* **1990**, *93* (12), 9002–9011.
5. Sliwinska-Bartkowiak, M.; Dudziak, G.; Sikorski, R.; Gras, R.; Radhakrishnan, R.; Gubbins, K. E. Dielectric spectroscopy and molecular simulation. *J. Chem. Phys.* **2001**, *114* (2), 950–962.
6. Christenson, H. K. *J. Phys.: Condens. Matter* **2001**, *13* (11), R95–R133.
7. Jackson, C. L.; McKenna, G. B. *Rubber Chem. Technol.* **1991**, *64* (5), 760–768.
8. Hoei, Y.; Ikeda, Y.; Sasaki, M. *J. Phys. Chem. B* **1999**, *103* (25), 5353–5360.
9. Gibbs, J. W. *Collected works of J. W. Gibbs*; Dover, NY, 1928.
10. Zou, Q.; Habermann-Rottinghaus, S. M.; Murphy, K. P. *Proteins: Struct., Funct., Genet.* **1998**, *31* (2), 107–115.
11. Nandi, P. K.; Robinson, D. R. *Biochemistry* **1984**, *23* (26), 6661–6668.
12. Wells, D.; Drummond, C. J. *Langmuir* **1999**, *15* (14), 4713–4721.
13. Stumpe, M. C.; Grubmuller, G. *Comput. Biol.* **2008**, *4* (11), 1–10.
14. Stumpe, M. C.; Grubmuller, G. *J. Am. Chem. Soc.* **2007**, *129*, 16126–16131.
15. Stumpe, M. C.; Grubmuller, G. *Biophys. J.* **2009**, *96*, 3744–3752.
16. Frank, H. S.; Franks, F. *J. Chem. Phys.* **1968**, *48* (10), 4746–4757.
17. Nozaki, Y.; Tanford, C. *J. Biol. Chem.* **1963**, *238*, 4074–4081.
18. Sun, X.; Zhu, L.; Wang, D. Latex based adhesives derived from soybeans. U.S. Patent, No. 35091-PCT, pending, filed 2006.
19. Sun, X.; Wang, D. Surface active and interactive protein polymers. U.S. Patent, PCTUS2006/015943, filed 2006.
20. Della Gatta, G.; Ferro, D. *Thermochim. Acta* **1987**, *122*, 143–152.
21. Vogel, L.; Schuberth, H. Some physicochemical data of urea near the melting point. *Chem. Tech. (Leipzig)* **1980**, *32*, 143–144.
22. Gunawan, L.; Johari, G. P. *J. Phys. Chem. C* **2008**, *112* (51), 20159–20166.
23. Lu, H. M.; Wen, Z.; Jiang, Q. *J. Phys. Org. Chem.* **2007**, *20* (4), 236–240.

24. Jiang, Q.; Tong, H. Y.; Hsu, D. T.; Okuyama, K.; Shi, F. G. *Thin Solid Films* **1998**, *312* (1,2), 357–361.
25. Jiang, Q.; Aya, N.; Shi, F. G. *Appl. Phys. A: Mater. Sci. Process.* **1997**, *64* (6), 627–629.
26. Jiang, Q.; Shi, H. X.; Zhao, M. *Acta Mater.* **1999**, *47* (7), 2109–2112.
27. Sartor, G.; Mayer, E.; Johari, G. P. *Biophys. J.* **1994**, *66*, 249–258.
28. Salvetti, G.; Tombari, E.; Mikheeva, L.; Johari, G. P. *J. Phys. Chem. B* **2002**, *106*, 6081–6087.

Chapter 6

Extraction and Characterization of Sugar Beet Polysaccharides

Marshall L. Fishman,* Peter H. Cooke,† and Arland T. Hotchkiss Jr.

Crop Science and Engineering Research Unit and Microbial Biophysics and Residue Chemistry and Core Technology Research Unit, Eastern Regional Research Center, Agricultural Research Center, U.S. Department of Agriculture, 600 East Mermaid Lane, Wyndmoor, Pennsylvania 19038
*Marshall.fishman@ars.usda.gov
†Present address: Electron Microscopy Laboratory, New Mexico State University, PO Box 30001, MSC 3EML, Las Cruces, NM 88003

Sugar Beet Pulp (SBP), contains 65 to 80% (dry weight) of potentially valuable polysaccharides. We separated SBP into three fractions. The first fraction, extracted under acid conditions, was labeled pectin, the second was comprised of two sub fractions solubilized under alkaline conditions and was labeled alkaline soluble polysaccharides (ASP) and part of the remaining fraction was solubilized by derivatizing with carboxy methyl groups (DWCM). We have studied the global structure of these fractions after microwave-assisted extraction (MAE) from fresh sugar beet pulp. MAE was employed to minimize the disassembly and possibly the degradation of these polysaccharides during extraction. Fractions were characterized by their carbohydrate composition, by HPSEC with on-line molar mass and viscosity detection and by imaging with Atomic Force Microscopy (AFM). AFM revealed that pectin formed integrated networks comprised of spheres and strands. ASP aggregated but did not form networks.

Introduction

In the U.S., each year it is estimated that the extraction of sugar from sugar beets produces about two million tons of sugar beet pulp (SBP) (*1*). SBP has been used primarily as a low valued animal feed. Curently, demand and utilization of sugar beet pulp for animal feed is high. Nonetheless, on a dry weight basis, SBP is a rich source of carbohydrates in that it contains 65 to 80% of potentially high valued polysaccharides (2). Typically, polysaccharides are obtained by sequentially separating SBP into three fractions. The first fraction, extracted under acid conditions, has been labeled pectin. Pectins are polysaccharides in which (1→4) linked α-D-galacturonates and its methyl ester predominate (3). In addition to these "smooth" homogalaturonan (HG) regions, also common to all pectins are "hairy" regions which are comprised of rhamnogalacturonan (RG) units (3). HG units may contain OH protons substituted with acetyl groups or may be substituted with xylose. The (1→2) linked α –L- rhamnose units in the RG region may be substituted with galactan, arabinan or arabinogalactan side chains in the 4 position.

Pectin from sugar beets often differs from pectin from other sources in that it tends to have a higher degree of acetylation and a higher neutral sugar content (4). Furthermore, unlike pectins from many other sources, sugar beet pectin contains feruloyl groups (4). Moreover, unlike pectin from citrus peels or apple pomace, sugar beet pectin does not gel when high concentrations of sugar and acidic conditions are present. The poor gelling properties of sugar beet pectin have been attributed to the presence of acetyl groups and relatively low molar mass (5, 6). It also has been suggested that the relatively high amounts of neutral sugar side chains in sugar beet pectin prevent it from gelling (7).

There have been numerous studies on the structural characteristics and physico-chemical properties of pectin from sugar beet pulp (8). Also there have been many studies involving the structural characteristics and physico-chemical properties of enzymatically and chemically modified sugar beet pectins from pulp (9). Most of these studies were concerned with the fine structure of sugar beet pectin and how it might affect its functional properties. Few if any of these studies were directed toward understanding the global structure of sugar beet pectin. Recently, sugar beet pectin extracted by acid from cell walls isolated from fresh sugar beet roots and characterized by atomic force microscopy (AFM) revealed linear and branched molecules attached to globular molecules (10). The linear molecules were attributed to polysaccharides whereas the globular particles were attributed to proteins.

The second fraction, solubilized under alkaline conditions, is labeled alkaline soluble polysaccharides (ASP). This fraction may be associated in part with pectin (11). The third fraction (i.e. the extracted residue) contains cellulose microfibrils (12). The insoluble fraction was partially characterized after solubilizing by introducing carboxy methyl groups. The objectives of the study was to minimize the disassembly and possibly the degradation of these polysaccharides during extraction and to obtain all three fractions from the same matrix in order to assess their properties for value added applications. Therefore, we chose extraction conditions which minimized the disassembly and possibly the degradation of the extracted polysaccharides.

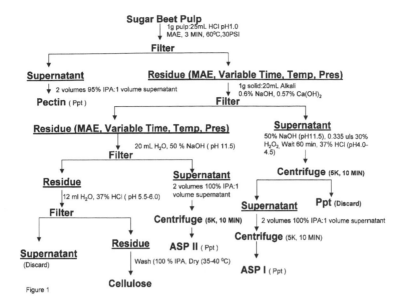

Figure 1

Figure 1. Flow diagram for the extraction of polysaccharides from sugar beet pulp.

Experimental

Materials

Partially dewatered SBP with sugar removed was a gift from American Crystal Sugar, Moorhead, MN. SBP was shipped frozen and stored at -20 °C until prepared for extraction. The frozen sample was dried and ground to approximately 20 mesh.

Microwave Assisted Extraction (MAE)

The flow diagram for the extraction of the various polysaccharides is illustrated in Figure 1. Details of the extractions of pectin (*13*) and alkali soluble polysaccharides (ASP) (*14*) have been described elsewhere. Briefly, microwave heating was performed in a CEM Corporation, model Mars X microwave sample preparation system equipped with valves and tubing which permitted the application of external pressure to each of the extraction vessels via nitrogen from a tank equipped with a pressure gauge. Samples were irradiated with 1200 watts of microwave power at a frequency of 2450 MHz.

Carboxymethylation (CM) of the SBP Residue and Determination of Degree of Substitution (DS)

Derivitization of the insoluble fraction was done by the method of Heinze and Pfeiffer (*15*). One gram of cellulose was slurried in iso-propanol, sodium hydroxide solution (various concentrations) was added to the mixture and stirred

73

for an hour. Then sodium monochloroacetate was added to the mixture which was heated for 2 to 5 hours at 55°C. The resultant product was then isolated and dried. FTIR spectra of CMC were used to calculate its degree of substitution (DS) according to the method and parameters of Pushpamalar et al. (16). FTIR spectra were collected with a Thermo Electron Nexus 670 FTIR. Two hundred fifty mg of the CMP was dissolved in 5 mL of water with stirring overnight. The resulting solution was centrifuged at 50,000g in a Sorvall RC-5B centrifuge. Ten drops of the supernatant were placed on a CaF_2 plate, allowed to air dry overnight, and then vacuum dried for thirty minutes. The films were scanned from 1200 cm^{-1} to 3700 cm^{-1} with a resolution of 4 cm^{-1}. Each sample underwent an average of 64 scans. The degree of substitution was calculated from the methine absorbance at 2920 cm^{-1} and the carboxyl absorbance at 1605 cm^{-1}.

Chromatography

Dry sample (2 or 4 mg/ml) was dissolved in mobile phase (0.05 M $NaNO_3$), centrifuged at 50,000 g for 10 minutes and filtered through a 0.22 or 0.45 micrometer Millex HV filter (Millipore Corp., Bedford, MA). The flow rate for the solvent delivery system, model 1100 series degasser, auto sampler and pump (Agilent Corp.), was set at 0.7 mL/min. The injection volume was 200 µL. Samples were run in triplicate. The column set consisted of two PL Aquagel OH-60 and one OH-40 size exclusion columns (Polymer Laboratories, Amherst, MA) in series. The columns were in a water bath set at 35 °C. Column effluent was detected with a Dawn DSP multi-angle laser light scattering photometer (MALLS) (Wyatt Technology, Santa Barbara, CA), in series with a model H502 C differential pressure viscometer (DPV) (Viscotek Corp., Houston TX) and an Optilab DSP interferometer (RI) (Wyatt Technology). Electronic outputs from the 90 degree light scattering angle, DPV and RI were sent to one directory of a personal computer for processing with TRISEC software (Viscotek Corp.). Electronic outputs from all the scattering angles measured by the MALLS, DVP and RI were sent to a second directory for processing with ASTRA™ software (Wyatt Technology).

Atomic Force Microscopy (AFM)

AFM of pectin (13) and ASP (14) have been described previously. Typically, dry sample was dissolved in HPLC grade water and serially diluted to the desired concentration. Two µLs of the solution was pipetted onto a freshly cleaved 10 mm diameter disk of mica and air dried. The mica was mounted in a Multimode Scanning Probe microscope with a Nanoscope IIIa controller, operated as an atomic force microscope in the Tapping Mode (Veeco Instruments, Santa Barbara, CA). The thin layer of pectin adhering to the mica surface was scanned with the AFM operating in the intermittent contact mode using etched silicon probes (TESP). The instrument was operated in soft tapping mode. Images were analyzed by software version 5.12 rev. B which is described in the Command Reference Manual supplied by the manufacturer.

Table 1. Compositional Analysis of Sugar Beet Pectin Extracted at 60 °C and for 3 Minutes

Pressure lb/in²	% PR	%AGA	%NS	%DM	%DA
Commercial		57(1)[a]	39(2)	83(9)	
Fresh[b]	33	38	37	54	28
25	13.0	52(3)	41(1)	99(6)	59(4)
30	11.3	48(2)	45(4)	114(13)	68(5)
75	9.2	60(3)	29(3)	77(4)	44(2)
Average[c]	11(2)	54(6)	38(8)	96(18)	57(12)

[a] Standard deviation of triplicate analysis.   [b] Data from reference (7).   [c] Does not include commercial or fresh sample

Compositional Analysis

Determination of anhydrogalacturonate content (%AGA), degree of methyl esterification (%DM), acetylation (%DA) neutral sugar content (%NS) and monosaccharide analysis have been described previously (*14*). Percentage of alcohol insoluble pectin was determined gravimetrically (%PR).

Results and Discussion

Characterization of Pectin

Compositional Analysis

Table 1 contains percentages of pectin recovered (%PR), anhydrogalacturanate (%AGA), neutral sugars (%NS), degree of methylesterification (%DM) and degree of acetylation (%DA) for microwave extractions at 60 °C . Heating time was 3 minutes and pressures ranged from 25 to 75 lbs/in². At 25 lb/in², values of percentage PR, AGA, NS, DM and DA were comparable to those found for commercial sugar beet pectin. By way of comparison (7) for fresh sugar beets extracted at pH 1, 75 °C for 30 minutes with HCl, %PR was 33, %AGA was 38, %NS was 37, %DM was 54, and %DA was 28. The lower values of %AGA, %DM and %DA and the higher value of %PR over those found for the 3 minute, 25 lb/in² extraction may indicate more pectin degradation occurred in the fresh sample extracted by conventional heating for 30 minutes.

Analysis of Molar Mass, Radius of Gyration and Intrinsic Viscosity

Pectin heated for 3 minutes gave weight average molar masses (M_w) that ranged from 517,000-1.2 million Daltons, radii of gyration (Rg_z) from about 35-40 nm, weight average intrinsic viscosity (η_w) from about 3.00-4.30 dL/g.

Table 2. Molecular Properties of Pectin from Sugar Beet Pulp Heated for 3 Minutes

Temp/Press. (°C, lb/in²)[a]	$M_w \times 10^{-3}$	R_{gz} (nm)	$[\eta_w]$ (dL/g)
Commercial	640(14)	40.4(2)	3.27(.03)
60/19	956(5)	40.1(.3)	4.23(0.05)
60/23	978(3)	35.2(.5)	3.81(.06)
60/25	1020(70)	37.3(1)	3.49(.08)
60/30	1200(6)	40.3(.3)	3.97(.02)
60/50	622(3)	37.6(1)	2.83(0.03)
60/60	517(7)	34.3(.4)	2.58(.05)
60/75	753(8)	40.8(2)	4.24(.07)
70/30	781(5)	36.3(.5)	3.45(.03)

[a] Sample concentration, 2 mg/ml

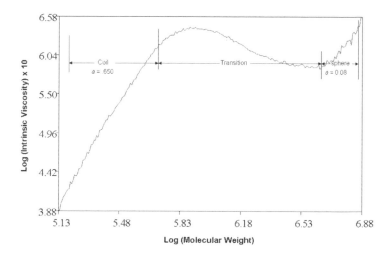

Figure 2. Mark-Houwink plot for sugar beet pectin. Heating time 3 minutes. Temperature 60 °C. Pressure 30 lb/in2.

We have plotted log intrinsic viscosity against log molar mass in Figure 2, the Mark Houwink (M-H) plot. The plot has been divided into three regions as marked by the vertical lines with the values of "a", the slope or Mark-Houwink exponent, indicated for each end section of the plot. The molecular properties for these three regions are given in Table 3. The high molar mass region had an M-H exponent of 0.076 which indicates an extremely compact molecule. This region comprised about 7.8% of the sample. The low molar mass region comprised about 34 % of the sample and had an M-H exponent of 0.65 which indicated a much less compact molecule than the high molar mass fraction.

Table 3. Molecular Properties of Sugar Beet Pectin Fractions[a]

Fraction	% recovery	$M_w \times 10^{-4}$	$[\eta]_w$ (dl/g)	Rg_z (nm)	a[b]
1. Sphere[c]	7.8(0.6)[d]	372(154)	3.8(0.1)	54(1.0)	0.076(0.02)
2. Trans. Region	60(1)	132(37)	4.1(0.1)	39(1)	---------
3. Random Coil[c]	34(1)	394(8)	3.7(0.1)	38(1)	0.650(0.06)

[a] (3/60/30) (time/temperature/pressure).   [b] Mark-Houwink exponent.   [c] Determined from "a".   [d] Standard deviation (triplicate analysis).

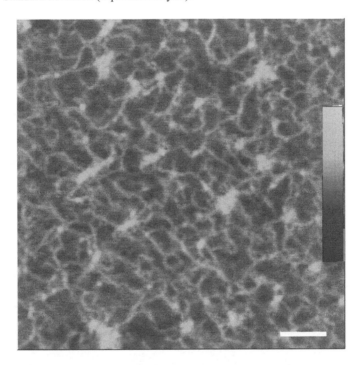

Figure 3. AFM image of sugar beet pectin network deposited from water at a concentration of 12.5 μg/mL. Scale bar is 100 nm, inset height scale, 0-2.5 nm. (see color insert)

AFM

To further elucidate its structure, images of sugar beet pectin were obtained by AFM. Figure 3 is an image of a sugar beet pectin network obtained by depositing a 2 μL drop of pectin dissolved in water at a concentration of 12.5 μg/mL and air drying it. The image revealed a number of spherical or odd shaped particles surrounded by strands. Many of the particles appear to be in contact with multiple strands thus appearing to be embedded in an open network structure. Dilution of the sugar beet pectin in water to 6.25 μg/mL produced the image in Figure 4. In that Figure the network has dissociated into particles attached to expanded strands. The strands appear to be comprised of rods, kinked rods, and segmented rods. Many of these appear to be aggregated end to end and/or side by side. Some of the strands have formed closed loops. The images in Figure 3 appear similar to peach pectin networks dissociated by dissolution in 5 mM NaCl and imaged by electron microscopy (*17*, *18*). Nonetheless, spherical or odd shaped particles were not observed in the peach pectin images.

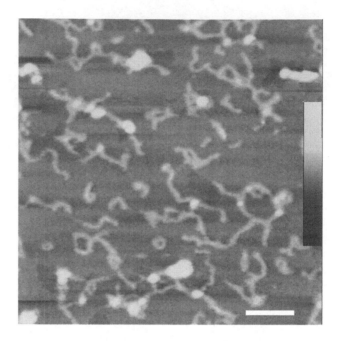

Figure 4. AFM image of sugar beet pectin network deposited from water at a concentration of 6.25 µg/mL. Scale bar is 100 nm, inset height scale, 0-5 nm. (see color insert)

Characterization of Alkaline Soluble Polysaccharides (ASP)

% Weight Recovery and Composition Analysis

The % wt recovery of ASP after 10 minutes of heating time in the microwave at various temperatures and pressures are contained in Table 4. Total % wt recovery of ASP ranged from 16.3 to 29.8%. Heating at 100 °C and 60 lb/in² gave the highest ASP I recovery and a somewhat smaller recovery of ASP II. Nonetheless the quantity of ASP II recovered was sufficient for further characterization studies.

Composition analysis of ASP (Table 5) revealed that in all cases that % anhydrogalacturonate (AGA), % degree of methyl esterification (DM) and % neutral sugar (NS) were appreciably lower than thiose values for either pulp or MAE pectin (sample 3/60/30 in Table 1).

Table 4. Weight percentage recovery of alkaline soluble polysaccharide (ASP) heated for 10 Minutes

	ASP I	ASP II	Total
100/30	4.2	21.6	25.8
100/40	12.3	15.4	27.7
100/60	16.0	13.8	29.8
105/50	12.8	12.8	25.6
105/90	3.3	15.0	16.3
120/90	14.6	10.2	24.8
Average	11(5)[a]	15(4)	25(5)

[a] Standard Deviation

Table 5. Compositon Analysis of Alkaline Soluble Polysaccharides

	% AGA		% DM		% NS	
Sample[a]	ASP I	ASP II	ASP I	ASP II	ASP I	ASP II
SB pulp[b]	34.4(4)[e]		70.4(9)		37.8(4)	
3/60/30[c,d]	48.3(2.3)		113.7(13.0)		44.7(3.5)	
10/80/30[d]	41.4(0.05)	26.6(0.3)	3.9(1)	9.0(2)	22.2(0.2)	22.8(0.04)
10/90/30[d]	21.6(0.2)	44.2(0.3)	11.8(1)	2.7(0.7)	29.6(0.2)	24.1(0.1)
10/100/30[d]	27.7(0.1)	29.8(0.2)	6.4(0.8)	7.5(1)	26.5(0.3)	13.9(0.1))
10/110/30[d]	23.2(0.2)	39.5(0.2)	11.5(0.6)	3.4(0.4)	25.7(0.1)	25.6(0.9)

[a] Time(min.)/temp(°C)/pres.(lb/in²). [b] Unfractionated; [c] Pectin. [d] Obtained by MAE. [e] Standard deviation of triplicate analysis.

Neutral Sugar Analysis

As indicated by their molar percentages in Table 6, the largest amount of neutral sugar present is in the form of arabinose followed by either rhamnose or galactose. For MAE sugar beet pectin, the order of neutral sugars detected was Ara>Gal>Rha>Glc>Xyl>Fuc. Recently reported research (*10*) on sugar beet pectin found that the order of neutral sugar detection was similar (Ara>Gal>Rha) when freshly harvested sugar beet roots were extracted. In that study only trace amounts of Glc, Xyl, Fuc and Man were found . In contrast, we detected the molar percentages of neutral sugars to be in the order Ara>Rha>Gal for ASP I and ASP II fractions extracted from sugar beet. We note that the ratio of uronate to neutral sugars in Table 6 is appreciably lower than expected for pectin. Possibly, this is a result of the hydrolysis being optimized for neutral sugar recovery. It is well known that the neutral sugars glycosidic linlages in sugar beet pectin are more readily hydrolyzed than (1→4)-α –D-galacturonate linkages. The presence of galacturonic acid in the ASP fractions posibly indicates that residual pectic polysaccharides may have been extracted with the ASP's or that they are covalently linked to the ASP's.

Table 6. Percentage Recovery of Neutral Sugar in Alkaline Soluble Polysaccharides (ASP) and in Acid Extracted Pectin

Sample[a]	Fraction	Ara	Gal	Rha	Glc	Xyl	Fuc	GalA	GlcA
SB pulp[b]	Control	41.29	11.53	6.31	15.12	6.76	0.46	16.96	1.57
3/60/30[c,d]	Pectin	46.26	20.47	7.74	7.57	4.01	0.63	11.11	2.3
10/80/30[d]	ASP I	41.81	10.7	19.1	0.6	0.29	0.1	26.1	1.35
	ASP II	54.4	11.0	22.6	0.02	0.11	0.03	10.7	1.2
10/90/30[d]	ASP I	54.1	11.7	20.6	0.15	0.01	0.05	12.2	1.2
	ASP II	44.3	9.4	19.8	0.55	0.19	0.07	24.5	1.2
10/100/30[d]	ASP I	52.4	11.7	20.7	0.13	0.07	0.05	13.4	1.6
	ASP II	44.9	10.5	15.7	1.4	1.3	0.09	24.6	1.6
10/110/30[d]	ASP I	52.5	11.4	20.9	0.23	0.09	0.05	13.5	1.33
	ASP II	37.7	8.0	13.0	0.22	0.23	0.03	40.1	0.81

[a] Time(min.)/temp(°C)/pressure (lb/in²).   [b] Unfractionated.   [c] Pectin.   [d] Obtained by MAE.

Table 7. Molecular Properties of ASP Heated for 10 Minutes

Sample	ASP I			ASP II		
Temp./ Press °C lb/in²	M_w x 10^{-3}	Rg_z (nm)	$[\eta_w]$ (dL/g)	M_w x 10^{-3}	Rg_z (nm)	$[\eta_w]$ (dL/g)
100/30	94(9)[a]	10.2(1)	0.31(0.01)	109(8)	15.9(2)	0.33(0.03)
100/40	92.1(0.9)	10.1(2)	0.37(0.01)	113(4)	16.7(1)	0.40(0.01)
100/60	92.2(0.3)	11.2(.4)	0.38(0.01)	92.6(0.3)	14.0(1)	0.34(0.01)
105/50	106(3)	23.8(1)	0.40(0.01)	122(7)	23.2(3)	0.41(0.03)
105/90	192(5)	23.1(1)	0.40(0.01)	252(4)	20.9(0.8)	0.43(0.01)
120/90	83.3(1)	14.4(4)	0.35(0.01)	88.6(0.7)	16.3(0.6)	0.36(0.01)
Average	109(40)	15.(6)	0.37(0.03)	130(61)	18(3)	0.38(0.04)

[a] Standard deviation of triplicate analysis.

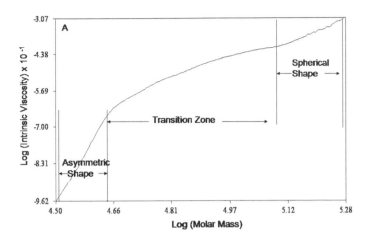

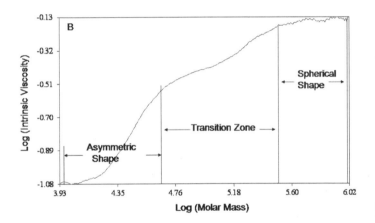

Figure 5. Mark-Houwink Plot. A.) ASP I. B.) ASP II. Sample 10/100/30.

Analysis of Molar Mass, Radius of Gyration and Intrinsic Viscosity

Table 7 contains the weight average molar mass, M_w, the z-average radius of gyration (Rg_z) and the weight average intrinsic viscosity ($[\eta_w]$) for ASP I and II. For ASP I, M_w values ranged between 83,000 Da and 192,000 Da and for ASP II between 92,000 Da and 252,000 Da. For ASP I, values ranged from about 10 to 24 nm for Rg_z and about 0.31 to 0.40 dL/g for $[\eta_w]$. For ASP II, values ranged from about 14 to 23 nm for Rg_z and about 0.33 to 0.43 dL/g for $[\eta_w]$. By way of contrast, sugar beet pectin values of M_w, Rg_z, and $[\eta_w]$ indicated that sugar beet pectins are larger and have higher molar masses than alkaline soluble polysaccharides (see Table 2).

Table 8. Molecular properties of alkaline soluble polysaccharide fractions

Region[a]	Frac.[b]	% Rec.[c]	$M_w \times 10^{-3}$	Rg_z nm	η_w dL/g	a
Whole[e]	ASP I	58 (2)	94(9)[d]	10(1)	0.31(.01)	.59(.04)
	ASP II	45(1)	109(8)	16(2)	0.33(.03)	.66(.04)
Spherical	ASP I	12(1)	205(31)	13(.5)	0.40(.01)	.27(.03)
	ASP II	5.0(.2)	512(75)	18(1)	0.65(.07)	.26(.05)
Trans[f]	ASP I	39(1)	74(5)	7(1)	0.30(.01)	.87(.2)
	ASP II	20(1)	91(5)	11(2)	0.34(.01)	.55(.07)
Asym[g]	ASP I	7.0(.6)	24(2)	3.9(.1)[f]	0.13(.01)	.86(.06)
	ASP II	20(1)	19(2)	3.5(.1)[f]	0.12(.01)	.89(.03)

[a] Region of the chromatogram. [b] Fraction extracted. [c] Percentage recovered after extraction. [d] Standard deviation of triplicate analysis. [e] Properties of entire sample 10/100/30 (heating time, minutes/temperature, °C, pressure, lb/in².) [f] Transition. [g] Asymmetric.

Figures 5a and 5b are M-H plots for ASP I and II respectively. We have divided both plots into three sections. The high molar mass end in which ASP I has an "a" value of 0.27 and ASP II has an "a" value of 0.26, a transition region in which the "a" values were 0.87 and 0.55 respectively and low molar mass region in which the "a" values are 0.86 and 0.89 (See Table 8.). Based on their M-H exponents, ASP in the high molar mass region tend toward a compact, possibly spherical, shape whereas those in the low molar mass region tend toward a more asymmetric shape. Presumably the transition region contains a mixture of both kinds of molecules with the asymmetric shapes being greater in number. The M-H exponents for the two ASP fractions at the ends of the distribution are comparable to one another but somewhat larger than M-H exponents obtained previously for sugar beet pectin which we extracted prior to extracting ASP (see Figure 2). In that case the high molar mass fraction had an "a" value of 0.08 and a low molar mass "a" value of 0.650. Because the transition region for sugar beet pectin in the M-H plot had a relative minimum and maximum, no "a" value was calculated.

AFM

Figure 6A contains ASP I molecules of sample 10/100/30 imaged from solution by AFM at a concentration of 0.125 µg/mL. As indicated by the inset height bars in the Figure, the higher the molecules rise above the surface, the brighter is their color. Size measurements at four concentrations on the fifty largest molecules revealed that the particles were asymmetric in shape and grew in size as their concentration increased (*14*). Remarkably, as shown in Figure 6B, at the concentration of 25 µg/mL many of the smaller compact molecules have aggregated into linear, circular or combined linear-circular skeletal structures. Nevertheless, many of the smaller compact molecules remain visible.

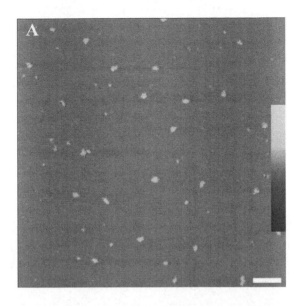

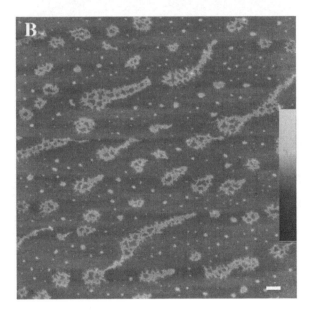

Figure 6. AFM images of ASP I deposited from water. Sample 10/100/30. A.) solution concentration 0.125 µg/mL. Scale bar is 250 nm; inset height scale is 0-5 nm. B.) Solution concentration 25 µg/mL. Scale bar is 250 nm; inset height scale is 0-10 nm (see color insert)

Figure 7 contains images of ASP II, extracted from sample 10/100/30 and deposited from solution at the concentration of 25 µg/mL. Unlike AFM images of sugar beet pectin (see Figure 3), network structures of ASP I and ASP II were not observed at 25 µg/mL. Figure 7 is comprised mostly of compact molecules with heights ranging from 0.8 to 5 nm and with a mean of about 2.9 nm, surrounded by fine strands (*14*).

Figure 7. AFM images of ASP II deposited from water. Sample 10/100/30. Solution concentration 25 µg/mL. Scale bar is 250 nm; inset height scale is 0-10 nm. (see color insert)

Partial Characterization of Carboxy Methyl Cellulose Extracted from the Residue

The insoluble residue which remained after removal of pectin, ASPI and II was partially solubilized by derivatizing with carboxy methyl groups (CWMG). Depending on reaction conditions, degree of carboxymethylation ranged from 0.59 to 1.38. High Performance Size Exclusion Chromatography with molar mass and viscometric detection revealed that solubilized material had M_w values ranging from about 84,000 to 206,000 Daltons, Rgz values ranging from 32 to 47 nm and [η_w] values ranging from 1.92 to 4.66 dL/g.

Conclusion

We have demonstrated that MAE of sugar beet pulp under moderate pressure and temperature could extract under acid conditions high molar mass, moderate viscosity pectin and alkaline soluble polysaccharides in minutes rather than hours as is required by conventional heating. HPSEC with online light scattering and viscosity detection revealed that both acid and alkaline soluble polysaccharides are bimodal distributions of high molar mass compact particles and lower molar mass less compact particles. AFM images of air dried solutions of sugar beet pectin corroborated this conclusion and further revealed that in the case of pectin strands and spherical particles were integrated into networks. In the case of ASP, only the compact spherical particles were visible when imaged by AFM at low concentrations. ASP I aggregated into skeletal structures with increasing concentration, ASP II only aggregated into larger compact particles. HPSEC and AFM in combination clearly showed the structural similarities and differences between pectin, ASP I and II. Initial results on their molecular properties indicate that sugar beet pulp has the potential to be a sustainable source of carboxymethyl cellulose.

Acknowledgments

We thank André White for his technical assistance. We thank Hoa K. Chau and David R. Coffin for their editorial and technical assistance. Mention of trade names or commercial products in this publication is solely for the purpose of providing specific information and does not imply recommendation or endorsement by the U.S. Department of Agriculture.

References

1. Schwartz, T. Beet Sugar Development Foundation, personal communication.
2. Oosterveld, A.; Beldman, G.; Schols, Henk A.; Voragen, A. G. J. *Carbohydr. Res.* **1996**, *288*, 143–153.
3. Carpita, N.; McCann, M. C. In *Biochemistry and Molecular Biology of Plants*; Buchanan, B., Ed.; American Society of Plant Physiologists: Rockville, MD, 2000; pp 52–108.
4. Rombouts, F. M.; Thibault, J.-F. *Carbohydr. Res.* **1986**, *154*, 177–187.
5. Roboz, E.; Van Hook, A. *Proc. Am. Soc. Sugar Beet Technol.* **1946**, *4*, 574–583.
6. Pippen, E. L.; McCready, R. M.; Owens, H. S. *J. Am. Chem. Soc.* **1950**, *72*, 813–816.
7. Levigne, S.; Ralet, M.-C.; Thibault, J.-F. *Carbohydr. Polym.* **2002**, *49*, 145–153.
8. Oosterveld, A.; Beldman, G.; Voragen, A. G. J. *Carbohydr. Polym.* **2002**, *48*, 73–81.
9. Bucholt, H. C.; Christensen, T. M. I. E.; Fallensen, B.; Ralet, M.-C.; Thibault, J.-F. *Carbohydr. Polym.* **2004**, *58*, 149–161.
10. Kirby, A. R.; MacDougall, A. J.; Morris, V. J. *Food Biophysics* **2006**, *1*, 51–56.
11. Sun, R.C.; Hughes, S. *Carbohydr. Polym.* **1999**, *38*, 273–281.
12. Dinand, E.; Chanzi, H.; Vignon, M.R. *Food Hydrocolloids* **1999**, *13*, 275–283.
13. Fishman, M. L.; , H. K.Chau; Cooke, P. H.; Hotchkiss, A. T., Jr. *J. Agric. Food Chem.* **2008**, *50*, 1471–1478.
14. Fishman, M. L.; Chau, H. K.; Cooke, P. H.; Yadav, M. P.; Hotchkiss, A. T. *Food Hydrocolloids* **2009**, *23*, 1554–1562.
15. Heinze, T.; Pfeiffer, K. *Angew. Makromol. Chem.* **1998**, *266*, 37–45.

16. Pushpamalar, V.; Langford, S. J.; Ahmad, M.; Lim, Y. Y. *Carbohydr. Polym.* **2006**, *64*, 312–318.
17. Fishman, M. L.; Cooke, P.; Levaj, B.; Gillespie, D. T. *Arch. Biochem. Biophys.* **1992**, *294*, 253–260.
18. Fishman, M. L.; Cooke, P.; Hotchkiss, A.; Damert, W. *Carbohydr. Res.* **1993**, *248*, 303–316.

Chapter 7

Novel Biobased Plastics, Rubbers, Composites, Coatings and Adhesives from Agricultural Oils and By-Products

Yongshang Lu and Richard C. Larock*

Department of Chemistry, Iowa State University, Ames, IA 50011
***larock@iastate.edu**

A remarkable range of exciting new plastics and rubbers have been made by the cationic, thermal, free radical and ring-opening metathesis polymerization of regular and modified vegetable oils with a number of readily available, commercial comonomers, including styrene, divinylbenzene, acrylonitrile and dicyclopentadiene. The bioplastics and rubbers possess excellent thermal and mechanical properties, plus outstanding damping and shape memory properties. Fillers, such as glass fibers, organic clays, and agricultural co-products, such as soybean hulls and spent germ, have been used to reinforce these vegetable oil polymer resins, resulting in biocomposites with significant improvement in their mechanical properties and thermal stability. A number of novel new vegetable oil-based waterborne polyurethane dispersions and polyurethane/ acrylics hybrid latexes have also been prepared and show promising applications as decorative/protective coatings and pressure sensitive adhesives.

Introduction

The utilization of fossil fuels for the manufacture of plastics accounts for about 7% of the worldwide use of oil and gas, which will arguably be depleted within the next one hundred years (1). Therefore, a change from fossil feedstocks to renewable resources is important for sustainable development into the future (2). The utilization of renewable resources can consistently provide raw materials for every day products, effectively avoiding further contribution to greenhouse gas effects, because of the minimalization of CO_2 emissions (3).

The renewable raw materials most widely used to replace petroleum are polysaccharides (mainly cellulose and starch), proteins, sugars, natural rubbers and plant oils (4, 5). Among these, vegetable oils are considered to be the most promising materials for the chemical and polymer industries, due to their superb environmental credentials, including their inherent biodegradability, low toxicity, avoidance of volatile organic chemicals, easy availability, and relatively low price (6). Vegetable oils have been used as an ingredient or component in many manufactured products, such as soaps, drying agents, paints, coatings, insulators, hydraulic fluids and lubricants. In recent years, the amounts of vegetable oils and fats produced have increased by approximately 4% per year, one third of which results from the growing industrial use of vegetable oils. Vegetable oils possess a triglyceride structure linked with different fatty acids as shown in Scheme 1. There are several positions that are amenable to chemical reactions: ester groups, C=C double bonds, allylic positions and the α-position of the ester groups. These functional groups can be used to directly polymerize triglycerides or to modify the triglyceride structure with more readily polymerizable groups to obtain thermosets (7).

Recently, a variety of new polymeric materials have been prepared from vegetable oils and derivatives, which possess industrially viable thermophysical and mechanical properties and thus may find many applications. One of the major efforts in this field has taken advantage of the carbon-carbon double bonds of the vegetable oils themselves for cationic polymerization (8) or the carbon-carbon double bonds of vegetable oil derivatives for free radical polymerization (9) or olefin metathesis polymerization (10). Other major processess have involved the conversion of vegetable oils or fats into epoxidized vegetable oils (EVO) (11) or polyols (12). The EVO can be cationically polymerized by latent thermal catalysts (11) or cured by amines (13) or anhydrides (14) to produce thermosetting epoxy resins, whereas the polyols can react with diisocyanates to produce vegetable oil-based polyurethane foams, thermosets (15, 16) or waterborne polyurethane dispersions (17, 18). In this review, we highlight the most recent advances made in novel polymers, biocomposites and nanocomposites based on vegetable oils, which have been subjected to cationic, free radical, olefin metathesis and step growth polymerizations.

Vegetable Oil-Based Bioplastics and Biocomposites from Cationic Polymerization

The carbon-carbon double bonds in vegetable oils are slightly more nucleophilic than those of ethylene and propylene and are susceptible to cationic polymerization (19). However, compared with ethylene, propylene and isobutylene, vegetable oils are a multifunctional monomer because of the multiple carbon-carbon double bonds in the triglycerides and the branching of the monomer, which results in crosslinked polymers with high molecular weights.

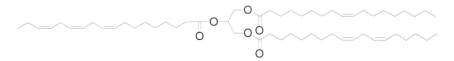

Scheme 1. The typical structure of vegetable oils

Protic acids and Lewis acids, such as $TiCl_4$, $ZnCl_2$ and $BF_3 \cdot OEt_2$ (BFE), are capable of initiating the cationic polymerization of vegetable oils under mild conditions (*20*). Among the Lewis acids, however, BFE has proved to be the most efficient initiator. Except for tung oil (*21*) and conjugated linseed oil, cationic homopolymerization of regular vegetable oils or the corresponding conjugated oils initiated by BFE affords only low molecular weight viscous oils or soft rubbery materials consisting of solid polymers and liquid oligomers in most cases. These materials generally possess limited utility. Therefore, alkene comonomers, such as styrene (ST), divinylbenzene (DVB), norbornadiene (NBD) and dicyclopentadiene (DCP), have usually been copolymerized with the vegetable oils (*22*). The cationic polymerization of a vegetable oil with ST and DVB is illustrated in Scheme 2. The cationic copolymerization of 50-60 wt % of various soybean oils [regular soybean oil (SOY), LowSat soybean oil (LLS) and conjugated LLS (CLS)] with DVB (25-35 wt %) initiated by BFE (1-5 wt %) provides thermosetting polymers ranging from soft rubbers to hard plastics, depending on the reagents, stoichiometry and initiators used (*21*). The room temperature moduli of these thermosets are approximately 400-1000 MPa, which are comparable to those of conventional plastics. However, due to the poor miscibility between the soybean oils and the initiator and the big difference in the reactivity of the soybean oils and DVB, heterogeneous reactions with a lot of solid white particles occur at the early stages of this copolymerization, leading to phase separated copolymers consisting of oil-rich phases and DVB-rich phases (*22*). It has been found that homogeneous copolymerization of the various soybean oils with DVB can be achieved by using ST as the main comonomer, instead of DVB, or using an initiator of BFE modified with Norway fish oil ethyl ester (NFO), since NFO is completely miscible with vegetable oils and DVB. The resulting soybean oil (SOY, LLS and CLS)-ST-DVB bulk polymers, consisting of ~ 45-50 wt % of the vegetable oil as a starting material are typically opaque materials with a glossy dark brown color, and have T_gs ranging from approximately 0 to 105 °C. The soybean oil-ST-DVB thermosets exhibit tensile stress-strain behavior ranging from soft rubbers through ductile to relatively brittle plastics (*23*). The Young's moduli of these polymers vary from 3 to 615 MPa, the ultimate tensile strengths vary from 0.3 to 21 MPa, and the elongation at break vary from 1.6 to 300%, depending on the stoichiometry and the soybean oil employed.

Material damping is one of the most effective solutions to the problem of vibration and noise. Viscoelastic polymers useful as damping materials have attracted considerable interest in recent years because of their high damping values around the glass transition temperature (T_g) (*24*). Soybean oil-ST-DVB polymers with appropriate compositions exhibit good damping properties over a broad temperature and frequency range with a loss factor maximum (tan δ)$_{max}$ of 0.8-4.3, an overall damping capacity value of 50-124 K after correcting the background,

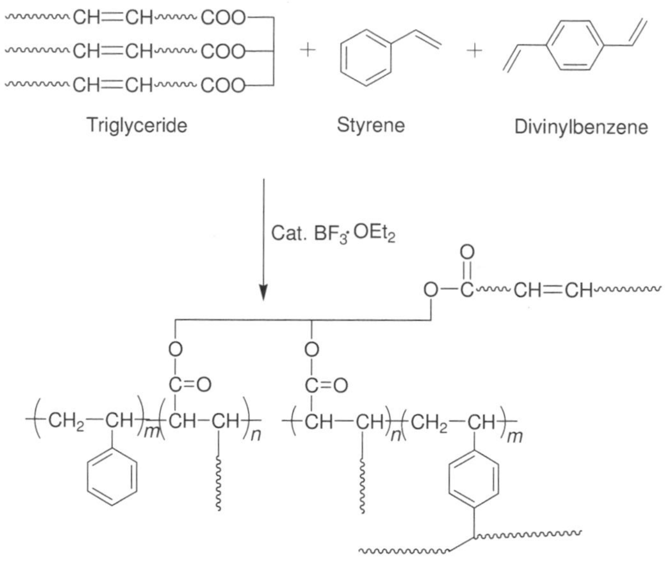

Scheme 2. Cationic copolymerization of soybean oils with ST and DVB.

and an 80-110 degree temperature range for high damping ($\tan \delta > 0.3$) (*25*). The high damping intensities observed in these materials can be attributed to the large number of ester groups present in the triglycerides directly attached to the polymer chains, whereas the broad damping region is due to the segmental inhomogeneities upon crosslinking.

In addition to good damping properties, these soybean oil-ST-DVB polymers can also be tailored to show good shape memory properties (*26*). It is found that a T_g well above ambient temperature and a stable crosslinked network are two prerequisites for these polymers to exhibit good shape memory effects. Through structural design of the polymer chain rigidity, the resuting soybean oil-based polymers possess excellent processability in the elastomeric state, being able to fix over 97% of their deformation at room temperature, and completely recover their original shape upon being reheated, making these materials particularly promising in applications where shape memory properties are desirable. In addition to ST and DVB, dicyclopentadiene (DCP) has also been used as a comonomer for cationic copolymerization with SOY or conjugated soybean oil (CSOY) initiated by BFE, resulting in a variety of novel thermosetting polymers ranging from tough and ductile to very soft rubbers (*27*). The SOY-DCP and CSOY-DCP bulk copolymers have T_gs ranging from -22.6 to 56.6 °C and are thermally stable below 200 °C.

New silicon-containing soybean-oil-based copolymers have been prepared from SOY, ST, DVB, and *p*-(trimethylsilyl)styrene by cationic polymerization using BFE as the initiator. The resulting thermosets exhibit glass transition temperatures ranging from 50 to 62 °C and limiting oxygen index (LOI) values from 22.6 to 29.7, suggesting that these materials may prove to be useful alternatives for current non-renewable-based flame-retardant materials (*28*).

Using soybean/corn oil-based resins as the matrix, a series of novel high performance biocomposites have been prepared by reinforcing the resins with glass fibers (*29, 30*). With increasing glass fiber content from 0 to 52 wt%, the composites exhibit a significant increase in Young's modulus from 150 to 2730 MPa and ultimate tensile strength from 7.9 to 76 MPa. Increasing the crosslinking (*e.g.* DVB) in the vegetable oil-based matrix results in composites with improved thermal and mechanical properties. When compared with biocomposites from a corn oil-based resin reinforced with glass fiber, the biocomposites based on the soybean oil resin possess higher thermal and mechanical properties (*30*), which is attributed to the higher crosslinking present in the soybean oil resin, due to the higher unsaturation of the soybean oil.

A montmorillonite clay modified with triethyl(4-vinylbenzyl)ammonium chloride (VTAC), abbreviated VMMT, has been used to reinforce corn (*31*) and soybean oil-based cationically polymerized resins (*32*), where the polymerizable vinyl groups of the VTAC can be incorporated into the polymer resin through chemical bonding. It has been found that a heterogeneous structure consisting of intercalation and partial exfoliation or an intercalation structure occurs in the resulting materials, depending on the amount of VMMT. The resulting nanocomposites with 1-2 wt % VMMT exhibit a significant improvement in their thermal stability, mechanical properties, and vapor barrier performance. For example, the modulus, strength and strain at failure for the CLS-based nanocomposites increase by 100-128%, 86-92% and 5-7%, respectively, when the VMMT loading is in the range of 1-2 wt % (*32*).

Vegetable Oil-Based Bioplastics and Biocomposites by Free Radical Polymerization

The carbon-carbon double bonds in the vegetable oils are capable of being polymerized through a free radical mechanism. However, due to the presence of chain transfer processes occurring at the allylic positions of the fatty acid chains, the free radical polymerization of triglyceride double bonds has received relatively little attention. The drying oils, such as tung oil, can react with atmospheric oxygen to form polymeric materials with a network structure. The oxidation of drying oils by air involves hydrogen abstraction from a methylene group between two double bonds in a polyunsaturated fatty acid chain (*33–35*), which leads to peroxidation, perepoxidation, hydroperoxidation, epoxidation, and crosslinking via radical recombination. Using this method, a variety of grafted copolymers with higher biodegradability and biocompatibility have been synthesized by the free radical polymerization of methyl methacrylate (MMA) or *n*-butyl methacrylate (nBMA) initiated by using polymeric peroxides prepared by the autooxidation of linseed oil (LIN) (*33*), soybean oil (SOY) (*34*), and linoleic acid (LIA) (*35*).

When heated above 100 °C, ST undergoes thermal polymerization, which involves the formation of a Diels-Alder dimer from two ST molecules and subsequent hydrogen transfer to styrene to yield two radicals that can initiate polymerization of a drying oil (*36*). Tung oil-based (30-70 wt %) thermosetting polymers have been synthesized by the thermal copolymerization of tung oil, ST and DVB (*37*). These fully cured thermosets, ranging from elastomers to tough and rigid plastics, possess T_gs of -2 to +116 °C, crosslink densities of 1.0×10^3

to 2.5×10^4 mol/m^3, coefficients of linear thermal expansion of 2.3×10^{-4} to 4.4×10^{-4}/ °C, compressive moduli of 0.02-1.12 GPa, and compressive strengths of 8-144 MPa. A series of samples with similar properties have also been prepared from conjugated linseed oil (CLIN), ST and DVB (*38*).

Conjugated carbon-carbon double bonds in vegetable oils are relatively easily attacked by free radicals (*8*). Much more viscous polymerized conjugated oils can be obtained by free radical polymerization of conjugated linseed oil (CLIN) or CLS initiated by either benzoyl peroxide (BPO) or *tert*-butyl hydroperoxide (TBHP) or combinations of these reagents. Some comonomers, such as DVB, acrylonitrile (AN) and DCP, have been successfully copolymerized with CLIN and CLS to obtain thermosetting polymers with good thermal and mechanical properties ranging from hard and brittle plastics to soft and rubbery materials (*39, 40*). For example, for the CLIN-AN-DVB thermosets, approximately 61-96 wt % of the CLIN has been incorporated into the final products. The T_gs determined from the tan δ peaks determined by DMA analysis range from 60 to 101 °C and the room temperature storage moduli vary from 160 MPa to 1.6 GPa, depending on the oil content in the formulation. These thermosets are thermally stable up to 300 °C. The wide range of properties attained with these materials makes them suitable for potential applications in which petroleum-based polymers are currently used (*39*).

Novel biocomposites have been prepared by the free radical polymerization of a tung oil-based resin using spent germ (SG), the co-product of wet mill ethanol production, as a filler (*41*). The composites produced are quite thermally stable with T_{max} values in the vicinity of 430 °C. In general, the thermal and mechanical properties of the composites are improved by decreasing the size of the filler. This is most likely the result of enhanced interfacial adhesion and filler-matrix interaction. As more SG is added to the composite, the mechanical properties tend to decrease, due to filler-filler agglomerations and an increase in voids expected when the amount of filler is increased. DVB is used as an effective crosslinker and, as expected, the thermal and mechanical properties of the composites increase as the concentration of DVB in the matrix increases.

In addition to tung oil-based biocomposites, biocomposites have been prepared by the free radical polymerization of a CSOY-based resin reinforced with soybean hulls (*42*). The resin consists initially of 50 wt % CSOY and varying amounts of DVB (5-15 wt %), DCP (0-10 wt %), and *n*BMA (25-35 wt %). Two soybean hull particle sizes have been tested (<177 and <425 μm) and two different filler/resin ratios have been compared (50 : 50 and 60 : 40). It has been observed that the mechanical properties of the composites tend to decrease when higher filler/resin ratios or larger particle sizes are used. This behavior is closely related to impregnation of the filler by the resin; whenever good dispersion of the filler in the matrix is compromised, the mechanical properties of the composites are negatively affected.

Besides moving the carbon-carbon double bonds into conjugation to improve reactivity, the incorporation of more reactive carbon-carbon double bonds through chemical modifications is another important approach to improving the overall reactivity of vegetable oils. Wool and co-workers have developed a number of chemical routes to vegetable oil derivatives useful in the preparation of polymer and composite materials via free radical polymerization (*43*, *44*). For example, acrylated epoxidized soybean oil (AESO), prepared by ring opening epoxidized soybean oil by acrylic acid, when diluted with ST affords AESO-ST thermosets suitable for structural applications after polymerization. The corresponding glass fiber-reinforced composites display a tensile modulus of 5.2 GPa and flexural modulus of 9 GPa with glass fiber content lower than 35 wt %. At higher glass fiber contents (50 wt %), the composites display tensile and compression moduli of 24.8 GPa each. In addition to glass fibers, natural fibers, such as flax and hemp, can be used to reinforce these AESO-based resins for preparation of composite materials. The properties exhibited by both the natural- and synthetic fiber-reinforced composites can be combined through the production of "hybrid" composites. These materials combine the low cost of natural fibers with the high performance of synthetic fibers, resulting in a wide range of mechanical properties.

Vegetable Oil-Based Bioplastics and Biocomposites from ROMP

Olefin metathesis is a reaction which involves the exchange of alkylidene groups catalyzed by metals, such as W, Mo, Re, and Ru (*45*). The olefin metathesis of unsaturated fatty acid esters was first examined by Boelhouwer *et al.* (*46*), who found that methyl oleate was converted into equimolar amounts of 9-octadecene and dimethyl 9-octadecenedioate in the presence of WCl_6 plus $(CH_3)_4Sn$. Using the same catalyst system, Kohashi and Foglia (*47*) examined the co-metathesis of methyl oleate with other unsaturated diesters. The WCl_6/Me_4Sn-catalyzed metathesis of soybean oil has been reported to produce an improved drying oil (*48*). However, the standard metathesis conditions utilized in this system suffer from a number of disadvantages, including the high sensitivity of the catalyst to moisture and oxygen, the use of a chlorinated solvent (chlorobenzene), and a relatively high catalyst load.

An alternative, very effective catalyst, Grubbs' ruthenium catalyst $(Cy_3P)_2Cl_2Ru=CHPh$ has been reported to affect the olefin metathesis of vegetable oils (*49*). The reaction proceeds in the absence of solvent, with very low catalyst concentrations (0.1 mole %) under moderate temperatures and low pressures. Furthermore, the higher molecular weight oligomers obtained can be easily separated from the unreacted oil and the lower molecular weight alkene by-products.

Olefin metathesis has also been employed in polymer synthesis, primarily through ring opening metathesis polymerization (ROMP) (*50*) and acyclic diene metathesis polymerization (ADMET) (*51*). The Grubbs' ruthenium catalyst has been used for the ADMET polymerization of soybean oil, resulting in a variety of materials ranging from sticky oils to rubbers (*52*). Through ROMP, a bicyclic castor oil derivative (BCO) prepared from the esterification of castor oil with the commercially available bicyclic anhydride bicyclo[2.2.1]hept-5-ene-2,3-dicarboxylic anhydride has been successfully copolymerized with cyclooctene to synthesize rubbery thermosets (*53*). The resulting rubbery thermosets possessing 55 to 85 wt % oil are transparent with a light tan hue, thermally stable above 200 °C, and have T_gs ranging from -14 to 1 °C, well below ambient temperature.

Dilulin is a commercially available vegetable oil derivative from Cargill, which is a mixture of norbornenyl-functionalized linseed oil and cyclopentadiene oligomers in a weight ratio of 95:5. Using DCP (54) or other polycyclic norbornene-based comonomers (55) as a crosslinker, a variety of unique Dilulin-based thermosetting resins have been successfully synthesized. The ROMP of pure Dilulin itself affords a soft and flexible rubber with a T_g of -29 °C. However, with an increase in DCP content from 0-70 wt %, the resulting yellow, transparent thermosets vary from soft and flexible to hard and strong. The Young's modulus and tensile strength of the thermoset with 70 wt % DCP are 525 MPa and 29 MPa, respectively, which makes this material comparable to high density polyethylene (56). After reinforcement with 56 wt % glass fibers, the Young's modulus and tensile strength of the resulting Dilulin-DCP biocomposites are significantly increased to 1576 and 168 MPa, respectively. These properties suggest promising applications as structural materials (54). Dilulin and polycyclic norbornene-based comonomers undergo ROMP with Grubbs' 1st generation catalyst to form a variety of thermosetting polymers in which the Dilulin content ranges from 50-100 wt % (55). However, the vegetable oil-based thermosets prepared from BCO and Dilulin exhibit a phase separation consisting of an oil-rich phase and a petroleum-based comonomer-rich phase, due to the difference in their ROMP reactivity and their structures.

To address the phase separation while maintaining good thermophysical and mechanical properties, four novel norbornenyl-functionalized fatty alcohols with different side chain structures, e.g. NMSA, NMDA, NMMA and NMCA, have been synthesized from soybean oil, Dilulin, ML189 and castor oil respectively by first reducing the vegetable oil triglycerides to fatty alcohols, which were then reacted with 5-norbornene-2,3-dicarboxylate anhydride (57). These norbornenyl-functionalized biorenewable monomers can easily undergo ring opening metathesis homopolymerization using Grubbs 2nd generation catalyst, leading to vegetable oil-based thermosets with good thermophysical and mechanical properties and no apparent phase separation (Scheme 3). The differences in the structures of the side chains and the viscosities of the monomers result in different properties for the final thermosets. Compared with polyNMSA, the polyNMDA and polyNMMA thermosets exhibit lower soluble fractions, and higher thermal stabilities and mechanical properties, because of successful incorporation of the side chain into the polymer matrix to form effective crosslinking. However, polyNMCA affords a higher soluble fraction, and lower thermal stability, resulting from incomplete polymerization of the highly viscous NMCA monomer. Note that polyNMDA and polyNMMA exhibit tensile stress-strain behaviors of ductile plastics with Young's moduli ranging from 155 to 310 MPa, ultimate tensile strengths ranging from 10 to 15 MPa, and percent elongation at break values ranging from 35 to 41%. The properties of these materials are comparable to petroleum-based plastics, like HDPE and poly(norbornene).

Scheme 3. ROMP samples from norbornenyl-functionalized fatty alcohols.

Vegetable Oil-Based Waterborne Polyurethane Dispersions and Their Hybrid Latexes for Coatings and Adhesives

Polyurethanes (PUs) are one of the most versatile polymeric materials with regard to both processing methods and mechanical properties (*58*). PUs that range from high performance elastomers to tough rigid plastics can be easily synthesized by the proper selection of reactants. This wide range of achievable properties makes PUs an indispensable component in coatings, binders, adhesives, sealants, fibers, and foams (*59, 60*). Therefore, academic and industrial researchers are increasingly devoting their attention and efforts to the possible use of vegetable oils as raw materials for the production of polyurethane products (*61*). With the exception of castor and lesquerella oils, hydroxyl groups must be introduced at the unsaturated sites to synthesize PU thermosets based on vegetable oils. This has been accomplished by hydroformylation, followed by hydrogenation (*62*); epoxidation, followed by oxirane opening (*63*); ozonolysis, followed by hydrogenation (*64*); and microbial conversion (*65*). However, the conventional PU products usually contain a significant amount of organic solvents and sometimes even free isocyanate monomers (*66*). The increasing need to reduce volatile organic compounds (VOCs) and hazardous air pollutants (HAPs) has led to increased efforts to formulate waterborne PUs for use as coatings, adhesives, and related end uses (*67*). Waterborne PUs present many advantages relative to conventional solvent-borne PUs, including low viscosity at high molecular weight and good applicability, and are now one of the most rapidly developing and active branches of PU chemistry (*68*).

Environmentally friendly, vegetable oil-based waterborne polyurethane dispersions with very promising properties have been synthesized (Scheme 4) from a series of methoxylated soybean oil polyols (MSOLs) with hydroxyl functionalities ranging from 2.4 to as high as 4.0 (*69*). The particle sizes of the dispersions and the structure and thermophysical and mechanical properties of the resulting soybean oil-based waterborne polyurethane (SPU) films containing 50-60 wt % of biorenewable MSOL depend strongly on the polyol functionality and the hard segment content. The particle size of the SPU dispersions increases from about 12 to 130 nm when the MSOL functionality is increased from 2.4 to 4.0, whereas the particle size decreases with an increase in the hard segment content of the SPU. Increasing the OH functionality of the MSOLs

95

can significantly increase the crosslink density of the SPUs, whereas increasing the hard segment content of the SPU can effectively improve the interchain interactions caused by hydrogen bonds. Typical tensile stress-strain behaviors of the SPU films are shown in Figure 1. The SPU-135, prepared from an MSOL with hydroxyl number of 135 mg KOH/g, shows a rubbery modulus of 7.7 MPa, a viable ultimate tensile strength of ~4.2 MPa, and an elongation at break of ~280%. When the OH number of the MSOL is further increased to 149 mg KOH/g, the Young's modulus and ultimate tensile strength of the resulting SPU-149 increase, but its elongation at break slightly decreases. It is worth noting that both SPU-135 and SPU-149 exhibit a strain recovery of 100% because of their relatively low crosslink densities. This is similar to the tensile test behavior of an elastomeric polymer. However, the film SPU-176 exhibits behavior that is typical of a ductile plastic with a clear yield point and shows a modulus and a tensile strength that are approximately 41 and 3.6 times higher, respectively, than those of SPU-135. The MSOL with a higher hydroxyl number of 190 mg KOH/g results in a relatively hard plastic, SPU-190, which exhibits yielding behavior, followed by strain softening. No strain hardening behavior is observed before the specimen breaks. The film SPU-200 from an MSOL with hydroxyl number of 200 mg KOH/g exhibits characteristics of a very rigid plastic. Its Young's modulus and ultimate tensile strength reach approximately 720 and 22 MPa, respectively. These changes in the mechanical behavior are a result of increases in the crosslink density in the resulting SPU films from the MSOLs with high average OH functionality.

PUs feature exceptional performance properties, such as excellent elasticity, mar resistance, toughness, adhesion, and superior low temperature impact properties, but suffer from poor hardness, water, and alkali resistance (70). In order to obtain vegetable oil-based waterborne polyurethane resins with good overall properties, a series of new waterborne polyurethane (PU)/acrylic hybrid latexes have been synthesized by the emulsion polymerization of acrylic monomers (butyl acrylate and methyl methacrylate) in the presence of a soybean oil-based waterborne PU dispersion using potassium persulfate as an initiator (71). The resulting PU/acrylic hybrid latexes, containing 15-60 wt % of a chlorinated soybean oil polyol (SOL) with 2.3 hydroxyl functionalities as a renewable resource, are very stable and exhibit a uniform particle size of 125 ± 20 nm. The particle sizes of the final hybrid latexes are not significantly affected by the urethane/acrylic ratio. The occurrence of extensive grafting of the acrylics onto the PU and interpenetration between the acrylics and the PU results in miscible hybrid latexes with enhanced thermal and mechanical properties. The tensile stress-strain curves of the PU/acrylic hybrid latexes as a function of the acrylic content is shown in Figure 2. The tensile strength and elongation at break of the PU film are 8.8 MPa and 200%, respectively. These values are very close to those of the waterborne PUs from castor oil- and rapeseed oil-based polyols (17, 18). For the hybrid latex films, the tensile strengths are significantly increased from 8.8 to 16.9 MPa with an increase in the acrylic content from 0 to 75 wt %. The elongation at break of the hybrid latex films first increases with increasing acrylic content and reaches a maximum of 365% at 50 wt % acrylic content. Hybrid latex films with good mechanical properties can be prepared over the entire range of

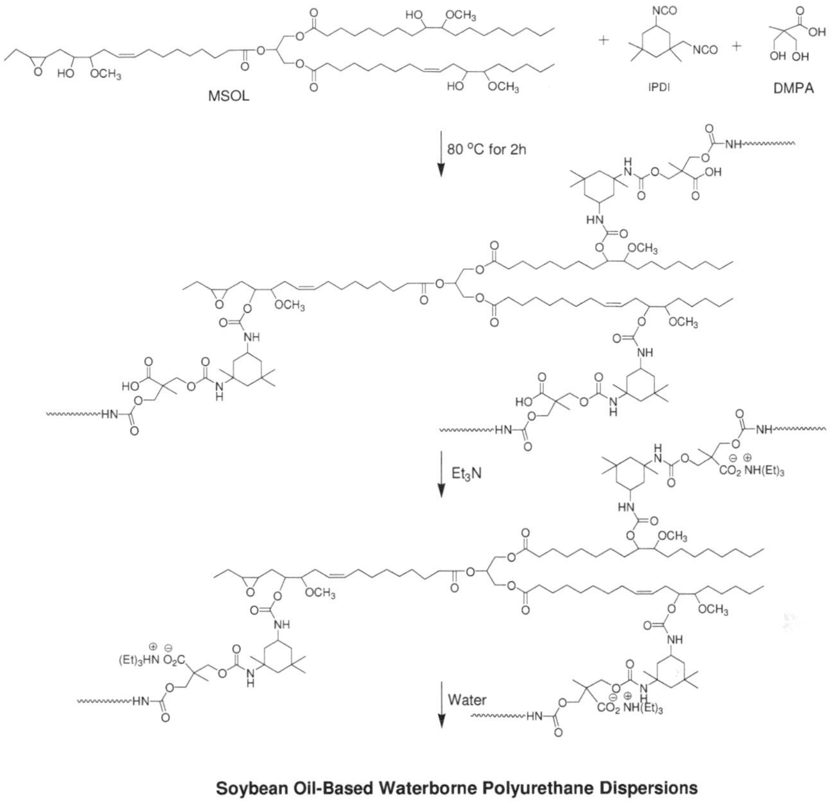

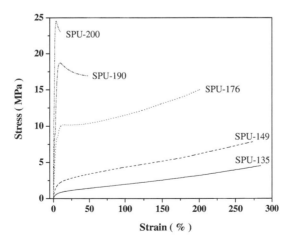

Soybean Oil-Based Waterborne Polyurethane Dispersions

Scheme 4. Synthesis of waterborne polyurethane dispersions from MSOLs.

Figure 1. Stress-strain curves of SPU films.

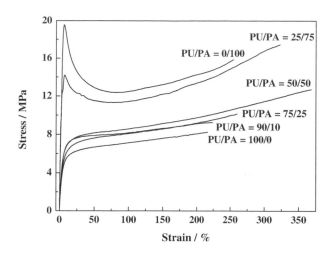

Figure 2. Tensile stress-strain curves of PU/PA hybrid latex films.

composition by hybrid emulsion polymerization technology. However, physically blended films from a PU dispersion and a PA emulsion are very brittle when the PA content is higher than 30 wt %, due to the poor miscibility between component polymers. The deformation behavior of the hybrid latex films at room temperature greatly depends on the acrylic content. When the acrylic content is less than 50 wt %, the latex films exhibit characteristics typical of soft and tough polymers that show uniform extension. However, the hybrid latex film containing 75 wt % acrylics displays behavior typical of rigid and tough plastics with a necking down across the width of the specimen, similar to the behavior of the PA copolymers.

It has been reported that graft copolymerization between vinyl-containing polyurethane (VPU) macromonomers and vinyl monomers, such as ST (*72*), and acrylics (*73*), can significantly increase the thermal stability, tensile strength, hardness, and water resistance of the resulting materials, because of the covalent linkage between the two components. Generally, the synthesis of a grafted VPU-vinyl copolymer is carried out in a two-stage reaction. First, excess isocyanate is allowed to react with the polyol to obtain a VPU prepolymer, followed by addition of 2-hydroxyethyl acrylate (HEA) or 2-hydroxyethyl methacrylate (HEMA) as an end-capping monomer. Then, the vinyl monomers are added to the VPU for graft copolymerization, leading to grafted urethane-vinyl copolymers with high performance.

Using AESO as a starting monomer, a novel bio-based vinyl-containing waterborne polyurethane (VPU) dispersion has been synthesized from the reaction of Toluene-2,4-diisocyanate (TDI), dimethanol propionic acid (DMPA) and a mixture of SOL and AESO (*74*). By polymerization of acrylic monomers of methyl methacrylate (MMA) and butyl acrylate (BA) in the presence of the VPU dispersion, a variety of VPU-acrylics grafted latexes with a relatively uniform particle diameter of 75 ± 15 nm have been prepared using emulsion polymerization. The resulting grafted latexes, containing 15-60 wt % of the SOL and AESO as renewable resources, exhibit enhanced thermal stability and mechanical properties, because of the occurrence of graft copolymerization between the acrylics and the VPU. Depending on the ratio of components, the grafted latex films exhibit characteristics typical of soft and tough polymers or rigid and tough plastics, whose mechanical properties are comparable to or better than those of petroleum-based urethane-acrylic hybrid latex films. Compared with a hybrid latex prepared from soybean oil-based PU without AESO, the grafted latex from the VPU exhibits a substantial improvement in thermal stability and mechanical properties, due to the extensive grafting, which has occurred in the grafted latexes.

By changing the structure and hydroxyl functionality of the soft segment, the type of hard segment and the molar ratio between hard and soft segments, as well as the ratio between the PU and the acrylics, these vegetable oil-based waterborne PU dispersions and hybrid latexes can be easily formulated into latex architecture coatings and pressure-sensitive adhesives with physical properties that can be comparable to or even better than petroleum-based analogues.

Conclusion

Vegetable oil-based plastics, rubbers, composites, latex coatings and adhesives have been prepared by cationic, free radical, ROMP, step-growth and emulsion polymerizations. These vegetable oil-based polymeric materials possess conventional, industrially useful characteristics, as well as unique properties, showing promising applications as alternatives to their petroleum-based counterparts. The utilization of vegetable oils as a starting raw material for the synthesis of valuble polymeric materials is becoming more and more important from a social, environmental and energy standpoint, with the increasing emphasis on waste disposal and the depletion of non-renewable resources. At the same time, research on the development and applications of vegetable oil-based polymers will broaden the scope of the synthetic materials available and give new polymeric materials.

References

1. Williams, C. K.; Hillmyer, M. A. *Polym. Rev.* **2008**, *48*, 1–10.
2. Meier, M. A. R.; Metzger, J. O.; Schubert, U. S. *Chem. Soc. Rev.* **2007**, *36*, 1788–1802.
3. Eissen, M.; Metzger, J. O.; Schmidt, E.; Schneidewind, U. *Angew. Chem., Int. Ed.* **2002**, *41*, 414–436.
4. Padma, L. N. *Polym. Rev.* **2000**, *40*, 1–21.
5. MacGregor, E. A.; Greenwood, C. T. *Polymers in Nature*; John Wiley & Sons: New York, 1980.
6. Erhan, S. Z., Ed. *Industrial Uses of Vegetable Oils*; AOCS Press: Champaign, IL, 2005.
7. Galià, M.; Montero de Espinosa, L.; Ronda, J. C.; Lligadas, G.; Càdiz, V. *Eur. J. Lipid Sci. Technol.* **2009**, *111*, in press.
8. Lu, Y.; Larock, R. C. *ChemSusChem* **2009**, *2*, 136–147.

9. Wool, R. P.; Sun, X. S. *Bio-Based Polymers and Composites*; Elsevier: Amsterdam, The Netherlands, 2005.

10. Kaminsky, W.; Fernandez, M. *Eur. J. Lipid Sci. Technol.* **2008**, *110*, 841–845.

11. Park, S.-J.; Jin, F.-L.; Lee, J.-R. *Macromol. Chem. Phys.* **2004**, *205*, 2048–2054.

12. Javni, I.; Zhang, W.; Petrović, Z. S. *J. Appl. Polym. Sci.* **2003**, *88*, 2912–2916.

13. Miyagawa, H.; Mohanty, A. K.; Misra, M.; Drzal, L. T. *Macromol. Mater. Eng.* **2004**, *289*, 636–641.

14. Miyagawa, H.; Mohanty, A. K.; Misra, M.; Drzal, L. T. *Macromol. Mater. Eng.* **2004**, *289*, 629–635.

15. Petrović, Z. S.; Guo, A.; Zhang, W. *J. Polym. Sci., Part A: Polym. Chem.* **2000**, *38*, 4062–4069.

16. Lligadas, G.; Ronda, J. C.; Galià, M.; Cádiz, V. *Biomacromolecules* **2007**, *8*, 1858–1864.

17. Lu, Y.; Tighzert, L.; Dole, P.; Erre, D. *Polymer* **2005**, *46*, 9863–9870.

18. Lu, Y.; Tighzert, L.; Berzin, F.; Rondot, S. *Carbohydr. Polym.* **2005**, *61*, 174–182.

19. Kennedy, J. P.; Marechal, E. *Carbocationic Polymerization*; John Wiley & Sons: New York, 1982; pp 31−55.

20. Marks, D.; Li, F.; Pacha, C. M.; Larock, R. C. *J. Appl. Polym. Sci.* **2001**, *81*, 2001–2012.

21. Andjelkovic, D. D.; Valverde, M.; Henna, P.; Li, F.; Larock, R. C. *Polymer* **2005**, *46*, 9674–9685.

22. Li, F.; Hanson, M. V.; Larock, R. C. *Polymer* **2001**, *42*, 1567–1579.

23. Li, F.; Larock, R. C. *J. Polym. Sci., Part B: Polym. Phys.* **2001**, *39*, 60–77.

24. Qin, C.-L.; Cai, W.-M.; Cai, J.; Tang, D.-Y.; Zhang, J.-S.; Qin, M. *Mater. Chem. Phys.* **2004**, *85*, 402–409.

25. Li, F.; Larock, R. C. *Polym. Adv. Technol.* **2002**, *13*, 436–449.

26. Li, F.; Larock, R. C. *J. Appl. Polym. Sci.* **2002**, *84*, 1533–1543.

27. Andjelkovic, D. D.; Larock, R. C. *Biomacromolecules* **2006**, *7*, 927–936.

28. Sacristán, M.; Ronda, J. C.; Galià, M.; Cádiz, V. *Biomacromolecules* **2009**, *10*, 2678–2685.

29. Lu, Y.; Larock, R. C. *Macromol. Mater. Eng.* **2007**, *292*, 1085–1094.

30. Lu, Y.; Larock, R. C. *J. Appl. Polym. Sci.* **2006**, *102*, 3345–3353.

31. Lu, Y.; Larock, R. C. *Biomacromolecules* **2006**, *7*, 2692–2700.

32. Lu, Y.; Larock, R. C. *Macromol. Mater. Eng.* **2007**, *292*, 863–872.

33. Singleton, D. A.; Hang, C.; Szymanski, M. J.; Meyer, M. P.; Leach, A. G.; Kuwata, K. T.; Chen, J. S.; Greer, A.; Foote, C. S.; Houk, K. N. *J. Am. Chem. Soc.* **2003**, *125*, 1319–1328.

34. Çakmakli, B.; Hazer, B.; Tekin, I. O.; Cömert, F. B. *Biomacromolecules* **2005**, *6*, 1750–1758.

35. Çakmakli, B.; Hazer, B.; Tekin, I. O.; Açıkgöz, Ş.; Can, M. *J. Am. Oil Chem. Soc.* **2007**, *84*, 73–81.

36. Yamada, B.; Zetterlund P. B. In *Handbook of Radical Polymerization*; Matyjaszewski, K., Davis, T. P., Eds.; John Wiley & Sons: New York, 2002; pp 117–186.

37. Li, F.; Larock, R. C. *Biomacromolecules* **2003**, *4*, 1018–1025.

38. Kundu, P. P.; Larock, R. C. *Biomacromolecules* **2005**, *6*, 797–806.

39. Henna, P. H.; Andjelkovic, D. D.; Kundu, P. P.; Larock, R. C. *J. Appl. Polym. Sci.* **2007**, *104*, 979–985.

40. Valverde, M.; Andjelkovic, D. D.; Kundu, P. P.; Larock, R. C. *J. Appl. Polym. Sci.* **2008**, *107*, 423–430.

41. Pfister, D. P.; Baker, J. R.; Henna, P. H.; Lu, Y. S.; Larock, R. C. *J. Appl. Polym. Sci.* **2008**, *108*, 3618–3625.

42. Quirino, R. L.; Larock, R. C. *J. Appl. Polym. Sci.* **2009**, *112*, 2033–2043.

43. Khot, S. N.; LaScala, J. J.; Can, E.; Morye, S. S.; Williams, G. I.; Palmese, G. R.; Küsefoğlu, S. H.; Wool, R. P. *J. Appl. Polym. Sci.* **2001**, *82*, 703–723.

44. Wool, R. P.; Küsefoğlu, S. H.; Palmese, G. R.; Zhao, R.; Khot, S. N. U.S. Patent 6,121,398, 2000.

45. Banks, R. L.; Bailey, G. C. *Ind. Eng. Chem. Prod. Res. Dev.* **1964**, *3*, 170–173.

46. Boelhouwer, C.; Mol, J. C. *J. Am. Oil Chem. Soc.* **1984**, *61*, 425–430.

47. Kohashi, H.; Foglia, T. A. *J. Am. Oil Chem. Soc.* **1985**, *62*, 549–554.

48. Erhan, S. Z.; Bagby, M. O.; Nelsen, T. C. *J. Am. Oil Chem. Soc.* **1997**, *74*, 703–706.

49. Refvik, M. D.; Larock, R. C.; Tian, Q. *J. Am. Oil Soc.* **1999**, *76*, 93–98.

50. Ivin, K. J. *Olefin Metathesis*; Academic Press: London, 1983.

51. Walba, D. M.; Keller, P.; Shao, R.; Clark, N. A.; Hillmyer, M. A.; Grubbs, R. H. *J. Am. Chem. Soc.* **1996**, *118*, 2740–2741.

52. Tian, Q.; Larock, R. C. *J. Am. Oil Chem. Soc.* **2002**, *79*, 479–88.

53. Henna, P. H.; Larock, R. C. *Macromol. Mater. Eng.* **2007**, *292*, 1201–1209.

54. Henna, P. H.; Kessler, M. R.; Larock, R. C. *Macromol. Mater. Eng.* **2008**, *293*, 979–990.

55. Mauldin, T. C.; Haman, K.; Sheng, X.; Henna, P.; Larock, R. C.; Kessler, M. R. *J. Polym. Sci., Part A: Polym. Chem.* **2008**, *46*, 6851–6860.

56. Mark, J. E. *Polymer Data Handbook*; Oxford University Press: New York, 1999.

57. Xia, Y.; Lu, Y. S.; Larock, Y. S. *Polymer* **2010**, *51*, 53–61.

58. Szycher, M. *Szycher's Handbook of Polyurethanes*; CRC Press: Boca Raton, 1999.

59. Zhang, L.; Jeon, H. K.; Malsam, J.; Herrington, R.; Macosko, C. W. *Polymer* **2007**, *48*, 6656–6667.

60. Liaw, D.-J.; Lin, S.-P.; Liaw, B.-Y. *J. Polym. Sci., Part A: Polym. Chem.* **1999**, *37*, 1331–1339.

61. Petrović, Z. S. *Polym. Rev.* **2008**, *48*, 109–155.

62. Petrović, Z. S.; Guo, A.; Javni, I.; Cvetković, I.; Hong, D. P. *Polym. Int.* **2008**, *57*, 275–281.

63. Guo, A.; Cho, Y.; Petrović, Z. S. *J. Polym. Sci., Part A: Polym. Chem.* **2000**, *38*, 3900–3910.

64. Kong, X. H.; Narine, S. S. *Biomacromolecules* **2007**, *8*, 2203–2209.

65. Hou, C. T. *Adv. Appl. Microbiol.* **1995**, *41*, 1–23.
66. Kim, B. S.; Kim, B. K. *J. Appl. Polym. Sci.* **2005**, *97*, 1961–1969.
67. Noble, K.-L. *Prog. Org. Coat.* **1997**, *32*, 131–136.
68. Lee, S. Y.; Lee, J. S.; Kim, B. K. *Polym. Int.* **1997**, *42*, 67–76.
69. Lu, Y.; Larock, R. C. *Biomacromolecules* **2008**, *9*, 3332–3340.
70. Wu, L.; You, B.; Li, D. *J. Appl. Polym. Sci.* **2002**, *84*, 1620–1628.
71. Lu, Y.; Larock, R. C. *Biomacromolecules* **2007**, *8*, 3108–3114.
72. Romero-Sánchez, M. D.; Pastor-Blas, M. M.; Martín-Martínez, J. M. *Int. J. Adhes. Adhes.* **2003**, *23*, 49.
73. Zhang, H. T.; Guang, R.; Yin, Z. H.; Lin, L. L. *J. Appl. Polym. Sci.* **2001**, *82*, 941.
74. Lu, Y.; Larock, R. C. *J. Appl. Polym. Sci.* **2009**, DOI:10.1002/app.29029.

Chapter 8

Enzyme-Nanotube-Based Composites Used for Chemical and Biological Decontamination

Cerasela Zoica Dinu,[1,2,*] Indrakant V. Borkar,[1] Shyam Sundhar Bale,[1] Guangyu Zhu,[1] Karl Sanford,[3] Gregg Whited,[3] Ravi S. Kane,[1] and Jonathan S. Dordick[1]

[1]Department of Chemical and Biological Engineering and Rensselaer Nanotechnology Center, Rensselaer Polytechnic Institute, Troy, NY 12180
[2]Department of Chemical Engineering, West Virginia University, Morgantown, WV 26506
[3]Genencor International, Palo Alto, CA 94304
*cerasela-zoica.dinu@mail.wvu.edu

"Smart" coatings capable of both detecting and actively eliminating hazardous agents are being investigated as new ways for combating chemical and biological contamination. Specifically, we took advantage of the unique surface properties of carbon nanotubes (e.g. high surface area, controllable morphology and size etc.) to immobilize a cocktail of enzymes and subsequently we incorporated the enzyme-carbon nanotube conjugates into latex-based paint. Operational properties of the enzymes were optimized to yield high loading and activity on the nanotube support, and long-term operational stability in the composite. This "green" namely enzyme-based technology, eliminates the risks associated with chemical decontamination that uses corrosive agents and generates substantial amounts of residual waste. Moreover, the "green"-based composites developed herein are cost-effective and decontaminate on contact without imposing logistical burden to the personnel.

Introduction

There is a critical need to develop and deploy safe and effective means to decontaminate chemical and biological warfare agents before they are used against military, civilian, agricultural, or other targets. To this end, we developed enzyme-based composites in the forms of paints capable to effectively recognize and neutralize the warfare agents. These composites are mechanically robust, safe, user and environmentally friendly, and stable under long-term operation and storage. Specifically, we took advantage of the unique surface properties of carbon nanotubes (high surface area, controllable morphology and size, increase in the number of active sites on their surface by user-directed functionalization) to immobilize a cocktail of enzymes (including perhydrolase S54V and chloroperoxidase). We then incorporated the resulting nanostructured enzyme-carbon nanotube conjugates into latex-based paint to generate active composites to be tested for decontamination against spore of the biological agent *B.cereus*, asimulant of *B. anthracis* and 2-chloroethyl ethyl sulfide (CEES), a mustard gas analog (Figure 1). Carbon nanotubes enabled high enzyme loading, extended stability without enzyme leaching, and structural reinforcement of the paints. The use of biocatalysts inherently makes this technology safe, environmentally benign, operational at mild conditions while requiring low energy input. Importantly, the availability of suites of enzymes with desired activities endows flexibility and multi-functionality to the proposed technology.

Experimental

Enzyme Coupling to MWNTs

Perhydrolase S54V (AcT) solution was provided by Genencor International Inc. (Palo Alto, CA). Chloroperoxidase (CPO) was purchased from Sigma, St. Louis, MO. AcT and CPO were covalently attached to acid functionalized multi walled carbon nanotubes (MWNTs) via a three-step process (*2*). First, carboxylic acid groups were created on MWNT (purity > 95%, outer diameter 15 ± 5 nm, length 5-20 μm, NanoLab, Inc., Newton, MA) by acid treatment. Typically, a ratio of 3:1 of sulfuric to nitric acid, v/v, total of 60 ml (H_2SO_4, 95-98%, HNO_3, 68%-70%, Fisher Scientific, Hampton, NH) were added to 100 mg MWNTs and the suspension was sonicated at room temperature for 6 h in a VWR ultrasonic cleaner (model 50T, 45 W). The functionalized MWNTs were then diluted in 200 ml Milli-Q water and filtered through a 0.2 μm filter membrane (Isopore type GTTP, polycarbonate, Millipore, Billerica, MA). The nanotubes on the filter membrane were further redispersed by sonication in Milli-Q water; the filtration was repeated several times to remove any residual acids and solubilized impurities. The functionalized MWNTs were subsequently used in the second reaction step, for spacer binding. Specifically, 2 mg functionalized MWNTs were dispersed in 2 ml of 2-(N-morpholino)ethanesulfonic acid sodium salt (MES) buffer (50 mM, pH 4.7) containing 160 mM 1-ethyl-3-[3-dimethylaminopropyl] carbodiimide hydrochloride (EDC, Acros Organics, Morris Plains, NJ) and 80 mM N-hydroxysuccinimide (NHS, Pierce, Rockford, IL) by brief sonication. After 15 min shaking at 200 rpm and room temperature, the EDC/NHS activated MWNTs were filtered, washed thoroughly with MES and redispersed in 10 ml of amino-dPEG$_{12}$-acid (1 mg/ml, Quanta Biodesign, Powell, OH) solution in potassium phosphate buffer (PB, 50 mM, pH 7.1). The mixture was further

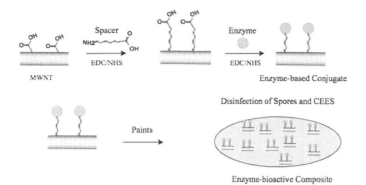

Figure 1. Enzymes are immobilized onto multi-walled carbon nanotubes using covalent binding and a spacer. The spacer assures the enhanced flexibility and greater substrate accessibility of the immobilized enzyme (1). The resulting enzyme-nanotube conjugates are incorporated into paint and lead to active biocomposites capable to generate agents for chemical and biological decontamination.

incubated for 3 h at room temperature with shaking at 200 rpm. Subsequently, the PEG-functionalized MWNTs were filtered, washed to remove any residual amino-dPEG$_{12}$-acid and used for the third reaction step, namely enzyme coupling reaction (again via the EDC/NHS as previously described). For the AcT, PEG-functionalized MWNTs were redispersed in 10 ml PB, 50 mM, pH 7.1 containing 4 mg AcT while for CPO, the PEG-functionalized MWNTs were redispersed in 10 ml citrate buffer (5 mM, pH 3.2) containing 4 mg CPO. Both enzyme-nanotube mixtures were incubated for at room temperature for 3 h with shaking at 200 rpm. The resulting conjugates were subsequently filtered and washed extensively with the corresponding buffers, while the supernatants were collected and used for the bicinchoninic acid (BCA, Pierce, Rockford, IL) protein assay in order to evaluate enzyme loading.

Preparation of Composites Containing Enzyme-MWNTs

Enzyme-nanotube latex-based composites were prepared by adding water suspension of enzyme-nanotube conjugates (different concentrations) into latex-based paint (typically 0.2 ml) in a glass vial (2.5 cm diameter). The two components were mixed thoroughly using a pipette tip and the mixture was air-dried under the hood for about 48 h.

Enzyme Activity Test

The activity of AcT was determined by measuring the concentration of peracidic acid (PAA) generated by the conjugates or composites (3). In a typical reaction, 10.6 µl hydrogen peroxide solution (H$_2$O$_2$, stock solution 30%, Sigma, St. Louis, MO, final concentration 100 mM) was added to a mixture of 0.8 ml propylene glycol diacetate (PGD, Sigma, St. Louis, MO, final concentration 100 mM in PB, 50 mM, pH 7.1) and 0.2 ml AcT solution (2.0 µg/ml final concentration for free AcT or equivalent concentration of AcT for AcT-nanotube conjugates). The mixture was incubated for 20 min at 200 rpm and room temperature. PAA assay was conducted by diluting 25 µl of reaction solution 400-times in Milli-Q water and subsequently mixing 100 µl of the diluted solution with 0.9 ml reagent assay (where the reagent was prepared by mixing 5 ml potassium citrate buffer, 125 mM, pH 5.0 with 50 µl 2,2'-azino-bis(3-ethylbenzthiazoline-6-sulphonic acid solution of 100 mM in Milli-Q water, Fisher Scientific, Hampton, NH and 10 µl potassium iodide, 25 mM solution in Milli-Q water). The mixture was then incubated at room temperature for another 3 min and the absorbance at 420 nm was measured on a UV-Vis spectrophotometer. PAA concentration was calculated as:

$$[PAA] \, (mM) = A_{420nm} \times 0.242 \times 400$$

where 0.242 is the calibrated constant correlation between concentration of PAA and absorbance at 420 nm and 400 is the dilution factor.

The activity of CPO was measured by following the conversion of monochlorodimedon to dichlorodimedon in the presence of H_2O_2 and potassium chloride, at 25°C, pH = 2.75, and 278 nm. Specifically, a reaction mixture containing 98 mM citric acid, 98 mM potassium phosphate, 0.096 mM monochlorodimedon (in potassium chloride buffer, 20 mM), 19 mM potassium chloride, 0.006% (v/v) H_2O_2 and equivalent of 0.01 - 0.05 unit of CPO (all final concentrations) were monitored at 278 nm on the UV-Vis spectrophotometer for approximately 5 minutes (all reagents were purchased from Sigma, St. Louis, MO).

Alternatively, for preliminary tests on chemical decontamination, 10 g/m^2 of 2-chloroethyl ethyl sulfide (CEES, Sigma, St. Louis, MO) were incubated with the paint composites containing different concentrations of CPO-based conjugates and CEES degradation was monitored using gas chromatography.

Spore Growth Conditions

Bacillus Cereus (B. cereus) 14737 was purchased from ATCC (Manassas, VA) and routinely cultured in nutrient broth (3 g/L beef extract, 5 g/L peptone) (Difco, Detroit, MI, USA) prepared in Milli-Q water for 48 h. The samples were next centrifuged at 3000 rpm for 3 min and sporulation was induced by resuspending the cells in GYC media at 30°C and 180 rpm for 48 h (4). All reagents were purchased from Sigma, unless otherwise mentioned. To end the sporulation, the solution was centrifuged and the sediment redispersed in Milli-Q water; the procedure was repeated 5 times. Subsequently, the washed sediment was redispersed in lysosome solution (2 mg/ml, Sigma, St. Louis, MO) and incubated at room temperature and 200 rpm for another 3 h. The spores were recovered by centrifugation at 3000 rpm for 3 min, washed and stored at 4°C in Milli-Q water. Spore purity was confirmed using Schaeffer-Fulton staining method (5). Spore concentration was estimated using standard plate count technique. The sporicidal efficiency of composite films was determined by using a diluted suspension containing 10^6 CFU/ml spore.

Results

Operational properties of the enzymes were optimized to yield high loading and activity and long-term stability in the composites. The loading of AcT-based conjugates was around 0.06 mg AcT/mg MWNTs, while the activity of these conjugates was about 24% of the activity of the free AcT. The capability of generating sufficiently high amount of PAA was investigated in reaction with *B.cereus* spores. For instance, following 20 min incubation of the AcT-based paint composite (0.16 % conjugate concentration in the paint) with *B. cereus* spore solution, the supernatant was capable of killing > 99% of spores initially charged at 10^6 CFU/ml. Furthermore, preliminary results on chemical decontamination of the composites containing 0.16 mg CPO/mg MWNTs (activity conjugates of about 30% when compared with free CPO activity) showed more than 98% degradation of 10 g/m^2 blister agent CEES (where 10 g/m^2 is considered a standard in chemical decontamination).

Conclusion

Water-soluble enzyme-nanotube conjugates were prepared by covalent attachment of enzymes onto carbon nanotube supports. Uniformly incorporation of the conjugates into latex-based paint led to composites. These composites contained "green" namely enzyme-based technology, were user friendly and generated reactive species capable of decontaminating chemical (CEES) and biological agents (spores).

Acknowledgments

This work was supported by a contract (W911SR-05-0038) from the U.S. Army under a subcontract from Genencor International, and a contract from DTRA (HDTRA1-08-1-0022).

References

1. Dinu, C. Z.; Zhu, G.; Bale, S. S.; Reeder, P.; Anand, G.; Kane, R. S.; Dordick, J. S. *Adv. Funct. Mater.* **2009**, in press.
2. Jiang, K.; Schadler, L. S.; Siegel, R. W.; Zhang, X.; Zhang, H.; Terrones, M. *J. Mater. Chem.* **2004**, *14*, 37–39.
3. Pinkernell, U.; Luke, H.-J.; Karst, U. **1997**, *122*, 567–571.
4. Kwon, S. J.; Lee, M. Y.; Ku, B.; Sherman, D. H.; Dordick, J. S. *ACS Chem. Biol.* **2007**, *2*, 419–25.
5. O'Mahony, T.; Rekhif, N.; Cavadini, C.; Fitzgerald, G. F. *J. Appl. Microbiol.* **2001**, *90*, 106–14.

New or Improved Biocatalysts

Chapter 9

Hemin-Binding Aptamers and Aptazymes

Mingzhe Liu, Hiroshi Abe, and Yoshihiro Ito*

Nano Medical Engineering Laboratory, RIKEN Advanced Science Institute,
2-1 Hirosawa, Wako-Shi, Saitama, 351-0198, Japan
*y-ito@riken.jp

Aptamers are oligonucleotides that have been engineered through repeated rounds of *in vitro* selection or by the process of Systematic Evolution of Ligands by EXponential enrichment (SELEX) to bind to various molecular targets such as small molecules, proteins, nucleic acids, and even cells and tissues. Aptamers offer similar molecular recognition properties to the more commonly used antibodies. In addition to their ability to discriminate their target, aptamers offer some advantages over antibodies. For example, aptamers can be engineered completely in a test tube and are readily produced using chemical processes. On the other hand, some aptamers that have been developed exhibit enzyme activity after binding to their target molecule, and function as 'aptazymes'. Numerous aptamers and aptazymes have been used in biotechnology, diagnostics and therapy. In this study, we demonstrate that natural DNA/RNA aptamers produced by selection *in vitro*, not only bind to hemin, but also show peroxidase activity when in such complexes (i.e., they function as aptazymes).

1. *In Vitro* Selection

In vitro ligand selection, or the systematic evolution of ligands by exponential enrichment (SELEX) was developed by the Gold and Szostack groups in 1990 (*1*, *2*). They successfully isolated functional oligonucleotides from randomized nucleic acid libraries *in vitro* that could bind selectively to target molecules. The SELEX method has become a general method for molecular engineering in the past twenty years. Aptamers are the oligonucleotides (DNA or RNA) isolated by the SELEX method. They offer similar molecular recognition properties to the more commonly used antibody approach. The overall process of SELEX is illustrated in Figure 1. Initially, a random library of oligonucleotides is prepared (usually 10^{10} to 10^{16}). Subsequently, affinity selection is performed in which the library is passed through a matrix with immobilized target molecules and then the few oligonucleotides that can bind to the target are collected. These are amplified by polymerase chain reaction (PCR) and the resulting products are applied to the next round of selection. So far, numerous DNA/RNA aptamers that can bind various molecules such as small molecules, proteins, nucleic acids, and even cells and tissues have been developed. Some aptamers also function as 'aptazymes', in which the aptamers exhibit enzyme activities after binding their target molecules (3–7). The aptamers not only have similarly high target affinity to antibodies, but also offer some advantages. For example, compared with antibodies, aptamers can be engineered completely *in vitro* and are readily produced using chemical process. Furthermore, the target range of aptamers is wider than that of antibodies. Therefore, aptamers have became rivals of antibodies and are widely used in many areas such as biotechnology, diagnostics, nanotechnology and therapy (3, 8–11).

2. Hemin-Binding Aptamers and Aptazymes

Heme is an essential molecule that plays critical roles in numerous biological phenomena. Free heme can act as an intracellular messenger by playing a role in the regulation of gene expression or in ion channel signal transduction (12, 13). Moreover, some hemoproteins containing a heme group also play critical roles such as in electron transfer and catalytic reactions. One typical group of hemoproteins is that of the peroxidases, which utilize heme as a cofactor to catalyze the oxidation of various substrates (14, 15). Peroxidases have been used as biocatalysts for bioreactors and biosensors (16–19). However, several intrinsic properties of these enzymes, such as their limited thermostability and high cost, limit the applications of their functions as natural catalytic copolymers. Therefore, the development of heme-binding artificial receptors or heme-containing artificial enzymes is an attractive and important theme in biotechnology. Many groups have tried to construct artificial heme receptors or mimics of peroxidases using rational design methods (20–25). For example, Takahashi et al. prepared a polyethylene glycol modified hemin, a chloride form of heme, that exhibited peroxidase activity in several organic solvents, although the enzyme activity was low (25). Hayashi *et al.* reported that horse heart myoglobin reconstituted with a modified heme showed enhanced peroxidase activity (20). On the other hand, several groups have reported catalytic antibodies with peroxidase activity (26–29). For example, Cochran et al. used *N*-methyl mesoporphyrin IX (NMM) as a transition state analogue to obtain antibodies (26, 30). They found that the obtained antibodies could also bind hemin and that their complexes exhibited peroxidase activity.

In contrast to the above hemin derivatives or hemin-binding protein, nucleic acid aptamers would be good candidates as hemin-binding receptors, because aptamers have a simpler structure than protein, are easy to produce by chemical engineering and might also provide higher peroxidase activity than hemin

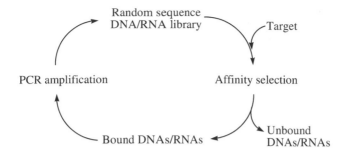

Figure 1. Schematic illustration of the in vitro systematic evolution of ligands by exponential enrichment selection (SELEX).

Table 1. *In vitro* selection of nucleic acid aptamers for binding to various porphyrins[a]

Type of aptamers	Target	Binding affinity to hemin	Peroxidase activity	References
RNA aptamer	NMM	NR	NR	*(31)*
DNA aptamer	NMM	Yes	Yes	*(32, 36)*
DNA aptamer	HPIX	NR	NR	*(34)*
Non-natural RNA aptamer	NMM	Yes	Yes	*(35)*
RNA aptamer	MPIX	Yes	NR	*(33)*

[a] NR, not reported

derivatives. So far, several porphyrin-binding nucleic acid aptamers based on *in vitro* selection have been reported (Table 1) *(31–36)*. The porphyrins used are shown in Figure 2. Travascio et al. reported on a DNA aptamer–hemin complex that exhibited peroxidase activity: the first example of a nucleic acid catalyst with peroxidase activity *(36)*. We have also reported an artificial RNA aptamer carrying amino groups that bound hemin and exhibited peroxidase activity *(35)*. However, each of these aptamers, which also function as aptazymes, was selected using indirect methods. We further developed selection using columns with immobilized NMM as a target.

We have used hemin as target to isolate hemin-binding natural DNA/RNA aptamers by *in vitro* selection. The DNA/RNA aptamers thus selected not only bound to hemin, but also showed peroxidase activity when complexed with hemin. That is, they functioned as aptazymes. Both the binding affinity and peroxidase activity for the aptamers selected were evaluated *(37, 38)*.

3. *In Vitro* Selection of Hemin-Binding DNA Aptamer

In vitro selection of hemin-binding DNA aptamer was performed as shown in Figure 3. The DNA library was synthesized chemically using phosphoramidite chemistry. About 10^{13} single-stranded (ss) DNAs were used in the first round

113

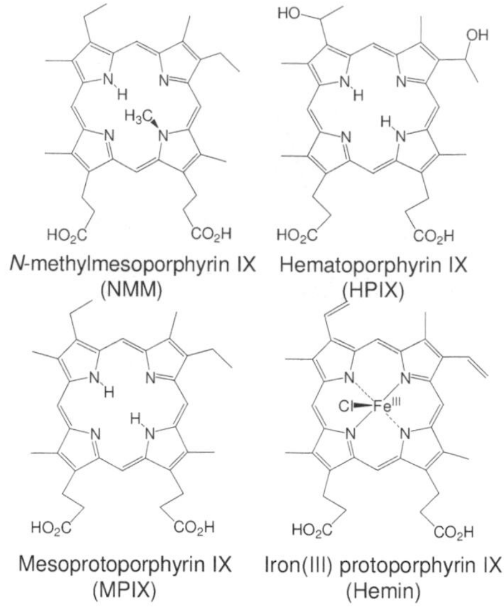

Figure 2. Porphyrins used for in vitro selection.

of selection *in vitro*. The sequence of each member of the ssDNA library consisted of a 60-nucleotide (nt) random region and two constant primer regions located at the 5' and 3' sites (5'–TAGGGAATTCGTCGACGGATCC–N60–CTG CAGGTCGACGCATGCGCCG–3'). In the first round, the ssDNA library was heated at 90 °C and then annealed at room temperature for 1 h. The folded DNAs were then loaded onto a hemin–agarose column and then incubated with hemin. Unbound DNAs were washed out. Substrate elution was performed using a hemin-saturated binding buffer to allow recovery of even the tightest binding aptamers. The eluted DNAs were amplified by PCR using primers labeled with biotin at the 5' termini. After an additional nine PCR cycles, the amplified biotin-labeled double-stranded (ds) DNAs were purified and reacted with streptavidin-coated magnetic beads. After treating with NaOH, the separated unlabeled ssDNAs were isolated and purified. The amount of each ssDNA was measured by ultraviolet-visible (UV-VIS) spectrophotometry and then used for the next round of selection. Absorbance at 260 nm of the DNA loaded in the column, or of the eluted bound DNA was measured and was subsequently used for calculating the binding percentage of DNA in each selection round.

Compared with the first round, a significant increase in the binding percentage of ssDNA was observed after the fourth round. Therefore, we amplified and cloned the ssDNAs collected from the fourth round. Seven clones were picked randomly and then sequenced using the dideoxy method. The resulting sequences of the random region are shown in Table 2. We chemically synthesized three kinds of G-rich sequences (4c15, 4c19 and 4c21) and one kind of G-poor sequence (4c20) and investigated their binding affinities to hemin. We found that 4c15 exhibited the highest binding affinity. Therefore, the secondary structure of 4c15 was predicted using software reported by Zuker (Figure 4) (*39*).

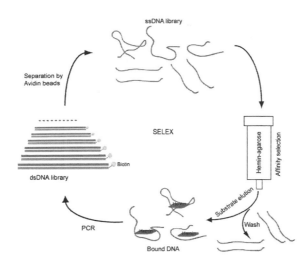

Figure 3. In vitro selection of hemin-binding DNA aptamer.

Notably, no featured secondary structure was found in 4c15. Because several G-rich DNAs that bind porphyrin have been reported (*40–42*), we synthesized a G-rich motif (4c15-s; enclosed by a dashed frame in Figure 4) to investigate the binding property in detail.

A conformational analysis of the interaction between 4c15-s and hemin was performed using the acquisition of circular dichroism (CD) spectra (Figure 4). The 4c15-s molecule exhibited a strong positive band at 260–265 nm and a negative band at 235–240 nm, indicating that it exhibits a parallel G-quartet structure, as described (*34, 41, 43*). When hemin was added to the solution in large excess to 4c15-s, no significant change was observed in the spectra and the characteristic G-quartet signature was retained. This suggests that the parallel G-quartet structure remained, even after binding of hemin. Okazawa et al. (*34*) also reported that hematoporphyrin did not induce significant changes in the conformation of selected DNA aptamers.

Next, the interaction of the truncated 4c15-aptamer (4c15-s) with hemin was investigated by UV-VIS spectrometric analysis (Figure 6a). Absorption at the Soret band of hemin increased with the increase of the concentration of 4c15-s. An increase of absorbance at the Soret band of hemin indicates an increase in the hydrophobicity of the hemin environment, which is directly proportional to the binding affinity (*20, 25, 36, 44*). This indicated that 4c15-s could bind to hemin. The molar extinction coefficient of the 4c15-s–hemin complex at 404 nm (ε = 165000 M^{-1} cm^{-1}) was determined. A Scatchard plot for the titration of hemin with a tetramer of 4c15-s was obtained by analyzing Figure 6a following a previous report (Figure 6b) (*45*). The dissociation constant was 45 nM at a stoichiometric ratio of 1:1 between hemin and the 4c15-s tetramer. This finding suggested that the 4c15-s tetramer formed a parallel G-quartet structure upon binding to hemin. The binding affinity of 4c15-s tetramer was similar to that of a PS2.M aptamer–hemin complex (K_d = 27 ± 3 nM) described by Travascio et al., although they reported that the binding stoichiometry was 1:1 (*36*).

Table 2. Sequences of the random regions of *in vitro* selected DNAs.
Copyright 2009 The Chemical Society of Japan

No.	Random sequence region	G
4c2	GTAGAAAGATCAGGTTGCTAGTTGGGCTGTAGCGTCGTTCAGCTCTTACGTGCCGCGTTC	19
4c11	CCGTTAGGCTAGTTTGGGGGTGGGCTGTTACGGACGGAT*TGAGTTAAGAGGGGGCAGTAA*	26
4c15	*AGGTGGGGAGGAGCGGGGTG*CTAGGCTCTTATGGAGTCAGCGCAAAAGGGTTGTTGGAGC	28
4c19	TTGCTGCCCCTGCCCCAACGGGAGGTTGGCGGCGGAGTG*CGGCTCTGGGGCTGGGGCTGT*	27
4c20	CTGAGCGAGTCACAGCCATCGGCCTAGAAGCGCCCACCCTCATGAAACTATTAGGGTGCT	15
4c21	GCAGCCATGGTCC*GTGGGGGGCTGGGAGCACGG*TGTGGAATTGCGTTT*CCGGGGGTGGT*	29
4c27	CGGGCCGTAATAACTTGACGAAACGGCGCTTGTACTCC*GAGGGGTCGCAGTGT*	18

4. *In Vitro* Selection of Hemin-Binding RNA Aptamer

In vitro selection was performed as shown in Figure 7 (*46*). The library of single-strand ssDNAs (103 nt) containing a random region of 59 bases (5′–TA GGGAATTCGTCGACGGATCC–N59–CTGCAGGTCGACGCATGCGCCG–3′) was synthesized and amplified using PCR using the 5′ primer containing the T7 promoter sequence, and the 3′ primer. The amplified DNA was transcribed using T7 RNA polymerase. The product of the transcription reaction was purified by denaturing polyacrylamide gel electrophoresis (PAGE). For the first round of selection, a random sequence RNA pool (about 10^{11} molecules) was annealed and then loaded on the hemin-immobilized agarose column. Unbound RNAs were washed away. Bound RNAs were eluted with the binding buffer containing hemin. The collected RNAs were quantified by UV absorption at 260 nm and then amplified by reverse transcription (RT)–PCR. The amplified DNA was transcribed and applied to the hemin-immobilized column again for the next round of selection. After the fourth round of selection, the ability of the RNAs to bind to immobilized hemin increased significantly. We cloned the fourth and sixth rounds of amplification of the DNA pool and sequenced them. After testing the affinity to hemin with the collected clones, we obtained an RNA aptamer (6c5 RNA) with the highest binding affinity to hemin. The predicted secondary structure of the 101-nt full length RNA and the random region of the 6c5 RNA aptamer are shown in Figure 8 (*39*).

Interactions between this RNA aptamer and hemin were also analyzed by measuring the Soret band of hemin in the absence or in the presence of the aptamer. Absorption increased with the increasing concentration of 6c5 RNA. The absorbance changes at 404 nm depicted in Figure 9a were plotted against the concentration of 6c5 RNA (Figure 9b). The apparent dissociation constant (K_d) determined was 0.8 µM (*38, 47*). This is similar to that of RNA aptamers reported by other groups, in which the K_d was in the submicromolar order (*33, 35, 36*). Scatchard analysis gave a value of 1.3 moles of RNA bound per mole of hemin, suggesting that the binding stoichiometry was 1:1 (data not shown). The predicted secondary structure of either the 101-nt full-length or the random region of 6c5 RNA suggested that the conserved domain enclosed by the dashed frame was the active domain binding to hemin, as shown in Figure 8. Notably, the 6c5 RNA sequence does not contain a G-rich motif, which has been observed in other hemin-binding RNA aptamers (*33, 35, 48*).

4c15　*AGGTGGGGAGGAGCGGGGTGC*TAGGCTCTTATGGAGTCAGCGCAAAAGGGTTGTTGGAGC
4c15-s　*AGGTGGGGAGGAGCGGGGTGC*

(b)

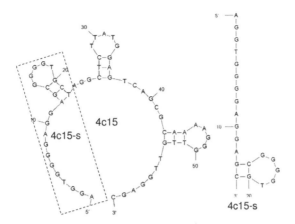

Figure 4. Sequences of DNA aptamers (a) and its predicted secondary structure (b). Copyright 2009 The Chemical Society of Japan.

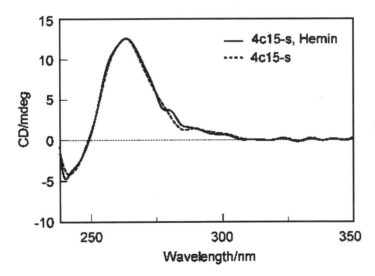

Figure 5. CD spectra of 4c15-s in the absence (dashed line) and presence (solid line) of hemin. The 4c15s and hemin concentrations were 2.0 µM and 5.0 µM, respectively, in 40KT buffer (50 mM 2-morpholinoethanesulfonic acid hydrate pH 6.5; 100 mM Tris acetate; 40 mM potassium acetate; 1% DMSO; 0.05% Triton X-100) (36). Copyright 2009 The Chemical Society of Japan.

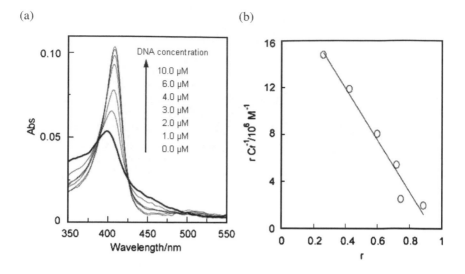

Figure 6. (a) UV-VIS spectra around the Soret band of hemin with increasing concentrations of the 4c15s DNA. Concentrations of 4c15s DNA were 0, 1.0, 2.0, 3.0, 4.0, 6.0 and 10.0 μM. Titration of hemin with 4c15s DNA was carried out by incubating 4c15s DNA with hemin (0.67 μM) in 40KT buffer at room temperature. (b) Scatchard plot for the titration of hemin with 4c15-s. $r = C_b/[(4c15-s)_4]$ where C_b is the concentration of bounded hemin and C_f is the concentration of unbounded hemin. Copyright 2009 The Chemical Society of Japan.

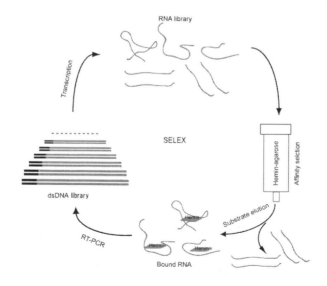

Figure 7. In vitro selection of hemin-binding RNA aptamer.

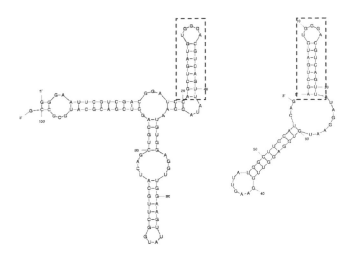

Figure 8. The predicted secondary structure of the 101-nt full length (a) and the random region (b) of RNA aptamer (6c5 RNA). Copyright 2009 Elsevier.

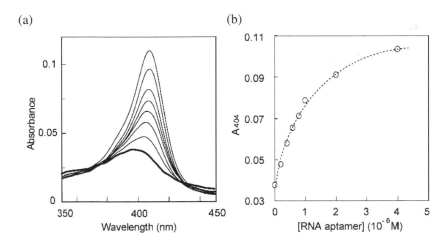

Figure 9. (a) Absorption of Soret band of hemin with increasing concentrations of 6c5 RNA. The concentrations of 6c5 RNA were 0, 0.2, 0.4, 0.6, 0.8, 1.0, 2.0 and 4.0 µM. Titration of hemin with 6c5 RNA was carried out by incubating 6c5 RNA with hemin (0.5 µM) in 40KT buffer at room temperature. (b) Absorbance at 404 nm versus varying concentrations of 6c5 RNA. The data are from Figure 9a. Copyright 2009 Elsevier.

5. Hemin-Binding Aptazymes

We successfully isolated hemin-binding DNA and RNA aptamers. As shown in Table 1, several nucleic aptamers that could bind hemin also exhibited peroxidase activity (i.e., they functioned as aptazymes). Therefore, we used the 4c15-s DNA and 6c5 RNA aptamers to investigate whether they exhibited peroxidase activity when complex with hemin (Figure 10).

(a)   (b)

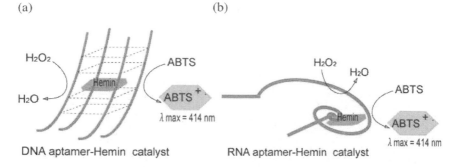

DNA aptamer-Hemin catalyst   RNA aptamer-Hemin catalyst

Figure 10. Illustration of peroxidase reactions catalyzed by DNA type (a) or RNA type (b) aptazymes. Both the DNA aptamer–hemin and RNA–aptamer complexes showed enhanced peroxidase activity. Copyright 2009 The Chemical Society of Japan. Copyright 2009 Elsevier.

(a)   (b)

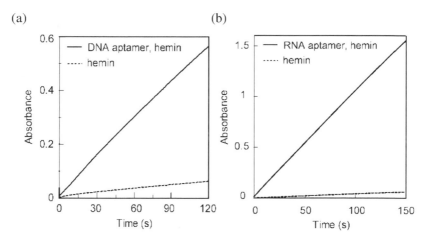

Figure 11. Time-dependent absorbance (414 nm) changes upon analysis of peroxidase activity of hemin alone (dashed line) and aptamer–hemin complexes (solid line). In all experiments, the reactions consisted of the 4c15-s DNA aptamer or the 6c5 RNA aptamer at 2 μM; 0.33 μM hemin, 2.5 mM 2,2'-azinobis (3-ethylbenzothiazoline-6-sulfonic acid) (ABTS) and 0.75 mM H_2O_2 in 40KT buffer.

The peroxidase activity was investigated by the addition of ABTS and H_2O_2 to the solution used in the binding assay. If the aptamer–hemin complexes exhibit peroxidase activity, the reduction of H_2O_2 will induce ABTS oxidation, resulting in the formation of its radical cation. This has an absorption maximum at 414 nm ($\varepsilon = 36000$ M^{-1} cm^{-1}) (49). The rate of ABTS oxidation by hemin in the presence or absence of the aptamer was measured by monitoring the increase of absorbance at 414 nm. As shown in Figure 11, in the case of hemin alone the rate of ABTS oxidation was very slow. In contrast, in the presence of aptamers (both 4c15-s DNA and 6c5 RNA), the rates of ABTS oxidation were enhanced significantly. This indicated that the aptamer–hemin complexes exhibited peroxidase activity and that the aptamers functioned as aptazymes. With the DNA aptamer the rate was 8-fold that of hemin alone (Figure 11a) and with the RNA aptamer the rate was 10-fold that of hemin alone.

Kinetic assay was also performed using the Michaelis–Menten equation, in which the ABTS oxidation rate was a function of the concentration of H_2O_2. The results are summarized in Table 3. When either the 6c5 RNA or 4c15-s DNA were used as an aptazyme, there were considerable increases in k_{cat} values compared with using hemin alone. In contrast, the K_m values of RNA/DNA aptazymes were higher than that of hemin alone. The difference in K_m values suggests that the access of H_2O_2 to the reaction center of the aptamer–hemin complex may be more difficult than that to hemin. However, compared with hemin alone, significant enhancement in catalytic efficiency (k_{cat}/K_m) was achieved using 6c5 RNA/4c15-s DNA as an aptazyme. In addition, although the catalytic efficiency of aptazymes was considerably lower than that of horseradish peroxidase (HRP), it was higher than the efficiency of some hemoproteins that have peroxidase activity such as wild-type ferrimyoglobin ($k_{cat}/K_m = 540$ M^{-1} s^{-1}) and a catalytic antibody ($k_{cat}/K_m = 233$ M^{-1} s^{-1}) (48). We achieved a 10-fold enhancement in the catalytic efficiency of the 6c5 RNA-aptazyme compared with hemin alone. This enhancement by 6c5 RNA-aptazyme was 2-fold higher than that induced by the non-natural aptazyme reported previously by our group and was 1.7-fold of that induced by 4c15-s DNA-aptazyme, although the binding affinity of 6c5 RNA was lower than that of the non-natural RNA aptamer and 4c15-s DNA (35, 37). This result suggests that peroxidase activity depends not only on the binding affinity to hemin, but also on the conformation of the aptamer–hemin complexes such as axial coordination for hemin and the microenvironment surrounding hemin.

Table 3. Kinetic parameters for peroxidation by hemin and aptamer–hemin complexes (aptazymes)

Catalysts	k_{cat} (min^{-1})	K_m (mM)	k_{cat}/K_m (M^{-1} s^{-1})
6c5 RNA-aptazyme	650	9.0	1203
4c15-s DNA-aptazyme	302	7.0	719
Hemin	9	1.3	115

6. Conclusion

Here we reported hemin-binding aptamers and aptazymes. The DNA/RNA aptamers developed by *in vitro* selection not only bind to hemin but also show peroxidase activity when complex with hemin. Therefore, these aptamers also function as aptazymes. We successfully determined an active motif of the DNA aptamer (4c15-s) just 21 nt long. Moreover, we clarified that the 4c15-s tetramer formed a parallel quaduplex to bind hemin. We also predicted the active domain of the RNA aptamer. Thus, these aptamers could be used as key components of biosensors and could serve as biocatalysts by forming complexes with hemin (*50–54*).

References

1. Ellington, A. D.; Szostak, J. W. *Nature* **1990**, *346*, 818–822.
2. Tuerk, C.; Gold, L. *Science* **1990**, *249*, 505–510.
3. Breaker, R. R. *Nature* **2004**, *432*, 838–845.
4. Famulok, M.; Hartig, J. S.; Mayer, G. *Chem. Rev.* **2007**, *107*, 3715–3743.
5. Ito, Y.; Abe, H.; Wada, A.; Liu, M. Z. *ACS Symp. Ser.* **2008**, *999*, 194–215.
6. Sen, D.; Geyer, C. R. *Curr. Opin. Chem. Biol.* **1998**, *2*, 680–687.
7. Wilson, D. S.; Szostak, J. W. *Annu. Rev. Biochem.* **1999**, *68*, 611–647.
8. Lee, J. F.; Stovall, G. M.; Ellington, A. D. *Curr. Opin. Chem. Biol.* **2006**, *10*, 282–289.
9. Lu, Y.; Liu, J. W. *Curr. Opin. Biotechnol.* **2006**, *17*, 580–588.
10. Navani, N. K.; Li, Y. F. *Curr. Opin. Chem. Biol.* **2006**, *10*, 272–281.
11. Phillips, J. A.; Lopez-Colon, D.; Zhu, Z.; Xu, Y.; Tan, W. H. *Anal. Chim. Acta* **2008**, *621*, 101–108.
12. Hou, S. W.; Reynolds, M. F.; Horrigan, F. T.; Heinemann, S. H.; Hoshi, T. *Acc. Chem. Res.* **2006**, *39*, 918–924.
13. Ogawa, K.; Igarashi, K.; Nishitani, C.; Shibahara, S.; Fujita, H. *J. Health. Sci.* **2002**, *48*, 1–6.
14. Dawson, J. H. *Science* **1988**, *240*, 433–439.
15. Sono, M.; Roach, M. P.; Coulter, E. D.; Dawson, J. H. *Chem. Rev.* **1996**, *96*, 2841–2887.
16. Dai, L. H.; Klibanov, A. M. *Biotechnol. Bioeng.* **2000**, *70*, 353–357.
17. Grabski, A. C.; Grimek, H. J.; Burgess, R. R. *Biotechnol. Bioeng.* **1998**, *60*, 204–215.
18. Lin, T. Y.; Wu, C. H.; Brennan, J. D. *Biosens. Bioelectron.* **2007**, *22*, 1861–1867.
19. Rao, S. V.; Anderson, K. W.; Bachas, L. G. *Biotechnol. Bioeng.* **1999**, *65*, 389–396.
20. Hayashi, T.; Hitomi, Y.; Ando, T.; Mizutani, T.; Hisaeda, Y.; Kitagawa, S.; Ogoshi, H. *J. Am. Chem. Soc.* **1999**, *121*, 7747–7750.
21. Hayashi, T.; Sato, H.; Matsuo, T.; Matsuda, T.; Hitomi, Y.; Hisaeda, Y. *J Porphyrine Phthalocyanines* **2004**, *8*, 255–264.
22. Kamiya, N.; Goto, M.; Furusaki, S. *Biotechnol. Bioeng.* **1999**, *64*, 502–506.
23. Ozaki, S.; Matsui, T.; Watanabe, Y. *J. Am. Chem. Soc.* **1997**, *119*, 6666–6667.
24. Ozaki, S.; Yang, H. J.; Matsui, T.; Goto, Y.; Watanabe, Y. *Tetrahedron: Asymmetry* **1999**, *10*, 183–192.
25. Takahashi, K.; Matsushima, A.; Saito, Y.; Inada, Y. *Biochem. Biophys. Res. Commun.* **1986**, *138*, 283–288.
26. Cochran, A. G.; Schultz, P. G. *J. Am. Chem. Soc.* **1990**, *112*, 9414–9415.
27. Kawamura-Konishi, Y.; Asano, A.; Yamazaki, M.; Tashiro, H.; Suzuki, H. *J. Mol. Catal. B: Enzym.* **1998**, *4*, 181–190.

28. Quilez, R.; deLauzon, S.; Desfosses, B.; Mansuy, D.; Mahy, J. P. *FEBS Lett.* **1996**, *395*, 73–76.

29. Takagi, M.; Kohda, K.; Hamuro, T.; Harada, A.; Yamaguchi, H.; Kamachi, M.; Imanaka, T. *FEBS Lett.* **1995**, *375*, 273–276.

30. Cochran, A. G.; Schultz, P. G. *Science* **1990**, *249*, 781–783.

31. Conn, M. M.; Prudent, J. R.; Schultz, P. G. *J. Am. Chem. Soc.* **1996**, *118*, 7012–7013.

32. Li, Y. F.; Geyer, C. R.; Sen, D. *Biochemistry* **1996**, *35*, 6911–6922.

33. Niles, J. C.; Marletta, M. A. *ACS Chem. Biol.* **2006**, *1*, 515–524.

34. Okazawa, A.; Maeda, H.; Fukusaki, E.; Katakura, Y.; Kobayashi, A. *Bioorg. Med. Chem. Lett.* **2000**, *10*, 2653–2656.

35. Teramoto, N.; Ichinari, H.; Kawazoe, N.; Imanishi, Y.; Ito, Y. *Biotechnol. Bioeng.* **2001**, *75*, 463–468.

36. Travascio, P.; Li, Y. F.; Sen, D. *Chem. Biol.* **1998**, *5*, 505–517.

37. Liu, M. Z.; Kagahara, T.; Abe, H.; Ito, Y. *Bull. Chem. Soc. Jpn.* **2009**, *82*, 99–104.

38. Liu, M. Z.; Kagahara, T.; Abe, H.; Ito, Y. *Bioorg. Med. Chem. Lett.* **2009**, *19*, 1484–1487.

39. Zuker, M. *Nucleic Acids Res.* **2003**, *31*, 3406–3415.

40. Yi, X.; Pavlov, V.; Gill, R.; Bourenko, T.; Willner, I. *ChemBioChem* **2004**, *5*, 374–379.

41. Sugimoto, N.; Toda, T.; Ohmichi, T. *Chem. Commun.* **1998**, 1533–1534.

42. Mikuma, T.; Ohyama, T.; Terui, N.; Yamamoto, Y.; Hori, H. *Chem. Commun.* **2003**, 1708–1709.

43. Balagurumoorthy, P.; Brahmachari, S. K.; Mohanty, D.; Bansal, M.; Sasisekharan, V. *Nucleic Acids Res.* **1992**, *20*, 4061–4067.

44. Slamaschwok, A.; Lehn, J. M. *Biochemistry* **1990**, *29*, 7895–7903.

45. Scatchard, G. *Ann. N. Y. Acad. Sci.* **1949**, *51*, 660–672.

46. Kawazoe, N.; Teramoto, N.; Ichinari, H.; Imanishi, Y.; Ito, Y. *Biomacromolecules* **2001**, *2*, 681–686.

47. Wang, Y.; Hamasaki, K.; Rando, R. R. *Biochemistry* **1997**, *36*, 768–779.

48. Travascio, P.; Bennet, A. J.; Wang, D. Y.; Sen, D. *Chem. Biol.* **1999**, *6*, 779–787.

49. Childs, R. E.; Bardsley, W. G. *Biochem. J.* **1975**, *145*, 93–103.

50. Elbaz, J.; Shlyahovsky, B.; Li, D.; Willner, I. *ChemBioChem* **2008**, *9*, 232–239.

51. Elbaz, J.; Shlyahovsky, B.; Willner, I. *Chem. Commun.* **2008**, 1569–1571.

52. Kolpashchikov, D. M. *J. Am. Chem. Soc.* **2008**, *130*, 2934–2935.

53. Li, D.; Shlyahovsky, B.; Elbaz, J.; Willner, I. *J. Am. Chem. Soc.* **2007**, *129*, 5804–5805.

54. Li, T.; Dong, S. J.; Wang, E. K. *Chem. Commun.* **2007**, 4209–4211.

Chapter 10

Synthesis of Covalently Linked Enzyme Dimers

Sanne Schoffelen,[1] Loes Schobers,[1] Hanka Venselaar,[2] Gert Vriend,[2] and Jan C. M. van Hest[1,*]

[1]Radboud University Nijmegen, Department of Organic Chemistry, P.O. Box 9010, 6500 GL Nijmegen, The Netherlands
[2]Centre for Molecular and Biomolecular Informatics, Radboud University Nijmegen, P. O. Box 9101, 6500 HB Nijmegen, The Netherlands
*j.vanhest@science.ru.nl

The covalent linkage of enzymes that collaborate in a multistep reaction may have a beneficial effect on substrate conversion. Site-specific modification strategies ensure absolute control over the site of coupling. We show that well-defined enzyme architectures can be prepared via site-specific incorporation of novel functionalities into the model enzyme *Candida Antarctica* Lipase B (CalB). Different kinds of dimers were produced in which the covalent connection brought both active sites or both N-termini in proximity of each other. Dimers were as active as their respective monomers, both towards monofunctional substrates and molecules containing two substrate moieties.

Introduction

The organization of proteins into complexes is of great importance for their function (*1*). Several classes of protein receptors, such as growth factor receptors and ion channels, need to form dimers or even larger oligomeric structures to effectuate transfer of a biochemical signal across the cell membrane. Specific families of eukaryotic transcription factors bind to DNA as dimers through recognition of palindromic DNA sequences (*2*). Additionally, many enzymes are organized into macromolecular complexes. They cooperate in the biosynthesis of a wide range of molecules. Their spatial organization establishes that the (unstable) intermediates are transferred between the catalytic sites without diffusion into the bulk phase of the cellular environment (*3*).

The amount of examples in literature describing well-defined syntheses of artificial protein oligomers is limited. Native chemical ligation (NCL) has been used to form dimers of expressed polypeptides, and to attach multiple fluorescent protein molecules to a dendrimer (*4*, *5*). Furthermore, the NCL method has been exploited to introduce an azide or alkyne moiety at the N-terminus of a protein for dimerization via the Cu(I) catalyzed azide-alkyne cycloaddition (*6*).

An alternative approach has been followed for the production of a heterodimer of bovine serum albumin (BSA) and *Thermomyces Lanuginosa* lipase (TLL). The proteins were linked via the free, solvent-accessible cysteine of BSA and the only free lysine of a TLL mutant using 3-azidopropyl-1-maleimide and 4-pentynoic acid as additional reagents. In this case, BSA functioned as an anchor to immobilize the lipase on a surface for single enzyme activity studies (*7*). In a more recent study a heterodimer of BSA and streptavidin was formed using a biotin-maleimide heterotelechelic polymer (*8*).

To our knowledge, no studies have been reported that aimed at linking multiple enzymes together that collaborate in a biocatalytic process. As stated earlier, the co-localization of multiple enzymes can have a beneficial effect on substrate conversion. In that case it would be desirable to have control over the relative orientation of the active sites. Thus, a site-specific bioconjugation strategy is required that allows variation in the site of conjugation. Here, we describe a method that meets this requirement resulting in the preparation of well-defined enzyme architectures. We show that the catalytic activity of the produced enzyme dimers does not significantly change upon dimer formation.

Site-Specific Coupling of *Candida Antarctica* Lipase B

To demonstrate the abilty to make well defined enzyme dimers a suitable biocatalyst and coupling strategy has to be chosen. As a model enzyme we have selected *Candida Antarctica lipase B* (CalB). This lipase is known to be very stable over a broad pH range and at elevated temperatures. It is active in organic solvents while exhibiting broad substrate specificity. This has resulted in CalB being a very popular biocatalyst in both the industrial and academic field (*9*).

As reported previously, CalB can be modified at a single position via the incorporation of azidohomoalanine (*10*). Whereas the recombinant enzyme contains five azide moieties upon residue-specific replacement of methionine, only one out of five is available for conjugation via the Cu(I)-catalyzed azide-alkyne cycloaddition, which was shown to be the moiety located near the N-terminus of the protein. This enabled the site-specific immobilization onto polymersome surfaces and conjugation to elastin-like polypeptides (*11*, *12*). It was shown that the enzyme remained active while the function of the polymer-like materials was retained as well. Additionally, fast and efficient PEGylation of the azide-functionalized enzyme was achieved via the strain-promoted (3+2) cycloaddition using a newly developed aza-dibenzocyclooctyne (*13*).

The highly selective and bio-orthogonal azide-alkyne cycloaddition, which was used in the studies mentioned above, will also be exploited in linking two CalB molecules together. Both azide and alkyne functionalities can be site-specifically introduced into proteins via the incorporation of non-natural analogues of

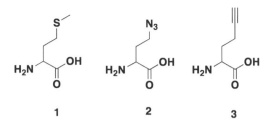

*Scheme 1. Structure of methionine **1** and its analogues azidohomoalanine **2** and homopropargylglycine **3**.*

methionine (*14*, *15*). The structures of methionine **1**, azidohomoalanine **2** and homopropargylglycine **3** are depicted in Scheme 1.

Via site-directed mutagenesis a methionine residue can be introduced in an enzyme at any location of choice provided that the side-chain is solvent-accessible and the amino acid replacement does not lead to structural changes. Using this method, conjugation is not limited to a specific site as is the case with for example native chemical ligation.

Different kinds of CalB dimers can be formed.   First of all, an azide-functionalized CalB can directly be linked to a CalB variant containing an alkyne moiety. Additionally, short dialkynyl spacers 1,18-diethynyl tetraethylene glycol **4** and 1,4-diethynyl benzene **5** (Scheme 2) will be used to link two CalB molecules that contain a solvent-accessible azide at either the N-terminus or near the active site. Scheme 3 shows the schematic representations of the different constructs.

Results and Discussion

Preparation of CalB Dimers Linked via the N-Termini

The most straightforward approach towards the formation of CalB dimers was to directly link an azide-functionalized enzyme molecule to one with an alkyne moiety. The azide moiety was introduced into CalB using the methionine analogue azidohomoalanine **2** (AHA). According to the same protocol, the alkyne handle was introduced via the incorporation of homopropargylglycine **3** (HPG). We tested whether the azide and alkyne variants of the wildtype protein (wt(AHA)CalB and wt(HPG)CalB) could be covalently linked.   In a 1:1 ratio wt(AHA)CalB and wt(HPG)CalB were reacted overnight at room temperature in presence of 50 equivalents Cu(I). The reaction product was purified by size-exclusion chromatography (see Figure 1a). The separation of dimer from unreacted CalB was verified by SDS-PAGE (see Figure 1b). As expected, the first peak showed to be CalB dimer. The second peak contained both unreacted CalB and an impurity in the protein sample which was known to be YodA, a native protein from *E. coli* (*16*). The relative amount of formed dimer compared to unreacted CalB was quantified by measuring the fluorescence of the corresponding protein bands, which were stained by Coomassie Brilliant Blue. 40% of the total amount of CalB had formed dimers.

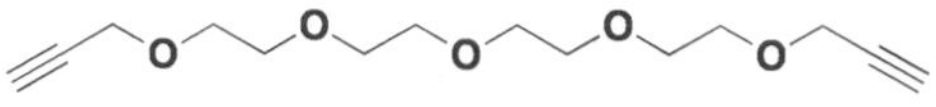

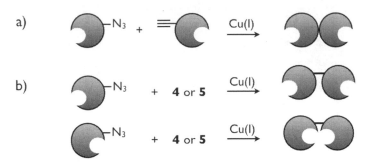

Scheme 2. The structures of 'clickable' spacers **4** en **5** which will be used for dimerization of azide-functionalized CalB.

Scheme 3. The different kinds of CalB dimers whose synthesis and activity studies are described in this paper. a) Direct linkage of azide-functionalized CalB to CalB containing an alkyne moiety by addition of Cu(I). b) Formation of CalB dimers using a short cross-linker and different variants of CalB resulting in dimers that have either their N-termini or their active sites in proximity of each other.

The activity of the wtCalB dimer was compared to the activity of the unreacted wt(AHA)CalB and wt(HPG)CalB. *p*-Nitrophenyl butyrate (pNPB) was used as a substrate. Different concentrations of pNPB were added to a dilution of the protein in phosphate buffer containing 0.1% triton. As shown in Figure 2 the activity of the dimer was not significantly different from the monomer. Apparently, the linkage of the two molecules to each other did not affect their structure and/or activity.

Production of CalB Mutants with a Functional Handle near the Active Site

In order to produce a so-called 'kissing' dimer, *i.e.* a dimer in which the active sites of the enzyme molecules are facing each other, an additional methionine codon had to be introduced. The 3D structure of the protein was examined *in silico* to select amino acids near the active site that could be replaced by methionine. Four different amino acids were chosen being located in the two loops surrounding the catalytic cleft, which were leucine(147), valine(149), leucine(219) and valine(221) (see Figure 3). Like methionine these residues are hydrophobic. Furthermore they seemed to be accessible to the solvent.

A multiple sequence alignment among the lipase family was performed to check for conservation of these residues. Since they appeared to be hardly conserved, we expected that replacement by methionine or its non-natural analogue would not harm the enzyme's structure and function.

To prevent functionalization at two positions, the solvent-accessible methionine codon located near the N-terminus was replaced by glycine. In this

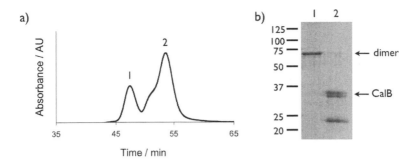

Figure 1. Dimerization of wt(AHA)CalB and wt(HPG)CalB. a) Size-exclusion chromatogram of the reaction mixture. The shoulder in the second peak is caused by the presence of a native E. coli protein which co-elutes with CalB. b) SDS-PAGE gel on which a fraction of both peaks was loaded.

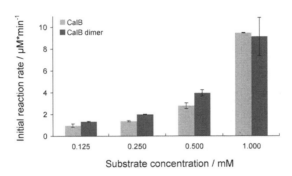

Figure 2. Hydrolytic activity of CalB dimer linked via the N-termini. Its activity was compared to the activity of CalB that had remained unreacted. Different concentrations of p-nitrophenyl butyrate were used. The slope in the curves obtained by measuring absorbance at 405 nm for 20 minutes were plotted in this chart. The assay was performed in duplo.

way only the newly introduced azide moiety would be available for conjugation. Upon expression and purification the different mutants were analyzed by mass spectrometry. The detected masses corresponded with the calculated ones indicating that the correct mutations had been introduced (see Table 1).

In the first column of Table 1, wt stands for wildtype. MG refers to the Met(1)Gly mutation and 147, 149, 219 and 221 refer to the additional sites that were mutated in these protein variants.

Conjugation of alkyne-functionalized poly(ethylene glycol) (Mw 5000) or dansyl to the protein mutants proved that the newly introduced azides were indeed solvent-accessible. No conjugation product was detected using the variant containing only the Met(1)Gly mutation. This indicated that the four other azide moieties that are buried inside the protein had remained unaccessible, meaning that the overall structure of CalB had not significantly changed.

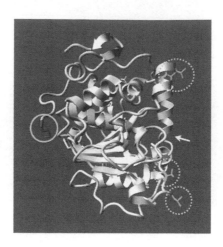

Figure 3. Picture of CalB indicating in green with dashed circles the amino acids that were changed into methionine. The entrance of the channel leading to the active site is indicated by the arrow and the N-terminus is indicated in red with a solid-line circle.

Table 1. Overview of the different CalB mutants that were produced

Name	Mutation	Calculated mass (Da)	Detected mass (Da)
wt(AHA)CalB	-	34244.5	34245.3
MG(AHA)CalB	Met(1)Gly	34175.4	34175.5
147(AHA)CalB	Met(1)Gly + Leu(147)Met	34188.4	34189.9
149(AHA)CalB	Met(1)Gly + Val(149)Met	34202.4	34203.9
219(AHA)CalB	Met(1)Gly + Leu(219)Met	34188.4	34191.7
221(AHA)CalB	Met(1)Gly + Val(221)Met	34202.4	34203.7

Preparation of CalB Dimers Using a Dialkynyl Linker

Dimers of the CalB mutants were formed in two steps. This appeared to be necessary as no dimer formation was observed when a mutant functionalized with HPG was reacted with a mutant containing AHA. Probably the newly introduced functional handles were less accessible than the one near the N-terminus. A dialkynyl linker was used to circumvent the problem of steric hindrance.

First, azide-functionalized protein was incubated with 50 equivalents of the linker molecule **4** or **5** in presence of Cu(I). After washing away the excess of linker, fresh azide-protein was added to allow the reaction with the alkyne-functionalized protein to take place. Using this approach, 147CalB dimers were prepared (147-dimer). As a control, dimers of wtCalB were produced in a similar way (wt-dimer).

Dimers that were produced using **4** were separated from the unreacted monomer by size-exclusion chromatography (see Figure 4a). The peak areas in the chromatogram were calculated to determine the relative amount of protein in the different fractions. Using **4**, 27% of the total 147CalB protein amount had been converted into dimers. In the case of wtCalB this was 33%.

As shown in Figure 4b, the 147-dimer runs higher on a (denaturing) SDS-PAGE gel than the wt-dimer. This may be explained by taking into account the position at which the two protein chains are coupled. For the wt-dimer, this is at the N-terminus of the chains, which results in a totally linear construct upon unfolding. Unfolding of the 147-dimer results in a branched structure because the two chains are linked in the middle. Apparently, such a branched structure results in a lower electrophoretic mobility.

Dimers were also formed using spacer **5**. According to the size-exclusion chromatograms, 34% of the total amount of 147AHACalB and 41% of wt(AHA)CalB had formed dimers (see Figure 5a). The SDS-PAGE gel showed that dimer and monomer were well-separated. Again, the 147-dimer had a lower electrophoretic mobility than the wt-dimer (see arrows in Figure 5b).

The Effect of Dimerization on the Activity of the 'Kissing' Dimer

We investigated the effect of dimerization on the activity of the 'kissing' 147-dimer. As a model reaction the esterification of 1-octanol using vinyl acetate was studied (see Scheme 4). This reaction, which proceeds in dry organic solvents, has been employed before to study the activity of different CalB formulations (*17*).

We compared the activity of 147-dimers with the activity of their corresponding monomers. These monomers had undergone the same reaction procedures, such as the click reaction with dialkynyl linker **4** or **5** in presence of Cu(I), followed by dialysis to remove the excess of linker, incubation with freshly added Cu(I) and purification via size-exclusion chromatography. As the click reaction during the first step did not result in 100% conjugation, a part of these monomers was functionalized with the linker whereas the remaining portion was not changed.

Based on the absorbance signal in the size-exclusion chromatograms, equal amounts of protein were taken from the first (dimer) and second (monomer) peak fractions. These aliquots were lyophilized followed by the addition of a mixture of 1-octanol and vinyl acetate in dried THF. Conversion into octyl acetate was followed in time by GC-MS. Both the decrease in the amount of 1-octanol and the increase in the amount of octyl acetate were detected and plotted against time (Figure 6).

Figure 6a shows the plot that was obtained from the 147-dimer linked by spacer **4**. The slope of the curves in the first three hours was taken as a measure for the activity. In Figure 6b these initial reaction rates of all four analysed samples are shown. Hardly any difference between dimer and monomer was detected. Bar 1 and 2 represent the activities of the 147-dimer linked by **4** and its unreacted monomer. Bar 3 and 4 correspond to the activities of the 147-dimer linked by **5** and its monomer.

The activity of 147(AHA)CalB that had been reacted with **5** is higher. However, this is probably due to higher amounts of protein that were present in both the monomer and dimer fractions of this variant. As spacer **4** and **5** are both very small in comparison to the protein, no difference in activity would be expected between monomer conjugated to **4** (bar 2) and monomer conjugated to **5** (bar 4) if the amounts of protein were the same.

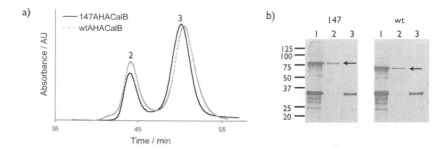

Figure 4. Dimerization of wt(AHA)CalB and 147(AHA)CalB mediated by dialkynyl linker 4. a) Size-exclusion chromatograms obtained upon injection of the reaction products on a column without any pre-purification. The numbers above the peaks correspond to the lanes in figure b). b) SDS-PAGE gel containing aliquots of the crude reaction mix (lane 1) and of the purified fractions (lane 2 and 3). The arrows indicate the dimers.

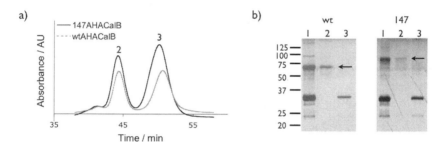

Figure 5. Dimerization of wt(AHA)CalB and 147(AHA)CalB mediated by dialkynyl linker 5. a) Size-exclusion chromatograms obtained from both reactions after injection onto a Superdex 75 column. The numbers above the peaks correspond to the lanes in figure b). b) SDS-PAGE gel containing aliquots of the crude reaction mix (lane 1) and of the purified fractions with the dimer in lane 2 and monomer in lane 3.

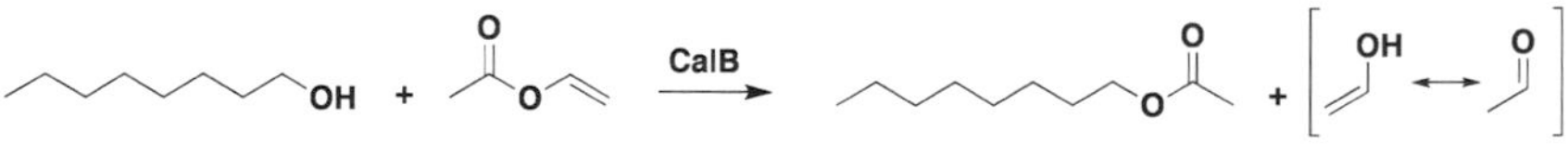

Scheme 4. The reaction between 1-octanol and vinyl acetate catalyzed by CalB. Due to rearrangement of vinyl alcohol to acetaldehyde the reaction is irreversible.

The Activity of 'Kissing' Dimers Towards Alkyl Diols

We hypothesized that the activity of the so-called 'kissing' dimers towards substrates containing two hydroxyl or carboxyl moieties could be higher than the activity of CalB monomer or N-terminally linked CalB dimers, due to the close proximity and right orientation of the active sites. In the esterification of alkyl diols with vinyl acetate we therefore expected to see a faster conversion of the monoacetylated intermediate into the diacetylated product.

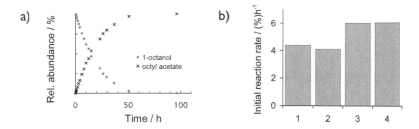

Figure 6. Esterification of 1-octanol catalyzed by 147-dimer and its unreacted monomers. a) The conversion of 1-octanol into octyl acetate catalyzed by the 147-dimer which was coupled by 4. b) The initial reaction rates of the esterification of 1-octanol by (1) 147-dimer linked by 4 and (2) its monomer, and (3) 147-dimer linked by 5 and (4) its monomer.

This hypothesis was tested in an assay using 1,8-octanediol and vinyl acetate. The activity of the 147-dimer was compared with the activity of its corresponding monomer and the activity of the wt-dimer. Both dimers were formed by using **4** as linker.

Figure 7 shows the plots that were obtained by following the conversion of 1,8-octanediol in time. The starting compound, the monoacetylated intermediate and the diacetylated product were detected by GC-MS. By comparing Figure 7a with Figure 7b it looks like the 147-dimer is more active than the 147-monomer. On the other hand, the wt-monomer seems to be more active than the wt-dimer (Figure 7c, d). Furthermore, the curves of the 147-dimer look quite similar to those of the wt-dimer. Apparently, the formation of diacetylated product (C) does not proceed significantly faster for the 147-dimer compared to the wt-dimer.

Since it could be that the effect we expected could not be detected with 1,8-octanediol, because this substrate was relatively short compared to the distance between the two active sites, some test reactions were performed with longer substrates, *i.e.* 1,10-decanediol and 1,12-dodecanediol. Unfortunately, using the reaction and analysis set-up that was optimized for 1,8-octanediol the results that were obtained with these substrates were not reliable. This was due to solubility problems and the fact that the compounds could not be well-seperated on the GC column that was available.

The Activity of 'Kissing' Dimers Towards Poly(ethylene glycol) Diols

The distance between the two active sites of a 'kissing' dimer was difficult to predict as it not only depends on the depth of the catalytic clefts and the length of the cross-linker, but also on the flexibility of the dimer and the substrates. Therefore, it was hard to estimate the particular length of a difunctional substrate that would fit in between the active sites of the dimer in such a way that conversion would proceed faster.

We therefore decided to investigate the esterification of dihydroxy-poly(ethylene glycol) (HO-PEG-OH). The polydispersity of commercially available PEG would allow us to test multiple lengths at the same time. If one of the lengths would fit better between the active sites of a 'kissing' CalB dimer

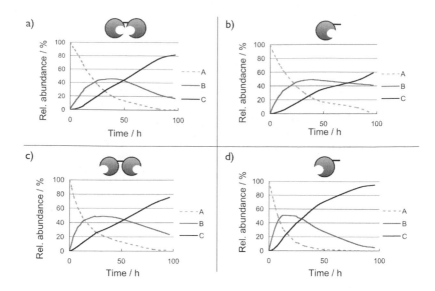

Figure 7. Esterification of 1,8-octanediol catalyzed by (a) 147-dimer, (b) the respective monomer, (c) wt-dimer and (d) its corresponding monomer. Dimers were prepared with 4 as linker. The relative amounts of (A) 1,8-octanediol, (B) octyl acetate and (C) the diacetylated product as determined by GC-MS are plotted against time.

than the other lengths, then esterification of that specific length would proceed faster. We expected that this effect would be visible in mass spectra of the crude reaction mixture.

We performed an assay using HO-PEG300-OH and vinyl 2-chloroacetate. The amount of ethylene glycol units in PEG300 varies from approximately 5 to 11 units. The lengths of the molecules will therefore vary between 17.5 and 38.5 Å. The catalytic cleft of CalB is 12 Å. Taking this into account, we hypothesized that PEG300 would be well suited to determine if there was an optimal substrate length with respect to rate of conversion by the 'kissing' dimer.

The shortest of the two available spacers (**5**) was used to prepare the 147-dimer and wt-dimer. Acetylation of PEG by these dimers and their respective monomers was investigated. As in earlier assays, equal amounts of dimer and monomer were taken based on the peak areas in the size-exclusion chromatograms. Lyophilization of these fractions was followed by addition of a mixture containing PEG300 and vinyl 2-chloroacetate in dry THF. At different time points aliquots of the reaction mixtures were analysed by mass spectrometry.

Figure 8 shows some typical mass spectra that were obtained. The signals could be assigned to the different PEG lengths and their corresponding acetylated products. However, no clear preference for one of the lengths was visible in the reaction performed by the 147-dimer. The fact that both NH_4^+ and Na^+ adducts were present in varying degree resulted in peaks being split up. Furthermore, as mass spectrometry on its own is not quantitative, minor effects could not be detected.

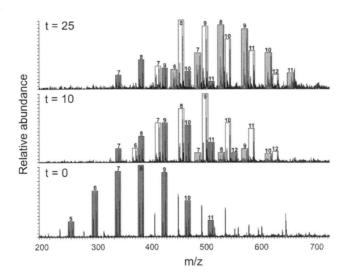

Figure 8. Esterification of poly(ethylene glycol) diol catalyzed by 147-dimer. The acetylation of HO-PEG300-OH was followed in time. Mass spectra of three different time points (in hours) are shown. The peaks surrounded by the dark grey bars belong to the starting compound, white bars indicate the monoacetylated intermediates and light grey bars correspond to the diacetylated products. The numbers indicate the amount of ethylene glycol units.

Another analytic technique is required to verify whether the 'kissing' dimer exhibits some special activity towards molecules with two substrate moieties. RP-HPLC analysis may enable the separation of the individual PEG oligomers to quantify their relative abundance (*18*).

Furthermore, it could be that the orientation of the two active sites with respect to each other is not optimal. Even though the enzyme molecules are linked near the catalytic clefts, these domains may point in another direction and not face each other. Therefore, it may be interesting to study the activity of the dimers in aqueous solution instead of organic solvent. It is known that the conformational dynamics of enzymes is higher in an aqueous environment (*19*). It could be that such a higher degree of flexibility is required for the 'kissing' dimer to enable the active sites to cooperate.

Experimental

Protein Production and Purification

The methionine auxotrophic strain B834(DE3)pLysS *E. coli* (Novagen) was used for protein production. Competent cells were transformed with pET22 plasmid containing the CalB gene of interest. As described previously, protein expression was induced in presence of azidohomoalanine or homopropargyl glycine (Chiralix) followed by purification via affinity and size-exclusion chromatography (*10*). The protein concentration of the purified samples was determined by measuring absorbance at 280 nm using a NanoDrop 2000

(Thermo Scientific). The whole protein was analysed by electrospray ionization time-of-flight (ESI-TOF) on a JEOL AccuTOF. CalB (10 µM) in formic acid (0.1 %) was injected. Deconvoluted spectra were obtained using Magtran software.

Site-Directed Mutagenesis

The structure of the protein was available from PDB file 1tca (*20*). The Yasara & WHAT IF Twinset was used for visualization and analysis of the protein to select amino acids that could be changed into methionine residues (*21*).

A standard mutagenesis protocol was used to produce the different CalB mutants (*22*). The following primers (Biolegio) were used: Met(1)Gly forward: G CCG GCG ATG GCC GGG GGA CTA CCT TCC GG. Met(1)Gly reverse: CC GGA AGG TAG TCC CCC GGC CAT CGC CGG C. Leu(147)Met forward: CCT CTC GAT TCA ATG GCG GTT AGT GC. Leu(147)Met reverse: GC ACT AAC CGC CAT TGC ATC GAG AGG. Val(149)Met forward: GAT GCA CTC GCG ATG AGT GCA CCC TCC. Leu(149)Met reverse: GGA GGG TGC ACT CAT CGC GAG TGC. Leu(219)Met forward: GTG TGT GGG CCG ATG TTC GTC ATC GAC. Leu(219)Met reverse: GTC GAT GAC GAA CAT CGG CCC ACA CAC. Val(221)Met forward: GGG CCG CTG TTC ATG ATC GAC CAT GC. Val(221)Met reverse: GC ATG GTC GAT CAT GAA CAG CGG CCC.

Cu(I)-Catalyzed Click Reactions

For every new experiment a fresh stock solution of $CuSO_4$ (20 mM) and sodium ascorbate (25 mM) in MQ was prepared and premixed in a 1:1 ratio with a 40 mM stock solution of tristriazole ligand (*23*) in MeCN. This mixture will be referred to as 'click mix'.

For the direct linkage of wt(AHA)CalB and wt(HPGCalB), 44 nmol Cu(I) (4.4 µL click mix) was added to 0.88 nmol of each protein variant in an aqueous medium containing 50 mM NaH_2PO_4 and 150 mM NaCl (pH 7.0). The final reaction volume was 45 µL. The reaction was incubated overnight at room temperature while gently shaken.

For the dimerization of (AHA)CalB using a dialkynyl linker, 10 mM stock solutions were prepared of **4** (in MQ) and **5** (in DMSO). 25 nmol of linker **4** or **5** and 25 nmol Cu(I) (2.5 µL click mix) were mixed with 0.44 nmol wt(AHA)CalB or 147(AHA)CalB resulting in a final reaction volume of 12.5-20 µL. After overnight incubation at room temperature the reaction was dialyzed against 0.5 mL of an aqueous solution containing 50 mM NaH_2PO_4 and 150 mM NaCl (pH 7.0) using a 10 kDa MWCO centrifugal filter unit (Millipore). 0.44 nmol fresh wt(AHA)CalB or 147(AHA)CalB was added together with 25 nmol Cu(I) resulting in a final volume of 25-33 µL. The reaction was left at room temperature for an additional night.

Size-Exclusion Chromatography

Dimerization and oligomerization reactions were analyzed on a *Pharmacia SMART* fast performance liquid chromatography (FPLC) system equipped with a

Superdex 75 PC 3.2/30 column (GE Healthcare Life Sciences) using an aqueous solution containing 50 mM NaH_2PO_4 and 150 mM NaCl (pH 7.0) as an eluent. The flow rate was typically 25 μL/min and fractions of 25 μL were collected. Only for the analysis of the directly linked wt(AHA)CalB and wt(HPG)CalB a flow rate of 20 μL/min was used.

SDS-PAGE Analysis

Protein samples were analysed by electrophoresis on 12% (w/v) polyacrylamide gels followed by Coomassie Brilliant Blue or silver staining. The Odyssey Infrared Imaging System (LI-COR Biosciences) was used to quantify the intensity of the Coomassie stained protein bands.

Activity Assay with *p*-Nitrophenyl Butyrate

Hydrolysis of *p*-nitrophenyl butyrate (pNPB) was detected by monitoring the absorbance at 405 nm using a Multicounter Wallac Victor² (PerkinElmer Life Sciences). A 96-wells plate well was filled with 45 μL of an aqueous solution containing 50 mM NaH_2PO_4, 150 mM NaCl (pH 7.0) and 0.1% Triton-X100. 2.5 μL of enzyme solution in an aqueous solution containing 50 mM NaH_2PO_4 and 150 mM NaCl (pH 7.0) was added. Directly before the measurement was started 2.5 μL substrate (concentrations varying from 2.5, 5, 10 and 20 mM in isopropanol) was added. Autohydrolysis of the substrate was monitored by adding 2.5 μL of an aqueous solution containing 50 mM NaH_2PO_4 and 150 mM NaCl (pH 7.0) without enzyme. Absorbance was measured every two minutes for at least 60 minutes while the temperature was kept at 25 °C. The slope of the curve in the first 20 minutes was taking as a measure of hydrolytic activity.

Esterification Assay with Vinyl Acetate and 1-Octanol, 1,8-Octanediol and PEG.

Esterification of 1-octanol or 1,8-octanediol using vinyl acetate was performed in dry THF. A 2x stock was prepared containing 2.2 M vinyl acetate and 0.38 M 1-octanol or 1,8-octanediol in THF. The compounds were dried in advance using molecular sieves (3 Å). 500 μL stock solution and 500 μL THF were added to an eppendorf tube containing the lyophilized enzyme. Lyophilization was performed using a solution of the enzyme in an aqueous solution containing 50 mM NaH_2PO_4 and 150 mM NaCl (pH 7.0) without addition of any cryoprotectant such as PEG-6000.

The reaction mix was incubated at room temperature while gently shaken. 20μL-aliquots were withdrawn from the reaction mixture at specific time points. They were analysed by GC-MS after 10x dilution in acetone. The temperature of the GC oven was increased by 30 °C min⁻¹ from 100 to 190 °C. The retention times for 1-octanol and the corresponding ester were 3.00 and 3.53 min, respectively. For 1,8-octanediol, a GC-gradient was used in which the oven was heated by 30 °C min⁻¹ from 160 to 270 °C. This resulted in retention times of 2.82, 3.22 and 3.60 min, for diol, monoester and diester, respectively.

The acetylation assay of PEG300 was performed in a similar way using a stock solution of 2.2 M vinyl 1-chloroacetate and 0.38 M PEG300. 5-µL aliquots were withdrawn from the reaction mixture and analyzed by LR-MS upon 10,000x dilution in MeOH. For PEG-6 the following masses were detected: Before incubation with CalB; LRMS (ESI+) m/z calcd. for $C_{12}H_{26}O_7$ [M+NH$_4$]$^+$ 300.2, found: 300.1; [M+Na]$^+$ 305.2, found: 305.3. After incubation with CalB; LRMS (ESI+) m/z calcd. for $C_{16}H_{28}Cl_2O_9$ [M+NH$_4$]$^+$ 452.3, found: 452.1; [M+Na]$^+$ 457.3, found: 457.1.

Conclusion

A series of CalB dimers have been produced. Via the incorporation of non-natural amino acid analogues of methionine, CalB was produced with either an azide or alkyne at the N-terminus. They were directly linked by the addition of Cu(I). Dimerization did not affect hydrolytic activity.

Mutants of CalB were produced with an accessible azide near the active site. Using a dialkynyl linker, dimers were prepared in which both active sites were brought in proximity of each other. Remarkably, these dimers behaved differently on a SDS-PAGE gel compared to dimers that were linked at their N-termini.

A model substrate with one alcohol moiety was converted as efficiently by the dimers as by the CalB monomers. Also in case of molecules containing two substrate moieties no difference in rate was detected. A more quantitative assay may give more insight in the question whether in particular the 'kissing' dimer can convert the latter kind of substrates faster. A hydrolytic assay in an aqueous environment will be investigated in the near future as some degree of flexibility may be necessary for the 'kissing' dimer to function in a cooperative way.

Acknowledgments

The authors would like to thank L. Canalle for providing the 'clickable' linkers, and P. van Galen for assistance with the mass spectrometry analysis. NWO is thanked for the financial support of this research.

References

1. Marianayagam, N.; Sunde, M.; Matthews, J. *Trends Biochem. Sci.* **2004**, *29*, 618.
2. Lee, K. A. W. *J. Cell Sci.* **1992**, *103*, 9.
3. Winkel, B. S. J. *Annu. Rev. Plant Biol.* **2004**, *55*, 85.
4. Ziaco, B.; Pensato, S.; D'Andrea, L. D.; Benedetti, E.; Romanelli, A. *Org. Lett.* **2008**, *10*, 1955.
5. van Baal, I.; Malda, H.; Synowsky, S. A.; van Dongen, J. L. J.; Hackeng, T. M.; Merkx, M.; Meijer, E. W. *Angew. Chem., Int. Ed.* **2005**, *44*, 5052.
6. Xiao, J.; Tolbert, T. J. *Org. Lett.* **2009**, *11*, 4144.

7. Hatzakis, N. S.; Engelkamp, H.; Velonia, K.; Hofkens, J.; Christianen, P. C. M.; Svendsen, A.; Patkar, S. A.; Vind, J.; Maan, J. C.; Rowan, A. E.; Nolte, R. J. M. *Chem. Commun.* **2006**, *19*, 2012.

8. Heredia, K. L.; Grover, G. N.; Tao, L.; Maynard, H. D. *Macromolecules* **2009**, *42*, 2360.

9. Anderson, E. M.; Karin, M.; Kirk, O. *Biocatal. Biotransform.* **1998**, *16*, 181.

10. Schoffelen, S.; Lambermon, M. H. L.; van Eldijk, M. B.; van Hest, J. C. M. *Bioconjugate Chem.* **2008**, *19*, 1127.

11. van Dongen, S. F. M.; Nallani, M.; Schoffelen, S.; Cornelissen, J. J. L. M.; Nolte, R. J. M.; van Hest, J. C. M. *Macromol. Rapid Commun.* **2008**, *29*, 321.

12. Teeuwen, R. L. M.; van Berkel, S. S.; van Dulmen, T. H. H.; Schoffelen, S.; Meeuwissen, S. A.; Zuilhof, H.; de Wolf, F. A.; van Hest, J. C. M. *Chem. Commun.* **2009**, *27*, 4022.

13. Debets, M. F.; van Berkel, S. S.; Schoffelen, S.; Rutjes, F. P. J. T.; van Hest, J. C. M.; van Delft, F. L. *Chem. Commun.* **2010**, *46*, 97.

14. Kiick, K. L.; Saxon, E.; Tirrell, D. A.; Bertozzi, C. R. *Proc. Natl. Acad. Sci. U.S.A* **2002**, *99*, 19.

15. van Hest, J. C. M.; Kiick, K. L.; Tirrell, D. A. *J. Am. Chem. Soc.* **2000**, *122*, 1282.

16. Bolanos-Garcia, V. M.; Davies, O. R. *Biochim. Biophys. Acta, Gen. Subj.* **2006**, *1760*, 1304.

17. Secundo, F.; Carea, G.; Soregaroli, C.; Varnelli, D.; Morrone, R. *Biotechnol. Bioeng.* **2001**, *73*, 157.

18. Trathnigg, B.; Gorbunov, A.; Skvortsov, A. *J. Chromatogr., A* **2000**, *890*, 195.

19. Serdakowski, A. L.; Dordick, J. S. *Trends Biotechnol.* **2008**, *26*, 48.

20. Uppenberg, J. H.; Hansen, M. T.; Patkar, S.; Jones, T. A. *Structure* **1994**, *2*, 293.

21. Krieger, E.; Koraimann, G.; Vriend, G. *Proteins* **2002**, *47*, 393.

22. According to the instruction manual of QuikChange Site-Directed Mutagenesis Kit (Stratagene).

23. van Kasteren, S. I.; Kramer, H. B.; Gamblin, D. P.; Davis, B. G. *Nat. Protoc.* **2007**, *2*, 3185.

Chapter 11

Biotransformations Using Cutinase

Peter James Baker[a] and Jin Kim Montclare[a,b,*]

[a]Department of Chemical and Biological Sciences, Polytechnic Institute
of NYU, Brooklyn, NY 11201, USA
[b]Department of Biochemistry, SUNY Downstate Medical Center, Brooklyn,
NY 11203, USA
*jmontcla@poly.edu

There is a growing interest in sustainable and environmentally-friendly solutions for the industrial manufacturing of chemicals to replace non-renewable fossil fuel-based feedstocks. Biotransformations provide an alternative methodology to traditional reactions by taking advantage of the biochemical diversity of microorganisms to provide a chemo-, regio- and enantioselectivity, which are not always available via traditional synthetic approaches. Cutinases are enzymes secreted from phytopathogens and have been proven to be useful for several different biotransformations. This review will investigate the structure and function of cutinase. Further, it will detail cutinase activity towards both natural and non-natural substrates as well as methods employed to impart stability. Finally, we will describe the different role that cutinase has played in biotransformation reactions for biotechnology applications.

Introduction

The cuticle serves as a lipophillic cell wall barrier covering the aerial surface of plants. The cutical surface is between 0.1 and 10 μm thick. It is composed of two layers: a waxy waterproofing layer and an aliphatic insoluble layer that provides structural support. The formation of these layers is thought to be a key adaptation in aerial plant evolution (*1*, *2*). This cuticle surface is chiefly comprised of cutin — an insoluble, crosslinked, lipid-polyester matrix comprised of n-C_{16}, n-C_{18} ω-hydroxy and epoxy fatty acids (Figure 1) (*3*). The composition of cutin differs depending upon the plant species. In general, fast growing plants contain cutin bearing C_{16} monomers, while slower growing plants contain cutin with both C_{16} and C_{18} monomer families (3). Remarkably, the plant cuticular barrier outperforms many engineered polymer films of comparable thickness as well as resists the penetration of pesticides used in crop protection and foreign molecules from the atmosphere (4). However, several phytopathogens have been identified that can degrade this barrier and infect plants.

Pioneering studies by Kolattukudy on the cellular and molecular mechanism of the fungal pathogen *Fusarium solani*, revealed a 25,000 Da secreted enzyme able to hydrolyze the cutin surface (5, 6). This enzyme was aptly named cutinase. An up-regulation of cutinase expression was observed upon contact of the fungi with the cuticle, suggesting that the cutin monomer serves as the signaling molecule (7). Several species of fungi, including *F. solani*, were able to grow on cutin as their sole carbon source (8–10). Through the use of antibodies, inhibitors and cutinase-deficient fungal mutants, it was confirmed that cutinase played a critical role in fungal pathogenesis (11).

Several other phytopathogenic fungi were identified to produce cutinolytic enzymes (12). In addition to cutin hydrolysis, cutinases were used to hydrolyze *in vitro* many other industrially important esters including soluble synthetic esters and insoluble long-chain triglycerides (triolein and tricaprylin) (Table 1) (6, 13). The hydrolysis of *p*-nitrophenyl esters resulted in the production of a spectrophotometrically detectable *p*-nitrophenyl group, providing a rapid assay to detect cutinolytic activity (6). Furthermore, cutinases were exploited for synthetic reactions such as: esterification, transesterification and polymer synthesis (Tables 2, 3) (14).

To date *Fusarium solani* cutinase (FsC) has provided the bulk of the information regarding cutinase biocatalysis because its three-dimensional structure was the first to be solved (Figure 2a) (15). Recently, the three-dimensional structure of cutinases from *Aspergillus oryzae* (AoC) and *Glomerella cingulata* (GcC) have been solved, revealing different surface charges and active site geometries potentially opening up new areas of investigation (16, 17). Aside from fungal cutinases, these enzymes have also been found in several species of nitrogen fixing bacteria and pollen (18, 19). As our knowledge of biodiversity expands and we accumulate biochemical data from different species, potential applications for new biotransformations will increase.

Here we will review the structural data of cutinase and the role it plays in enzyme activity. Furthermore, we will survey the different means by which cutinases have been immobilized and how that has assisted in its use in non-traditional media. Finally, we will discuss the potential biotransformation applications in polymer degradation, modification and synthesis.

Cutinase Structure

Cutinases are approximately 200 amino acid residues long and have molecular weights between 20,000 and 22,000 daltons. They are members of the α/β hydrolase family, which is hallmarked by three highly conserved features: (*i*) the α/β fold; (*ii*) a catalytic triad; and (*iii*) an oxyanion hole (Figure 2a)

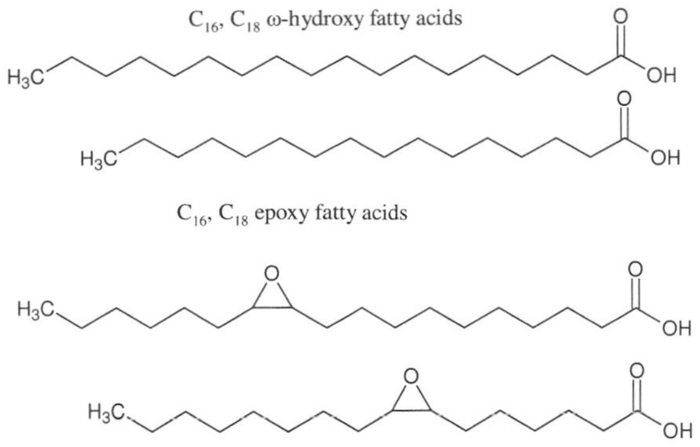

Figure 1. C_{16}, C_{18} ω-hydroxy and epoxy fatty acids are the major components that form cutin.

Table 1. Hydrolytic and synthetic activity of FsC towards triglycerides and esters (n.d. = not determined)

		Substrate (Chain length)	K_m μM	Ref.
Hydrolysis	Triglycerides	Triolein (C_8)	1.32×10^5	[20]
		Trilaurin (C_{12})	1.51×10^5	[20]
		Tricaprylin (C_{18})	3.95×10^5	[20]
	Esters	p-nitrophenylacetate (C_1)	0.67	[16]
		p-nitrophenylbutyrate (C_3)	1.26	[16]
		p-nitrophenylvalerate (C_4)	1.48	[16]
		p-nitrophenylhexanoate (C_5)	1.5	[16]
Synthesis	Esters	1-diol	n.d.	[21]
		2-diol	n.d.	[21]
		3-diol	n.d.	[21]
		4-diol	n.d.	[21]
		butyric acid (C_4)	n.d.	[22]
		valeric acid (C_5)	n.d.	[22]
		octonic acid (C_8)	n.d.	[22]
		decanoic acid (C_{10})	n.d	[22]

(*23*). The α/β hydrolase family of enzymes are important for the biotechnology industry as they exhibit chemoselectivity, can be expressed in high yields, do not require cofactors and remain enzymatically active in organic solvents (*24–26*). In addition to cutinases, lipases, esterases and proteases are members of this family of enzymes.

The α/β fold is found in a majority of hydrolytic enzymes regardless of their phylogenetic origin or substrate specificity (*27*). The fold is composed of central β-sheets, in either a parallel or anti-parallel orientation, surrounded by α-helices. This α/β fold is the most frequently fold found in nature and provides a stable scaffold for the active site of a wide variety of enzymes (*28*). Cutinases possess a central β-sheet of 5 parallel strands and 5 α-helices, with the active site residues on top of the sheets (Figure 2a) (*15–17, 23*).

The catalytic triad of α/β hydrolase is composed of a single acidic Asp/Gln residue, a basic His residue as well as a residue containing a hydroxyl group (*29*). Originally, identified in α-chymotrypsin, this triad has been further observed in lipases, trypsin, subtilisin, and several other α/β hydrolases (*29–31*). Backbone overlay of the active site shows structural homology despite the low overall sequence homology, suggesting a specific geometric arrangement is required for hydrolase activity (*31*). In lipases, an α-helical hydrophobic lid often covers the active site. Modeling of lipase at the lipid/water interface shows that the lid is displaced thereby activating the enzyme (*32*). The rearrangement of this lid allows for the substrate to enter the active site and the product to leave it (*28*).

Cutinases differ from lipases as this lid is absent, allowing for a more solvent exposed active site as well as preventing interfacial activation (*15, 33*). The structural geometry of these two enzymes allows for different reactivity; cutinases show a preference for substrates with a shorter acyl chain while lipases prefer longer chains (*34*). Furthermore, cutinases hydrolyze water-soluble esters whereas lipases are efficient at hydrolyzing insoluble esters (*35, 36*).

The cutinase catalytic triad of is comprised of highly conserved Ser, Asp and His residues, where Ser is located at the C-terminal end of one of the β-strands in an extremely sharp turn toward the next helix, creating a so-called 'nucleophile elbow' (Figure 2a) (*13, 15*). The imidazole ring of the His residue serves as a general base catalyst abstracting the proton from the hydroxyl group of the Ser. The Ser oxygen performs a nucelophilic attack on the acyl carbonyl carbon of the substrate (Figure 2b). The role of the Asp residue has been an issue of controversy. In the initially proposed charged relay hypothesis, the carboxyl anion moiety of the Asp residue serves as a second general base by removing a proton from the imidazole group from the His (*40, 41*). This hypothesis has been challenged by Bruice *et. al.*, who suggests that this mechanism is thermodynamically unfavorable because Asp is a much poorer base than the His (*39*). In other members of the α/β hydrolase family, it has been suggested that the Asp residue forms a low-energy hydrogen bond with the His residue (Figure 2b) (*38*).

Pauling postulated that the effectiveness of enzymatic catalysis is due to the protein's ability to bind the substrate in a conformation resembling the transition state; by lowering the activation energy, the reaction rate is increased (*42*). The oxyanion hole of α/β hydrolase accomplishes this by stabilizing the transition state via hydrogen bonds with two of the backbone chain nitrogens (Figure 2b) (*31*). The oxyanion hole in cutinase is composed of a Ser and Gln residue. Specifically, these two residues serve to stabilize the developing negative charge on the substrate ester or amide carbonyl oxygen during the formation of the tetrahedral intermediate for acyl transfer (Figure 2b) (*43*).

Comparing the apo- FsC crystal structure to that of the covalently inhibited by diethyl-*p*-nitrophenyl phosphate (E600) structure, reveals little change in the overall conformations. The residues in the oxyanion hole, Ser 42 and Gln 121, occupy the same positions in both structures, indicating that unlike other lipases, FsC has a preformed oxyanion hole (Figure 2) (*43*). Simulation studies demonstrate that Oγ of Ser 42 is stabilized by hydrogen bonding with Gln 121

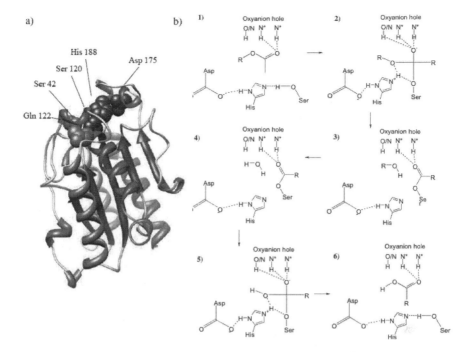

Figure 2. a) FsC (pdb ID = 1CEX) in 3D ribbon representation (23). The catalytic residues and oxyanion hole are highlighted (15, 23). b) Schematic representation of cutinase ester- cleavage: (1) non-covalent Michaelis complex; (2) first tetrahedral transition state; (3) formation of the acyl-enzyme and release of an alcohol molecule; (4) attack of water molecule on the acyl enzyme; (5) second tetrahedral transition state; and (6) release of the fatty acid. N* represent the backbone residues and hydrogen bonds are depicted in grey (37–40).

Nε2 and Asn 84 Nδ2 with distances of 3.15 Å and 3.06 Å, respectively (*37*). NMR studies show a greater flexibility, contrary to the crystallographic studies in which the helical flap opens and closes the binding site in the both the apo-enzyme and the cutinase-inhibitor complex (*44*). The Ser 42 Ala mutations results in a 1000-fold decrease in activity; indicating the importance of this residue in catalysis (*37*).

The crystallographic analysis of AoC reveals a deep continuous hydrophobic groove along the active site, which topologically favors more hydrophobic substrates (Figure 3a). Comparing the activity of AoC to FsC, AoC has higher affinity and catalytic efficiency toward *p*-nitrophenyl esters with longer acyl chains (*16*).The crystal structure of cutinase from GcC showed an active site that is capable of cycling between conformations, rendering the enzyme in either an active or inactive state (Figure 3b). In this structure, the catalytic His residue (His 204) is swung out of the active site to a distance (11Å) where is it unable to participate in activity. GcC must undergo a geometric rearrangement where the solvent exposed His 204 shifts into a buried conformation, switching to an active state (*17*). The catalytic His residues from the other structures predict no such rearrangement.

Immobilized Cutinase for Industrial Applications

Cutinase hydrolyzes a broad range of substrates including: *p*-nitrophenyl esters of fatty acids and triglycerides, lending its usefulness for applications in industry (Table 1) (*6, 45, 46*). These enzymes offer advantages to the dairy industry by providing a low energy methodology for modification of milks as well as cheeses to diversify flavors and smells (*14*). Cutinases provide alternatives for grease or stain removal processes in laundry and dishwashing detergents (*47*). These enzymes are routinely employed by the chemical industry for stereoselective esterification reactions of alcohols (*21, 45*).

Cutinases, like most enzymes are marginally stable macromolecules with a T_m between 50 to 70°C, at pH 8.0. In acidic conditions the structure is maintained but the activity is lost due to the ionization state of the His residue in the catalytic triad (Figure 2b) (*48*). Furthermore, the soluble nature of enzymes presents many problem for biotechnological applications such as end-product contamination, poor stability, and limited reuse (*49*). Below we investigate several different techniques that have assisted in cutinase stability and activity in aqueous and non-aqueous environments.

Lyophilization is a straightforward technique where the purified enzyme is exchanged into water and sublimated, by reducing the surrounding pressure and decreasing the temperature, resulting in a freeze-dried enzyme (*50*). Lyophilized enzymes are then suspended in the appropriate organic solvent. In particular, lyophilized cutinases have been used either free in solution or immobilized on a solid support. After lyophilization, FsC has been used to examine hydrolysis at the triolein-water interface; a linear relationship between FsC concentration and loss of tension at the oil-water interface exists, suggesting that the FsC is hydrolyzing triolein at the interface (Table 1) (*51*). The lyophilized FsC in organic solvent has been used to elucidate the selectivity and stereospecificity for both degradative and synthetic reactions (*45, 52, 53*). Further, esterification reactions of acids with increasing chain lengths and alcohol carried out by FsC in iso-octane show a preference for butyric (C_4) and valeric (C_5) acids over octonic (C_8) and decanoic acids (C_{10}) (Table 1) (*22*). Lyophilization has proven to be a useful method for demonstrating cutinase activity and specificity in organic solvents, however, many industrial reactions require the catalysts to be easily separated from the products, which is not possible with free enzymes (*54*).

The immobilization of cutinase offers many potential enhancements over the use of free enzymes. The advantages include: increased conformational stability in both traditional and non-traditional media; reusability; and ease in product recovery (*55*). The different methods by which cutinase have been immobilized are physical adsorption, covalent attachment, entrapment in reverse micelles and gel encapsulation. Here we will discuss the pros and cons of each of these methods (Table 2, Figure 4).

Physical adsorption uses weak interactions such as electrostatic, hydrogen bonding and hydrophobic interactions to immobilize proteins non-specifically on solid supports (Table 2, Figure 4). FsC immobilized on zeolites or microporous aluminosilicate minerals have been shown to have substantially higher stability as compared to other polymeric supports (*vide infra*) (*71*) . The enzyme exhibits

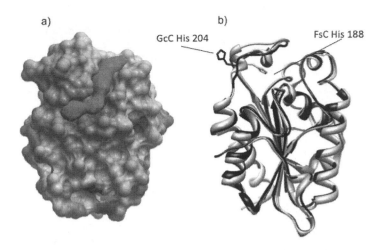

Figure 3. a) Electrostatic surfaces of AoC (pdb ID = 3GBS) as rendered by ICM. The pronounced grey solid density generated by PocketFinder function of ICM illustrates the groove in the active site surface (16). b) Overlay of the 3D ribbon structures of FsC (grey) and GcC (black) (pdb ID = 3DCN).The His residue backbones are shown in stick form (17, 23).

an enhanced stability when it is physically adsorbed to NaY as compared to those covalently attached to polymeric supports. Further, mutational analysis by the introduction of a charged residue (Leu 153 Gln) reveals enhanced stability suggesting the interaction between the solid support and the enzyme is largely electrostatic (*58*). Analysis of transesterification reactions on immobilized zeolites have demonstrated that cutinases prefer immobilization on NaY in terms of activity and stability (Table 2) (*72*). Although zeolites improve the stability and the rate of transesterification reactions by FsC, there is loss of conformation upon immobilization; this may limit the enzyme's activity towards other substrates or in other environments (*63, 72*). Cutinases immobilized on Lewatit resins have been particularly useful for polymer synthesis (Table 4, Figure 4) (*56, 57*). Physical absorption of enzymes has been proven to be an excellent technological development for biotransformations, however, two major drawbacks are orientation and reversibility. Nonspecific adsorption of the enzyme often leaves the enzyme in an orientation where the active site is inaccessible, rendering the enzyme inactive (Figure 4). Furthermore, these interactions are reversible, resulting in the enzyme leaching off the solid support (*73*).

Covalently linking the enzymes to the solid support is an irreversible process where the orientation of the enzyme may be controlled (Figure 4). Covalent immobilization can ensure a solvent exposure of the active site orientation, reduce enzyme leaching and enhance stability (*74*). FsC covalently immobilized on a dextran solid support has been found to have an enhanced thermal stability in aqueous environments, resulting in a long-term stability of the enzyme (Table 2). However, the rate of tricaprylin hydrolysis is significantly decreased (*63*). A major limitation of this method is the loss of enzyme activity following covalent immobilization due to increased structural rigidity (*75*). In order to achieve higher

Table 2. Cutinase mediated reactions carried out under different immobilized methods

Method of Immobilization	Enzyme	Support	Reaction	Ref.
Physical adsorption	FsC, HiC	Lewatit beads Accurel EP100 Zeolite	Hydrolysis Esterification Transesterification Polymerization	(56–60)
Covalent modification	FsC, HiC	Zeolite Dextran	Hydrolysis Transesterification	(58, 59, 61–63)
Reverse micelles	FsC, HiC μHiC	AOT CTAB	Hydrolysis Esterification Transesterification	(20, 62, 64–68)
Gel encapsulation	FsC	Sol-gel	Transesterification	(69, 70)

activity while also improving enzyme stability without rigidification and inactivity, alternative approaches of immobilization have been investigated.

The development of nanoscale structures capable of forming spheres or scaffolds has provided a new means of encapsulating enzymes for biotransformations (74). These structures provide reduced diffusion limitations, maximize the functional surface area and enhance particle mobility, providing a significant technological advancement over physical and covalent immobilization methods (74). Here we will discuss two such means of encapsulation: reverse micelle and sol-gel encapsulation (Figure 4).

Reverse micelles or microemulsions are nanometer-sized water droplets dispersed in organic media by the action of surfactants (76). Structurally, reverse micelles can be thought of as inverse analogues of micelles where the hydrophilic head is turned in and the hydrophobic tails are exposed to the surrounding organic solvent (Table 2, Figure 4) (75). Microemulsions are optically transparent and thermodynamically stable mixtures (75). They are generally evaluated by the critical micelle concentration (CMC), which is the minimum surfactant concentration required to reach the lowest surface tension. Above the CMC, the surfactant molecules form supramolecular structures (77). These microemulsions offer many favorable properties for biotransformations by overcoming substrate solubility, enhancing the stereo-selectivity, activity and stability of enzymes. Lipases have been one of the most thoroughly investigated enzymes in these microemulsions because of their enhanced stability and activity in this medium (78). Cutinases have also been investigated in reverse micelles by several different groups (20, 64, 66, 76, 79, 80).

The anionic surfactant sodium-di-2-ethylhexylsulfosuccinate, commercially available under the name Aerosol OT (AOT), is commonly used to form reverse micelles in organic media (Figure 4) (64). Microencapsulated FsC shows a decrease in activity towards triolein hydrolysis and stability as compared to FsC in aqueous environments. The activity is dependent on physical parameters such as pH, temperature and water content, suggesting the anionic surfactant inhibits the hydrolytic activity in a concentration dependent manner (Table 2) (20, 64).

NMR spectra of the aromatic and amide spectral regions of FsC in AOT reverse micelles indicate the loss of structure (64). Further, the encapsulation of cutinase in AOT reverse micelles results in a time-dependent unfolding process due to interfacial interactions (65).

To improve the stability of cutinase in reverse micelles, two different approaches have been undertaken: media engineering and protein engineering (20, 67, 68, 76). Media engineering is the substitution of aqueous reaction media by non-conventional media; it also implies the adaptation of the microenvironment of the biocatalyst by immobilization or by introduction of additives for stabilization (81). Protein engineering utilizes recombinant DNA technology to make single or multiple changes to the primary amino acid sequence to improve the function or stability of biocatalysts.

Media engineering has improved the thermal stability and activity of cutinase in reverse micelles (76). Using the cationic surfactant hexadecyltrimethyl-ammonium bromide (CTAB) there is an improvement in the stability of FsC however, the surfactant still serves as an inhibitor to the hydrolysis of p-nitrophenyl esters (Table 2) (82). In the presence of 1-hexanol, a co-surfactant, the AOT micelles are 10-fold larger, resulting in new interfacial characteristics that does not cause protein denaturation (76, 83). The stabilization of the FsC-AOT reverse micelle by 1-hexanol allows for an operational half-life of 674 days (76). Both the wild-type FsC and *Humicola insolens* cutinase (HiC) sequences demonstrate similar sensitivity to AOT, however a mutant variant of HiC (μHiC) displays a significantly improved stability (64).

Protein engineering has been employed to improve the stability of FsC: saturation mutagenesis libraries of FsC has resulted in variants with increased stability in the presence of AOT without compromising activity. The resulting variants suggest the presence of hot spots in the cutinase that contribute to the inhibitory effects of AOT on cutinase activity (67). The Ser 54 Asp mutant has been further investigated and demonstrated to have a similar secondary structure to the wild-type in aqueous environments as well as a significant resistance to AOT denaturation (68). Often these types of engineering are empirical and require time consuming experimentation; gel encapsulation has proven to be a useful alternative to reverse micelle entrapment (69, 70).

The encapsulation of enzymes in inorganic silica matrices of metal or semi-metal oxides (sol-gel) through the aqueous processing of hydrolytically labile precursors has become an efficient method for biocatalysis (Figure 4) (84, 85). This method restricts the conformational mobility of the enzyme, enhancing stability and preventing the enzyme from leaching during catalysis (85). FsC encapsulated in sol-gel matrices exhibits decreased activity towards transesterification reactions in hexane, when compared to other immobilization techniques (Table 2) (69). However, this decrease in activity is restored by the fact that this method allows for twice the amount of FsC to be immobilized. FsC immobilization onto zeolites combined with sol-gel encapsulation shows increased transesterification rates, while maintaining the increased number of FsC molecules (Table 2) (70). Moreover, sol-gel encapsulation shows improved activity and stability of cutinases and lipases in organic solvents and supercritical fluids, which will be further discussed in the next section (69).

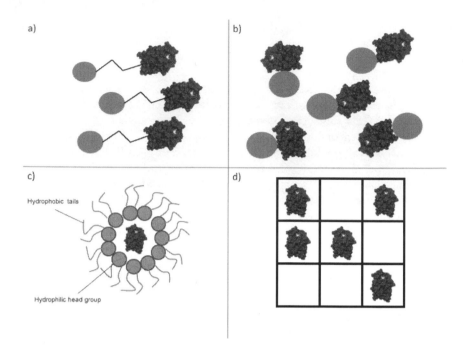

Figure 4. Cartoon representation of different cutinase immobilization techniques including: a) physical adsorption, b) covalent attachment, c) reverse micelle and d) sol-gel encapsulation. Shown is FsC, highlighted in light grey is the active site. The dark grey balls represent the solid support matrices and the black lines represent the sol-gel matrix.

Table 3. Reactions conducted in non-traditional media

Enzyme	Non-traditional media	Reaction	Ref.
FsC, TfC	Organic solvent, Supercritical fluid	Hydrolysis	(*61, 62*)
FsC	Organic solvent	Esterification	(*58, 80*)
FsC	Organic solvents, Supercritical fluid	Transesterification	(*62*)
HiC	Organic solvent	Polymerization	(*56, 57*)

Immobilization has been widely investigated as a method to stabilize cutinase against different kinds of inactivation. The approaches described above not only improve cutinase stability and activity but also expands its function in non-biological environments. This allows cutinases to be employed for a wider range of industrial applications.

Cutinase Activity in Non-Aqueous Environments

In order to maximize the full biotechnological potential of cutinase, it has been investigated in a wide array of non-traditional media (22, 62, 71). Enzyme selectivity is often different in non-aqueous environments than the more traditional aqueous environments (86). Specifically, enzymes have been employed in organic solvents (OS) and supercritical fluids (SCFs) (Table 3). In these media, cutinases are reactive in both the hydrolytic and synthetic directions in a water-dependent manner.

It has been demonstrated that in OS, enzymes are capable of mediating synthetic reactions (87, 88). There are several advantages offered by OS such as: increased solubility of organic substrates, enhanced thermostability, and ease in recovery of substrates (89) . Further, these solvents allow for unique conformational changes of the enzyme, energetic of substrate desolvation and transition state stabilization, conformational mobility and pH memory (49) . Thus activity of cutinase in OS has provided useful industrial biotransformations.

FsC, immobilized on zeolites or Accurel PA-6 through physical adsorption, is capable of mediating transesterification reactions between vinyl butyrate and 2-phenyl-1-propanol in acetonitirile , in water-dependent manner (Tables 2, 3) (90). When the water activity (a_w) is 0.2, the initial rates for the transesterification reaction are higher, however, when the a_w value is increased to 0.7, the rates for the hydrolytic reactions are favored (90). Switching the reaction media from acetonitirile to either ethane or n-hexane in the presence of acidic/basic buffers increases the rate of the transesterification reaction (62). These rates depend on the pH; at pH 8.5 (optimal pH for FsC) the rates are the highest, while at slightly acidic or basic pH the reaction rates decrease, indicating a 'pH memory' (91, 92). Transesterification reactions of 1-4 diols and one prochiral acyclic phenyalkanediolas with vinyl acetate in dichloromethane, using physically adsorbed FsC, exhibit a chain length preference (Table 3). Progress curves show 80% consumption diol 1 while diols 2-4 are unreacted demonstrating the preference FsC has for short chained substrates (Table 1, Table 3) (21). Despite the successes achieved with OS, there are several drawbacks with this media that include toxicity, flammability and substrate contamination (60).

Supercritical fluids (SCFs) technology is rapidly becoming a green-alternative to biocatalytic mediated reactions in organic solvents (93). At critical pressures and temperature, they exhibit properties of both liquids and gas. The solvent properties of SCFs are tailored by controlling the pressure and/or temperature (94). These fluids have been used for a host of polymer synthesis and transesterification reactions (95, 96).

FsC immobilized onto the hydrophobic polypropylene resin has been investigated for esterification reactions in supercritical CO_2 (Tables 2, 3). These experiments demonstrate that FsC mediates the esterification of hexanoic acid with hexanol (97). FsC immobilized onto zeolites shows a strong dependence on water. These studies reveal FsC has a stronger selectivity towards the (R)-enantiomer of 1-phenylethanol (Tables 2, 3) (62). Although SCFs demonstrates a significant advancement in cutinase biotechnology, this technology requires special equipment and high costs.

Table 4. Biotransformations mediated by cutinases

Biotransformation	Enzyme	Substrate	Ref.
Degradation	AoC, FsC, HiC, TfC	PCL, PBS, PBSA, PTT PET	(99–102)
Modification	FsC, HiC, PmC	PVAc, PET	(103)
Polymerization	HiC	ε-caprolactone δ-valerolactone ω-pentadecalactone β-butyrolactone	(56, 57)

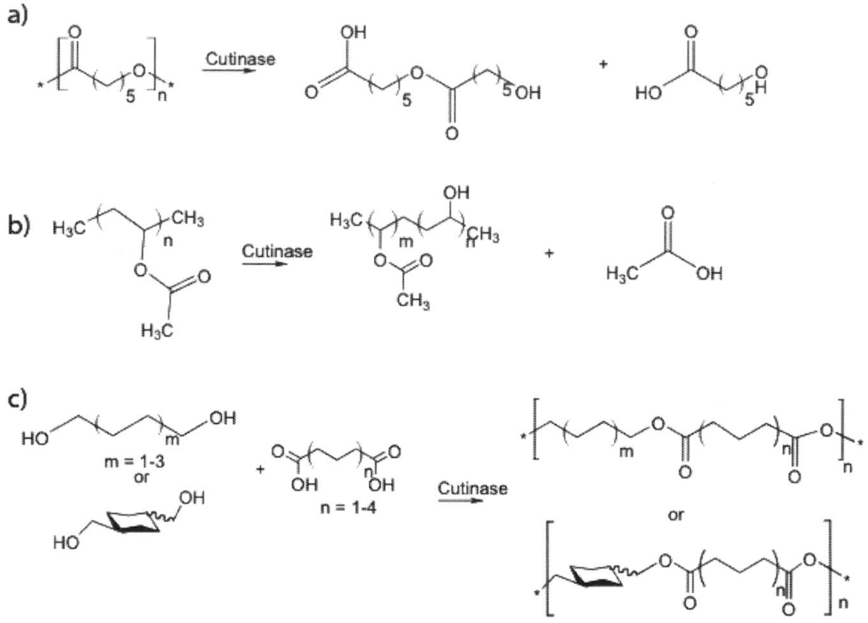

Figure 5. Schematic representation of cutinase biotransformations for polymer applications: a) degradation of polycaprolactone (PCL) (16); b) modification of polyvinyl acetate (PVAc) (103); and c) polycondensation between diols and diacids (56).

Biotransformations for Polymer Applications

Plastics effect our environment on two ecological fronts. The feedstocks for plastic synthesis are from petroleum, oil, gas and coal. Secondly, plastic waste contributes to over 67 million tons of the generated annual waste, comprising one-third of all the municipal waste (98). The development of microorganism and enzymes for the degradation, surface modification and synthesis of plastics has been a long-standing industrial goal. In this section we will survey how cutinase mediates the biotransformations of all three of those reactions.

Biodegradable plastics such as polycarpolactone (PCL), polylactic acid (PLA), polyhydroxybutyrate (PHB), poly-(butylenes succinate) (PBS), and poly-(butylenes succinate-co-adipate) (PBSA) are polymeric materials capable of undergoing decomposition into CO_2, methane, water in the presence of microorganisms and/or enzymes (98). Several different biodegradable plastics including: PCL and PBSA have long been shown to be degraded by *F. solani* and *A. oryzae* in a cutinase-dependent manner (Table 4, Figures 2a, 3a, 4) (99, 100). The purified recombinant AoC and FsC are able to degrade PCL films in which AoC is more effective in degradation (87% vs. 30% in 6 hours). The crystal structure of AoC shows a deep hydrophobic groove along the active site making it more accessible to hydrophobic polymeric surfaces as compared to FsC (Figure 3a) (16). Further, the AoC enzyme has high hydrolytic activity towards PBS/PBSA films, completely degrading them in 6 hours (Table 4) (100). Cutinase derived from thermophilic organisms such as *Humilica insolens* (HiC) and *Thermobifida fusca* (TfC) are effective at degrading synthetic polymers. Poly(ethylene terephthalate) (PET) films treated with HiC at 70°C showed 97 ± 3% weight loss in 96 hours, corresponding to a loss of film thickness of 30 μm per day (Table 4) (101). TfC is able to hydrolyze the aromatic polyester poly(trimethylene terephthalate) (PTT) in both the film and fiber form (Table 4) (102). These studies suggest that cutinase will continue to play an important role in the enzymatic degradation of both biodegradable and synthetic plastics.

The enzymatic modification of synthetic and natural polymers provides a highly specific and environmentally-friendly alternative to the harsh chemical methods currently employed. Lipases, proteases and cutinases are the most predominantly used enzymes for modification reactions (104). PET fibers treated with cutinase from several different species of tropical fungi, display increased absorption and enhanced surface hydropholicity (105). Variants of FsC have been tailored to react with PET (Table 4) (106). Specifically, Leu 81 Ala and Leu 182 Ala is proven to be four- and five-fold, increased in the active site towards PET fibers as compared to the wild type , respectively (106).

Poly(vinyl acetate) (PVAc) is another important industrial synthetic polymer used in adhesives and textiles (107). Cutinase from three different species (HiC, Pseudomonas *medocina* (PmC) and FsC) have been analyzed for their ability to deacetylate PVAc surface (Table 4, Figure 5) (103). FsC demonstrates poor affinity towards PVAc and possesses low catalytic activity, losing 93% of its activity within 24 hours while HiC and PmC retains activity after 192 hours. Due to the thermophilic nature of HiC, it shows the highest thermostability and catalytic efficiency (103).

Polycondensation, transesterification and ring-opening polymerization (ROP), have been performed using hydrolases, including cutinase (108, 109). HiC physically absorbed on Lewatit resin demonstrates activity for polymer synthesis of a series of diols and diacids using toluene as the solvent (Tables 3, 4 and Figure 5) (57). By increasing the diol and diacid lengths from C_8 to C_{10} results in higher molecular weight polymers (57). Further, immobilized HiC also mediates the catalysis for ROPs of ε- caprolactone, ω- pentadecalactone and 1,4-cyclohexanedimethanol (1,4-CHDM) at 70°C (Tables 2, 4) (56). These results show that immobilized HiC is capable of mediating a broad range of polymer synthesis reactions.

Conclusion

In recent years, cutinases have been investigated for biotransformations. Developments in stabilization and use in non-traditional media expands their reactive potential. This in conjunction with the identification of new cutinase sequences with different reactivates and stabilities will facilitate the widespread use of cutinases in biotechnology. We anticipate that further understanding of the relationship between sequence, structure and activity will lead to the development of several new biotransformation applications of cutinase.

References

1. Kolattukudy, P. E. *Lipids* **1973**, *8* (2), 90–2.
2. Kunst, L.; Samuels, L. *Curr. Opin. Plant Biol.* **2009**.
3. Pollard, M.; Beisson, F.; Li, Y.; Ohlrogge, J. B. *Trends Plant Sci.* **2008**, *13* (5), 236–46.
4. Riederer, M.; Schreiber, L. *J. Exp. Bot.* **2001**, *52* (363), 2023–32.
5. Purdy, R. E.; Kolattukudy, P. E. *Biochemistry* **1975**, *14* (13), 2824–31.
6. Purdy, R. E.; Kolattukudy, P. E. *Biochemistry* **1975**, *14* (13), 2832–40.
7. Woloshuk, C. P.; Kolattukudy, P. E. *Proc. Natl. Acad. Sci. U.S.A.* **1986**, *83* (6), 1704–1708.
8. Purdy, R. E.; Kolattukudy, P. E. *Arch. Biochem. Biophys.* **1973**, *159* (1), 61–9.
9. Sweigard, J. A.; Chumley, F. G.; Valent, B. *Mol. Gen. Genet.* **1992**, *232* (2), 174–82.
10. Kolattukudy, P. *Ann. Rev. Phytopathol.* **1985**, *23*, 223–50.
11. Fett, W. F.; Gerard, H. C.; Moreau, R. A.; Osman, S. F.; Jones, L. E. *Appl. Environ. Microbiol.* **1992**, *58* (7), 2123–2130.
12. Baker, C. J.; Bateman, D. F. *Phytopathology* **1978**, *68*, 1577–1584.
13. Egmond, M. R.; de Vlieg, J. *Biochimie* **2000**, *82* (11), 1015–21.
14. Pio, T. F.; Macedo, G. A. *Adv. Appl. Microbiol.* **2009**, *66*, 77–95.
15. Martinez, C.; De Geus, P.; Lauwereys, M.; Matthyssens, G.; Cambillau, C. *Nature* **1992**, *356* (6370), 615–8.
16. Liu, Z.; Gosser, Y.; Baker, P. J.; Ravee, Y.; Lu, Z.; Alemu, G.; Li, H.; Butterfoss, G. L.; Kong, X. P.; Gross, R.; Montclare, J. K. *J. Am. Chem. Soc.* **2009**.
17. Nyon, M. P.; Rice, D. W.; Berrisford, J. M.; Hounslow, A. M.; Moir, A. J.; Huang, H.; Nathan, S.; Mahadi, N. M.; Bakar, F. D.; Craven, C. J. *J. Mol. Biol.* **2009**, *385* (1), 226–35.
18. Shayk, M.; Kolattukudy, P. E. *Plant Physiol.* **1977**, *60* (6), 907–915.
19. Sebastian, J.; Chandra, A. K.; Kolattukudy, P. E. *J. Bacteriol.* **1987**, *169* (1), 131–6.

20. Melo, E. P.; Aires-Barros, M. R.; Cabral, J. M. *Appl. Biochem. Biotechnol.* **1995**, *50* (1), 45–56.
21. Borreguero, C.; Carvalho, C. M.; Cabral, J. M. S.; Sinisterra, J. V.; Alcantara, A. R. *J. Mol. Catal. B: Enzym.* **2001**, *11*, 613–622.
22. de Barros, D. P. C.; Fonseca, L. P.; Fernandes, P.; Cabral, J. M. S.; Mojovic, L. *J. Mol. Catal. B: Enzym.* **2009**, *60*, 178–185.
23. Longhi, S.; Czjzek, M.; Lamzin, V.; Nicolas, A.; Cambillau, C. *J. Mol. Biol.* **1997**, *268* (4), 779–99.
24. Maugard, T.; Remaud-Simeon, M.; Monsan, P. *Biochim. Biophys. Acta* **1998**, *1387* (1–2), 177–83.
25. Zaks, A.; Klibanov, A. M. *Science* **1984**, *224* (4654), 1249–51.
26. Kwon, M. A.; Kim, H. S.; Yang, T. H.; Song, B. K.; Song, J. K. *Protein Expression Purif.* **2009**, *68* (1), 104–9.
27. Ollis, D. L.; Cheah, E.; Cygler, M.; Dijkstra, B.; Frolow, F.; Franken, S. M.; Harel, M.; Remington, S. J.; Silman, I.; Schrag, J.; et al. *Protein Eng.* **1992**, *5* (3), 197–211.
28. Nardini, M.; Dijkstra, B. W. *Curr. Opin. Struct. Biol.* **1999**, *9* (6), 732–7.
29. Matthews, B. W.; Sigler, P. B.; Henderson, R.; Blow, D. M. *Nature* **1967**, *214* (5089), 652–6.
30. Huber, R.; Kukla, D.; Bode, W.; Schwager, P.; Bartels, K.; Deisenhofer, J.; Steigemann, W. *J. Mol. Biol.* **1974**, *89* (1), 73–101.
31. Kraut, J. *Annu. Rev. Biochem.* **1977**, *46*, 331–58.
32. Reis, P.; Holmberg, K.; Watzke, H.; Leser, M. E.; Miller, R. *Adv. Colloid Interface Sci.* **2009**, *147–148*, 237–50.
33. Brzozowski, A. M.; Derewenda, U.; Derewenda, Z. S.; Dodson, G. G.; Lawson, D. M.; Turkenburg, J. P.; Bjorkling, F.; Huge-Jensen, B.; Patkar, S. A.; Thim, L. *Nature* **1991**, *351* (6326), 491–4.
34. Anthonsen, H. W.; Baptista, A.; Drablos, F.; Martel, P.; Petersen, S. B.; Sebastiao, M.; Vaz, L. *Biotechnol. Annu. Rev.* **1995** (315–371).
35. Petersen, M. T. N.; Fojan, P.; Petersen, S. B. *J. Biotechnol.* **2001**, *85*, 115–147.
36. Shah, D. B.; Wilson, J. B. *J. Bacteriol.* **1965**, *89*, 949–53.
37. Nicolas, A.; Egmond, M.; Verrips, C. T.; de Vlieg, J.; Longhi, S.; Cambillau, C.; Martinez, C. *Biochemistry* **1996**, *35* (2), 398–410.
38. Frey, P. A.; Whitt, S. A.; Tobin, J. B. *Science* **1994**, *264* (5167), 1927–30.
39. Bruice, T. C. *Annu. Rev. Biochem.* **1976**, *45*, 331–73.
40. Blow, D. M.; Birktoft, J. J.; Hartley, B. S. *Nature* **1969**, *221* (5178), 337–40.
41. Blow, D. M. *Biochem. J.* **1969**, *112* (3), 261–8.
42. Zhang, X.; Houk, K. N. *Acc. Chem. Res.* **2005**, *38* (5), 379–85.
43. Martinez, C.; Nicolas, A.; van Tilbeurgh, H.; Egloff, M. P.; Cudrey, C.; Verger, R.; Cambillau, C. *Biochemistry* **1994**, *33* (1), 83–9.
44. Audit, M.; Barbier, M.; Soyer-Gobillard, M. O.; Albert, M.; Geraud, M. L.; Nicolas, G.; Lenaers, G. *Biol. Cell* **1996**, *86* (1), 1–10.
45. Mannesse, M. L.; Cox, R. C.; Koops, B. C.; Verheij, H. M.; de Haas, G. H.; Egmond, M. R.; van der Hijden, H. T.; de Vlieg, J. *Biochemistry* **1995**, *34* (19), 6400–7.

46. Longhi, S.; Cambillau, C. *Biochim. Biophys. Acta* **1999**, *1441* (2–3), 185–96.

47. Egmond, M. R.; van Bemmel, C. J. *Methods Enzymol.* **1997**, *284*, 119–29.

48. Petersen, S. B.; Jonson, P. H.; Fojan, P.; Petersen, E. I.; Petersen, M. T.; Hansen, S.; Ishak, R. J.; Hough, E. *J. Biotechnol.* **1998**, *66* (1), 11–26.

49. Klibanov, A. M. *Nature* **2001**, *409* (6817), 241–6.

50. Wasserman, A. E.; Hopkins, W. J. *Appl. Microbiol.* **1958**, *6* (1), 49–52.

51. Flipsen, J. A.; van der Hijden, H. T.; Egmond, M. R.; Verheij, H. M. *Chem. Phys. Lipids* **1996**, *84* (2), 105–15.

52. Rogalska, E.; Douchet, I.; Verger, R. *Biochem. Soc. Trans.* **1997**, *25* (1), 161–4.

53. Mannesse, M. L.; de Haas, G. H.; van der Hijden, H. T.; Egmond, M. R.; Verheij, H. M. *Biochem. Soc. Trans.* **1997**, *25* (1), 165–70.

54. Klibanov, A. M. *Anal. Biochem.* **1979**, *93* (1), 1–25.

55. Klibanov, A. M. *Science* **1983**, *219* (4585), 722–727.

56. Hunsen, M.; Abul, A.; Xie, W.; Gross, R. *Biomacromolecules* **2008**, *9* (2), 518–22.

57. Hunsen, M.; Azim, A.; Mang, H.; Wallner, S. R.; Ronkvist, A.; Xie, W.; Gross, R. A. *Macromolecules* **2007**, *40*, 148–150.

58. Costa, L.; Brissos, V.; Lemos, F.; Ribeiro, F. R.; Cabral, J. M. *Bioprocess Biosyst. Eng.* **2009**, *32* (1), 53–61.

59. Costa, L.; Brissos, V.; Lemos, F.; Ribeiro, F. R.; Cabral, J. M. *Bioprocess Biosyst. Eng.* **2008**, *31* (4), 323–7.

60. Sereti, V.; Stamatis, H.; Kolisis, F. N. *Biotechnol. Tech.* **1997**, *11*, 661–665.

61. Figueroa, Y.; Hinks, D.; Montero, G. *Biotechnol. Prog.* **2006**, *22* (4), 1209–14.

62. Garcia, S.; Vidinha, P.; Arvana, H.; Gomaes da Silva, M. D. R.; Ferreira, M.; Cabral, J. M. S.; Macedo, E. A.; Harper, N.; Barreiros, S. *J. Supercrit. Fluids* **2004**, *35*, 62–69.

63. Goncalves, A. P.; Cabral, J. M.; Aires-Barros, M. R. *Appl. Biochem. Biotechnol.* **1996**, *60* (3), 217–28.

64. Ternstrom, T.; Svendsen, A.; Akke, M.; Adlercreutz, P. *Biochim. Biophys. Acta* **2005**, *1748* (1), 74–83.

65. Melo, E. P.; Carvalho, C. M.; Aires-Barros, M. R.; Costa, S. M.; Cabral, J. M. *Biotechnol. Bioeng.* **1998**, *58* (4), 380–6.

66. Melo, E. P.; Costa, S. M.; Cabral, J. M. *Ann. N. Y. Acad. Sci.* **1995**, *750*, 85–8.

67. Brissos, V.; Eggert, T.; Cabral, J. M.; Jaeger, K. E. *Protein Eng., Des. Sel.* **2008**, *21* (6), 387–93.

68. Brissos, V.; Melo, E. P.; Martinho, J. M.; Cabral, J. M. *Biochim. Biophys. Acta* **2008**, *1784* (9), 1326–34.

69. Vidinha, P.; Augusto, V.; Almeida, M.; Fonseca, I.; Fidalgo, A.; Ilharco, L.; Cabral, J. M.; Barreiros, S. *J. Biotechnol.* **2006**, *121* (1), 23–33.

70. Vidinha, P.; Augusto, V.; Nunes, J.; Lima, J. C.; Cabral, J. M.; Barreiros, S. *J. Biotechnol.* **2008**, *135* (2), 181–9.

71. Cabral, J. M.; Aires-Barros, M. R.; Pinheiro, H.; Prazeres, D. M. *J. Biotechnol.* **1997**, *59* (1–2), 133–43.

72. Serralha, F. N.; Lopes, J. M.; Aires-Barros, M. R.; Prazeres, D. M.; Cabral, J. M. S.; Lemos, F.; Ramoa Ribeiro, F. *Enzymes Microb. Technol.* **2002**, *31*, 29–34.

73. Hodneland, C. D.; Lee, Y. S.; Min, D. H.; Mrksich, M. *Proc. Natl. Acad. Sci. U.S.A.* **2002**, *99* (8), 5048–52.

74. Betancor, L.; Luckarift, H. R. *Trends Biotechnol.* **2008**, *26* (10), 566–72.

75. Orlich, B.; Schomacker, R. *Adv. Biochem. Eng. Biotechnol.* **2002**, *75*, 185–208.

76. Melo, E. P.; Baptista, R. P.; Cabral, J. M. S. *J. Mol. Catal. B: Enzym.* **2003**, *22*, 299–306.

77. Carvalho, C. M.; Cabral, J. M. S. *Biochimie* **2002**, *82*, 1063–1085.

78. Stamatis, H.; Xenakis, A.; Kolisis, F. N. *Biotechnol. Adv.* **1999**, *17* (4–5), 293–318.

79. Carvalho, C. M.; Aires-Barros, M. R.; Cabral, J. M. *Biotechnol. Bioeng.* **1999**, *66* (1), 17–34.

80. Sebastiao, M. J.; Cabral, J. M.; Aires-Barros, M. R. *Biotechnol. Bioeng.* **1993**, *42* (3), 326–32.

81. Vermue, M. H.; Tramper, J. *Pure Appl. Chem* **1995**, *67*, 345–373.

82. Goncalves, A. M.; Serro, A. P.; Aires-Barros, M. R.; Cabral, J. M. *Biochim. Biophys. Acta* **2000**, *1480* (1-2), 92–106.

83. Melo, E. P.; Costa, S. M.; Cabral, J. M.; Fojan, P.; Petersen, S. B. *Chem. Phys. Lipids* **2003**, *124* (1), 37–47.

84. Ellerby, L. M.; Nishida, C. R.; Nishida, F.; Yamanaka, S. A.; Dunn, B.; Valentine, J. S.; Zink, J. I. *Science* **1992**, *255* (5048), 1113–5.

85. Eggers, D. K.; Valentine, J. S. *Protein Sci.* **2001**, *10* (2), 250–61.

86. Klibanov, A. M. *Trends Biotechnol.* **1997**, *15* (3), 97–101.

87. Klibanov, A.; Samokhin, G. P.; Martinek, K.; Berezin, I. V. *Biotechnol. Bioeng.* **1977**, *19* (9), 1351–61.

88. Klibanov, A. M. *Trends Biochem. Sci.* **1989**, *14* (4), 141–4.

89. Zaks, A.; Klibanov, A. M. *Proc. Natl. Acad. Sci. U.S.A.* **1985**, *82* (10), 3192–6.

90. Vidinha, P.; Harper, N.; Micaelo, N.; Lourenco, N. M.; Gomaes da Silva, M. D. R.; Cabral, J. M. S.; afonso, C. A.; Soares, C. M.; Barreiros, S. *Biotechnol. Bioeng.* **2003**, *85*, 442–449.

91. Costantino, H. R.; Griebenow, K.; Langer, R.; Klibanov, A. M. *Biotechnol. Bioeng.* **1997**, *53* (3), 345–8.

92. Klibanov, A. M. *Nature* **1995**, *374* (6523), 596.

93. Karmee, S. K.; Casiraghi, L.; Greiner, L. *Biotechnol. J.* **2008**, *3* (1), 104–11.

94. Ramsey, E.; Sun, Q.; Zhang, Z.; Zhang, C.; Gou, W. *J. Environ. Sci. (China)* **2009**, *21* (6), 720–6.

95. Kao, F. J.; Ekhorutomwen, S. A.; Sawan, S. P. *Biotechnol. Tech.* **1997**, *11*, 849–852.

96. Oliveria, M. V.; Rebocho, S. F.; Ribeiro, A. S.; Macedo, E. A.; Loureiro, J. M. *J. Supercrit. Fluids* **2009**, *50*, 138–145.

97. Kamat, S. V.; Iwaskewycz, B.; Beckman, E. J.; Russell, A. J. *Proc. Natl. Acad. Sci. U.S.A.* **1993**, *90* (7), 2940–4.

98. Song, J. H.; Murphy, R. J.; Narayan, R.; Davies, G. B. *Philos. Trans. R. Soc. London, B* **2009**, *364* (1526), 2127–39.

99. Murphy, C. A.; Cameron, J. A.; Huang, S. J.; Vinopal, R. T. *Appl. Environ. Microbiol.* **1996**, *62* (2), 456–60.

100. Maeda, H.; Yamagata, Y.; Abe, K.; Hasegawa, F.; Machida, M.; Ishioka, R.; Gomi, K.; Nakajima, T. *Appl. Microbiol. Biotechnol.* **2005**, *67* (6), 778–88.

101. Ronkvist, A.; Xie, W.; Lu, W.; Gross, R. A. *Macromolecules* **2009**, *42*, 5128–38.

102. Eberl, A.; Heumann, S.; Kotek, R.; Kaufmann, F.; Mitsche, S.; Cavaco-Paulo, A.; Gubitz, G. M. *J. Biotechnol.* **2008**, *135* (1), 45–51.

103. Ronkvist, A.; Lu, W.; Feder, D.; Gross, R. A. *Macromolecules* **2009**, *42*, 6086–97.

104. Gubitz, G.; Paulo, A. *Curr. Opin. Biotechnol.* **2003**, *14* (6), 577–582.

105. Nimchua, T.; Eveleigh, D. E.; Sangwatanaroj, U.; Punnapayak, H. *J. Ind. Microbiol. Biotechnol.* **2008**, *35* (8), 843–50.

106. Araujo, R.; Silva, C.; O'Neill, A.; Micaelo, N.; Guebitz, G.; Soares, C. M.; Casal, M.; Cavaco-Paulo, A. *J. Biotechnol.* **2007**, *128* (4), 849–57.

107. Matama, T.; Vaz, F.; Gubitz, G. M.; Cavaco-Paulo, A. *Biotechnol. J.* **2006**, *1* (7–8), 842–9.

108. Gross, R. A.; Kalra, B.; Kumar, A. *Appl. Microbiol. Biotechnol.* **2001**, *55* (6), 655–60.

109. Gross, R. A.; Kumar, A.; Kalra, B. *Chem. Rev.* **2001**, *101* (7), 2097–124.

Syntheses of Polyesters and Polycarbonates

Chapter 12

Biosynthesis of Polyhydroxyalkanoates from 4-Ketovaleric Acid in Bacterial Cells

Jian Yu*

Hawaii Natural Energy Institute, School of Ocean & Earth Science & Technology, University of Hawaii, Honolulu, Hawaii 96822, USA
*jianyu@hawaii.edu

4-Ketovaleric acid (4KVA) is a promising platform chemical derived from biomass for renewable fuels, chemicals and polymers. Two problems are encountered in microbial synthesis of polyhydroxyalkanoates (PHA) from 4KVA, the toxicity of the organic acid to microbial cells and the composition control of biopolyesters formed from 4KVA. A laboratory strain of *Ralstonia eutropha* that has high metabolic activity on short chain organic acids exhibits high tolerance to 4-ketovaleric acid and can use it as the sole carbon source for cell growth and PHA formation. The biopolyesters formed on different carbon substrates have the similar molecular weight and polydispersity, but different monomeric composition and material properties. A terpolyester with two primary monomers, 3-hydroxybutyrate (3HB) and 3-hydroxyvalerate (3HV), and one minor monomer, 4-hydroxyvalerate (4HV), is formed on 4KVA. The composition varies depending on co-substrates such as glucose and 4KVA. The minor monomer 4HV does not increase with increased supply of precursors, 4-hydroxyvaleric acid or 4-ketovaleric acid. The presence of 4HV in PHA backbone is attributed to random errors of a reductase and a synthase, two key enzymes involved in PHA biosynthesis.

Introduction

With increased concerns on peak oil and global warming, lignocellulosic biomass has attracted renewed interest as a renewable and CO_2-neutral feedstock for production of transportation fuels, chemicals and materials (*1, 2*). Biorefinery, a concept analogous to petroleum refinery, has emerged in bio-based manufacturing where different technologies are used to make various products from biomass for maximized values (*2*). In biomass refining, the complex lignocellulosic matrix must be broken down to small platform chemicals that can be further converted into fuels, chemicals and polymers in a way competitive with petroleum-based manufacturing process. 4-Ketovaleric acid (4KVA), or levulinic acid, is a promising platform chemical for the advanced biomass refining (*3*). It can be derived from hexoses, starch, cellulose, wheat straw and other lignocellulosics via thermal hydrolysis in aqueous solution (*4–8*). The reaction is catalyzed with protons. Because of its two reactive functional groups that allow a great number of synthetic transformations, numerous chemicals can be derived from 4-ketovaleric acid such as 5-aminolevulinic acid, 4-hydroxypentanoic acid, γ-valerolactone (GVL) and 1,4-pentanediol (*3, 9–11*).

4-Ketovaleric acid is also a monomer precursor of biodegradable polyesters that can be formed via chemical or biological synthesis. In chemical synthesis, sequential steps are needed including purification and cyclic ester formation of 4KVA, followed by ring-opening polymerization to poly(4-hydroxyvalerate) (*12, 13*). In biological synthesis, microbial cells can conduct polyester synthesis directly in the aqueous solution of biomass hydrolysates without the expensive purification and cyclic esterification. The formed biopolyesters are polyhydroxyalkanoates (PHA) that are accumulated in the cells for carbon and energy storage. The individual PHA granules (0.2-0.5μm in diameter) can be seen with transmission electron microscope as shown in Figure 1. In contrast to the homopolyester formed in chemical synthesis, the biopolyesters formed from 4-ketovaleric acid are copolymers and the monomeric composition changes with carbon substrates, microbial species and environmental conditions, which finally affects the material properties of biopolyester (*14*). For example, small monomers, such as 3-hydroxybutyrate (3HB) and 3-hydroxyvalerate (3HV), of short-chain length P3HAs result in a stiff material with high crystallinity, high tensile modulus, and low ductility, while large monomers (>C6) of mcl-P3HAs result in an elastic material with low crystallinity, low melting temperature, and high ductility (*14, 15*). The material properties can also be changed by incorporating longer monomers, such as 4-hydroxybutyrate (4HB) and 4-hydroxyvalerate (4HV), in the PHA backbone. For instance, as 4HB content in a co-polyester, poly(3-hydroxybutyrate-co-4-hydroxybutyrate) (P3HB4HB) increases, the crystallinity of the copolymer declines and its ductility increases (*16*).

Few studies have been conducted so far on 4-ketovaleric acid as a carbon source for PHA biosynthesis, probably because of the high toxicity of the acid (*17–20*). Another technical challenge is how to control the monomeric composition of biopolyesters formed on different carbon substrates under specific environmental conditions. Cultivated on glucose and 4-ketovaleric acid, a strain of *Alcaligenes* sp. formed poly(3-hydroxybutyrate-co-3-hydroxyvalerate) (P3HB3HV) and the content of 3-hydroxyvalerate (3HV) depended on the concentrations of glucose and 4KVA. The highest content of 3HV was around 40 mol% (*17*). The cell growth and PHA synthesis were inhibited at acid

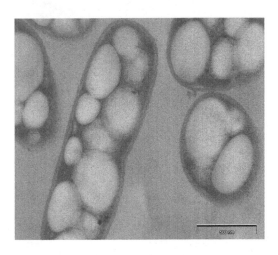

Figure 1. The inclusion bodies of PHA granules stored in bacterial cells. The bar is 500 nm.

concentration of 1 g/L or above, which indicated that the toxicity of 4-ketovaleric acid was actually higher than that of propionic acid, a popular precursor substrate for 3HV in PHA production (*21*). Grown on fructose and 4-ketovaleric acid, a strain of *Ralstonia eutropha* (formerly *Alcaligenes eutropha*) also formed copolyesters of 3HB and 3HV (*18*). The content of 3HV was about 35 mol% or lower, depending on the concentrations of sugar and acid. The cells were inhibited by 4-ketovaleric acid ranging from 0.5 g/L to 8 g/L. The substrate inhibition at a low concentration level (0.5 to 1 g/L) is not feasible to industrial manufacturing because of a high operation cost.

The early studies also observed an interesting phenomenon that no 4-hydroxyvaleric acid (4HV) was incorporated into the PHA backbone even though 4-ketovaleric acid was a precursor of 4-hydroxyvaleric acid. Little is known on how 4-ketovalerate is converted into 3-hydroxyvalerate in PHA biosynthesis. The content of 4-hydroxyvalerate (4HV) has a significant effect on the material properties of biopolyesters (*20*). With a recombinant strain of *Pseudomonas putida* harboring PHA-biosynthesis genes *phaC* and *phaE* of *Thiocapsa pfennigii* grown on 4-ketovaleric acid, a terpolyester, poly(3-hydroxybutyrate-co-3-hydroxyvalerate-co-4-hydroxyvalerate) (P3HB3HV4HV), was formed. The PHA has a high mol% of 4HV (19-30 mol%) and a low mol% of 3HB (0.7-0.8 mol%) (*20*). The biopolyester exhibited the properties of an elastic material such as low crystallinity, slow solidification, high stickiness, and long relaxation time with increased shear frequency. It actually has a poor performance in convential polymer processing such as injection molding and melt spinning. The similar effect has also be observed with 4-hydroxybutyrate (4HB) (*16*). It seems that 4HV can efficiently introduce irregularity into highly stereo-regular PHA matrix, resulting in a reduced crystallinity and increased ductility of short-chain-length PHAs such as P3HB and P3HB3HV (*22*).

Table 1. Cell growth and yield on 4-ketovaleric acid and valeric acid in a nutrient medium

Initial acid (g/L)	Valeric acid		4-Ketovaleric acid	
	Cell density (g/L)	Cell yield (g/g acid fed)	Cell density (g/L)	Cell yield (g/g acid fed)
0*	2.50	-*	2.45	-*
5	3.85	0.27	4.63	0.43
10	6.20	0.37	7.05	0.46
15	5.65	0.21	5.55	0.21
20	2.65	0.01	2.15	0.0

* Note: the controls do not contain organic acids and their cell densities are used as the base for the apparent cell yields from organic acids.

Cell Performance on 4-Ketovaleric Acid

Short-chain organic acids such as acetic, propionic, and valeric acids are metabolic products of microbes under anaerobic conditions, and may inhibit the cells' activity with increased concentration. 4-Ketovaleric acid is not a natural metabolite and may exhibit high toxicity if it is slowly used and accumulated within the cells (23). By using a laboratory strain of *Ralstonia eutropha* that shows high tolerance to short chain organic acids, the cells' metabolic activity on 4-ketovaleric acid and valeric acid are compared. Valeric acid (pentanoic acid) is used by *R. eutropha* and other bacteria as a precursor of 3-hydroxyvaleric acid (3HV) in short chain length PHAs (24). The strain is cultivated in a nutrient-rich medium containing (per liter): 10 g yeast extract, 10 g peptone, 5 g meat extract and 5 g ammonium sulfate. Any possible poor performance of the cells is therefore attributed to the inhibition of organic acids, rather than the limitation of nutrients. Work solutions of organic acids are adjusted to pH 6.8-7.0, if necessary, with 10 M NaOH, and autoclaved before use. Pre-determined amounts of organic acids are added aseptically into the medium (200 mL) in 500 mL baffled flasks. The flask cultures are shaken at 30 °C and 200 rpm on an orbital rotary incubator. The strain is first cultivated in the absence of organic acids for about 24 hours till the cell density reaches about 2 g/L. Aliquots of acid solutions are then added into the flasks for additional 48 hour cultivation under the same conditions. The cells are harvested with centrifugation at 5,000 g for 10 min and the wet pellets are freeze-dried for later use.

Table 1 gives the dry cell mass concentrations and apparent cell yields. The strain exhibits a high metabolic activity on both 4-ketovaleric acid and valeric acid at concentrations of up to 10-15 g/L in comparison with the controls. This acid tolerance is much higher than those reported in the literature (17–19) and is feasible for large scale PHA biosynthesis in industrial medium. Interestingly, 4-ketovaleric acid is a better carbon substrate than valeric acid, as indicated by the cell yield that is determined from the gain of dry cell mass on the amount of organic acid added. The high cell yield of 0.4-0.5 on 4KVA implies that most organic acid

Table 2. The weight-averaged molecular mass (Mw), polydispersity (Mw/Mn) and monomeric composition of PHAs formed on glucose, valeric acid and 4-ketovaleric acid

Substrate	Mw (kDa)	Mw/Mn (-)	3HB (mol%)	3HV (mol%)	4HV (mol%)
Glucose	1,500	2.3	100	0	0
Valeric acid	1,600	2.8	21.8	78.2	0
4-Ketovaleric acid	1,570	2.3	55.5	43.2	1.3

is utilized for cell growth and PHA formation. The inhibitory effect of organic acids is observed at the initial concentration of 15 g/L, and little cell mass and PHA is formed at 20 g/L in comparison with the controls. Both 4-ketovaleric acid and valeric acid can completely inhibit the cell growth at 20 g/L or above.

PHA Biosynthesis on C5 Acids

The content and composition of PHA in the cell mass are determined after methanolysis of the freeze-dried cell mass in methanol (3 wt% sulfuric acid) at 100 °C for 15 to 18 hours. The methyl hydroxyalkanoates are further hydrolyzed into hydroxyalkanoic acids by adding 10 M NaOH solution for HPLC analysis (25). The biopolyesters are also extracted from the freeze-dried cell mass in hot chloroform followed by precipitation with methanol. The molecular weight distribution of biopolyesters is measured, after dissolving the polymers in hot chloroform, with a SEC (Shimadzu) equipped with a RI detector and two Shodex mixed bed K805L columns in series (Showa Denko). The molecular mass is calibrated with narrow-cut polystyrene standards and the weight-average molecular mass are calculated from the distribution curves with SEC software (26). Table 2 gives the analysis results of PHA formed on valeric acid and 4-ketovaleric acid. For comparison, PHA formed on glucose in the same nutrient medium is also provided.

Although the average molecular weight and polydispersity of the biopolyesters formed on different carbon substrates are quite similar, the monomer composition varies. As expected, a homopolyester P3HB with 100 mol% of 3-hydroxybutyrate (3HB) is formed on glucose. From valeric acid, a copolyester P3HB3HV is formed in which 3-hydroxyvalerate (3HV) is the predominant monomer (~78 mol%) and 3HB the minor monomer (~22 mol%). A terpolyester is formed on 4-ketovaleric acid. The two primary monomers are 3HB (~55 mol%) and 3HV (~43 mol%), while 4-hydroxyvalerate (4HV) is a minor monomer (~1 mol%). The content of 4HV in P3HB3HV4HV cannot be increased even when the precursor 4-hydroxyvaleric acid is provided as the sole carbon source (data not shown here). This fact implies that the small amount of 4HV in terpolyesters is not restricted by the supply of precursor 4HV, but by the selectivity of PHA synthase. The key enzyme in PHA biosynthesis prefers 3-hydroxyacyl over 4-hydroxyacyl precursors. The enzyme, however, incorporates a small amount of 4HV into the PHA backbone probably because of random error.

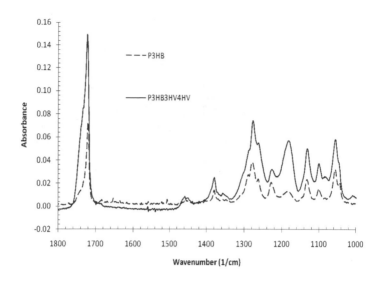

Figure 2. FTIR absorption spectra of P3HB and P3HB3HV4HV.

IR Absorption

The polyesters are dissolved in chloroform, cast into thin films on clean glass surface and aged in ambient conditions for more than 30 days. Figure 2 is the FTIR absorption spectra of P3HB and P3HB3HV4HV formed on glucose and 4-ketovaleric acid, respectively (see Table 2). The spectra show the characteristic IR absorption of chemical bonds, functional groups as well as polymer matrix (*27*). In particular, the IR absorption intensity at wave number 1180 cm-1 is related to the crystallinity of PHA matrix (*27, 28*). A high crystalline gives a low absorption intensity, vice versa (*28*). The IR absorbance at wave number 1380 cm-1 is assigned to the vibration energy of methyl group (-CH$_3$) of biopolyesters and is often used as a reference for a relative absorbance (e.g. ABS$_{1180}$ /ABS$_{1380}$) (*27*).

P3HB is a well-known brittle PHA because of its high stereo-regularity and high crystallinity (~60%). Its relative IR absorbance at 1180 cm-1 (ABS$_{1180}$ /ABS$_{1380}$) is 0.93. In comparison, the relative absorbance of P3HB3HV4HV at 1180 cm-1 is 2.38, indicating a lower crystallinity of the terpolyester. This change in crystallinity is attributed to its high content of 3HV (43 mol%) as well as the irregularity introduced by the minor monomer 4HV. The ethyl group of 3HV has a larger size than methyl group of 3HB and provides to some extent the irregularity of PHA chains, resulting in a lower crystallinity of P3HB3HV. The size effect of ethyl group on PHA crystallization, however, is not very significant because of co-crystallization of 3HB and 3HV (*29*). A small amount of 4HV in PHB backbone may introduce substantial irregularity and hence low crystallinity of P3HB3HV4HV. Indeed, the cast film of terpolyester exhibits a very high elongation at break (~500%) in comparison with P3HB (<5%) and P3HB3HV (<50%).

166

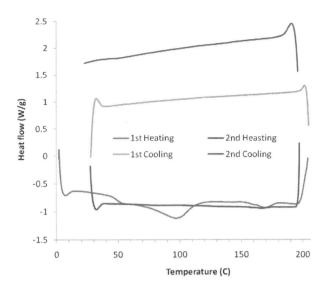

Figure 3. Heating-cooling-heating scans of P3HB3HV4HV (see color insert)

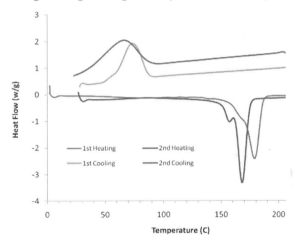

Figure 4. Heating-cooling-heating scans of P3HB (see color insert)

Thermal Analysis

A modulated mode DSC (TA Instruments 2920 Modulated DSC) equipped with a refrigerated cooling system is used for thermal analysis of the biopolyesters. Figures 3 and 4 are the heating-cooling-heating curves of P3HB3HV4HV and P3HB, respectively. In the first heating scan of a terpolyester film, two exothermic events are observed. The first event starts at 43 °C, peaks at 97 °C, and completes at 115 °C. The heat of melting is about 40 J/g. The second event starts at 156 °C, peaks at 169 °C, and completes at 178 °C. The heat of melting is around 4 J/g. The second event is very close to the melting of P3HB and may reflect the presence of a small amount of P3HB in the sample. In the cooling scan, however, no heat of crystallization is measured, and the two exothermic events are not observed in the second heating scan (Figure 3). Most likely, the terpolyester has a low crystallinity as well as slow crystallization rate that is not detected when the scan is performed at ±5 °C/min.

In contrast to the terpolyester, P3HB has a substantial heat of melting (92.8 J/g in the first heating, and 90.7 J/g in the second heating) and hence a high crystallinity (~60%). The exothermic event starts at 149-155 °C, peaks at 157- 178 °C, and completes at 174-185 °C in two heating scans. Estimated with the heat of melting of the second event mentioned above, the amount of P3HB in the terpolyester is less than 5%. P3HB also shows a clear heat of crystallization starting at 87 °C. Both DCS and FTIR measurements show a consistent result that the terpolyester has a much lower crystallinity than P3HB does.

Composition and Metabolic Pathway of Co-Polyesters

Polyester Composition on Co-Substrates

The material properties of PHA are very much dependent on their monomeric composition that can be controlled by using co-substrates. When biopolyesters were synthesized on co-substrates of 4-ketovaleric acid and glucose in nutrient medium, variation in mole composition of 3HB and 3HV is observed (Table 3). With increase in glucose (5 to 15 g/L) and correspondingly decrease in 4-ketovaleric acid (15 to 5 g/L), the content of 3HB increased from 45 to 82 mol% while the 3HV decreased from 53 to 17 mol%. It implies that glucose provides the precursor of 3HB and 4-ketovaleric acid the precursors of both 3HV and 3HB. It is interesting to note that the content of 4HV varies in a small range of 1-2 mol%, indicating that its presence in terpolyesters was more or less dependent on the error of PHA synthase, not restricted by the supply of precursor 4-ketovaleric acid. Even at a low concentration of 4KVA (e.g. 5 g/L), its incorporation into PHA backbones is still dependent on the probability of enzyme error. Except the contents of 3HB and 3HV, the polymer content, molecular mass and polydispersity were quite similar, not influenced very much by the composition of co-substrates.

In a mineral solution, the effect of co-substrates on the composition of copolyesters is not as straightforward as in the nutrient medium above. The change in polymer composition on glucose and 4-ketovaleric acid looks opposite to those found in nutrient medium (Table 4). With increase of 4KVA from 5 to 10 g/L and correspondingly decrease of glucose from 10 to 5 g/L, the 3HV content actually declines from 14 to 9 mol% while 3HB increases from 83 to 87 mol%. This phenomenon may be attributed to the higher toxicity of 4KVA in the mineral solution than in nutrient medium (data not shown here), which results in the reduced utilization of 4KVA at high acid concentration. Because of possible nutrient limitation in the mineral medium, the cells accumulate more PHA (52-62 wt%) than the low PHA content (27-36 wt%) formed in nutrient medium. The high PHA content was most likely contributed by 3HB derived from glucose, and hence the content of 3HV is diluted. It is interesting to note that the content of 4HV formed in the mineral media is in line with those observed in nutrient

Table 3. Weight-averaged molecular mass (Mw), polydispersity (Mw/Mn) and monomeric composition of PHA formed on co-substrates of 4-ketovaleric (4KVA) and glucose in nutrient-rich medium

4KVA (g/L)	Glucose (g/L)	Mw (kDa)	Mw/Mn (-)	3HB (mol%)	3HV (mol%)	4HV (mol%)
15	0	1,530	1.99	53.5	45.3	1.2
15	5	1,800	2.26	45.3	53.0	1.7
10	10	1,530	2.34	75.6	21.7	2.7
5	15	1,610	1.79	81.8	17.1	1.1

Table 4. PHA biosynthesis on 4KVA and glucose in a mineral solution*

4KVA (g/L)	Glucose (g/L)	PHA (wt%)	3HB (mol%)	3HV (mol%)	4HV (mol%)
5	10	62.2	83.7	13.8	2.5
7.5	7.5	58.7	85.6	10.6	3.8
10	5	51.6	87.4	9.4	3.2

* Mineral solution (per liter) contains: 2.8 g K_2HPO_4, 2 g NaH_2PO_4, 0.5 g $MgSO_4.7H_2O$, 0.5 g $NaHCO_3$, 1 g $(NH_4)_2SO_4$, 0.05 g ferric ammonium citrate and 1 mL trace solution (*30*). After inoculated, the cultures of 200mL are shaken in 500 mL baffled flasks for 48 hours at 30 °C and 200 rpm.

medium: small amount of 4HV in the presence of 4-ketovaleric acid while no 4HV is formed on valeric acid (Table 2).

Metabolic Pathway of PHA Synthesis

A possible metabolic pathway of PHA biosynthesis from 4-ketovaleric acid and co-substrates is presented in Figure 5. The reaction mechanism of PHA formation from glucose and valeric acid are well documented (*14, 31*). In brief, glucose is converted into two pyruvates via glycolysis and further into 3-hydroxybutyryl-CoA with a β-ketothiolase (*phaA*) and a NADPH-dependent reductase (*phaB*). The 3HB precursor is incorporated into a growing PHA backbone with a PHA synthase (*phaC*) or polymerase. A two-site working model of PHA synthase for chain elongation via transesterification is presented in Figure 6 (*32*). As expected, a homopolyester, P3HB, is formed from glucose. Valeric acid, after activation with CoA, is converted to 3-ketovaleryl-CoA via β–oxidation, and reduced to 3-hydroxyvaleryl-CoA, a precursor of 3HV. Similarly, it is incorporated with the PHA synthase into a growing PHA backbone as shown in Figure 6. Under the conditions with little PHA formation for carbon storage, the β–oxidation primarily generates acetyl-CoA and propionyl-CoA that can be used by cells for growth and energy supply. The ketothiolase, an enzyme responsible for the split reaction, may also be responsible for condensation of acetyl-CoA and propionyl-CoA into 3-ketovaleryl-CoA (*33*). At least three β-ketothiolases with different substrate specificity have been identified in *R.*

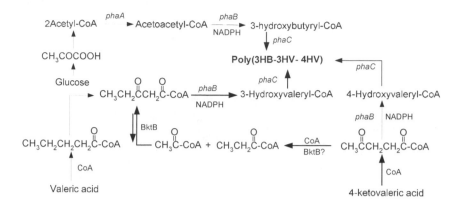

Figure 5. A possible metabolic pathway of PHA biosynthesis from co-substrates 4-ketovaleric acid, glucose and valeric acid. Key enzymes: β-ketothiolase (phaA), NADPH-dependent acetoacetyl-CoA reductase (phaB), PHA synthase (phaC), Co-enzyme A (CoA), β-ketothiolase B (BktB).

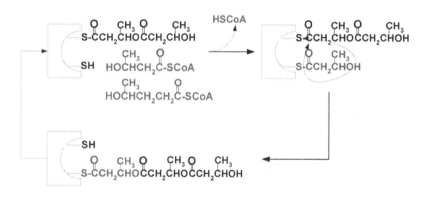

Figure 6. A two-site working mode of PHA synthase with chain elongation via transesterification. The synthase prefers 3-hydroxyacyl-CoA to 4-hydroxyacyl-CoA that is occasionally incorporated into the backbone because of enzyme's error. (see color insert)

eutropha. The reversible reaction with ketothiolases explains the formation of P3HB3HV with a low mol% of 3HB from valeric acid, as well as the formation of P3HB3HV with a moderate mol% of 3HV from acetic acid and propionic acid (*30*).

After a free 4-ketovaleric acid is activated with CoA, it may have two fates. It may be reduced to 4-hydroxyvaleryl-CoA by a NADPH-dependent acetoacyl-CoA reductase (phaB) and then incorporated into the PHA backbone by PHA synthase (phaC). Since the PHA synthase prefers 3-hydroxyvaleryl-CoA, a low carbon flow though this pathway has been observed and described in Figure 6. The selectivity of PHA synthase is confirmed by the fact that the mol% of 4HV in PHA backbone is not increased on 4-hydroxyvaleric acid (Table 3). Furthermore, the activity of reductase on the reduction of 4-ketovaleric acid may also not be high because no

4-hydroxyvaleric acid is accumulated. By random errors of these two enzymes, however, a small amount of 4HV appears in PHA backbone.

The second fate of 4-ketovaleric acid must have a major carbon flux for cell growth and PHA biosynthesis as shown in Table 1. There are a few possible initial reactions in using 4-ketovaleryl-CoA as the sole carbon source. In one reaction, it might be converted to 3-ketovaleryl-CoA via reduction and oxidation on γ and β carbons. This biochemical transformation, however, has a low possibility because the enzymatic activity on γ carbon (C4) is quite low as discussed above. Since very different contents of 3HB and 3HV are formed on 4-ketovaleric acid and valeric acid, respectively, as shown in Table 2, 3-ketovaleryl-CoA is not the first intermediate in the conversion of 4-ketovaleric acid. In a second possible reaction, 4-ketovaleryl-CoA may be converted into pyruvate and acetyl-CoA via β-oxidation (*14*). This reaction, however, would result in a very low 3HV content because of the strong oxidative decarboxylation of pyruvate to acetyl-CoA in this obligate aerobe (*30*). In a most possible reaction, 4-ketovaleryl-CoA is split into acetyl-CoA and propionyl-CoA with β-ketothiolase B (BktB) (*33*). The formed acetyl-CoA and propionyl-CoA could be re-condensed into 3-ketovaleryl-CoA for 3HV as happened with valeric acid. Two acetyl-CoA may also be converted to 3HB, which explains more 3HB and less 3HV are formed from 4-ketovaleric acid than from valeric acid. More importantly, acetyl-CoA and propionyl-CoA can easily enter the main metabolic pathways of cell growth and maintenance (*30*), explaining the high cell yield obtained on 4KVA as the sole carbon source. This pathway will be clarified in future work.

Conclusion

Bacterial strains that exhibit high metabolic activity on short chain organic acids can use 4-ketovaleric acid as a sole or co-substrate carbon for cell growth and biopolyester synthesis. The cells perform very well showing fast growth and high PHA yield at acid concentration levels feasible for industrial manufacturing. The terpolyester P3HB3HV4HV formed on 4-ketovaleric acid has a low crystallinity, low melting point and high ductility in comparison with homopolyester P3HB and copolyester P3HB3HV. 4-Hydroxyvalerate is a minor monomer of terpolyester, and may have a significant effect on the material properties of biopolyester. The molar composition and hence the material property of biopolyesters can therefore be changed or controlled by cultivating the cells on co-substrates.

References

1. Kamm, B.; Kamm, M.; Gruber, P. R.; Kromus, S. In *Biorefineries-industrial processes and products: status quo and future directions*; Kamm, B., Gruber P. R., Kamm, M. Wiley-VCH: Weinheim, 2006; *Vol.* 1, pp 3–40.
2. Yu, J.; Chen, L. *Environ. Sci. Technol.* **2008**, *42*, 6961–6966.
3. Bozell, J. J.; Moens, L.; Elliott, D. C.; Wang, Y.; Neuenscwander, G. G.; Fitzpatrick, S. W.; Bilski, R. J.; Jarnefel, J. L. *Resour., Conserv. Recycl.* **2000**, *28*, 227–239.

4. Girisuta, B.; Janssen, L. P. B. M.; Heeres, H. J. *Chem. Eng. Res. Des.* **2006**, *84* (A5), 339–349.

5. Cha, J. Y.; Hanna, M. A. *Ind. Crops Prod.* **2002**, *16*, 109–118.

6. Girisuta, B.; Janssen, L. P. B. M.; Heeres, H. J. *Ind. Eng. Chem. Res.* **2007**, *46*, 1696–1708.

7. Chang, C.; Cen, P.; Ma, X. *Bioresour. Technol.* **2007**, *98*, 1448–1453.

8. Girisuta, B.; Danon, B.; Manurung, R.; Janssen, L. P. B. M.; Heeres, H. J. *Bioresour. Technol.* **2008**, *99*, 8367–8375.

9. Sasaki, K.; Tanaka, T.; Nishizawa, Y.; Hayashi, M. *J. Ferment. Bioeng.* **1991**, *71*, 403–406.

10. Martin, C. H.; Prather, K. L. J. *J. Biotechnol.* **2009**, *139*, 61–67.

11. Hayes, D. J.; Fitzpatrick, S. W.; Hayes, M. H. B.; Ross, J. R. H. In *Biorefineries – industrial processes and products: status quo and future directions*; Kamm, B.; Gruber, P. R. Kamm, M., Eds.; Wiley-VCH: Weinheim, 2005; *Vol.* 1, pp 139–164.

12. Manzer, L. E. *Appl. Catal., A* **2004**, *272*, 249–256.

13. Saiyasombat, W.; Molloy, R.; Nicholson, T. M.; Johnson, A. F.; Ward, I. M.; Poshyachinda, S. *Polymer* **1998**, *39* (23), 5581–5585.

14. Sudesh, K.; Abe, H.; Doi, Y. *Prog. Polym. Sci.* **2000**, *25*, 1503–1555.

15. Kellerhals, M. B.; Kessler, B.; Witholt, B.; Tchouboukov, A.; Brandl, H. *Macromolecules* **2000**, *33*, 4690–4698.

16. Zhu, Z.; Dakwa, P.; Tapadia, P.; Whitehouse, R. W.; Wang, S.-Q. *Macromolecules* **2003**, *36*, 4891–4897.

17. Jang, J.-H.; Rogers, P. L. *Biotechnol. Lett.* **1996**, *18* (2), 219–224.

18. Chung, S. H.; Choi, G. G.; Kim, H. W.; Rhee, Y. H. *J. Microbiol.* **2001**, *39* (1), 79–82.

19. Keenan, T. M.; Tanenbaum, S. W.; Stipanovic, A. J.; Nakas, J. P. *Biotechnol. Prog.* **2004**, *20*, 1697–1703.

20. Gorenflow, V.; Schmack, G.; Vogel, R.; Steinbuchel, A. *Biomacromolecules* **2001**, *2*, 45–57.

21. Du, G.; Chen, J.; Yu, J.; Lun, S. *Biochem. Eng. J.* **2001**, *8*, 10–108.

22. Saito, Y.; Nakamura, S.; Hiramitsu, M.; Doi, Y. *Polym. Int.* **1996**, *39*, 169–174.

23. Yu, J.; Wang, J. *Biotechnol. Bioeng.* **2001**, *73* (6), 458–464.

24. Page, W.J.; Manchak, J.; Rudy, B. *Appl. Environ. Microbiol.* **1992**, *58* (9), 2866–2873.

25. Yu, J.; Plackett, D.; Chen, L. X. L. *Polym. Degrad. Stab.* **2005**, *89*, 289–299.

26. Yu, J.; Chen, L. X. L. *Biotechnol. Prog.* **2006**, *22*, 547–553.

27. Bloembergen, S.; Holden, D. A.; Hamer, G. K.; Bluhm, T. L.; Marchessault, R. M. *Macromolecules* **1986**, *19*, 2865–2871.

28. Xu, J.; Guo, B. H.; Yang, R.; Wu, Q; Chen, G. Q.; Zhang, Z. M. *Polymers* **2002**, *43* (25), 6893–6899.

29. Doi, Y.; Kitamura, S.; Abe, H. *Macromolecules* **1995**, *28*, 4822–4828.

30. Yu, J.; Si, Y. *Biotechnol. Prog.* **2004**, *20*, 1015–1024.

31. Madison, L. L.; Huisman, G. W. *Microbiol. Mol. Biol. Rev.* **1999**, *63*, 21–42.

32. Stubbe, J.; Tian, J. *Nat. Prod. Rep.* **2003**, *20*, 445–457.

33. Slater, S.; Houmiel, K. L.; Tran, M.; Mitsky, T. A.; Taylor, N. B.; Padgette, S. R.; Gruys, K. J. *J. Bacteriol.* **1998**, *180*, 1979–1987.

Chapter 13

Synthesis of Functional Polycarbonates from Renewable Resources

Kirpal S. Bisht[1,*] and Talal F. Al-Azemi[2]

[1]Department of Chemistry, University of South Florida, 4202 East Fowler
Avenue, Tampa, Florida 33620, USA
[2]Department of Chemistry, Kuwiat University, PO Box 5969, Safat 13060,
Kuwait
*kbisht@cas.usf.edu

Enantiomerically pure functional polycarbonates were
synthesized from a novel seven-membered-cyclic carbonate
monomer derived from naturally occurring L-tartaric acid in
three steps. The polymerization and copolymerization with
ε-CL were investigated using enzyme and chemical catalysts.
Immobilized *Candida antarctica* lipase -B (Novozym-435)
was found to be the most efficient lipase catalyst to carry out
the ring opening polymerization of ITC. Novozym-catalyzed
polymerizations led to formation of homopolymers (Mn=15500
g/mol) and suggested that during the polymerization the chain
transfer reactions were minimal and it showed characteristic of
non-terminating chain polymerization. Of the three chemical
catalysts, $Sn(Oct)_2$ was found to be the most efficient catalyst
(poly(ITC), M_n= 23000-26000 g/mol; PDI= 1.6; $[\alpha]_D^{20}$=+77.8)
and followed first order rate law. Optically active AB block
copolymers with various feed ratios were also synthesized
by 'one-shot feeding' of ITC with ε-CL. The deprotection of
the ketal groups led to polycarbonates with pendant hydroxy
groups with minimal degradation in the polymer chain.

Introduction

The commercial thermoplastic polymers are mostly hydrophobic and
biologically inert and their non-biodegradability is a major drawback, because

of the many environmental problems associated with their disposal. The growth in the use of synthetic polymers has a remarkable parallel to the growth of solid waste. Beside solution such as incineration and recycling; much research effort has been devoted to the design and development of renewable, functional and biodegradable polymers. Of the different natural resources, carbohydrates because of their natural abundance and functional diversity stand out as highly convenient raw material. In recent years, many carbohydrates and amino acids based polymers from renewable resources have been reported in the literature (1, 2). The key advantage to such polymers is that these can be converted, by microbial activity in a biologically active environment, to biomass and biocompatible products upon disposal. The environmental potential of polymers from renewable plant sources which are carbon neutral cannot be understated. The utility of the biocompatible polymers is also highlighted in biomedicine for degradable scaffoldings and drug delivery applications.

Poly lactic acid (PLA) is one of the most commercially viable biodegradable polymer, produced from lactic acid. PLA has good mechanical and physical properties and is prodcuced on large commerciall scale. PLA has found use in disposable consumer products and in biomedical applications such as sutures, stents, bone screws, and in long term delivery of drugs. However, despite its many applications, PLA is unfunctionalized and hydrophobic and hence has limitation as to designing 'on-demand' degradation and targeted drug delivery applications. It is, therefore, important to develop degradable polymers which can be functionalized with biologically relevant molecules and in which the degradability can be 'dialed-in'. It is well-known that incorporation of hydrophilic functional groups enhances the biodegradability of the polymers and hydroxyl (3), amine (4), and carboxyl (5) pendent functional groups have been reported. Water soluble poly(hydroxyalkylene carbonate)s and the polycarbonate based on 1,4:3,6-dihydrohexitrols and L-tartaric acid derivatives are examples, which are reported to exhibit high biodegradability in vitro and vivo hydrolysis (6).

L-tartaric acid is an optically pure and relatively inexpensive natural resource widely available from a large variety of fruits. It has been extensively utilized in organic synthesis as a source of chirality (7). Owing to the presence of two secondary hydroxyls and two carboxylic acids groups L-tartaric acid has been used, although somewhat limited, in synthesis of polyamide (8), polyesters (9), polyurethanes (10), and polycarbonate (6). Interestingly, the incorporation of tartaric acid or its derivative in degradable polymers has only been through condensation polymerization. The condensation polymerization is limited by the removal of the condensate, i.e., water or alcohol and efficient removal of the condensate is required to shift the equilibrium to the polymerization. Therefore, condensation polymerizations are driven by the use of vacuum, high temperature or gas to remove the condensate. However, the monomers are often thermally unstable and side reactions such as dehydration or decarboxylation are observed in aliphatic polymers. The ROP of cyclic monomers is more efficient as no leaving group is involved and unlike condensation polymerization, can be performed at much lower temperature and is hence energy efficient. Synthesis of a cyclic monomer derived from tartaric acid or its ring opening polymerization has not been reported.

176

Aliphatic polycarbonates have attracted much attention as biodegradable biomedical, and nontoxic material. Homopolymers and copolymers of trimethylene carbonate (TMC) and 5,5-dimethyl trimethylene carbonate (DTC), have been widely used in drug delivery, soft tissue implantation, and tissue regeneration (*11–15*). Although aliphatic ring-opening polymerization of larger ring size carbonates have been reported in the literature (*16, 17*), five and six membered cylic carbonate monomers are the most commonly polymerized via ring-opening polymerization (*18, 19*). There are only a few examples of polymerization of a seven-membered cyclic carbonates. The homopolymerization and copolymerization of 1,3-dioxepan-2-one (7CC), a seven-membered cyclic carbonate monomer, with valerolactone (VL), ε-Caprolactone (ε-CL) has been reported by both anionic and cationic catalysts ((*16a,b*), (*20*)). Various modification strategies have been employed to enhance the hydrophilicity of aliphatic polycarbonates to improve their biodegradation rates (*19*). Copolymerization and graft polymerization with other polar groups is an obvious approach. For example, amphiphilic copolymers [poly(PEG-*b*-DTC) and poly(PEG-*b*-TMC)] synthesized by the polymerization of DTC and TMC with the hydroxyl end group of methoxy-terminated poly(ethylene glycol) (PEG) showed much higher weight loss in phosphate buffer compared to the polycarbonate homopolymers of DTC and TMC, presumably because the hydrophilic PEG segments promoted water permeation into the copolymer matrix (*17*). Alternatively, introduction of hydroxyl or carboxylgroups not only increase the hydrophilicity and accelerate the degradation rates but also be conducive to post-polymerization modification (*5, 21*).

An alternative approach to chemical polymerization is to use biocatalysts, such as lipases, to catalyze the ring opening of cyclic monomers. This approach is especially useful as enzymes are versatile catalysts with demonstrated ability to carry out a wide range of transformation such as the polymerization of ring systems which are otherwise difficult to polymerize by conventional catalysis ((*5a,b*), (*18*)). Enzyme catalyzed ring opening polymerization of cyclic carbonates has also been studied and the polymerization is known to proceeds without any decarboxylation. Our earlier results on lipase catalyzed polymerization of 5-methyl-5-benzyloxycarbonyl-1,3-dioxan-2-one (MBC) and 5-methyl-5-carboxyl-1,3-dioxan-2-one (MCC) led to the first example of pendant carboxyl group polycarbonates. Random copolymers of trimethylene carbonate (TMC) were also synthesized with MBC and MCC using lipase catalyzed ROP (*5*).

In this chapter we summarize our efforts (*22–24*) towards the synthesis, homopolymerization and copolymerization of a new optically pure functional seven-membered carbonate monomer derived from naturally occurring L-tartaric acid. As an extension of our ongoing research efforts on biodegradable polymer synthesis based on renewable resources, novel polycarbonates containing pendant hydroxyl group has now been synthesized from homopolymerization and copolymerization of (5*S*,6*S*)-Dimethyl 5,6-*O*-isopropylidene-1,3-dioxepin-2-one (ITC, **3**). The ITC monomer was synthesized in three steps starting from L-tartaric acid. The ROP of the ITC monomer was investigated using enzymatic and chemical catalyst. Together with the use of renewable resources, the utilization of the enzyme-catalyzed ROP brings a "green-chemistry" appeal to this report. Copolymerization of ITC with ε-caprolactone was carried out using stannous octanoate [Sn(Oct)$_2$] and diblock copolymers with different feed ratios were synthesized, importantly in one shot feeding. The homo-and-copolymers were characterized by detail spectral and thermal analyses. Deprotection of the ketal groups resulted in optically active polycarbonates with free hydroxyl groups in the polymer backbone.

Results and Discussion

Monomer Synthesis

Enantiomerically pure seven membered-cyclic carbonate monomer, dimethyl 5,6-O-isopropylidene-1,3-dioxepin-2-one (ITC, **3**) was prepared from L-tartaric acid in three steps (Scheme 1).

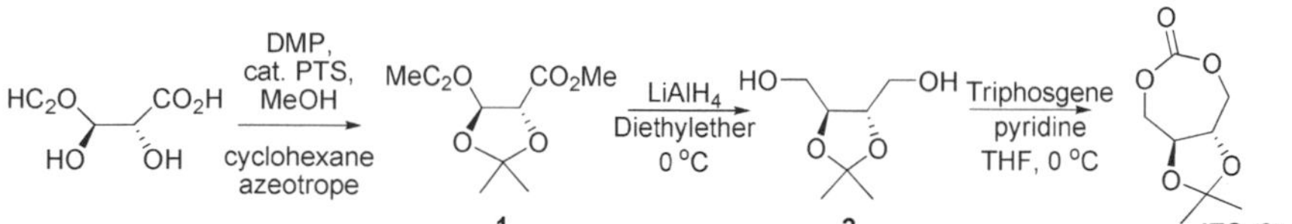

Scheme 1. Synthesis of seven-member cyclic carbonate monomer from L-Tartaric acid

The commercially available L-tartaric acid was treated with 2,2-dimethoxy propane in methanol in presence of catalytic amount of PTSA to result in formation of the ketal diester, **1**. The subsequent reduction of the methyl diester groups in **1** using LAH led to the diol **2** which was cyclized using triphosgene in presence of pyridine to yield (57%) the 5,6-O-isopropylidene-1,3-dioxepin-2-one (ITC, **3**, Figure 1). The monomer was recrystallized from hexane: dichloromethane (1:10) to yield colorless crystals; mp 75 °C, $[\alpha]_D^{20}$ =+84.79°, HRESIMS m/z [M+H]$^+$= Calcd. for C$_8$H$_{13}$O$_5$: 189.07630. Found 189.07593).

Enzymatic Ring-Opening Polymerization

The results obtained for the enzymatic polymerizations screen at 80 °C in bulk are listed in Table 1. Importantly, no polymerization was observed under the identical conditions in absence of the lipase or when a thermally deactivated lipase was used, which indicates the ROP of ITC was catalyzed by the active lipase. The four commercially available lipases screened for their ability to polymerize the ITC monomer (**3**) showed considerable variation in monomer conversion and molecular weight. For example, lipase from *Candida antarctica* (CAL-B immobilized on acrylic resin- Novozyme-435), *Pseudomonas fluorescens* (AK), and *Pseudomonas cepacia* (PS-30) accepted ITC as substrate but porcine pancreatic lipase (PPL) showed no activity and the monomer was recovered. Lipase AK showed the highest monomer conversion, though, with low molecular weights polymers, Mn= 1800-2300 g/mol and polydispersities of 1.5-1.6. Lipase PS-30 showed 72% monomer conversion with moderate Mn= 9500 after 24 h. With Novozyme-435, 88% monomer conversion was obtained after 24h with Mn = 15500 g/mol and polydispersity of 1.7. Therefore, the ring opening polymerization of the cyclic carbonate monomer ITC was further investigated using 50 wt % of the Novozym-435 at 80 °C in bulk.

In Figure 2 percent ITC monomer conversion as a function of polymerization time is plotted. The ITC monomer conversion reached 88% at 24h and increased slightly to 91% at 48 h. The slow increase in monomer conversion, from 88% - 91%, in 24h was attributed to the increased viscosity in the bulk polymerization. At high monomer conversion, especially in bulk polymerization, the increased viscosity of the reaction mixture limits monomer access to the lipase active site (25). We have observed similar phenomenon during bulk polymerization of MBC (5).

The absolute molecular weight of the poly(ITC) was measured using ^{1}H-NMR and compared to those measured by GPC calibrated with polystyrene standards. A very good agreement, within experimental error, was found between the molecular weight measured by the two techniques e.g., Mn 16000 g/mol by ^{1}H-NMR and 15500 g/mol by GPC. Therefore, further measurements were obtained using the more convenient GPC method. The number-average molecular weight (Mn) of the polymer showed a linear increase with percent monomer conversion (Figure 2), suggesting a fast initiation process and in which the propagating reactive centers are non-terminating. The polydispersity index (PDI) increased from 1.4 to 1.8 at 50% monomer conversion, suggesting new chain initiation events along with chain propagation during the initial stages of the polymerization. Importantly, beyond 50% monomer conversion, the PDI remained unchanged (within experimental error) with increasing polymer molecular weight suggesting no new initiation events and only chain propagation. Although, the transesterification reactions among polymer chains have been observed during the lipase catalyzed polymerizations (25), the linear increase in Mn with percentage monomer conversion suggests that the chain transfer reactions occurring during the polymerization, if any, were minimal.

The plot of $-\ln([M]/[M]_0)$ as a function of reaction time is shown in Figure 3. The correlation coefficient (R^2) of 0.973 from linear regression analysis shows linearity, which suggests that Novozym-435 catalyzed ROP of ITC monomer follows first-order rate law. This further confirms that throughout the chain propagation the number of growing chains is constant and termination is low and that ROP catalyzed by Novozym-435 show characteristic of non-terminating chain polymerization. The apparent rate constant (K_{app}) for the polymerization was found to be 5.29×10^{-2} h^{-1}.

Chemical Ring-Opening Polymerization of ITC

Three known catalysts, Sn(Oct)$_2$, Al(OiPr)$_3$ and ZnEt$_2$-H$_2$O, were also tested for the ring-opening polymerization of monomer (ITC, **3**). The screening homopolymerizations were investigated in bulk at 120 °C for 12 h (Table 2). The monomer-to-catalyst ratio (M/C) was varied to test the efficiency of the catalysts.

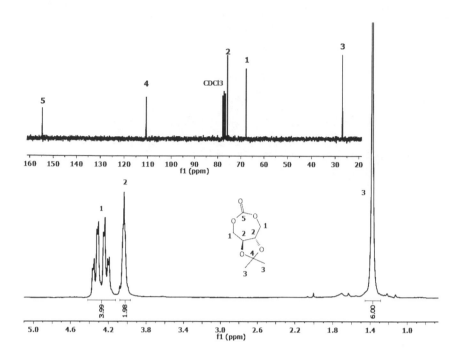

Figure 1. ¹H-NMR spectrum of (ITC, 3) monomer. Inserted in the figure ¹³C-NMR spectrum (62.9 MHz, CDCl₃).

Both Sn(Oct)₂, and ZnEt₂-H₂O catalysts showed 100 % monomer conversion in 24h. The monomer conversions for the polymerizations catalyzed by Al(OiPr)₃ were lower. The number-average molecular weight (Mn) was dependent upon the catalytic system used. In general, Sn(Oct)₂ catalyzed polymerizations had the highest Mn = 26000 g/mol and higer monomer conversion at all M/C ratio used. Therefore it was chosen for further investigation of the copolymerization of ITC and CL.

The monomer conversion vs the reaction time plot (Figure 4) for Sn(Oct)₂ catalyzed polymerization of ITC at 120 °C (M/C= 200) shows that the conversion increased linearly with time and reached 89% conversion by 4h; after 6h all of the ITC monomer was consumed. The molecular weight increased linearly with monomer conversion (Figure 4) and the correlation coefficient (R^2) of 0.99 suggests that no transfer reaction occurred during the course of the polymerization. The narrow molecular weight distribution advocates absence of chain initiation and transfer reactions occurring after 23% monomer conversion and that the main event is chain propagation. A plot of ln([M]₀/[M]) as a function of polymerization time (Figure 5, $R^2 = 0.9933$) suggested a first order rate law for Sn(Oct)₂ catalyzed the polymerization of ITC in bulk at 120 °C. These results are in agreement with general ROP mechanism of non-terminating chain polymerization. The apparent rate constant of propagation step was calculated, $K_{app}= 1.05 \times 10^{-3}$ s^{-1}.

Table 1. Enzymatic Ring-Opening Polymerization of ITC (3) in Bulk at 80 °C

Entry	Lipase [a]	Time (h)	Conversion [b] (%)	Mn [c] (g/mol)	Mw/Mn [c]
1	Novozyme-435	12	51	10000	1.8
2		24	88	15500	1.7
3	AK	12	80	1800	1.6
4		24	97	2300	1.5
5	PS	12	40	5500	1.7
6		24	72	9500	1.6
7	PPL	12	NR[d]	NA[e]	NA
8		24	NR	NA	NA

[a] Enzyme (source): Novozym-435 (Candida antarctica), AK (Pseudomonas fluorescens), PS-30 (Pseudomonas cepacia), and PPL (porcine pancreas). [b] Determined from ^{1}HNMR. [c] Determined from GPC. [d] no reaction (monomer recovered). [e] not applicable.

NMR Characterization of the Poly(ITC)

The structure of the poly(ITC) was analyzed from its ^{1}H- and ^{13}C-NMR (Figure 6) spectral data. The optical rotation measurement for the polymer was $[\alpha]_D^{20} = +77.8$. The absence of any diastereotopic resonances in its ^{13}C- NNMR spectrum suggested complete retention of its absolute stereochemistry.

The assignments in Figure 6 were based on comparison with NMR spectra of the monomer, DEPT-135, two dimensional ^{1}H-^{1}H COSY and ^{1}H-^{13}C HSQC experiments. The acetonide dimethyl hydrogens (H-8) were observed at 1.43 ppm; the repeat unit methylenes (H-6), owing to their diastereotopic relationship, were observed as 4.51 and 4.20 ppm; the methine (H-7) were at 4.20 ppm. End groups H-6′, H-6″ and H-7′ were observed at 3.72, 4.0, and 3.82 ppm, respectively. Interestingly, in the HSQC experiment (Figure 7), the resonance at 67.5 (C-6) ppm correlated to two proton signals, confirming that H6 hydrogens at 3.72 and 4.0 ppm, were diastereotopic. No proton resonances were found around 3.5 ppm for the ether linkages, which indicated that no decarboxylation in the main chain occurred during the polymerization. In the ^{13}C-NMR spectrum, the acetonide carbon (C-8) was at 28.7 ppm, the C-6 was at 67.5 ppm and the C-7 was at 77.2 ppm. The end group carbons, C-6′ and C-7′ were at 64.8 and 75.4 ppm, respectively.

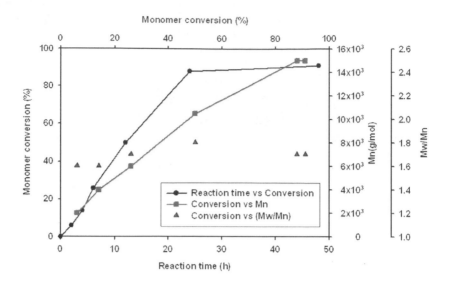

Figure 2. Plots of Novozym-435 catalyzed ring-opening polymerization at 80 °C in bulk for 48h.

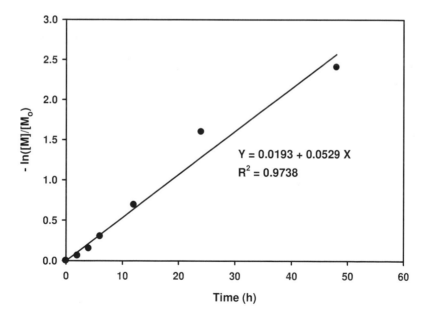

$$Y = 0.0193 + 0.0529\ X$$
$$R^2 = 0.9738$$

Figure 3. First-order kinetic plot for Novozyme-435 catalyzed ROP of ITC monomer at 80 °C for 48h. $[M]_0$ is the initial monomer concentration, and $[M]$ is the monomer concentration at time (t). The monomer conversions were calculated from the ^1HNMR spectra.

One-Shot Block Copolymerization

Copolymers of ε-caprolactone with carbonates provide opportunities for development of interesting biomaterial for their biocompatibility and bioresorbability. Because the ester bonds are more sensitive to hydrolysis than carbonates, the copolymerization of ε-caprolactone with carbonates enables tuning of the physical properties of the copolymers and thereby facilitates access to the tailor made biopolymers.

Typically copolymerization of a mixture of two monomers results in a random copolymer and sometime an alternating copolymer or mixture of two homopolymer is also formed. However, formation of block polymer has also been reported mostly through the sequential addition of the monomers that polymerize via living or controlled polymerization mechanism (27). An interesting approach to block copolymer synthesis known as 'one-shot feeding', in which the two monomers are fed together (28, 29), is advantageous as the process is much simpler compared to the sequential monomer feeding. However, dissimilar monomers that are polymerized by fundamentally different chemistries (e.g. ATRP, cationic or anionic ROP, and free radical polymerization) have only been reported. For example, there have been no reports of one shot block copolymerization of cyclic carbonates or lactones, which polymerize following similar chemistries. Importantly, in our case, copolymerization of ITC with ε-caprolactone led to diblock copolymers with different feed ratios in 'one-shot feeding'.

Copolymerization of ITC and ε- CL was carried out by $Sn(Oct)_2$ in bulk at 120 °C for 12h (Table 3). The copolymer was obtained in 92% isolated yield after 12h, when the feeding molar ratio of the monomers is 50:50 (Table 3, entry 4). All the copolymers had a unimodal molecular weight distribution with polydispersity index ranging from 1.4 to 1.6 indicating the polymers obtained were pure copolymers without homopolymers of ε-caprolactone and ITC. The number-average molecular weights (M_n) were in the range of 23000-26000 g/mol. Interestingly, the ^{1}H- and ^{13}C- NMR spectra of the copolymers did not contain resonances expected of diad and triad sequences, suggesting formation of AB block copolymers. The copolymerization of the optically pure ITC monomer with ε-caprolactone, a non chiral monomer, was evident as the specific rotation decreased with decreasing ITC content in the copolymers.

Formation of the AB block copolymers, as stated above, is traditionally accomplished through sequential addition of monomers such that the prepolymer from the first monomer initiates polymerization of the second monomer and inter- and intra chain transfer reaction are kept to a minimum. One shot block copolymerization involves successive polymerizations of each of the two monomers that are fed simultaneously. To understand the mechanism, copolymerization of ITC and ε-CL by $Sn(Oct)_2$ was further investigated (Figure 8). The progress of the copolymerization of equimolar amount of ITC and ε-CL was carefully monitored using ^{1}H NMR spectra collected at predetermined time intervals, and it could be observed that the rate of polymerization of ε-CL was much higher than that of ITC.

Table 2. Ring-Opening Polymerization of ITC Monomer in Bulk at 120 °C

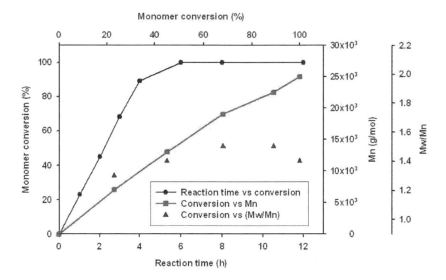

Entry	Catalyst	M/C[a]	Conversion (%)[b]	M_n[c] (g/mol)	M_n[d] (g/mol)	PDI[d]
1	Sn(Oct)$_2$	50	100	9400	9000	1.4
2		100	100	18800	15500	1.4
3		200	100	37600	26000	1.5
4	ZnEt$_2$-H$_2$O	50	100	9400	7500	1.4
5		100	100	18800	13500	1.4
6		200	100	37600	20500	1.4
7	Al(O^{i}Pr)$_3$	50	74	6960	5500	1.5
8		100	75	14100	11200	1.5
9		200	72	27100	16500	1.6

[a] Monomer/catalyst ratio (mol/mol) = 200.   [b] Calculated from ^{1}HNMR.   [c] Theoretical M_n calculated using the monomer conversion.   [d] Determined from GPC.

Figure 4. Plots of Sn(Oct)$_2$ catalyzed ring-opening polymerization at 120 °C in bulk for 12h.

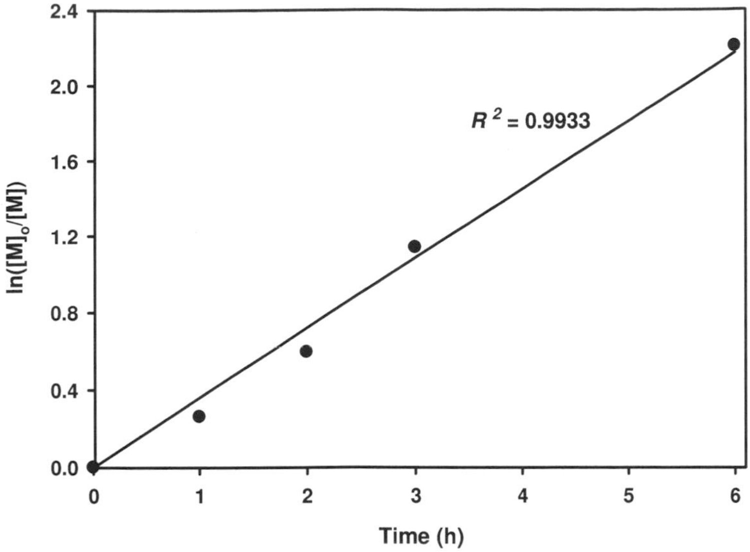

Figure 5. Plot of ln([M]₀/[M]) as function of polymerization time (h) for Sn(Oct)₂ catalyzed ROP of ITC monomer in bulk at 120 °C.

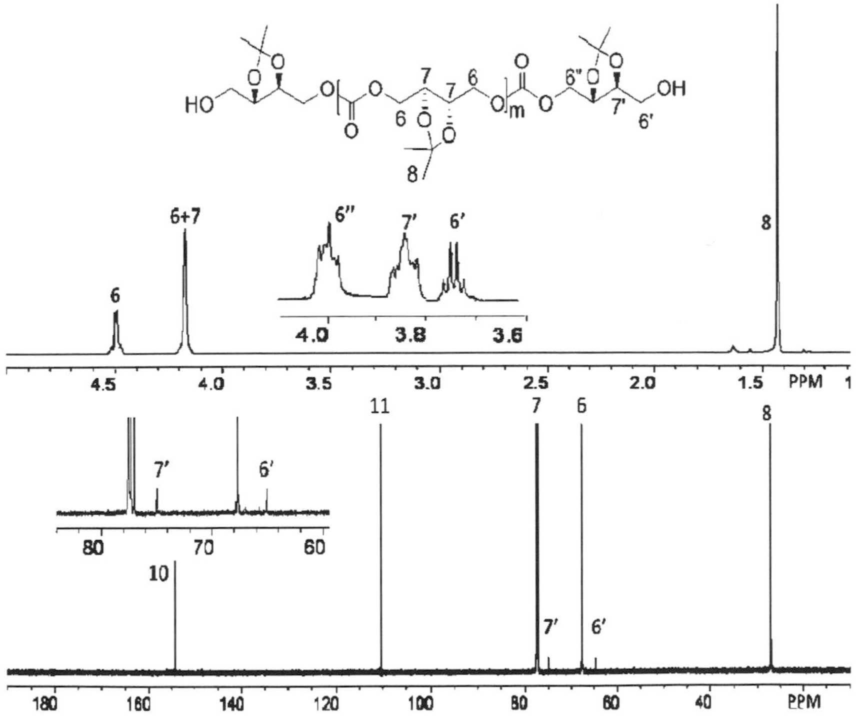

Figure 6. ^{1}H- and ^{13}C-NMR spectra of poly(ITC) obtained by $Sn(Oct)_2$ catalyzed ROP in bulk at 120 °C for 12h [Table 2, entry 3].

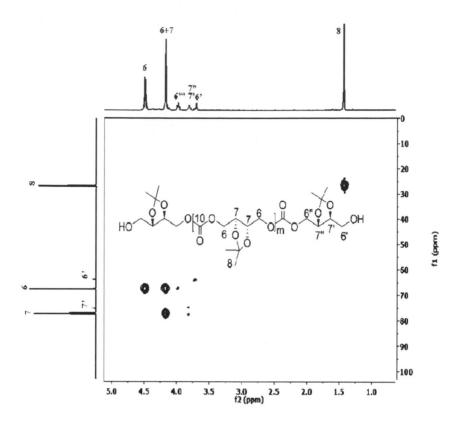

Figure 7. ¹H-¹³C HSQC-NMR (500 MHz, CDCl₃) spectrum of Poly(ITC) [Table 2, entry 1].

In the first 60 min, ε-CL conversion reached >98%, but none of ITC monomers had been consumed; in 80 min, ε-CL was completely consumed and ITC conversion was <10% (Figure 8). Clearly, ε-CL was polymerized first, and the polymerization of ITC was initiated only after all of the ε-CL monomer had been consumed (Figure 8). During the copolymerization the GPC curves (Figure 9) were unimodal and the polymer molecular weight increased with monomer conversion. Importantly, the GPC curves continue to be unimodal and molecular weight increased even after the consumption of ε-CL was complete (Figure 9).

The copolymerization of ITC with ε-CL catalyzed by Sn(Oct)₂ at 120 °C, therefore, resulted in formation of AB block copolymer, poly(ε-CL)-*block*-poly(ITC), in which the poly(ε-CL) prepolymer from the fast reacting monomer, ε-CL, initiated the ROP of the slow reacting ITC monomer. This is the first example, to the best of our knowledge, of the one shot block polymerization of ε-CL with a carbonate monomer.

Table 3. Sn(Oct)$_2$ catalyzed Ring-Opening Co-polymerizations of ITC and ε-CL* at 120 °C for 12h in bulk

#	Monomer feed ratio[a] [ITC:CL]	Copolymer molar composition[b] [ITC:CL]	Yield[c] (%)	M_n[d] (g/mol)	PDI[d]	T_m[e] (°C)	ΔH_m[e] (J/g)	$[\alpha]_D^{20}$[f]
1	100:0		87	24000	1.6	58.8	62.8	+ 77.8
2	90:10	81:19	88	23000	1.6	57.0	68.1	+ 62.5
3	70:30	62:38	91	24000	1.6	56.5	64.9	+ 47.9
4	50:50	44:56	92	24000	1.6	54.6	60.2	+ 33.8
5	30 :70	25:75	92	25000	1.5	54.1	59.2	+ 19.3
6	10:90	5:95	95	26000	1.4	52.7	56.2	+ 3.3
7	0:100		95	26000	1.4	52.7	57.1	

* Reactions were carried out using monomer/catalyst mole ratio (M/C) = 200. [a] Monomer feed ratio in mol/mol. [b] Calculated from ^{1}H-NMR spectrum. [c] Insoluble portion in methanol. [d] Determined from GPC. [e] Measured from DSC. [f] Specific rotation measured in CH$_2$Cl$_2$ (c =1.0).

NMR Characterization of the Poly(ITC-block-CL) Copolymers

The copolymers were characterized by detailed analyses of the ^{1}H-NMR, ^{13}C-NMR, ^{1}H-^{1}H-COSY, ^{1}H-^{13}C-HMQC, and by comparison with the polyCL and poly ITC spectra. The copolymer molar compositions were calculated from relative peak area of the H1 (-COC$\underline{H}_2$- at 2.32 ppm) and H6 resonances (-OC$\underline{H}_2$- at 4.17 and 4.50 ppm) in the ^{1}HNMR spectrum for ε-CL and ITC repeat units, respectively (Figure 10). The diastereotopic protons (H6$_{a,b}$) were observed at 4.17 and 4.50 ppm; the assignments were based on ^{1}H-^{13}C correlation HMQC experiment. The proton resonances belonging to the acetonide methyl (H8) in ITC and H3 of ε-CL repeat unit (-C$\underline{H}_2$-) overlapped at 1.33 ppm. Upon comparison to the ^{1}H-NMR spectrum of the poly(ITC), the low intensity resonances in the ^{1}H-NMR spectrum of the copolymer at 3.72, 3.82, and 4.0 ppm were assigned to H6′ (–C$\underline{H}_2$OH), H7′ (–C$\underline{H}$O–) and H6″ (–C$\underline{H}_2$OCOO), respectively, of the ITC block [insert in Figure 7]. The resonances of the methyleneoxy hydrogens (–C$\underline{H}_2$O–) belonging to the CL-ITC link were observed at 3.66 (H-5″, $\underline{CL}$-ITC), and 4.28 & 4.36 (H6‴, CL-$\underline{ITC}$) ppm.

187

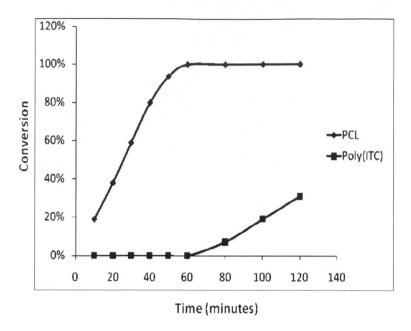

Figure 8. Plot of monomer conversion (%) as a function of reaction time (min) for Sn(Oct)$_2$ catalyzed the copolymerization of ITC monomer (♦) with ε-Caprolactone (■) at 120 °C in bulk [1:1 feed ratio].

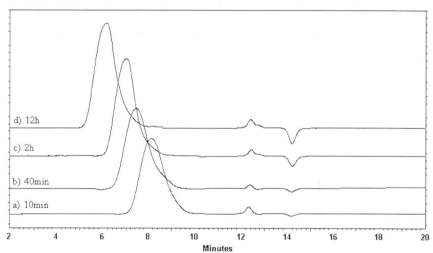

Figure 9. GPC chromatograms of poly[ITC-block-CL] at different reaction times catalyzed by Sn(Oct)$_2$ at 120 °C in bulk (M/I = 200): a) M_n = 1700 g/mol, PDI = 1.5. b) M_n = 6400 g/mol, PDI = 1.6. c) M_n = 13000 g/mol, PDI = 1.6.

In the ^{13}C-NMR spectrum of the copolymer the high intensity peaks were assigned to the Poly(ε-CL) and Poly (ITC) resonances. The low intensity peaks were assigned to the end groups and the PCL-PITC link carbons (Figure 11). The

PCL end groups C-1' (-CH₂COOH) and C-9' (-COOH) were observed at 32.8, and 177.0 ppm, respectively. The resonance at 62.8 ppm was assigned to the C-5" (-CH_2O) of the _PCL_-PITC link. The polyITC end group hydroxymethylene (C-6') was observed at 64.8 ppm and the C-7' was at 75.2 ppm. The carbonate carbonyl (C-10') linking the CL and ITC block was observed at 155.0 ppm. Interestingly, the C-6" and 6''' carbon were observed at 66.3 and 67.2 ppm owing to their proximity to the end group or the PCL-PITC link.

In order to confirm the end group assignment, the copolymer was subjected to acetylation of the hydroxyl end group and the ¹H-NMR spectra before and after acetylation is shown in Figure 10 insert. The new acetate group resonance was observed at 1.98 (s, CH_3CO) and its integral suggested that only one acetate group was attached, confirming only one hydroxyl end group in the copolymer (HOOC-pCL-pITC-OH). A careful comparison of the data before and after acetylation was used to confirm the assignments in the ¹H-NMR. Importantly, only the H-1' was deshielded by > 0.3 ppm suggesting an acetate formation at C-1'. As expected, the resonances arising from the CL-ITC link were unaffected by the end group acetylation. The ¹³C-NMR spectrum of the acetylated copolymer was also acquired and showed a new peak at 17.0 ppm for the acetate methyl and the C-6' (-CH_2OCOCH_3) resonance was shifted downfield to 67.4 ppm (Figure 11 insert). The downfield shift only in the resonance position of C-6' confirmed the structure of the block copolymer and that only one hydroxy end group was present. The AB block copolymer structure was thus established and was in line with observation of only unimodal peaks in the GPC analyses.

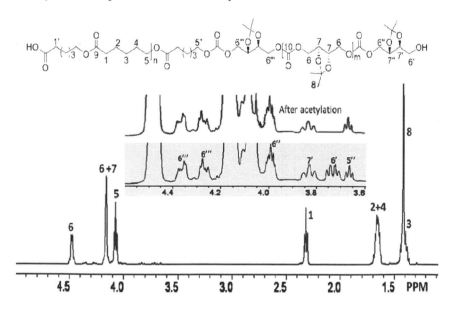

Figure 10. ¹H-NMR spectra (500 MHz, CDCl₃) of poly(ITC-block-CL) [Table 3, entry 4]. Insert in the figure shows a comparison of the data before and after acetylation.

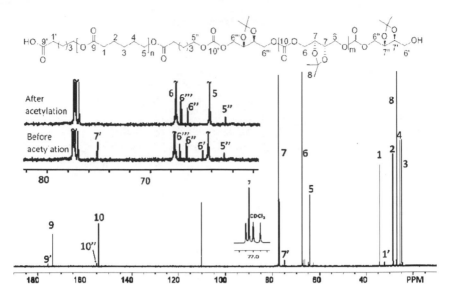

Figure 11. ^{13}C-NMR (125 MHz, CDCl₃) of poly(ITC-block-CL) for Sn(Oct)₂ catalyzed the copolymerization in bulk at 120 °C for 12h [Table 3, entry 4]. Insert in the figure shows expanded region before and after acetylation.

Deprotection of the Ketal Groups

An important aspect of this work is to introduce functional groups in the polymer backbone, which is expected to enhance the biodegradability of the polymer and the pendant groups provide opportunity for post polymerization modifications. Trifluoroacetic acid has been used as an efficient deprotecting agent for the removal of the ketal groups in polycarbonates with minimal degradation in the polymer chain (*26*). The Mn of homo polymer before the deprotection was 15500 with PDI of 1.7. After 5 minutes reaction time 43% removal of the ketal groups was observed; calculated from the new resonance at 3.6 ppm (-C*H*-OH), and broad peak at 5 ppm (–O*H*) in its ¹H-NMR spectrum (see Figure 12a). In 15 minutes, the conversion reached 91% (Mn= 12500 g/ mol) with slight increase in PDI from 1.7 to 1.8. A complete removal of the acetonide protecting groups was achieved after 20 minutes. The unimodal nature of the CPC trace confirmed minimal degradation, if any, of the polymer backbone (Figure 13). The specific rotation ([α]$_D^{20}$) of poly(ITC) before and after deprotection is + 77.8 (*c* = 1, CH₂Cl₂) and +56.0 (*c* = 1, EtOH), respectively.

Table 4 summarizes the results obtained for the deprotection of the model copolymer, poly(44%ITC)-*block*-poly(56% ε-CL) in various reaction times. After 5 min reaction time 51% of the ispropylidene protective were removed, the M_n dropped to 22500 which mainly due the lost of acetonide groups. The conversion was calculated from the new resonance at 3.7 ppm of the methine hydrogens (–C*H*-OH) in the ¹H-NMR spectrum (Figure 12b). After 15 minutes reaction 99% deprotection was achieved (M_n = 21,000 g/mol) with slight increase in PDI to 1.9.

The specific rotation ($[\alpha]_D^{20}$) of poly(44%ITC)-*block*-poly(56% ε-CL) before and after deprotection is +33.8 [CH_2Cl_2, c =1] and +10.9 [EtOH, c = 1], respectively.

Thermal Analysis of the ITC Block CL Copolymers

The thermal properties of the polymers before and after deprotection were examined by DSC analyses under helium atmosphere. The samples were scanned from -100 to 300 °C in rate of 10°C/min. The glass transition temperature (T_g) was not observed for poly(ITC), however, sharp melting temperature (T_m) peak was observed at 58.8 °C with heat of enthalpy $\Delta H_f = 62.28$ J/g.

Two glass transition temperatures (T_g) were observed for poly(44%ITC)-*block*-poly(56% ε-CL) (Table 3, entry 4) at -59.1 and -37.2 °C for the polyCL and the polyITC block, respectively, confirming the diblock nature of the copolymers (Figure 14). PCL is reported to have a T_g of -60 °C and T_m of 60 °C (*25*). The DSC thermogram also showed a sharp exothermic peak (T_m) for the poly[44%ITC-*block*-56%CL] at 54.6 °C with meting enthalpy (ΔH_m) of 60.23 J/g. The melting temperature (T_m) of PCL (M_n 24000 g/mol; PDI =1.4; Table 4, entry 7) is found to be 52.7 °C which was lower than the reported T_m value of 60 °C for a similar M_n PCL (*25*).

The DSC thermogram of the deprotected poly(ITC) is very comparable to the protected polymer. De-protected poly(ITC) showed T_m of 60.2 °C and $\Delta H_f = 69.56$ J/g. The slight increase in the Tm indicated that the crystallinity of free hydroxy polymer is slightly higher. Similar to that of the polyITC, a Tg was not found. For the poly[44%ITC-*block*-56%CL], after deprotection the T_m and ΔH_m of poly[44%ITC-*block*-56%CL] increased to 59.4 °C from 54.6 °C and 78.84 from 60.23 J/g, respectively. The ΔH_m increased by 30 % after the deprotection which indicates higher crystallinity of the free hydroxy copolymer.

Experimental Procedures

Materials

All reagents were used without further purification unless specified otherwise. L-tartaric acid (99%), *p*-Toluenesulfonic acid (98.5%) and stannous 2-ethyl-hexanoate (stannous octanoate, 95%), triisopropyl aluminum Al(O^iPr)_3, diethyl zinc monohydrate $ZnEt_2 \cdot H_2O$, and Triphosgene were purchased from the Aldrich Chemical Company. 2,2-Dimethoxypropane, Triethylamine (99%) and Lithium aluminium hydride (95%) were purchased from Acros Chemical Co. Diethyl ether and tetrahydrofuran (THF) were dried over Na before use. Porcine pancreatic lipase (PPL) Type II Crude (activity = 61 units/mg protein) was purchased from Sigma Chemical Co. Lipase PS-30 from *Pseudomonas cepacia* (20,000 units/g), and Lipase AK were obtained from Amano Enzymes Co., Ltd. The carrier fixed lipase Novozym 435 (from *Candida antarctica*, fraction B; *specified activity at pH 7.0 is 10,000 units/g*) was a gift from Novo Nordisk Inc.

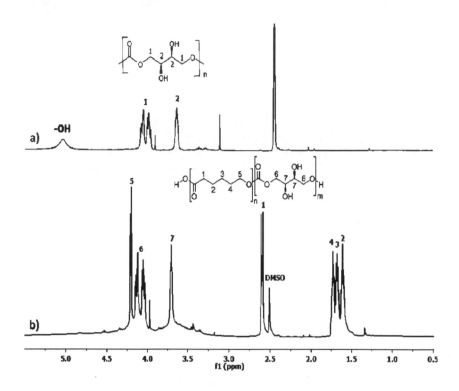

Figure 12. ¹HNMR (500MHz, DMSO-d₆) spectra:(a) Poly(ITC) after de-protection.(b) Poly[ITC-block-CL] after deprotection of [Table 3, entry 4].

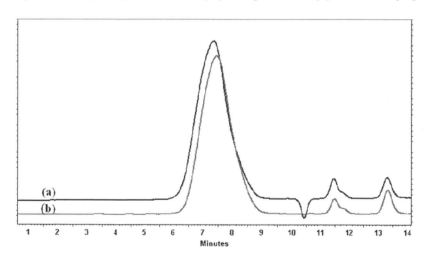

Figure 13. GPC chromatograms of Poly(ITC); (a) before de-protection (Mn = 15500 g/mol). (b) After de-protection (Mn = 12500 g/mol).

Measurements

Molecular weights were measured by gel permeation chromatography (GPC) using a Shimadzu HPLC system equipped with a model LC-10ADvp pump, model SIL-10A auto injector, model RID-10A refractive index detector (RI), model SPD-10AV UV-Vis detector, and waters HR 4E styragel column. $CHCl_3$ (HPLC grade) was used as an eluent at a flow rate of 1.0 mL/min. The sample concentration and injection volumes were 0.5 % (w/v) and 100 μL, respectively. EzChrome Elite (Scientific Software Inc.) was used to calculate molecular weights based on a calibration curve generated by narrow molecular weight distribution polystyrene standards (5.00×10^2, 8.00×10^2, 2.10×10^3, 4.00×10^3, 9.00×10^3, 1.90×10^4, 5.00×10^4, 9.26×10^4, 2.33×10^5, and 3.00×10^5 g/mol, Perkin-Elmer). ^{1}H- and ^{13}C-NMR spectra were recorded on a Bruker DPX-250, Varian inova-400 and 500 spectrometers. Sample concentrations were about 10% (w/v) in $CDCl_3$ containing 1% TMS as an internal reference. Monomer conversions were calculated from ^{1}H-NMR spectra upon integration of area of peaks for -2C$\underline{H}_3$ of the monomer at 1.38 ppm and for the polymer at 1.43 ppm. The degree of polymerization (DP) calculated from the ^{1}HNMR spectrum by determining the area under the repeat unit methylenes and methines (H-1 and H-2, 4.63 and 4.25 ppm) resonances and the end group C$\underline{H}_2$OH resonance (3.68 ppm) were in good agreement with the molecular weight obtained using GPC.

Optical rotations were measured on an Autopol IV (Rudolph Instruments) automated polarimeter at 20 °C in $CHCl_3$/MeOH at a concentration of 1.0. Thermal analyses were preformed on a Dupont DSC 2920 TA instrument attached to a Thermal Analyst 2000 TA instrument computer. Indium was used as the standard for the temperature calibration and the analyses were made under constant stream of nitrogen with a heating rate 10 °C/min and cooling rate of 40 °C/min.

Synthesis of (5*S*, 6*S*)-Dimethyl 5,6-*O*-Isopropylidene-1,3-dioxepin-2-one (ITC, 3) (*22*)

Triphosgene (0.01mol, 2.97g) was dissolved in dry THF (100mL) and the solution is added dropwise to a mixture of 2,3-Di-Oisopropylidene- L-Threitol (prepared in two steps from L-tartaric acid, (0.02mol,3.24g), and Pyridine (0.0633mol, 4.99g) dissolved in 200 mL tetrahydrofuran (THF) at 0 °C over a period of 30 minutes. The reaction mixture was stirred at room temperature for 6 hours. Precipitated pyridine hydrochloride was filtered off, and the filtrate was concentrated under reduced pressure. The product was purified by chromatography on silica gel using 20% ethyl acetate/hexanes as a solvent mixture. White solid, mp 75 °C (0.0114 mol, 2.14g, 57%); $[\alpha]_D$ [20] =+84.79° (CH_2Cl_2, c =1); HRMS m/z $[M+H]^+$. Calcd for $C_8H_{13}O_5$: 189.07630. Found 189.07593; ^{1}H NMR ($CDCl_3$) δ: 1.38 (s, 6 H, 2CH$_3$), 4.04 (m, 2H, -CH-O), 4.29 (dd, 5 and 12.5 Hz, 4H, -CH$_2$-O); ^{13}C NMR ($CDCl_3$) δ: 26.9, 67.7, 75.7, 110.7, 154.4.

Table 4. Removal of the acetonide groups in poly(44%ITC)-block-poly(56% ε-CL) #

entry	time (min)	conversion (%)[a]	yield (%)[b]	M_n (g/mol) [c]	PDI [c]
1	0			24000	1.6
2	5	51	89	22500	1.7
3	10	79	87	22000	1.8
4	15	99	83	21000	1.9
5	20	100	75	18000	2.2

$CH_2Cl_2/CF_3COOH/H_2O$ at room temperature (see experimental section). [a] Calculated from ^{1}HNMR. [b] Insoluble portion in methanol. [c] Determined from GPC.

General Procedure for the Enzymatic Polymerization (*22*)

All reactions were carried out in bulk. The lipase was dried (in a drying pistol over P_2O_5, at 50 oC/0.1 mm Hg; 15 h) in 6 mL sample vials. In a glove bag, maintained under nitrogen atmosphere, the monomer was transferred to a 6 mL reaction vial and the pre-weighed enzyme (94 mg/mmol of carbonate) was added. The reaction vial was capped with a rubber septum and placed in a constant temperature oil bath maintained at 80oC for predetermined times. Reactions were terminated by dissolution of the contents of the reaction vial in chloroform and removal of the enzyme (insoluble) by filtration (glass fritted filter, medium pore porosity). The filtrates were combined, solvents were removed *in vacuo* and the crude products were analyzed by proton (^{1}H) NMR and gel permeation chromatography (GPC). Polymers were purified by precipitation in methanol.

General Procedure for the Chemical Homopolymerization (*23*)

In a nitrogen atmosphere, ITC (11.6 mmol, 2g) was charged into dried, freshly silanized 15 mL schlenk glass tube and 2 x 10-4 mol of stannous octanoate (2% dry toluene solution) per mol of total monomer was added as a solution in sodium-dried toluene (1.48 x 10-2M). Subsequently, the toluene was removed by evacuation. The schlenk tubes were purged three times with dry nitrogen placed in an oil bath preheated to the polymerization temperature. After a predetermined time the schlenk tubes were quenched to room temperature (120 °C) and the polymer was dissolved in dichloromethane. Samples were taken for determination of the monomer conversion by ^{1}H NMR spectroscopy. For purification, the obtained polymers were dissolved in chloroform, filtered through a sintered glass filter and precipitated into an excess of ice cold methanol. The precipitated polymers were collected, washed with fresh methanol and dried at room temperature (RT) under reduced pressure.

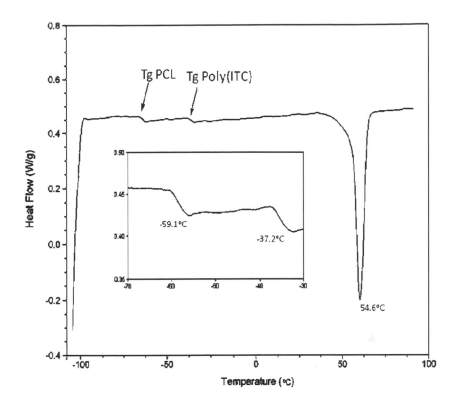

Figure 14. DSC thermogram of poly(44%ITC)-block-poly(56% ε-CL) (Table 3, entry 4).

General Procedure for Copolymerization (*23*)

In a nitrogen atmosphere, a mixture of ITC and CL (10 g scale) was charged into dried, freshly silanized 15 mL glass schlenk tubes. The monomer mixture was gently warmed and vigorously shaken in order to obtain a homogeneous mixture of the monomers. To the monomer mixture 2×10^{-4} mol of stannous octanoate (2% dry toluene solution, 1:200 catalyst/monomer ratio) per mol of total monomer was added as a solution in sodium-dried toluene (1.48×10^{-2}M). Subsequently, the toluene was removed by evacuation. The schlenk tubes were purged three times with dry nitrogen and placed in an oil bath preheated to the polymerization temperature (120 °C). After a predetermined time the ampoules were quenched to room temperature and the copolymer was dissolved in dichloromethane. Samples were taken for determination of the monomer conversion by ^{1}H NMR spectroscopy. For purification, the obtained copolymers were dissolved in chloroform, filtered through a sintered glass filter and precipitated into an excess of ice cold methanol. The precipitated polymers were collected, washed with fresh methanol and dried at room temperature (RT) under reduced pressure.

General Procedure for the Removal of Isopropylidene Protective Groups (*22, 23*)

Polymer (100mg) was dissolved in 1 mL of CH_2Cl_2. Then 1 mL CF_3COOH (80%) was added into the CH_2Cl_2. After stirring at room temperature for a predetermined time, the resulting solution was poured into 10mL iced cold methanol. The polymer was collected by vacuum filtration, and dried in a vacuum.

Conclusion

Enantiomerically pure seven-membered cyclic carbonate (ITC) synthesized from naturally occurring L-tartaric acid in three steps was investigated for it polymerization and copolymerization with ε-CL. Immobilized *Candida antarctica* lipase -B (Novozym-435) was found to be the most efficient catalyst to carry out the ROP of ITC. The study of the Novozym-catalyzed polymerization suggested that the chain transfer reactions occurring during the polymerization were minimal and that it showed characteristic of non-terminating chain polymerization. NMR examination of the polymers revealed hydroxy end groups at both terminals. Of the three chemical catalysts, namely stannous octanoate $[Sn(Oct)_2]$, triisopropoxide aluminum $Al(O^iPr)_3$, and diethyl zinc monohydrate $ZnEt_2$-H_2O screened for the polymerization of the monomer ITC, $Sn(Oct)_2$ was found to be the most efficient catalyst. The homopolymerization of ITC catalyzed by $Sn(Oct)_2$ followed first order rate law. The results were in agreement with general mechanism of the ROP of a non-terminating chain polymerization. Optically active copolymers with various feed ratios were synthesized by 'one-shot feeding' of ITC with ε-CL comonomers catalyzed by $Sn(Oct)_2$ at 120 °C for 12h, in bulk. Detailed investigation of the copolymers revealed them to be AB block copolymers. The deprotection of the ketal groups using trifluoroacetic acid offered polycarbonate with pendant hydroxy groups with minimal degradation in the polymer chain. The presence of hydroxy groups is expected to enhance the biodegradability, and the hydrophilicity of the polymers.

Acknowledgments

Financial support from the American Lung Association, American Cancer Society and the Herman Frasch Foundation (510-HF02) is greatly appreciated.

References

1. (a) Velter, I.; La Ferla, B.; Nicotra, F. *J. Carbohydr. Chem.* **2006**, *25*, 97. (b) Varma, A. J.; Kennedy, J. F.; Galgali, P. *Carbohydr. Polym.* **2004**, *56*, 429. (c) Qun Wang, Q.; Dordick, J. S.; Linhardt, R. *J. Chem. Mater.* **2002**, *14*, 3232.
2. (a) Torma, V.; Gyenes, T.; Szakacs, Z.; Noszal, B.; Nemethy, A.; Zrinyi, M. *Polym. Bull.* **2007**, *59*, 311. (b) Mori, H.; Iwaya, H.; Nagai, A.; Endo, T. *Chem. Commun.* **2005**, *38*, 4872. (c) Chung, Il-D.; Britt, P.; Xie, D.; Harth,

E.; Mays, J. *Chem. Commun.* **2005**, *8*, 1046. (d) Bentolila, A.; Vlodavsky, I.; Haloun, C.; Domb, A. J. *Polym. Adv. Technol.* **2000**, *11*, 377.

3. (a) Wang, X. L.; Zhuo, R. X.; Liu, L. J.; He, F.; Liu, G. *J. Polym. Sci., Part A: Polym. Chem.* **2002**, *40*, 70. (b) Vandenberg, E. J.; Tian, D. *Macromolecules* **1999**, *32*, 3613. (c) Ray, W. C., III; Grinstaff, M. W. *Macromolecules* **2003**, *36*, 3557. (d) Acemoglu, M.; Bantle, S.; Mindt, T.; Nimmerfall, F. *Macromolecules* **1995**, *28*, 3030.

4. Sanda, F.; Kamatani, J.; Endo, T. *Macromolecules* **2001**, *34*, 1564.

5. (a) Al-Azemi, T. F.; Bisht, K. S. *Macromolecules* **1999**, *32*, 6536. (b) Al-Azemi, T. F.; Harmon, J. P.; Bisht, K. S. *Biomacromolecules* **2000**, *1*, 493. (c) Lee, R. S.; Yang, J. M.; Lin, T. F. *J. Polym. Sci., Part A: Polym. Chem.* **2004**, *42*, 2303.

6. (a) Acemoglu, M.; Bantle, S.; Mindt, T.; Nimmerfall, F. *Macromolecules* **1995**, *28*, 3030. (b) Yokoe, M.; Aoi, K.; Okada, M. *J. Polym. Sci., Part A: Polym. Chem.* **2005**, *43*, 3909.

7. Jacek Gawronski, J.; Gawronska, K. *Tartaric and Malic Acids in Synthesis: A Source Book of Building Blocks, Ligands, Auxiliaries, and Resolving Agents*; 1st edition, Wiley-Interscience: 1999.

8. (a) Bou, J. J.; Rodriguez-Galin A.; Munoz-Guerra, S. *Macromolecules* **1993**, *26*, 5664. (b) Bou, J. J.; Iribarren, I.; Munoz-Guerra, S. *Macromolecules* **1994**, *27*, 5263. (c) Mathakiya, I. A.; Rakshit, A. K. *Int. J. Polym. Mater.* **2004**, *53*, 405.

9. Kimura, H.; Yoshinari, T.; Takeishi, M. *Polym. J.* **1999**, *31*, 338.

10. (a) Villuendas I.; Iribarren J. I.; Munoz-Guerra, S. *Macromolecules* **1999**, *32*, 8015. (b) Alia, A.; Rodriguez-Galfin, A.; Martinez de Ilarduya, A.; Munoz-Guerra, S. *Polymer* **1997**, *38*, 4935.

11. Hu, B.; Zhuo, R. X.; Fan, C. L. *Polym. Adv. Technol.* **1998**, *9*, 145.

12. Liu, J.; Zeng, F.; Allen, C. *J. Controlled Release* **2005**, *103*, 481.

13. Rokicki, G. *Prog. Polym. Sci.* **2000**, *25*, 259.

14. Khan, I.; Smith, N.; Jones, E.; Finch, D. S.; Cameron, R. E. *Biomaterials* **2005**, *26*, 621.

15. Chen, X. L.; Mccarthy, S. P.; Gross, R. A. *J. Appl. Polym. Sci.* **1998**, *67*, 547.

16. (a) Matsuo, J.; Sanda, F.; Endo, T. *Macromol. Chem. Phys.* **1998**, *199*, 97. (b) Matsuo, J.; Sanda, F.; Endo, T. *J. Polym. Sci., Part A: Polym. Chem.* **1997**, *35*, 1375. (c) Morikawa, H.; Sudo, A.; Nishida, H.; Endo, T. *Macromol. Chem. Phys.* **2005**, *206*, 592. (d) Tomita, H.; Sanda, F.; Endo, T. *J. Polym. Sci., Part A: Polym. Chem.* **2001**, *39*, 4091. (e) Shibasaki, Y.; Sanda, F.; Endo, T. *Macromol. Rapid Commun.* **2000**, *21*, 489.

17. (a) Matsuo, J.; Sanda, F.; Endo, T. *Macromol. Chem. Phys.* **2000**, *201*, 585. (b) Tomita, H.; Sanda F.; Endo, T. *J. Polym. Sci., Part A: Polym. Chem.* **2001**, *39*, 4091.

18. (a) Kobayashi, S.; Uyama, H.; Kimura, S. *Chem Rev.* **2001**, *101*, 3793. (b) Gross, R. A.; Kumar, A.; Kalra, B. *Chem. Rev.* **2001**, *101*, 2097. (c) Van, L.; Helmich, F.; De Bruijn, R.; Vekemans, J. A. J. M.; Palmans, A. R. A.; Meijer, E. W. *Macromolecules* **2006**, *39*, 5021. (d) Bisht, K. S.; Henderson, L. A.; Gross, R. A.; Kaplan, D. L.; Swift, G. *Macromolecules* **1997**, *30*,

2705. (e) Duda, A.; Kowalski, A.; Penczek, S.; Uyama, H.; Kobayashi, S. *Macromolecules* **2002**, *35*, 4266. (f) Albertsson, A.-C.; Srivastava, R.K. *Adv. Drug Delivery Rev.* **2008**, *60*, 1077.

19. (a) Kricheldrof, H. R.; Dunsing, R.; Serra I Albet, A. *Makromol.Chem.* 1987, *188*, 2453. (b) Kricheldorf, H. R.; Jenssen, J. *J. Macromol. Sci., Part A: Pure Appl. Chem.* **1989**, *26*, 631. (c) Kuhling, S.; Keul, H.; Hocker, H. *Macromol. Chem. Suppl.* **1989**, *15*, 9. (d) Kuhling, S.; Keul, H.; Hocker, H. *Macromol. Chem.* **1990**, *191*, 1611. (e) Keul, H.; Bacher, R.; Hocker, H. *Macromol. Chem.* **1990**, *191*, 2579. (f) Ariga, T.; Takata, T.; Endo, T. *J. Polym. Sci., Part A: Polym. Chem.* **1993**, *31*, 581. (g) Qiao, L.; Gu, Q.-M.; Cheng, H. N. *Carbohydr. Polym.* **2006**, *66*, 135. (h) Sahoo, S. K.; Nagarajan, R.; Roy, S.; Samuelson, L. A.; Kumar, J.; Cholli, A. L. *Macromolecules* **2004**, *37*, 4130.

20. (a) Nagai, D.; Yokota, K.; Ogawa, T.; Ochiai, B.; Endo, T. *J. Polym. Sci., Part A: Polym. Chem.* **2008**, *46*, 733. (b) Hachemaoui, A.; Belbachir, M. *Mater. Lett.* **2005**, *59*, 3904. (c) Shibasaki, Y.; Sanada, H.; Yokoi, M.; Sanda, F.; Endo, T. *Macromolecules* **2000**, *33*, 4316. (d) Lee, R.; Lin, T.; Yang. J. *J. Polym. Sci., Part A: Polym. Chem.* **2003**, *41*, 1435.

21. Li-Li, M.; Guo-Ping, Y.; Xiang-Hua, Y.; Si-Xue, C.; Jiang-Yu, W. *J. Appl. Polym. Sci.* **2008**, *108*, 93–98.

22. Wu, R.; Al-Azemi, T. F.; Bisht, K. S. *Biomacromolcules* **2008**, *9* (10), 2921–2928.

23. Wu, R.; Al-Azemi, T. F.; Bisht, K. S. *Macromolecules* **2009**, *42*, 2401–2410.

24. Wu, R.; Al-Azemi, T. F.; Bisht, K. S. *Polym. Prepr.* **2009**, *50* (2), 63–64.

25. (a) Kumar, A.; Kalra, B.; Dekhterman, A.; Gross, R. A. *Macromolecules* **2000**, *33*, 6303. (b) Kumar, A.; Gross, R. A. *Biomacromolecules* **2000**, *1*, 133.

26. Shen, Y.; Chen, X.; Gross, R. A. *Macromolecules* **1999**, *32*, 3891.

27. For example, see (a) Abhikari, R.; Michler, G. H. *Prog. Polym. Sci.* **2004**, *29*, 946. (b) Riess, G. *Prog. Polym. Sci.* **2003**, *28*, 1107. (c) *Ring-Opening Polymerization*; Ivin, K. J., Saegusa, T., Eds.; Elsevier: London, 1984; Vol. 1, p 218. (d) Covie, J. E. In *Comprehensive Polymer Science*; Eastmond, G. C., Ledwith, A., Russo, S., Sigwalt, P., Eds.; Pergamon: Oxford, 1989; Vol. 3, p 17; (e) Aoki, S.; Harita, Y.; Tanaka, Y.; Mandai, H.; Otsu, T. *J. Polym. Sci., Part. A: Polym. Chem.* **1968**, *6*, 2585.

28. (a) Raquez, J. M.; Dege´e, P.; Narayan, R.; Dubois, P. *Macromol. Rapid Commun.* **2000**, *21*, 1063. (b) Veld, P. J. A. I.; Velner, E. M.; DeWitte, P. V.; Hamhuis, J.; Dijkstra, P.J.; Feijen, J. *J. Polym. Sci., Part A: Polym. Chem.* **1997**, *35*, 219.

29. (a) Duxbury, C. J.; Wang, W.; De Geus, M.; Heise, A.; Howdle, S. M. *J. Am. Chem. Soc.* **2005**, *127*, 2384. (b) Huang, C.-F.; Kuo, S.-W.; Lee, H.-F.; Chang, F.-C. *Polymer* **2005**, *46*, 1561. (c) Didier, B.; Craig, J. H.; Elbert, E. H.; Zhiqun, L.; Thomas, P. R. *Macromolecules* **2000**, *33*, 1505. (d) Takagi, K.; Tomita, I.; Endo, T. *Chem. Commun.* **1998**, 681. (e) Azuma, N.; Sanda, F.; Takata, T.; Endo, T. *J. Polym. Sci., Part A: Polym. Chem.* **1997**, *35*, 1007. (f) Ariga, T.; Takata, T.; Endo, T. *Macromolecules* **1993**, *26*, 7106. (g) Feast, W. J.; Gibson, K. J.; Ivin, K. J.; Khosravi, E.; Kenwright, A. M.; Marshall, E. L.; Mitchell, J. P. *Makromol. Chem.* **1992**, *193*, 2103. (h) Saegusa, T.;

Chujo, Y.; Aoi, K.; Miyamoto, M. *Makromol. Chem. Makromol. Symp.* **1990**, *32*, 1.

Chapter 14

Polymers from Biocatalysis: Materials with a Broad Spectrum of Physical Properties

Mariastella Scandola,[a],* Maria Letizia Focarete,[a] and Richard A. Gross[b]

[a]Department of Chemistry 'G. Ciamician', University of Bologna, Via Selmi 2, 40126 Bologna (Italy)
[b]NSF-I/UCRC Center for Biocatalysis and Bioprocessing of Macromolecules, Department of Chemical and Biological Sciences, Polytechnic University, Six Metrotech Center, Brooklyn, New York 11201
*mariastella.scandola@unibo.it

Copolymers of ω-pentadecalactone (PDL) with ε-caprolactone, valerolactone, dioxanone and trimethylenecarbonate synthesized by biocatalysis show rather uncommon crystallization behavior, namely cocrystallization of the monomer units that leads to highly crystalline copolymers over the whole composition range. Hydrophilic/hydrophobic balance can be adjusted by a suitable choice of the comonomer and of composition, leading to materials with tunable hydrolytic degradation rate for environmental and biomedical applications. Copolyestercarbonates, copolyesteramides and polyol-containing copolyesters synthesized by lipase-catalysed polycondensation show strongly composition dependent physical properties, that can be easily tailored by composition control and cover the whole range from hard solid materials down to gluelike substances.

Introduction

High molecular weight polymers that cannot be obtained by chemical routes are easily synthesized by lipase-catalyzed polymerization (*1*). Some lipases such as *Candida antarctica* Lipase B (CALB), when used in ring opening polymerization or polycondensation reactions, allow incorporation of different monomers along chains so that copolymers with defined composition and microstructure can be prepared. A number of such copolymers, i.e. copolyesters (2–7), copolyestercarbonates (8–10) and copolyesteramides (11), have been successfully synthesized and their solid-state properties have been characterized.

Copolymers show a wide range of thermomechanical properties that can be finely tuned through composition and microstructure control, which in turn can be tailored by playing on feed ratio and reaction conditions. Such a control is critical to the ultimate goal of adjusting physical, mechanical, and biological properties of the copolymers in view of their materials applications.

Metal-free catalysis, mild reaction conditions, presence of hydrolysable bonds along the chains, tunable hydrophilic/hydrophobic ratio suggest potential uses of these polymers as bioresorbable materials in the medical field (surgery, implants, drug delivery, etc). Additional interesting applications of the biosynthesized polymers regard the area of environmentally friendly biodegradable materials.

Experimental

Materials

All polymers were synthesized using *Candida antarctica* Lipase B (CALB, Novozyme-435) as described elsewhere (2, 4–6, 8, 10–12). Solvents used for electrospinning (chloroform, dichloromethane, 1,1,1,3,3,3-Hexafluoro-2-propanol) were Aldrich products used without further purification.

Instrumentation

Polymer solid-state properties were characterized by differential scanning calorimetry (TA DSC-Q100, equipped with LNCS low-temperature accessory), thermogravimetric analysis (TA Instruments TGA2950), dynamical mechanical analysis (DMTA, Polymer Labs., MKII), wide angle X-ray diffraction (WAXS, PANalytical X'Pert PRO), tensile testing (INSTRON 4465) and scanning electron microscopy (SEM, Philips 515). The apparatus used for electrospinning was made in house (13) as was the system employed for the production of scaffolds through supercritical carbon dioxide (sc-CO_2) foaming (14).

Results and Discussion

Poly(ω-pentadecalactone)

Poly(ω-pentadecalactone), PPDL, synthesized by biocatalysis is a high molecular weight (>100 KDa) polyester that crystallizes from the melt with very fast kinetics. It is highly crystalline and melts around 100°C. It shows a pseudo-orthorombic monoclinic unit cell with dimensions a=7.49(1), b=5.034(9), and c=20.00(4)Å (fiber axis), and α= 90.06(4)°, that hosts two monomeric units belonging to polymer chains with opposite orientation (15). The glass transition of this highly crystalline polymer, that is barely detectable by DSC, was determined by DMTA and lays around -30°C (12).

The mechanical properties of PPDL resemble those of low density polyethylene, LDPE (Figure 1). Both polymers exhibit a hard and tough behavior, the curve of PPDL showing a steeper slope in the initial linear part, i.e. higher elastic modulus, as well as higher stress at yield than LDPE. Good mechanical properties associated with the presence along the polymer chain of hydrolysable ester linkages render PPDL an interesting biodegradable material for diversified purposes, including biomedical applications. In this context PPDL is a good candidate as a bioresorbable material where long healing times are required, thanks to the long methylene sequence in the monomer unit (14 C atoms) that 'dilute' the hydrolizable ester bonds along the chain.

PPDL scaffolds for tissue engineering were fabricated by means of electrospinning technology. Biocompatibility of PPDL and ability of the scaffolds to support cell growth were demonstrated using embryonic rat cardiac H9c2 cells (13). Figure 2 shows SEM micrographs of an electrospun nonwoven mat of PPDL fibers (average diameter 600 nm) before and after 14 days of culture. The seeded cells are seen to propagate and spread over the PPDL mat surface while retaining their native morphology. Cytotoxicity tests confirmed biocompatibility of PPDL (13).

Highly Crystalline ω-Pentadecalactone Copolymers

Random copolyesters of pentadecalactone (PDL) were synthesized using as co-units caprolactone (CL), valerolactone (VL) and dioxanone (DO) over the whole range of molar compositions (2, 4). The solid-state behaviour of such copolymers is very peculiar because, despite the fact that the two monomers are randomly distributed along the chain, high crystallinity develops at all compositions (3, 4). However, while both poly(PDL-co-CL) and poly(PDL-co-VL) are isomorphic systems (3), i.e. they crystallize in a lattice that smoothly changes with copolymer composition from that of one homopolymer (PPDL) to that of the other (PCL or PVL), poly(PDL-co-DO) shows isodimorphic behavior (4). The latter case occurs when the crystal lattice of each homopolymer is able to host foreign comonomer units only up to a certain degree. As a consequence, two crystal phases develop: PPDL-type in PDL-rich copolymers and poly(dioxanone)-type at the other end of the composition range. In isodimorphic systems the melting temperature changes with composition and shows a minimum at the so-called pseudo-eutectic, where the two crystal phases may be found to coexist. In poly(PDL-co-DO) the pseudo-eutectic lays at 71mol% DO content, a composition where both PPDL-type and PDO-type crystals are revealed by WAXS (4). Figure 3 compares the melting behavior of copolymers of PDL with CL and with DO, i.e. of systems that show isomorphic and isodimorphic behavior respectively.

An unusual crystallization behavior, involving cocrystallization, was also observed in chemo-enzymatically synthesized methacrylate brush copolymers carrying a PPDL block linked through a PEG spacer to the side chain (16). It was found that in these methacrylate polymer brushes a PPDL-type crystal phase develops upon side-chain crystallization. WAXS analysis revealed that, unexpectedly, the PEG segments were incorporated into the PPDL crystal lattice

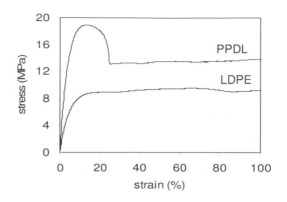

Figure 1. Stress-strain curves of PPDL and LDPE films (room temperature; strain rate: 10 mm/min)

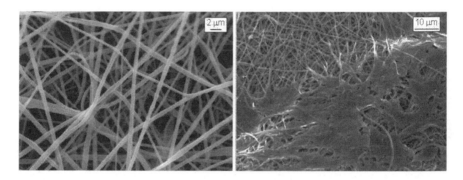

Figure 2. Left: Scanning electron micrograph of an electrospun PPDL fiber mat; right: PPDL fiber mat with H9c2 cells after 14 days of culture.

in the unusual extended zigzag conformation (PEG normally adopts a 7_2 helical conformation). This result further demonstrates the very peculiar ability of the long PDL unit to induce different moieties to enter the PPDL lattice via cocrystallization.

In all systems investigated it was found that copolymerization of smaller lactones with PDL is an effective way to enhance thermal stability of the corresponding homopolymers. This is illustrated in Figure 4 for a poly(PDL-co-VL) and a poly(PDL-co-DO) with near-equimolar PDL and VL content.

Given the high hydrophobicity of the PDL unit, copolymerization with more hydrophilic smaller lactones is also a means to tailor hydrolytic degradation rate of the resulting copolymers. Table 1 collects the results of water contact angle measurements on compression molded films of PPDL and of copolyesters with similar PDL molar content. Substitution of PDL units with CL or VL moieties decreases the contact angle, more markedly in the case of the shorter (VL) unit, as expected. Comparison of the results for the two five-chain-atom monomer units, VL and DO, highlights the effect of the presence of the oxygen atom in the DO repeat that additionally lowers contact angle. Copolymerization is

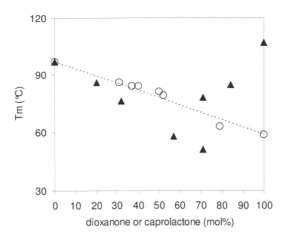

*Figure 3. Composition dependence of melting temperature of: (▲)
poly(PDL-co-DO), isodimorphic system, and (○) poly(PDL-co-CL), isomorphic
system.*

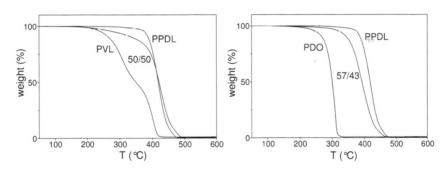

*Figure 4. Comparison of TGA curves of homopolymers PVL, PDO and PPDL
with those of poly(PDL-co-50mol%VL) and poly(PDL-co-57mol%DO)*

therefore a simple way to manipulate hydrophilicity and, consequently, hydrolytic degradation rate in view of potential biomedical applications of these polyesters.

The mechanical properties of PDL-copolymers reflect the presence of a rigid crystal phase associated with a mobile amorphous phase (all copolymers of PDL with CL, VL or DO have glass transition temperature, Tg, lower than room temperature). The amount of crystal phase, that depends on comonomer type and content, influences the stress-strain behavior. As an example, Figure 5 compares the stress-strain curves of two copolymers - poly(PDL-co-36mol%CL) and poly(PDL-co-53mol%DO) – with that of PPDL.

The three curves are remarkably different, mainly in terms of elastic modulus and strength. Table 2 reports tensile modulus and melting enthalpy (as an indication of crystallinity degree). Both parameters decrease in the same order, i.e. PPDL > poly(PDL-co-36mol%CL) > poly(PDL-co-53mol%DO), showing that the observed differences in mechanical properties are maily associated with different crystal phase content in the samples. Knowledge of structure-property

Table 1. Water Contact Angle of PPDL and of Copolyesters

sample	PDL (%mol)	PDL (%wt)	Contact Angle (°)	SD
PPDL	100	100	106	3
poly(PDL-co-CL)	50	79	95	2
poly(PDL-co-VL)	45	66	87	2
poly(PDL-co-DO)	47	68	76	3

relations in these copolymers (2–4, 7) teaches that crystallinity degree can be changed to a certain extent by playing on composition. Thus the material's properties can be adjusted to specific needs.

In summary, PDL copolymers can be biosynthesized with tunable crystallinity degree, hydrophilicity and mechanical properties suitable for a wide variety of applications. Among others, an important biomedical field of activity regards the use of scaffolds for cell expansion in tissue engineering. Scaffolds of PDL copolymers were therefore produced not only via electrospinning technology (13) as mentioned above, but also through sc-CO2 foaming (14). Figure 6 shows, as an example, the 3-D reconstruction by micro computer tomography (μ-CT) of a foamed porous scaffold of poly(PDL-co-36mol%CL). In order to optimize foaming of this highly crystalline material, a novel experimental procedure was setup that included controlled cooling during the depressurization stage (14). By this new method scaffolds with tunable porosity, pore size and interconnectivity were successfully obtained with mechanical properties suitable for applications in cartilage tissue regeneration.

A copolyester-carbonate containing the PDL unit was also synthesized using CALB (8).

The peculiar crystallizing ability of PDL and its tendency, when copolymerized with the cyclic trimethylenecarbonate monomer (TMC), to yield alternate comonomer sequences led to highly crystalline poly(PDL-TMC) copolymers. Worth noting is that the homopolymer poly(TMC) is totally amorphous. Nevertheless a new crystal phase, that melts at lower temperature than PPDL and is attributed to crystallization of PDL-TMC alternate sequences, was observed (9). Moreover the copolymers showed enhanced thermal stability compared with that of PTMC, as illustrated for the copolymer with 50mol%TMC in Figure 7.

Copolyestercarbonates, Copolyesteramides and Polyol-Containing Copolyesters with Strongly Composition Dependent Physical Properties

Poly(butylene carbonate-co-butylene succinate) random copolymers were synthesized by lipase catalysis over the whole range of comonomer content. Poly(BC-co-BS) copolymers varied from semicrystalline to near completely amorphous with changing composition (10). The semicrystalline poly(BC-co-BS) show only one crystal phase, either that of the homopolymer poly(BC) or that of poly(BS), indicating the inability of either crystal lattice to host different

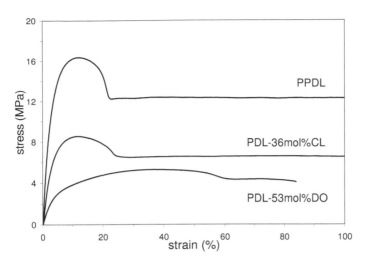

Figure 5. Comparison of stress-strain curves of poly(PDL-co-36mol%CL), poly(PDL-co-53mol%DO) and PPDL films (room temperature; strain rate: 10 mm/min)

Table 2. Mechanical properties of PPDL and of selected copolymers (strain rate: 10 mm/min)

sample	Tensile modulus (MPa)	Melting enthalpy [a] (J/g)
PPDL	530 ± 43	125
poly(PDL-co-36mol%CL)	230 ± 17	101
poly(PDL-co-53mol%DO)	85 ± 2	58

[a] From DSC

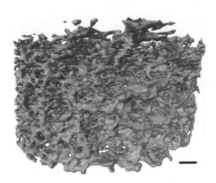

Figure 6. Three-dimensional reconstruction by μ-CT of a porous scaffold of poly(PDL-co-36mol%CL) obtained by sc-CO₂ foaming. Scale bar: 1 μm.

comonomer units. At particular copolymer compositions (around 70 mol % BC), neither BC units nor BS units are able to organize into a crystal structure. As a consequence of the disappearance of crystallinity the material, characterized by a low Tg (around -40°C irrespective of composition), changes from rigid to soft

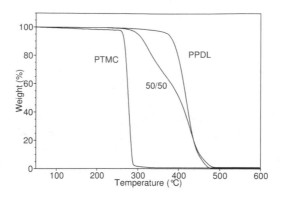

Figure 7. TGA curves of PTMC, PPDL and Poly(PDL-50mol%TMC)

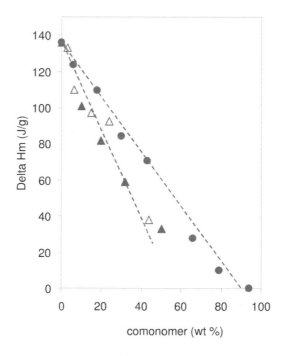

Figure 8. Composition dependence of melting enthalpy in copolymers of octamethylene adipate with: (△) glycerol adipate, (▲) sorbitol adipate and (●) silicone adipamide.

and sticky. It is reasonable to expect that the susceptibility to hydrolysis of these copolyestercarbonates will parallel the decrease of crystallinity.

Polyol-containing polyesters were also syhthesized using CALB (5, 6) and they were found to show a similar decrease of crystallinity with increasing amount of comonomer units (7). Figure 8 shows the decrease of melting enthalpy that is displayed by copolymers of octanediol adipate with either sorbitol adipate or glycerol adipate with increasing polyol content. Worth mentioning is that, concurrent with the decrease of crystallinity, the increasing amount of hydrophilic

Figure 9. Left: hard crystalline P(OA-co-10mol%SiAA); right: gluelike P(OA-co-50mol%SiAA)

polyol units in the polyester chain may enable fine tuning of the degradation rate of these copolyesters. Moreover, the hydroxyl functional groups on the polyol moieties along the chain may be exploited to link biomolecules, making this class of polymers attractive candidates for use as bioactive and bioresorbable materials.

CALB was also used to catalyse one-pot reactions between diethyl adipate (DEA), 1,8-octanediol (OD), and α,ω-(diaminopropyl)polydimethylsiloxane (SiNH$_2$) under mild conditions (*11*). The obtained silicone polyesteramides, that exhibit blocklike sequence distribution, show physical properties that strongly depend on the relative amount of amide and ester units along the polymer chain. High content of DEA-OA units leads to hard solid materials containing a well developed high-melting poly(octamethylene adipate)-type (POA) crystal phase, whose melting temperature and enthalpy change with composition. The decrease of melting enthalpy (i.e. of crystallinity) with increasing silicone adipamide content is shown in Figure 8, where comonomer content is reported in terms of weight fraction. When the DEA-SiAA units mole content is $\geq 33\%$ (66wt% in Figure 8), the material acquires a sticky appearance. As an example, Figure 9 compares the picture of hard crystalline P(OA-*co*-10mol%SiAA) with that of gluelike P(OA-*co*-50mol%SiAA), with the aim to illustrate the wide range of physical properties that these materials can display.

Conclusions

The copolymers of PDL, synthesized by biocatalysis, show a peculiar and rather uncommon crystallizing behavior, namely cocrystallization of the comonomer units leading to either isomorphic substitution or to isodimorphism. This behavior is associated with the peculiar crystallizing ability of the long PDL unit, that is able to incorporate into its polyethylene-type crystal lattice a number of different monomer units. These crystalline copolymers with tunable hydrolytic degradation rates can be used in medical applications, for example as scaffolds for tissue engineering. Such cell supporting structures have been produced using PPDL and PDL-copolymers both by electrospinning technology and by supercritical CO_2 foaming.

On the other hand, biosynthesized condensation copolymers, whose amount of crystal phase decreases with increasing comonomer content, display solid state properties that cover the whole range from hard solid materials down to gluelike substances. Such properties can be easily tailored by composition control and can be adjusted to meet the requirements of a broad range of applications.

Acknowledgments

Financial support from Italian Ministry of Foreign Affairs (Directorate General for Cultural Promotion and Cooperation - Significant Bilateral Project Italy-USA) is gratefully acknowledged.

References

1. Gross, R. A.; Kumar, A.; Kalra, B. *Chem. Rev.* **2001**, *101*, 2097–2124.
2. Kumar, A.; Kalra, B.; Dekhterman, A.; Gross, R. A. *Macromolecules* **2000**, *33*, 6303–6309.
3. Ceccorulli, G.; Scandola, M.; Kumar, A.; Kalra, B.; Gross, R. A. *Biomacromolecules* **2005**, *6*, 902–907.
4. Jiang, Z.; Azim, H.; Gross, R. A.; Focarete, M.L.; Scandola, M. *Biomacromolecules* **2007**, *8*, 2262–2269.
5. Kulshrestha, A.; Kumar, A.; Gao, W.; Gross, R. A. *Polym. Prepr., Am. Chem. Soc.* **2003**, *44*, 585–586.
6. Kumar, A.; Kulshrestha, A.; Gao, W.; Gross, R. A. *Macromolecules* **2003**, *36*, 8219–8221.
7. Fu, H.; Kulshrestha, A.S.; Gao, W.; Gross, R. A.; Baiardo, M.; Scandola, M. *Macromolecules* **2003**, *36*, 9804–9808.
8. Kumar, A.; Garg, K.; Gross, R. A. *Macromolecules* **2001**, *34*, 3527–3533.
9. Focarete, M.L.; Gazzano, M.; Scandola, M.; Gross, R. A. *Macromolecules* **2002**, *35*, 8066–8071.
10. Zini, E.; Scandola, M.; Jiang, Z.; Liu, C.; Gross, R. A. *Macromolecules* **2008**, *41*, 4681–4687.
11. Sharma, B.; Azim, A.; Azim, H.; Gross, R. A.; Zini, E.; Focarete, M. L.; Scandola, M. *Macromolecules* **2007**, *40*, 7919–7927.
12. Focarete, M. L.; Scandola, M.; Kumar, A.; Gross, R. A. *J. Polym. Sci.: Part B, Polym. Phys.* **2001**, *39*, 1721–1729.
13. Focarete, M. L.; Gualandi, C.; Scandola, M.; Govoni, M.; Giordano, E.; Foroni, L.; Valente, S.; Pasquinelli, G.; Gao, W.; Gross, R. A. *J. Biomater. Sci. (Polymer Ed.)* **2010**, in press.
14. Gualandi, C.; White, L. J.; Chen, L.; Gross, R. A.; Shakesheff, K. M.; Howdle, S. M.; Scandola, M. *Acta Biomater.* **2010**, *6*, 130–136.
15. Gazzano, M.; Malta, V.; Focarete, M. L.; Scandola, M.; Gross, R. A. *J. Polym. Sci.: Part B, Polym. Phys.* **2003**, *41*, 1009–1013.

16. Kalra, B.; Kumar, A.; Gross, R. A.; Baiardo, M.; Scandola, M. *Macromolecules* **2004**, *37*, 1243–1250.

Chapter 15

Lipase-Catalyzed Copolymerization of ω-Pentadecalactone (PDL) and Alkyl Glycolate: Synthesis of Poly(PDL-*co*-GA)

Zhaozhong Jiang* and Jie Liu†

Biomedical Engineering Department, Yale University, 55 Prospect Street, New Haven, Connecticut 06511
*Zhaozhong.jiang@yale.edu
†Current address: Advanced Biomaterials and Tissue Engineering Center, Huazhong University of Science and Technology, Wuhan 430074, China.

Candida antarctica lipase B (CALB) was found to be an efficient catalyst for copolymerization of ω–pentadecalactone (PDL) and ethyl glycolate (EGA) to form poly(PDL-*co*-glycolate) [poly(PDL-*co*-GA)] copolymers. The copolymerization reactions took place at 50-100 °C and were promoted by vacuum. The formed copolyesters contained up to 30 mol% glycolate (GA) units and had typical molecular weight (M_w) ranging from 18000 to 21000 and polydispersity values between 1.6 and 1.9. Compared to EGA content in the monomer feeds, the GA unit content in the corresponding copolymer products was found to be significantly lower primarily attributed to partial EGA elimination during the polycondensation under vacuum. NMR analyses, including statistical analysis on repeat unit sequence distribution, indicate that the PDL and GA unit arrangements in the copolymer chains are not completely random, but with a significant tendency toward an alternating structure. The synthesized poly(PDL-*co*-GA) copolyesters with various GA content were successfully transformed to nanoparticles with an average size between 170 and 190 nm.

Introduction

Although organometallic catalysts have been employed for synthesis of biodegradable aliphatic polyesters (*1*), use of enzymes (such as lipases) as alternative catalysts are preferred due to their higher activity and selectivity, and resultant high purity of products that are also metal-free (*2–6*). In contrast to metal catalysts which are known to be inefficient for polymerization of large lactones (*7*), lipases generally exhibit high activity toward these monomers (*8–10*). Nevertheless, lipases possess rather low activity in polymerization of short chain (e.g., C_2-C_4) substrates (*11*).

Copolyesters containing both hydrophobic and highly hydrophilic repeat units, such as poly(ω-pentadecalactone-*co*-glycolide) [poly(PDL-*co*-GA)] and poly(ω-pentadecalactone-*co*-*p*-dioxanone) [poly(PDL-*co*-DO)] (*12*), are of great interest because their degradation rates and physical properties can be varied over a wide range by adjusting the comonomer unit ratio in the polymers. However, synthesis of such copolymers is extremely challenging since the two types of monomers (e.g., PDL and glycolide) exhibit substantially different reactivity in the presence of lipase catalysts. This paper reports for the first time the synthesis of poly(PDL-*co*-GA) copolymers via copolymerization of PDL and alkyl glycolate under mild reaction temperatures using a lipase catalyst.

Experimental

Materials

Ethyl glycolate (EGA), ω-pentadecalactone (PDL), and diphenyl ether were purchased from Aldrich Chemical Co. in the highest available purity and were used as received. Immobilized *Candida antarctica* lipase B (CALB) supported on acrylic resin or Novozym 435, poly(vinyl alcohol) (PVA, M_w = 30,000-70,000, 88% hydrolyzed), toluene (99.8%, anhydrous), chloroform (HPLC grade), dichloromethane (99.5%), methanol (98%), and chloroform-*d* were also obtained from Aldrich Chemical Co. The lipase catalyst was dried at 50 °C under 2.0 mmHg for 20 h prior to use. Glycolide (99.9%) was purchased from Polysciences, Inc. and was used without further purification.

Instrumentation

^{1}H and ^{13}C NMR spectra were recorded on a Bruker AVANCE 500 spectrometer. The chemical shifts reported were referenced to internal tetramethylsilane (0.00 ppm) or to the solvent resonance at the appropriate frequency. The number and weight average molecular weights (M_n and M_w, respectively) of polymers were measured by gel permeation chromatography (GPC) using a Waters HPLC system. Chloroform was used as the eluent and polymer molecular weights were determined based on a conventional calibration curve generated by narrow polydispersity polystyrene standards from Aldrich Chemical Co. Surface morphology and size of nanoparticles were analyzed using a XL30 ESEM scanning electron microscope (FEI Company). Particle

samples were mounted on an aluminum stub using carbon adhesive tape and were sputter-coated with a mixture of gold and palladium (60:40) under low pressure argon using a Dynavac Mini Coater. The image-analysis application program, ImageJ (developed by Wayne Rasband, NIH), was used to measure particle diameters, to calculate average particle sizes, and to determine particle size distributions.

General Procedure for CALB-Catalyzed Copolymerization of Ethyl Glycolate (EGA) with ω-Pentadecalactone (PDL)

The ring-opening and condensation copolymerizations were performed in diphenyl ether solution using a parallel synthesizer connected to a vacuum line with the vacuum (± 0.2 mmHg) controlled by a digital vacuum regulator. In a typical experiment, EGA and PDL comonomers, Novozym 435 catalyst (10 wt% vs. total monomer), and diphenyl ether solvent (200 wt% vs. total monomer) were combined to form reaction mixtures. The copolymerization reactions were carried out in two stages: first stage oligomerization followed by second stage polymerization. During the first stage reaction, the reaction mixtures were stirred at 50-100 °C under 100-600 mmHg pressure for 15-45 h. Thereafter, the reaction pressure was reduced to 1-3 mmHg and the reactions were continued for an additional 40 to 80 h. Same temperature was employed for both oligomerization and polymerization steps. To monitor polymer chain growth, aliquots were withdrawn for analysis during the second stage polymerization. The aliquot samples were dissolved in HPLC-grade chloroform and filtered to remove the enzyme catalyst. The filtrates containing whole products were analyzed by GPC using polystyrene standards to measure polymer molecular weights. To determine polymer structures, aliquots were dissolved in chloroform-d. The resultant solutions were filtered to remove catalyst particles and then analyzed by ^{1}H and ^{13}C NMR spectroscopy.

PDL-EGA copolymer: ^{1}H NMR (CDCl$_3$; ppm) 1.25 (br., $-OCH_2-CH_2-$ $(CH_2)_{10}-CH_2-CH_2-CO-$), 1.61 (m, $-OCH_2-CH_2-(CH_2)_{10}-CH_2-CH_2-CO-$), 2.29/2.41 (t, $-OCH_2-CH_2-(CH_2)_{10}-CH_2-CH_2-CO-PDL/-OCH_2-CH_2-(CH_2)_{10}-$ $CH_2-CH_2-CO-GA$), 4.05/4.15 (t, PDL$-OCH_2-CH_2-(CH_2)_{10}-CH_2-CH_2-CO-$ /GA$-OCH_2-CH_2-(CH_2)_{10}-CH_2-CH_2-CO-$), 4.60/4.68, 4.73/4.82 (s, $-O-CH_2-$ CO$-$, PDL-GA*-PDL/PDL-GA*-GA, GA-GA*-PDL/GA-GA*-GA triads); ^{13}C NMR (CDCl$_3$; ppm) 24.77, 24.82, 24.99, 25.03, 25.76, 25.79, 25.95, 28.46, 28.50, 28.55, 28.67, 29.08, 29.17, 29.21, 29.27, 29.29, 29.45, 29.48, 29.54-29.64 (m), 33.74, 33.82, 34.40, 60.17, 60.56, 61.07, 64.38, 65.50, 167.15, 167.41, 168.00, 172.99, 173.11, 173.95, plus small absorptions at 173.86/173.36, 65.76/65.70, and 14.27/14.11 ppm presumably due to the presence of two different types of ethyl ester end groups $[-(CH_2)_{14}-COO-CH_2CH_3, -O-CH_2-COO-CH_2CH_3]$ and a small absorption at 60.12 ppm attributable to $-CH_2OH$ end groups.

215

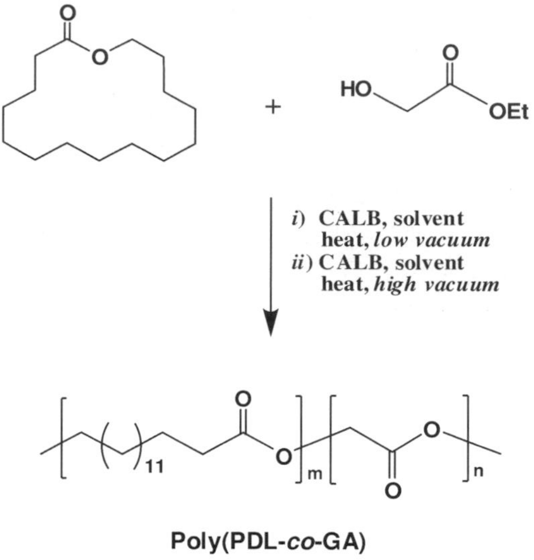

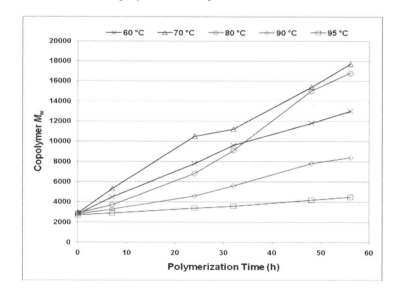

Scheme 1. Two-Stage Process for Poly(PDL-co-GA) Synthesis via
Copolymerization of PDL with EGA

Figure 1. Temperature effects on copolymerization of PDL with EGA
(polymerization conditions: 1:1 PDL/EGA, 1.6 mmHg pressure)

Preparation of Purified Poly(PDL-*co*-glycolate) [Poly(PDL-*co*-GA)] Copolymers with Varied PDL to Glycolate (GA) Unit Ratios

Four poly(PDL-*co*-GA) copolyesters with different unit ratios were synthesized via copolymerization of PDL with EGA following a procedure analogous to the one described above. The monomer molar ratios of PDL to EGA employed for the four reactions were 50:50, 60:40, 70:30, and 80:20, respectively. The first-stage oligomerizations were run at 80 °C under 600 mmHg for 18 h, and the subsequent second-stage polymerizations were carried out at 80 °C under 2.0 mmHg for 78 h. At the end of the reactions, each product mixture was dissolved in chloroform and the resultant chloroform solution was filtered to remove the enzyme catalyst. After being concentrated under vacuum, the filtrate was dropwise added to stirring methanol to cause precipitation of a white solid polymer. The obtained polymer was then filtered, washed with methanol three times, and dried under vacuum at 50 °C for 24 h.

Preparation of Poly(PDL-*co*-GA) Nanoparticles

The nanoparticles were fabricated using an oil-in-water single emulsion technique. In a typical experiment, 100 mg of purified poly(PDL-*co*-GA) was dissolved in 2 ml of methylene chloride in a glass tube. The resultant organic solution was dropwise added to 4 ml of 5 wt% PVA aqueous solution while vortexing. The mixture was subsequently sonicated three times (with each time lasting for 10 seconds) with a TMX 400 sonic disruptor (Tekmar, Cincinnati, Ohio) to yield a homogeneous oil-in-water emulsion. This emulsion was immediately poured into 100 ml of aqueous solution containing 0.3 wt% PVA and the whole mixture was magnetically stirred in an open beaker at room temperature for three hours. This process allows nanoparticles to form via gradual evaporation of the methylene chloride solvent. The formed nanoparticles were collected by centrifugation at 11,000g for 15 minutes, washed three times with deionized water, and then dried in a lyophilizer.

Results and Discussion

Temperature Effects on Copolymerization of PDL with EGA

Scheme 1 illustrates a general copolymerization reaction to form poly(PDL-*co*-GA) from PDL and EGA monomers. The ring-opening and polycondensation reactions were performed at different temperatures in diphenyl ether (200 wt% vs. total monomer) using 1:1 molar ratio of PDL to EGA and Novozym 435 (10 wt% vs. total monomer) as catalyst. The copolymerization reactions were carried out under 1.6 mmHg pressure for 56 h after initial oligomerization at 600 mmHg for 19 h and subsequently at 100 mmHg for 24 h. Figure 1 shows the polymer chain growth vs. polymerization time for the copolymerizations at 60, 70, 80, 90, and 95 °C. For all copolymerization reactions, continuous chain growth was observed during the 56 hour polymerization period. Among these reaction temperatures, the fastest chain growth occurred at 70-80 °C. For example, at 0, 7, 24, 32, 48, 56 h, the copolymerization at 70 °C yielded products with M_w (M_w/M_n shown in parentheses) values of 2900 (1.3), 5300 (1.3), 10500 (1.5), 11200 (1.6), 15400 (1.6), 17700 (1.6), respectively. For all products shown in Figure 1, the polydispersities of the copolymers were in the range between 1.2 and 1.6.

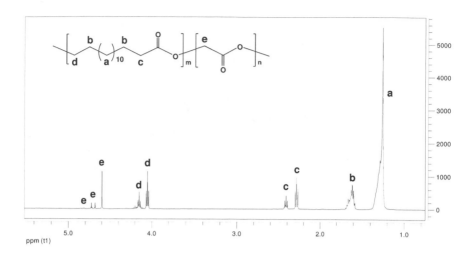

Figure 2. ¹H NMR spectrum of the poly(PDL-co-GA) (28 mol% GA unit content, M_w = 16800, M_n = 10500) formed at 80 °C after 56 h using 1:1 PDL/EGA monomer ratio (solvent: CDCl₃).

PDL-EGA copolymer composition and unit sequence distribution were analyzed by ¹H and ¹³C NMR spectroscopy. NMR resonance absorptions were assigned by comparing signals of PDL-EGA copolymers to those of two reference polymers, poly(PDL) (*9*) and poly(glycolide) (*13*), and by comparing signal intensities between the copolymers formed with different PDL/EGA monomer feed ratios.

The molar ratios of PDL to GA units in PDL-EGA copolymers were calculated from proton resonance absorptions: number of PDL units from total methylene absorptions at 2.29 and 2.41 ppm, and number of GA units from total absorptions at 4.60, 4.68, 4.73, and 4.82 ppm. Figure 2 shows ¹H NMR spectrum of the PDL-EGA copolymer formed at 80 °C after 56 h. The two triplets at 2.29 and 2.41 ppm due to the methylene groups adjacent to a carbonyl in PDL units are attributable to PDL*-PDL [–OCH₂–(CH₂)₁₂–C*H₂*–COO–CH₂–(CH₂)₁₂–CH₂–CO–] and PDL*-GA [–OCH₂–(CH₂)₁₂–C*H₂*–COO–CH₂–CO–] diad structures, respectively, in the polymer chains, whereas the two triplets at 4.05 and 4.15 ppm due to the –OCH₂– groups in PDL units are likely attributed to PDL-PDL* [–OCH₂–(CH₂)₁₂–CH₂–COO–C*H₂*–(CH₂)₁₂–CH₂–CO–] and GA-PDL* [–OCH₂–COO–C*H₂*–(CH₂)₁₂–CH₂–CO–] diad structures, correspondingly. On the other hand, the four singlets at 4.60, 4.68/4.73, and 4.82 ppm due to the methylene groups in GA units correspond respectively to PDL-GA*-PDL, PDL-GA*-GA/GA-GA*-PDL, and GA-GA*-GA triad structures in the copolymer. By using this ¹H NMR analysis method, the PDL-EGA copolymers formed at 60, 70, 80, 90, 95 °C after 56 h reaction time (Figure 1) were found to have GA unit content (vs. total of GA and PDL units) of 31, 30, 28, 27, and 24 mol%, correspondingly.

(A)

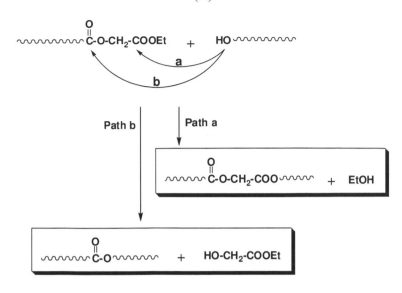

(B)

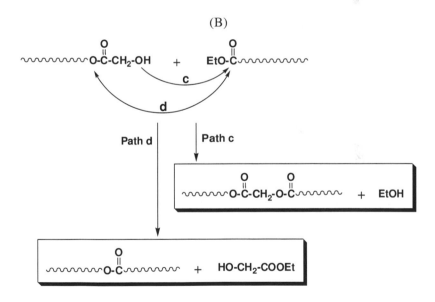

Scheme 2. Proposed Mechanism for Decreasing GA Content in Poly(PDL-co-GA) during Polymer Chain Growth

Table 1. Triad Distributions for Poly(PDL-*co*-GA) Copolyesters Formed via Copolymerization of PDL with EGA[a]

Reaction Temperature (°C)		60	70	80	90	95
Copolymer M_w		13000	17700	16800	8400	4500
GA Unit Content (mol%)[b]		31	30	28	27	24
Triad Distributions						
PDL-PDL*-PDL + GA-PDL*-PDL[b]	Meas.[c]	0.429	0.448	0.469	0.488	0.548
	Calc.[d]	0.476	0.490	0.518	0.533	0.578
PDL-PDL*-GA + GA-PDL*-GA[b]	Meas.[c]	0.266	0.258	0.251	0.242	0.216
	Calc.[d]	0.214	0.210	0.202	0.197	0.182
PDL-GA*-PDL	Meas.[c]	0.208	0.204	0.202	0.197	0.176
	Calc.[d]	0.148	0.147	0.145	0.144	0.139
PDL-GA*-GA/ GA-GA*-PDL	Meas.[c]	0.046/ 0.046	0.043/ 0.043	0.038/ 0.038	0.035/ 0.035	0.029/ 0.029
	Calc.[d]	0.066/ 0.066	0.063/ 0.063	0.056/ 0.056	0.053/ 0.053	0.044/ 0.044
GA-GA*-GA	Meas.[c]	0.004	0.003	0.003	0.003	0.001
	Calc.[d]	0.030	0.027	0.022	0.020	0.014

[a] Reaction conditions: 1:1 (mol/mol) PDL/GA; oligomerization under 600 mmHg for 19 h, and then under 100 mmHg for 24 h; polymerization under 1.6 mmHg for 56 h.   [b] (PDL-PDL*-PDL + GA-PDL*-PDL) and (PDL-PDL*-GA + GA-PDL*-GA) triads are shown as PDL*-PDL and PDL*-GA diads, respectively, in the ^{1}H NMR spectra.   [c] Measured from the ^{1}H NMR spectra.   [d] Calculated for random copolymers. Distribution of triad X-Y-Z = $f_X \times f_Y \times f_Z$, where X, Y and Z independently equal to PDL or GA; f_X, f_Y, and f_Z are molar fractions of X, Y, and Z units, respectively, in the polymer chains.

The above structural assignments of poly(PDL-*co*-GA) copolyesters on the basis of the ^{1}H NMR spectra are further supported by the ^{13}C NMR resonance absorptions of the copolymers.   The absorptions at 173.95, 173.11, 172.99 ppm are attributable to the carbonyl groups of PDL units due to PDL*-PDL [$-OCH_2-(CH_2)_{12}-CH_2-COO-CH_2-(CH_2)_{12}-CH_2-CO-$],   PDL*-GA-PDL [$-OCH_2-(CH_2)_{12}-CH_2-COO-CH_2-COO-CH_2-(CH_2)_{12}-CH_2-CO-$],   and PDL*-GA-GA [$-OCH_2-(CH_2)_{12}-CH_2-COO-CH_2-COO-CH_2-CO-$] structures, respectively, in the polymer chains.   The former is identical to the carbonyl resonance of poly(PDL) homopolymer (*9*) and the intensity of both absorptions at 173.11, 172.99 ppm decreased with decreasing GA content in the copolymer (the synthesis of PDL-EGA copolymers with various compositions is discussed in a latter section).   The resonance absorptions at 168.00, 167.41, and 167.15 are presumably attributed to the carbonyl groups of GA units (*13*) due to PDL-GA*-PDL, PDL-GA*-GA, and GA-GA*-PDL structures, correspondingly, in the chains.   The intensities of the latter two absorptions are comparable and substantially lower than that of the former absorbance for the copolymers with ≤ 30 mol% GA unit contents. The resonances at 64.38 and 65.50 ppm are ascribable to the –OCH$_2$– groups of PDL units due to PDL-PDL [–OCH$_2$–(CH$_2$)$_{12}$–CH$_2$–COO–CH$_2$–(CH$_2$)$_{12}$–CH$_2$–CO–] and GA-PDL [–OCH$_2$–COO–CH$_2$–(CH$_2$)$_{12}$–CH$_2$–CO–] structures in the copolymers. The former resonance is identical to that of poly(PDL) (9). The resonance absorptions at 60.56 and 60.17/61.07 ppm are presumably attributed to the –OCH$_2$– groups of GA units due to PDL-GA*-PDL [–OCH$_2$–(CH$_2$)$_{12}$–CH$_2$–COO–CH$_2$–COO–CH$_2$–(CH$_2$)$_{12}$–CH$_2$–CO–] and PDL-GA*-GA/GA-GA*-PDL [–OCH$_2$–(CH$_2$)$_{12}$–CH$_2$–COO–CH$_2$–COO–CH$_2$–CO–/ –OCH$_2$–COO–CH$_2$–COO–CH$_2$–(CH$_2$)$_{12}$–CH$_2$–CO–] structures, respectively, in the polymer chains. The latter two absorptions had a comparable intensity, but were substantially less intense than the former absorbance for the copolymers with low (≤ 30 mol%) GA contents.

Table 2. Yield and Structure of The Isolated Poly(PDL-*co*-GA) Copolyesters

PDL/EGA (Feed Molar Ratio)		50:50	60:40	70:30	80:20
Polymer Yield		87%	89%	91%	92%
PDL/GA (Unit Molar Ratio)		73:27	79:21	84:16	90:10
GA Unit Content in mol%		27	21	16	10
Copolymer	M_w	19000	21200	19500	17700
	M_w/M_n	1.9	1.9	1.9	1.8
Triad Distributions					
PDL-PDL*-PDL + GA-PDL*-PDL[a]	Meas.[b]	0.483	0.583	0.668	0.775
	Calc.[c]	0.533	0.624	0.706	0.810
PDL-PDL*-GA + GA-PDL*-GA[a]	Meas.[b]	0.250	0.210	0.172	0.122
	Calc.[c]	0.197	0.166	0.134	0.090
PDL-GA*-PDL	Meas.[b]	0.192	0.162	0.134	0.093
	Calc.[c]	0.144	0.131	0.113	0.081
PDL-GA*-GA/ GA-GA*-PDL	Meas.[b]	0.036/ 0.036	0.022/ 0.022	0.013/ 0.013	0.005/ 0.005
	Calc.[c]	0.053/ 0.053	0.035/ 0.035	0.022/ 0.022	0.009/ 0.009
GA-GA*-GA	Meas.[b]	0.004	0.001	0.000	0.000
	Calc.[c]	0.020	0.009	0.004	0.001

[a] (PDL-PDL*-PDL + GA-PDL*-PDL) and (PDL-PDL*-GA + GA-PDL*-GA) triads are shown as PDL*-PDL and PDL*-GA diads, respectively, in the ^{1}H NMR spectra. [b] Measured from the ^{1}H NMR spectra. [c] Calculated for random copolymers using the methods shown in Table 1.

Compared to the EGA content (50 mol%) in the monomer feeds, the lower GA unit contents (24-31 mol%) in the final PDL-EGA copolymers appear to result from partial elimination of ethyl glycolate during the second stage polymerization under high vacuum. Scheme 2 illustrates two possible pathways (path b of Scheme 2A and path d of Scheme 2B) accountable for decreasing glycolate content in poly(PDL-co-GA) during polymer chain growth. NMR and GPC analyses indicated that for the copolymerizations of PDL with EGA, PDL was completely reacted during the oligomerization step. Thus, polymer chain growth during the second stage polymerization took place predominantly via polycondensation reactions. Among these reactions, two transesterification reactions are anticipated: reaction between –COO–CH_2–COOEt and hydroxyl end groups (paths a and b, Scheme 2A), and reaction between –OOC–CH_2–OH and –COOEt end groups (paths c and d, Scheme 2B). The presence of two different types of ethyl ester end groups [–$(CH_2)_{14}$–COO–CH_2CH_3, –O–CH_2–COO–CH_2CH_3] is supported by small ^{13}C NMR resonances at 173.86/173.36, 65.76/65.70, and 14.27/14.11 ppm. Although the presence of –OOC–CH_2–OH end groups could not be confirmed due to peak overlaps, the observed small proton resonance (triplet) at 3.6 ppm and a small ^{13}C resonance absorption at 60.12 ppm are attributable to –OOC-$(CH_2)_{13}$-CH_2-OH termini. The transesterification paths a and c would lead to longer polymer chains via elimination of byproduct ethanol. In contrast, the reaction paths c and d would promote polymer chain growth via elimination of ethyl glycolate, which would, of course, reduce GA content in the polymer product. NMR analyses on aliquots withdrawn during the copolymerization of PDL and EGA at 80 °C (Figure 1) showed that the copolymers formed at 0, 7, 24, 32, 48, and 56 h had GA unit contents of 50, 40, 28, 28, 28, and 28 mol%, respectively. Thus, glycolate elimination takes place predominantly during the initial 24 h period of polycondensation. Thereafter, the glycolate unit content in the copolymers remained fairly constant. The proposed mechanism is also consistent with the observation that both ethanol and ethyl glycolate were present and condensed in the dry ice cold trap between the reactors and vacuum pump.

Although two different diads (PDL-PDL, PDL-GA) involving PDL unit and four different triads (PDL-GA*-PDL, PDL-GA*-GA, GA-GA*-PDL, GA-GA*-GA) involving GA unit were observed (Figure 2), the distribution of PDL and GA units in the copolymer chains does not appear totally random. Table 1 shows the distributions of (PDL-PDL*-PDL + GA-PDL*-PDL), (PDL-PDL*-GA + GA-PDL*-GA), PDL-GA*-PDL, PDL-GA*-GA, GA-GA*-PDL, and GA-GA*-GA triads for the polymer products formed via copolymerization of PDL with EGA after 56 h at 60, 70, 80, 90, and 95 °C. It needs to be noted that PDL*-PDL and PDL*-GA diads represent two sets of triads (PDL-PDL*-PDL + GA-PDL*-PDL) and (PDL-PDL*-GA + GA-PDL*-GA), respectively, in the ^{1}H NMR spectra due to near identical chemical shifts of proton resonances for the middle PDL units of the two triads in each set (Figure 2). For comparison, distributions of the triads were calculated for random copolymers with same compositions and the results are also included in Table 1. The data clearly indicate that regardless of the various reaction temperatures used, the PDL and GA unit arrangements in the polymer chains have a significant tendency toward an alternating structure.

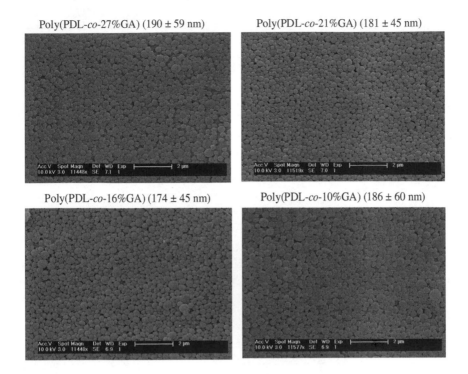

Poly(PDL-*co*-27%GA) (190 ± 59 nm)

Poly(PDL-*co*-21%GA) (181 ± 45 nm)

Poly(PDL-*co*-16%GA) (174 ± 45 nm)

Poly(PDL-*co*-10%GA) (186 ± 60 nm)

Figure 3. SEM micrographs of nanoparticles fabricated from purified poly(PDL-co-GA) copolymers. Scale bar: 2 μm

To determine whether the polymerization reactions were indeed catalyzed by CALB, control experiments were performed without the lipase. The control reaction was performed in diphenyl ether under identical conditions (stage 1: 1/1 PDL:EGA, 80 °C, 600 mmHg for 19 h, 100 mmHg for 24 h; stage 2: 80 °C, 1.6 mmHg for 56 h). Analysis of the resulting product by GPC showed that its M_w is below 800. Thus, CALB is indeed the catalyst for copolymerization of PDL and EGA.

To compare the reactivity between alkyl glycolate and the conventional monomer glycolide in copolymerization with PDL, a 1:2 (molar ratio) glycolide/PDL mixture in toluene (200 wt% vs. total monomer) was stirred in the presence of Novozym 435 (10 wt% vs. total monomer) at 50, 60, 70, 80, and 90 °C under 1 atmosphere of nitrogen for 24 h. GPC analysis showed that no polymers with $M_w > 800$ were formed during the reactions. Analyses of the products by NMR spectroscopy indicated that majority of PDL and glycolide comonomers remain unreacted at the end of the reactions. Since it is known that PDL ring-opening polymerization proceeds rapidly over Novozym 435 catalyst in toluene at the above temperatures (*8*), the current results suggest that glycolide is a strong inhibitor for CALB-catalyzed polymerization of lactones.

Preparation of Purified Poly(PDL-*co*-GA) Copolyesters with Various PDL to GA Unit Ratios and Their Nanoparticle Fabrication

Procedures were developed for preparation of pure poly(PDL-*co*-GA) copolymers with various compositions. The copolymers synthesized using 50:50, 60:40, 70:30, and 80:20 PDL/EGA monomer molar ratios were purified via reprecipitation in chloroform/methanol mixture. Table 2 summarizes the polymer yield, composition, molecular weight (M_w), and polydispersity for the four purified polymers. As shown in Table 2, the copolymers were obtained in good yields (87-92%) with M_w ranging from 17700 to 21200 and M_w/M_n between 1.8 and 1.9.

For the reactions at PDL/EGA feed molar ratio below 40:60, only low molecular weight oligomers ($M_w < 1000$) were formed. With the monomer feeds containing 50, 40, 30, and 20 mol% EGA, the PDL-EGA copolymerizations resulted in the formation of poly(PDL-*co*-GA) copolymers with 27, 21, 16, and 10 mol% GA unit contents, respectively, in the polymer chains (Table 2). The triad distributions for the purified copolymers, along with the calculated values for random copolymers with same compositions, are also shown in Table 2. The results again indicate that among all copolymers with different compositions, the PDL and GA unit arrangements in the polymer chains have a considerable tendency toward an alternating structure. Furthermore, nanoparticles with an average size between 170 and 190 nm were successfully prepared from all four purified copolymer samples (Figure 3). These particles are potentially useful carriers for delivery of therapeutic drugs.

Conclusions

This paper represents the first report on the synthesis of copolyesters containing both large lactone and GA units using an enzyme catalyst. The synthesized poly(PDL-*co*-GA) copolymers had significantly lower GA unit contents compared to EGA contents in the corresponding monomer feeds. This is primarily attributable to partial EGA elimination during the polycondensation under vacuum. NMR analyses, including statistical analysis on repeat unit sequence distribution, indicate that the PDL and GA unit arrangements in the copolymer chains are not completely random, but with a significant tendency toward an alternating structure. Free-standing nanoparticles were successfully fabricated from poly(PDL-*co*-GA) with various GA unit contents. The metal-free poly(PDL-*co*-GA) copolyesters are analogues of typical commercial copolyesters, such as poly(ε-caprolactone-*co*-GA), and are potentially important biodegradable materials suitable for medical applications.

Acknowledgments

The authors wish to thank Yale University for the financial support of this work.

References

1. Stridsberg, K. M.; Ryner, M.; Albertsson, A.-C. *Adv. Polym. Sci.* **2002**, *157*, 41–65.
2. Kobayashi, S. *Macromol. Rapid Commun.* **2009**, *30*, 237–266.
3. Uyama, H.; Kobayashi, S. *Adv. Polym. Sci.* **2006**, *194*, 133–158.
4. Matsumura, S. *Adv. Polym. Sci.* **2006**, *194*, 95–132.
5. Gross, R. A.; Kumar, A.; Kalra, B. *Chem. Rev.* **2001**, *101*, 2097.
6. Kobayashi, S.; Uyama, H.; Kimura, S. *Chem. Rev.* **2001**, *101*, 3793.
7. Nomura, R.; Ueno, A.; Endo, T. *Macromolecules* **1994**, *27*, 620.
8. Kumar, A.; Kalra, B.; Dekhterman, A.; Gross, R. A. *Macromolecules* **2000**, *33*, 6303.
9. Bisht, K. S.; Henderson, L. A.; Gross, R. A.; Kaplan, D. L.; Swift, G. *Macromolecules* **1997**, *30*, 2705–2711.
10. Uyama, H.; Kikuchi, H.; Takeya, K.; Kobayashi, S. *Acta Polym.* **1996**, *47*, 357.
11. Azim, H.; Dekhterman, A.; Jiang, Z.; Gross, R. A. *Biomacromolecules* **2006**, *7*, 3093–3097.
12. Jiang, Z.; Azim, H.; Gross, R. A.; Focarete, M. L.; Scandola, M. *Biomacromolecules* **2007**, *8*, 2262–2269.
13. Dobrzynski, P.; Kasperczyk, J.; Bero, M. *Macromolecules* **1999**, *32*, 4735–4737.

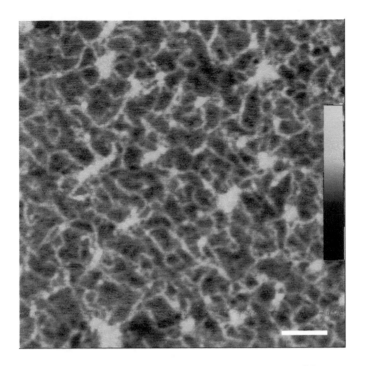

Figure 3. AFM image of sugar beet pectin network deposited from water at a concentration of 12.5 µg/mL. Scale bar is 100 nm, inset height scale, 0-2.5 nm.

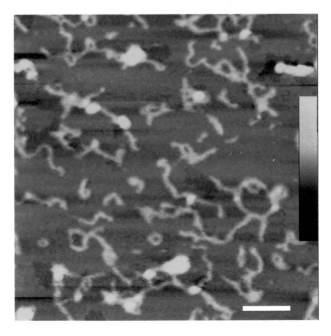

Figure 4. AFM image of sugar beet pectin network deposited from water at a concentration of 6.25 µg/mL. Scale bar is 100 nm, inset height scale, 0-5 nm.

Color Insert - 1

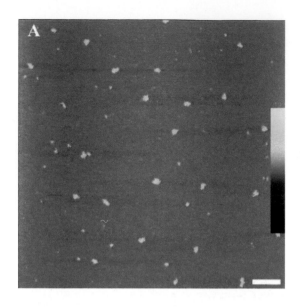

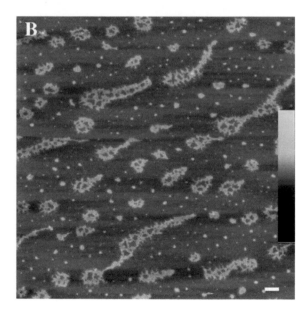

*Figure 6. AFM images of ASP I deposited from water. Sample 10/100/30. A.)
solution concentration 0.125 μg/mL. Scale bar is 250 nm; inset height scale is
0-5 nm. B.) Solution concentration 25 μg/mL. Scale bar is 250 nm; inset height
scale is 0-10 nm*

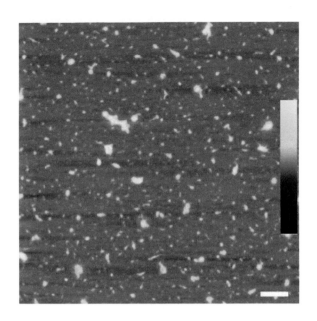

Figure 7. AFM images of ASP II deposited from water. Sample 10/100/30. Solution concentration 25 µg/mL. Scale bar is 250 nm; inset height scale is 0-10 nm.

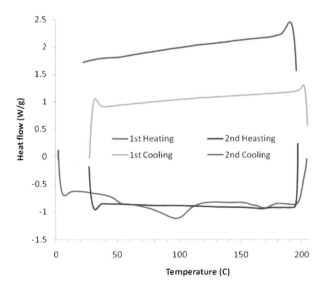

Figure 3. Heating-cooling-heating scans of P3HB3HV4HV

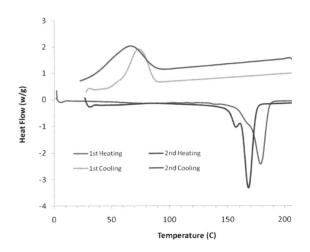

Figure 4. Heating-cooling-heating scans of P3HB

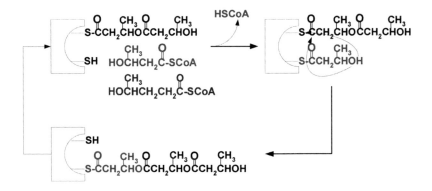

Figure 6. A two-site working mode of PHA synthase with chain elongation via transesterification. The synthase prefers 3-hydroxyacyl-CoA to 4-hydroxyacyl-CoA that is occasionally incorporated into the backbone because of enzyme's error.

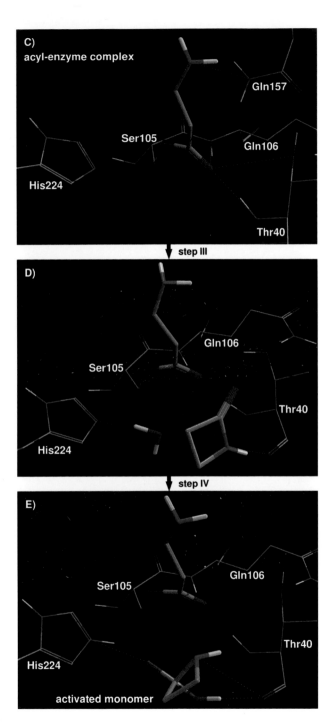

Figure 8. Covalent docking of acyl-enzyme complex (C), docking of second monomer in presence of catalytic water (D) and activation of β-lactam by hydroxyl group of this water molecule (step IV, E)

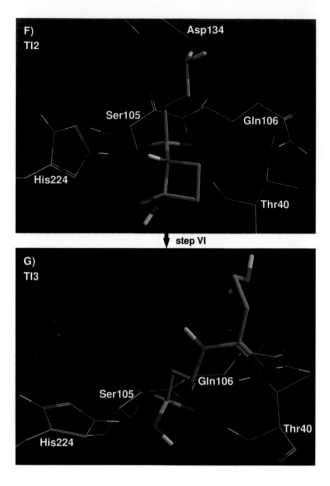

Figure 9. Rearrangement of short-lived intermediate (TI2) in which Ser105-O jumps from central carbon to terminal carbon yielding terminal bound tetrahedral intermediate (TI3).

A

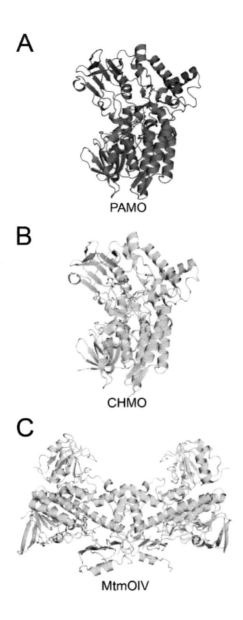

PAMO

B

CHMO

C

MtmOIV

Figure 2. Comparison of the three BVMO crystal structures. (A) Overall
structure of PAMO, shown in blue. (B) Overall structure of the closed
conformation of CHMO, shown in green. (C) Overall structure of MtmOIV, with
one protomer shown in yellow and the other shown in wheat.

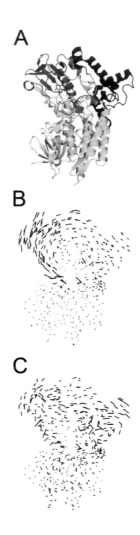

Figure 3. Domain rotations in Type I BVMOs. (A) Overall structure of the closed conformation of CHMO, colored by domain. The FAD-binding domain is shown in wheat, the NADPH-binding domain in blue, the helical domain in brown, the linker regions in green, the key mobile, catalytic loop in orange, and the BVMO signature sequence in yellow. (B) Equivalent atom representation of the open and closed conformations of CHMO. After aligning the structures using the FAD-binding domain, vectors were drawn between equivalent C-α atoms. (C) Same as in (B), but comparing the open conformation of CHMO with PAMO (37).

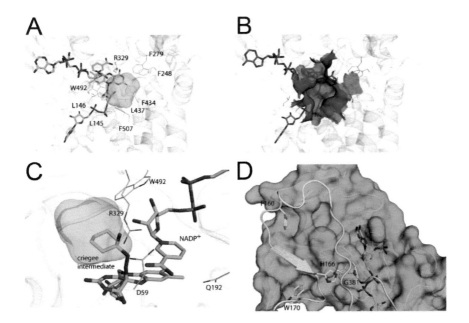

Figure 4. (A) The putative substrate binding cavity as seen in the closed conformation of CHMO, formed primarily by hydrophobic residues. (B) In comparison, the larger binding cavity in the open conformation caused by the movement of the loop formed by residues 487-504 can be seen. (C) Model of the Criegee intermediate based on the structures of CHMO. (D) The BVMO signature sequence (yellow) is shown here to be anchored by F160 and W170, which interacts with hydrophobic pockets in the NADPH-binding domain. H166 is shown to be interacting with G381, part of a key linker (cyan) that connects the FAD- and NADPH-binding domains (37).

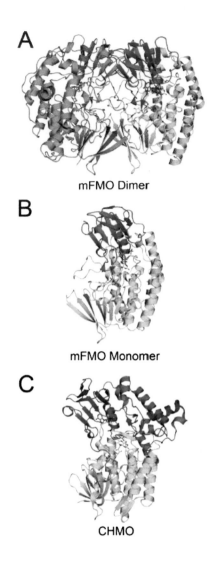

A

mFMO Dimer

B

mFMO Monomer

C

CHMO

Figure 5. Comparison of the crystal structure of mFMO with the open conformation of CHMO. (A)The overall structure of mFMO. The FAD- and NADPH-binding domains of one protomer are shown in wheat and blue, respectively. The second protomer in the dimeric mFMO is shown in grey. (B) One protomer of mFMO, using the same colour scheme as in (A). (C) The overall structure of the open conformation of CHMO, using the same colour scheme as in (A).

Chapter 16

Chemo-Enzymatic Syntheses of Polyester-Urethanes

Karla A. Barrera-Rivera,[a] Ángel Marcos-Fernández,[b] and Antonio Martínez-Richa[a],*

[a]Departamento de Química, Universidad de Guanajuato, Noria alta s/n. Guanajuato, Gto., 36050, México.
[b]Departamento de Química y Tecnología de Elastómeros, Instituto de Ciencia y Tecnología de Polímeros (CSIC), Juan de la Cierva No. 3, 28006 Madrid, Spain.
*richa@quijote.ugto.mx

The enzymatic synthesis of α-ω-telechelic polycaprolactone diols (HOPCLOH) and triblock copolymers was studied. Synthesis of α-ω-telechelic PCL diols was achieved by enzymatic ring opening polymerization with *Yarrowia lipolytica* lipase immobilized on a macroporous resin Lewatit VP OC 1026, and using diethylene glycol and poly(ethylene glycol) as initiators. Biodegradable linear polyester-urethanes were prepared from synthesized PCL diols and hexamethylenediisocyanate (HDI). Depending on the length of PCL in HOPCLOH, the polymers were amorphous or semicrystalline. Measured mechanical properties strongly depend upon the degree of crystallinity of HOPCLOH.

Introduction

The development of injectable materials to be used in non-invasive surgical procedures has triggered much attention in recent years. These materials are required to display low viscosity at insertion time, while a gel or solid consistency is developed *in situ*, later on. Block copolymers comprising poly(ethylene glycol) (PEG) segments and biodegradable polyester blocks such as poly(lactic acid), poly(glycolic acid) and poly(caprolactone) (PCL), have been described by various groups (*1–3*). The implementation of green chemistry in the field of polymer science includes the design and synthesis of sustainable plastics. Such plastics should be produced from feedstocks derived from renewable or biomass resources by environmentally benign processes, such as enzymatic and solvent free processes, and avoiding the use of hazardous materials. Moreover, they should be chemically recyclable and biodegradable. In addition, a high-performance material that leads to a reduction in its consumption is very important. A chemically recyclable and biodegradable polymer contains enzymatically cleavable linkages, such as ester and carbonate linkages. Such a polymer chain can be further broken down by environmental microbes during biodegradation. Also, it can be cleaved by a specific enzyme into oligomers or monomers that can be repolymerized in the reverse reaction of the enzyme (chemical recycling) (4). Polyurethane polymers are extremely important and versatile materials having numerous applications in foams, surface and textile coatings, adhesives and elastomers. They are used in a wide variety of industries such as furniture, construction, aircraft and automobile manufacture and mining equipment. These materials are manufactured from hydroxyl terminated polyester resins made by the high temperature Lewis acid catalyzed condensation of a diacid and diol, or hydroxyl terminated polyethers derived from propylene oxide, in both cases the subsequent reaction with diisocyanates produces the polyurethane polymer (5). Enzyme-catalyzed polymerization may become a versatile method for the production of sustainable polyurethanes, as lipase, for example, is a renewable catalyst with high catalytic activities (6). The most prominent advantage of using a hydrolysis enzyme for the production of polymers is the reversible polymerization-degradation reaction that allows chemical recycling (7).

The focus of this work is the synthesis of α, ω-telechelic poly(ε-caprolactone) diols (HOPCLOH), diblock and triblock copolymers using a one-step enzymatic method, useful for polyurethane synthesis.

Experimental

Materials

ε-CL (Aldrich) was distilled at 97-98 °C over CaH_2 at 10 mm Hg. Diethylene glycol (DEG) and poly(ethylene glycol) with different molecular weights (PEG200), (PEG400) and (PEG1000), Lewatit VP OC 1026 beads, stannous 2-ethylhexanoate, hexamethylenediisocyanate (HDI) and 1,2-dichloroethane anhydrous 99.8 % were purchased from Sigma Aldrich and used as received.

Instrumentation

Solution ^{1}H and ^{13}C-NMR spectra were recorded at room temperature on a Varian Gemini 2000 instrument. Chloroform-d ($CDCl_3$) was used as solvent. Matrix-assisted laser desorption ionization time-of-flight (MALDI-TOF) spectra were recorded in the linear mode by using a Voyager DE-PRO time-of-flight mass spectrometer (Applied Biosystems) equipped with a nitrogen laser emitting at λ= 337 nm with a 3 ns pulse width and working in positive-ion mode and delayed extraction. A high acceleration voltage of 20 kV was employed.

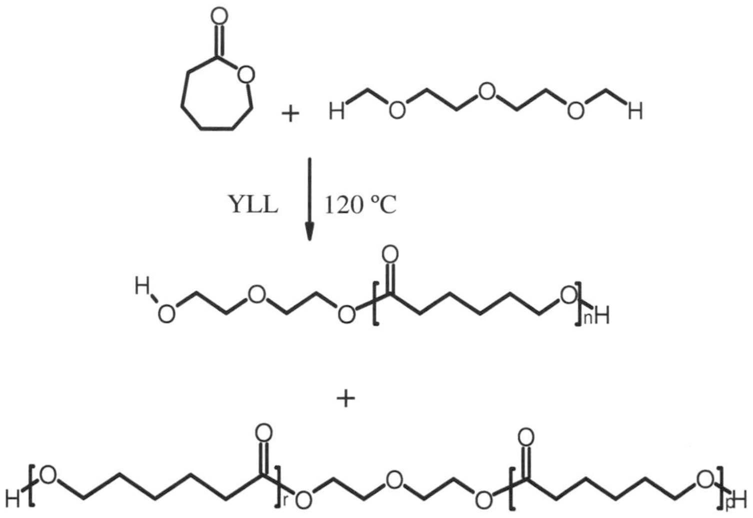

Figure 1. Synthesis of α, ω-telechelic poly(ε-caprolactone) diols (HOPCLOH).

2,5-dihydroxybenzoic acid (DHB) was used as matrix. Samples were dissolved in acetonitrile and mixed with the matrix at a molar ratio of approximately 1:100. DSC thermograms were obtained in a Mettler-Toledo 820e calorimeter using heating and cooling rates of 10 °C/min. Thermal scans were performed from 0 °C to 100 °C. FT-IR spectra were obtained with the ATR technique on films deposited over a diamond crystal on a Perkin-Elmer 100 spectrometer with an average of 4 scans at 4 cm^{-1} resolution. Gel permeation chromatography (GPC-MALLS) was used to determine molecular weights and molecular weight distributions, M_w/M_n, of macrodiols samples. The chromatographic set-up used consists of an Alliance HPLC Waters 2695 Separation Module having a vacuum degassing facility online, an auto sampler, a quaternary pump, a columns thermostat, and a Waters 2414 Differential Refractometer for determining the distribution of molecular weight. The temperature of the columns was controlled at 33 °C by the thermostat. Tensile properties were measured in a MTS Synergie 200 testing machine equipped with a 100 N load cell. Type 3 dumbbell test specimens (according to ISO 37) were cut from film. A crosshead speed of 200 mm/min was used. Strain was measured from crosshead separation and referred to 12 mm initial length. Five samples were tested for each polymer composition.

Lipase Isolation and Immobilization

Lipase production by *Yarrowia lipolytica* (YLL) was made as previously reported by Barrera *et al* (8). Before immobilization, Lewatit 1026 beads were activated with ethanol (1:10 beads: ethanol), washed with distilled water and dried under vacuum for 24 h at room temperature. The beads (1g) were shaken in a rotatory shaker in 15 mL of lipase solution with 0.1568 mg/mL of YLL at 4°C for

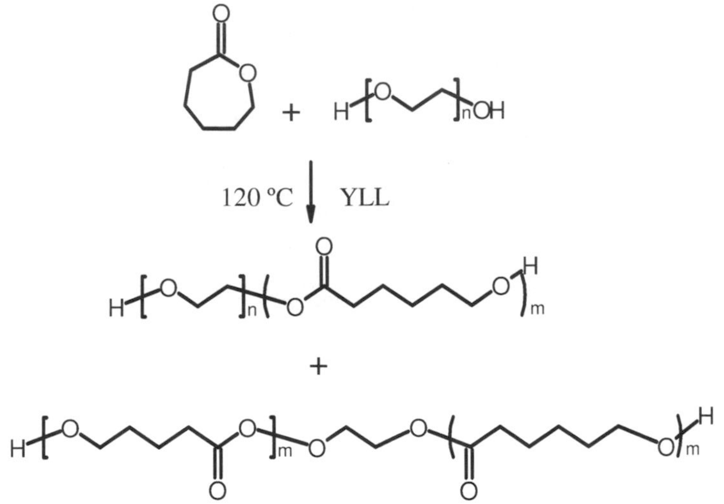

Figure 2. Synthesis of PEG-PCL diblock and triblock copolymers.

24 h. After incubation, the carrier was filtered off, washed with distilled water and then dried under vacuum for 24 h at room temperature.

Synthesis of α,ω-Telechelic Poly(ε-caprolactone) Diols (HOPCLOH)

Samples DEG1PCL, (10 mmol of ε-CL, 1 mmol of DEG), DEG2PCL (10 mmol of ε-CL, 0.5 mmol of DEG) and DEG3PCL (10 mmol of ε-CL,0.25 mmol of DEG) were placed in a 10 mL vial previously dried, and in all cases 12 mg of immobilized YLL was added. Vials were stoppered with a teflon silicon septum and placed in a thermostated bath at 120 °C for 6 h. No inert atmosphere was used. After the reaction was stopped, the enzyme was filtered off and the final polymer was analyzed by FT-IR, ^{1}H and ^{13}C-NMR, GPC-MALLS, DSC and MALDI-TOF.

Synthesis of the PEG-(CL)$_n$ Copolymers

The PEG-(CL)$_n$ copolymer was synthesized by a ring opening polymerization reaction. PEG200PCL (10 mmol of ε-CL, 1 mmol of PEG200), PEG400PCL (10 mmol of ε-CL, 1 mmol of PEG400) and PEG1000PCL (10 mmol of ε-CL, 0.1 mmol of PEG1000) were placed in a 10 mL vial previously dried, and then 12 mg of immobilized YLL was added. The reaction proceeded at 120 °C for 6 h. By varying the CL/PEG ratios and PEG molecular weight, caprolactone blocks of different length were produced. Depending on the molecular weight of the CL segment, soft waxes to hard solids were obtained.

PCL diols were dried at 70 °C *in vacuo* for 12 h, and stored at ambient temperature in a dessicator at vacuum until used.

230

Table 1. Molecular Weights of the synthesized poly(ε-caprolactone) diols

	Mn (GPC)	Mw/Mn (GPC)	Mn (MALDI)	Mn (Da) ¹H-NMR
PCLPEG200	3817	1.132	974	1066
PCLPEG400	4083	1.163	1120	1211
PCLPEG1000	4481	1.214	971	2504
PCLDEG1	4321	1.181	1363	836
PCLDEG2	5101	1.272	1978	1305
PCLDEG3	7426	1.531	2429	1780

Table 2. Percent of bisubstitution (% Bi (OH)), monosubstitution (% Mono (OH)) in the synthesized poly(ε-caprolactone) diols

	% Mono (OH)	% Bi (OH)	HAPCLOH (%)
PCLPEG200	47	53	10
PCLPEG400	44	56	13
PCLPEG1000	52	48	56
PCLDEG1	34	66	8
PCLDEG2	29	71	9
PCLDEG3	23	77	48

Synthesis of PCL Polyurethanes

Dry PCL diol (1.5 g) and HDI in the appropriate amount (OH:NCO ratio = 1:1) and 2 mL of 1,2-dichloroethane were charged into a round bottom flask. The catalyst, stannous 2-ethylhexanoate (1% mol by PCL diol moles) was added, and stirred for 4 h at 80 °C. The resulting solution was poured over a leveled glass. The solution was covered by a conical funnel to protect it from dust and to avoid the excessively fast solvent evaporation, and allowed to stand at ambient temperature for 24 h. The film was then released and dried in vacuum. Samples for physical characterization were cut from films; film thickness ranged from 50-80 μm.

Results and Discussion

Lipase Isolation and Immobilization

Lipase with a protein concentration of 0.1568 mg/mL was obtained. Lewatit VP OC 1026, a crosslinked polystyrene (macroporous) resin with a matrix active group di-2-ethylhexyl phosphate, was used. The saturation time for YLL absorption was 30 min. The enhanced adsorption rates of polystyrenic resins are attributed to stronger hydrophobic interactions between styrenic surfaces, functional groups and YLL. The dependence of adsorption rate on particle size is

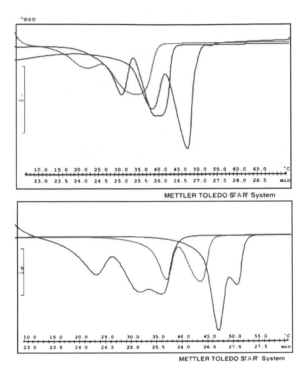

Figure 3. DSC second heating curves (20 °C/min) for the synthesized poly(ε-caprolactone) diols. Top: PCLPEG200 (black), PCLPEG400 (red) and PCLPEG1000 (blue). Bottom: PCLDEG1 (black), PCLDEG2 (red) and PCLDEG3 (blue).

due to the pore size that is limiting protein transport to the inside of the particles. Final resin with a protein content of 0.136 mg of protein/g of resin and a protein adsorption of 87 % was used in polymer synthesis.

Synthesis of α,ω-Telechelic Poly(ε-caprolactone) Diols (HOPCLOH)

The synthetic pathway is described in Figure 1. The first step was the formation of the PCL diols by the ring-opening polymerization of ε-CL and using DEG as initiator in the presence of immobilized YLL. Under the same conditions, CL and DEG were allowed to polymerize in the absence of enzyme as control. After precipitation, no corresponding polymers could be obtained, which indicate that the lipase enzymes actually catalyze the polymerization of CL and DEG. Results for the three synthesized PCL diols are shown in Tables 1 and 2.

Synthesis of the PEG-(CL)ₙ Copolymers

The synthetic pathway is described in Figure 2. Three PEG-PCL copolymers (shown in Tables 1 and 2) with different molecular weights and compositions were easily synthesized by changing the feed molar ratio of ε-CL/PEG in the presence

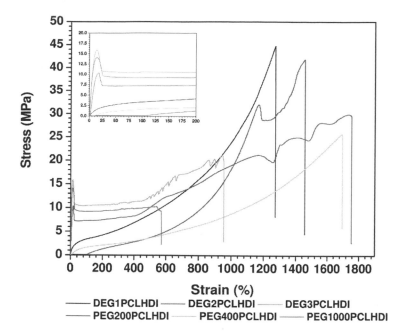

Figure 4. Stress–strain graph for the synthesized polyurethanes at room temperature.

of immobilized lipase. The apparent molecular weights and polydispersities of all samples were also measured by GPC. As shown in Table 1, the diblock copolymers exhibited a narrow molecular weight distribution.

In the last column of Table 2 we report the percentage of HAPCLOH present in the polymers. In some cases (PCLPEG1000 and PCLDEG3) in which the PEG1000 and DEG concentrations are lower (0.1 mmol and 0.25 mmol respectively) it reaches more than 50 % of the total polymer chains. These behavior could be associated to an increase in the chain length of PCL and a decrease in the initiator concentration, which causes that the proportional molecules of water present in the reaction medium (*9*) that initiates the reaction becomes larger.

DSC results show crystallization due to caprolactone segments, with an increase in the melting point when polymer length increases. Melting temperatures decrease in the polymers which have a lesser content of HAPCLOH chains. In the second scan a biphasic behavior can be observed, attributed to the existence of a multiphase morphology (Figure 3).

In the second stage, OH-terminated PCL was reacted with a stoichiometric amount of HDI (respect to HO hydroxyl groups), to form the PCL polyurethane. Finally, the chemical composition of the polymers was studied by different techniques. The FT-IR spectra of the polymers revealed the complete disappearance of the isocyanate peak 2268 cm[-1] and the appearance of urethane bands at 3320 and 2263 cm[-1]. The presence of urethane groups in the polymer was corroborated by ^{13}C-NMR, where the characteristic peak at 156.3 ppm was clearly observed, while the isocyanate band at 122.1 ppm totally disappeared.

Table 3. Mechanical properties of the synthesized polyurethanes

Polymer	Strain at yield (%)	Stress at break (MPa)	Strain at break (%)	Modulus (MPa)
PCLPEG200HDI	42 ± 4	43 ± 5	1678 ± 161	11 ± 2
PCLPEG400HDI	27 ± 1.3	26 ± 1.4	1698 ± 36	4 ± 0.2
PCLPEG1000HDI	16 ± 0.2	14 ± 0.1	20 ± 2	222 ± 7.3
PCLDEG1HDI	67 ± 8	45 ± 6	1434 ± 127	11 ± 2.2
PCLDEG2HDI	39 ± 2	32 ± 2	1779 ± 137.4	122 ± 47
PCLDEG3HDI	23 ± 3	21 ± 2.2	953 ± 373	228 ± 10

The stress-strain curves of the different polyurethanes are shown in Figure 4. Characteristic values derived from these curves are presented in Table 3.

Two very differentiated behaviors can be observed. When PCL segments crystallize extensively, the polymers behave as tough plastics and present a high modulus followed by yielding and high extension. A very small increase in the stress value is observed until the narrow part of the specimen is completely extended and reaches the wide part of the dumbbell; after that, the stress increases again until the sample breaks. When the crystallization of the PCL segments is limited, the polymer behaves as an elastomer with low modulus, and a steady increase of stress as a function of strain until rupture is seen. For samples PCLPEG400HDI and PCLPEG200HDI, the stress-strain diagrams are curved for practically the entire range of stress. PCLPEG400HDI shows a lower modulus than PCLPEG200HDI. This behavior can be associated to the PCL chain length, which is shorter for the PCLPEG400 diol. In both cases, PCL chains are amorphous and does not crystallize (as observed by DSC). PEG content in the polyurethane is larger in PCLPEG400HDI, and this reflects in a softer material. In both cases, it can be concluded that final polymers exhibit good mechanical performance.

Conclusion

A series of PCL–PEG–PCL block copolymer diols, and α,ω-telechelic poly (ε-caprolactone) diols were successfully synthesized for the first time, using immobilized lipase from *Yarrowia lipolytica* as catalyst. Results demonstrated that these diols are useful for polyurethane syntheses. The polymers based on them showed good mechanical behavior as tough plastics or rubbers depending on crystallization of the PCL segments.

Acknowledgments

Financial support by Consejo Nacional de Ciencia y Tecnología (CONACYT) Grant SEP-2004-C01- 47173E. We are indebted to Rosa Lebrón-Aguilar (CSIC) and Ricardo Vera-Graziano (IIM-UNAM) for obtaining MALDI-TOF spectra and GPC-MALLS data.

References

1. Cohn, D.; Younes, H. *J. Biomed. Mater. Res.* **1988**, *22*, 993.
2. Sawhney, S. A.; Hubbell, J. A. *J. Biomed. Mater. Res.* **1990**, *24*, 1397.
3. Penco, M.; Bignotti, F.; Sartore, L.; D'Antone, S.; D'Amore, A. *J. Appl. Polym. Sci.* **2000**, *78*, 1721.
4. Tokiwa, Y. *Biopolymers* **2002**, *9*, 323.
5. Chang, W. L.; Karalis, T. *J. Polym. Sci., Part A* **1993**, *31*, 493.
6. Kobayashi, S.; Uyama, H.; Kimura, S. *Chem. Rev.* **2001**, *101*, 3793.
7. Matsumura, S. *Adv. Polym. Sci.* **2006**, *194*, 95.
8. Barrera-Rivera, K. A.; Flores-Carreón, A.; Martínez-Richa, A. *J. Appl. Polym. Sci.* **2008**, *109* (2), 708.
9. Henderson, L. A.; Svirkin, Y. Y.; Gross, R. A.; Kaplan, D. L.; Swift, G. *Macromolecules* **1996**, *29* (24), 7759.

Chapter 17

Enzymatic Synthesis and Properties of Novel Biobased Elastomers Consisting of 12-Hydroxystearate, Itaconate and Butane-1,4-diol

Mayumi Yasuda, Hiroki Ebata, and Shuichi Matsumura*

Faculty of Science and Technology, Keio University, 3-14-1, Hiyoshi, Kohoku-ku, Yokohama 223-8522, Japan
***matumura@applc.keio.ac.jp**

A biobased elastomer was prepared by the lipase-catalyzed polymerization of methyl 12-hydroxystearate (12HS-Me), dimethyl itaconate and butane-1,4-diol using immobilized lipase from *Burkholderia cepacia* and subsequent thermal crosslinking. The molecular weight of the polyester was significantly increased by the ring-opening copolymerization of a cyclic butylene itaconate oligomer and 12HS-Me. The produced polyester with a M_w of 160000 g/mol consisted of randomly distributed 12HS units and butylene itaconate units. Thermal crosslinking of the polyester was carried out at 180 °C using a hot-press to obtain the transparent elastomer. The hardness of the elastomer increased with decreasing 12HS content. In contrast, Young's modulus of the elastomer decreased with increasing 12HS content.

Introduction

Biobased polymers produced from renewable biomass resources such as plant oil are attractive with respect to saving fossil carbon resources, reduction of CO_2 emission and energy minimization for sustainable development and the establishment of a sustainable society. Such green polymers should be produced by environmentally benign processes, e.g., enzymatic and solvent-free processes, and by avoiding the use and generation of hazardous materials. Unlike fossil resources, plant oil is regarded as an abundant renewable feedstock for next generation polymers (*1–6*). About one hundred million tons of plant oil are produced annually.

In recent years, soybean oil has been used as the feedstock for the production of paint, printing ink and biofuel. Furthermore, the application of plant oil as the chain extender for polyurethanes and crosslinkers for biodegradable polymers has been reported (*7–14*). However, as in the case of soybean oil, the price of plant oil is dependent on both its production and its demand in food supplies. Non-edible oils may not cause this conflict. Among plant oils, castor oil has attracted attention as a non-edible feedstock for the production of functional materials because of its relative abundance. Castor oil is obtained from the bean of the castor plant and 530,000 tons of it are produced annually. Approximately 85-90% of the triglyceride-derived fatty acid in castor oil is ricinoleic acid, 12-hydroxy-*cis*-9-octadecenoic acid. 12-Hydroxystearic acid (12HS) is produced industrially by the hydrogenation of ricinoleic acid. Thus, 12HS is attractive as a renewable biobased starting material for the production of various polymeric materials.

Elastomers show rubber elasticity at room temperature and are widely used in industrial applications. Elastomers are produced mainly from petroleum feedstocks and their consumption is increasing annually. The polybutadiene series and polybutadiene-acrylonitrile series elastomers are both widely used. A thermosetting poly(ricinoleic acid)-type elastomer was prepared by the enzyme-catalyzed polymerization of ricinoleic acid and subsequent crosslinking (*15, 16*). A thermoplastic elastomer composed of 12HS and 12-hydroxydodecanoate (12HD) was synthesized by lipase-catalyzed polymerization (*17*).

As biodegradable plastics, poly(butylene succinate) (PBS) is produced by the polymerization of succinic acid and butane-1,4-diol (BD). The former can be produced by the fermentation method and the latter can be produced by the reduction of succinic acid. Thus, PBS is regarded as a potentially biobased plastic. Itaconic acid is also produced by the fermentation method. Itaconic acid has a polymerizable C=C double bond in its structure (*18, 19*), thus it is an attractive monomer for the production of functional polyesters. However, the synthesis of polyesters involving itaconic acid has not been reported to date.

In this study, a novel biobased elastomer was prepared by the lipase-catalyzed copolymerization of 12HS, itaconic acid and BD with subsequent thermal crosslinking. Some physicochemical properties of the elastomer were also reported.

Experimental Part

Materials and Measurements

Methyl 12-hydroxystearate (12HS-Me) was purchased from Sigma (St. Louis, MO, USA). Dimethyl itaconate (IA-Me) and BD were purchased from Wako Pure Chemical Industries, Ltd. (Tokyo, Japan). 12HD was purchased from Aldrich (Milwaukee, WI, USA). 1-Hexanol was purchased from Junsei Chemical

Co. ltd. (Tokyo, Japan). 4-Decanol was purchased from Tokyo Kasei Kogyo Co. Inc. (Tokyo, Japan). Molecular sieves 4A (MS4A) were purchased from Junsei Chemical Co., Ltd. (Tokyo, Japan), and were dried at 150 °C for 2 h before use. Immobilized lipase from *Candida antarctica* (lipase CA: Novozym 435; 10000 PLU·g⁻¹ propyl laurate units; lipase activity based on ester synthesis) was kindly supplied by Novozymes Japan, Ltd. (Chiba, Japan). Lipase from *Burkholderia cepacia* immobilized on Diatomaceous Earth with 0.5 units·mg⁻¹ (lipase PS-D) was purchased from Wako Pure Chemical Industries, Ltd. (Tokyo, Japan). The immobilized enzymes were dried in a vacuum (3 mmHg) over P_2O_5 at 25 °C for 2 h.

The weight-average molecular weight (M_w), number-average molecular weight (M_n), polydispersity index (M_w/M_n) and monomer conversion were measured using size exclusion chromatography (SEC) with SEC column (Shodex K-805L, Showa Denko Co., Ltd., Tokyo, Japan) at 37 °C with a refractive index detector. Chloroform was used as the eluent at 1.0 mL·min⁻¹. The SEC system was calibrated with polystyrene standards with narrow molecular weight distribution. ¹H NMR spectra were recorded on a Varian MERCURY plus 300 spectrometer operating at 300 MHz. The ¹³C NMR and HMBC spectra were recorded with a Lambda 300 Fourier Transform Spectrometer (JEOL, Ltd., Tokyo, Japan) operating at 75 MHz. Matrix-assisted laser desorption ionization time-of-flight mass spectrometry (MALDI-TOF MS) was performed with a Bruker Ultraflex mass spectrometer equipped with a nitrogen laser. The detection was performed in the reflectron mode, 2,5-dihydroxybenzoic acid was used as the matrix, sodium bromide was used as the cation source, and positive ionization was used.

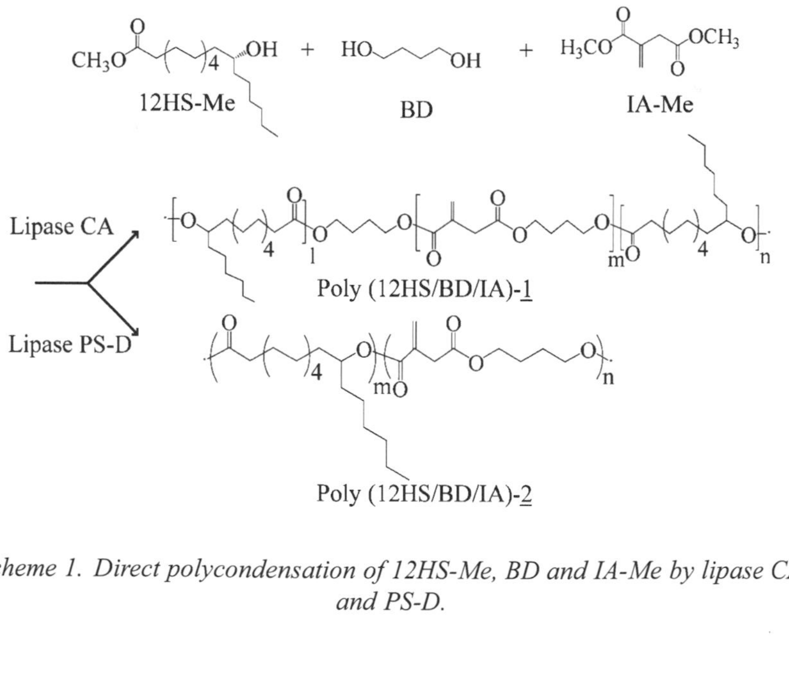

Scheme 1. Direct polycondensation of 12HS-Me, BD and IA-Me by lipase CA and PS-D.

The glass transition temperature (T_g), crystallization temperature (T_c), and melting temperature (T_m) of the polymer were determined by differential scanning calorimetry (DSC-60, Shimadzu Co., Kyoto, Japan). The heating rate was 10 °C·min⁻¹ within a temperature range of -150 to 60 °C. The polymer samples were heated in a nitrogen flow at a rate of 10 °C·min⁻¹ from 30 to 60 °C, cooled to -150 °C at a rate of -20 °C·min⁻¹, and then scanned at the same heating rate and over the same temperature range. The crosslinking behavior of the polymer was evaluated with a Scanning Vibrating Needle Curemeter (SVNC, RAPRA, Heisen Yoko Co., Ltd., Japan). The Young's modulus of elastomers was measured using an Autograph instrument (Shimadzu Co., Kyoto, Japan). The hardness of the elastomers was measured using a Haze meter (Nippon Denshoku Industries Co., Ltd., Japan).

General Enzymatic Polymerization Procedure

Direct Polycondensation of 12HS-Me, BD and IA-Me

The general procedure for the enzymatic polymerization of 12HS-Me, BD and IA-Me was carried out in a screw-capped vial with MS4A placed at the top of the vial to absorb the condensation byproducts, such as water or methanol. The preparation of poly(76.5% 12HS/BD/IA) with a M_w of 40300 is described as a typical example. A mixture of 12HS-Me (182 mg, 0.58 mmol), BD (6.5 mg, 0.072 mmol), IA-Me (11.4 mg, 0.072 mmol) and immobilized lipase PS-D (240 mg, 120 wt% relative to substrate) was stirred under a nitrogen atmosphere in an oil bath at 80 °C for 5 d. After the polymerization, the reaction mixture was dissolved in chloroform (20 mL), and the insoluble enzyme was removed by filtration. The solvent was then evaporated under reduced pressure to obtain the polymer. The crude polymer was reprecipitated from chloroform using methanol to remove any unreacted monomers. The molecular weight was determined using SEC. The molecular structure and monomer composition were determined by ^{1}H NMR, ^{13}C NMR and HMBC experiments. The spectral data of poly(76.5 mol% 12HS/BD/IA) is shown as being representative.

Poly(76.5 mol% 12HS/BD /IA)

^{1}H NMR (300 MHz, CDCl₃): δ 0.88 (t, 3H, -CH_3, J = 6.6 Hz), 1.25 (m, 22H, -CH_2-, 12HS), 1.49 [m, 4H, -CO(CH₂)₉CH_2-, -CH_2(CH₂)₄CH₃, 12HS], 1.60 (m, 2H, -COCH₂CH_2-, 12HS), 1.70 (m, 4H, -CH₂CH_2CH_2CH₂-, BD), 2.27 (t, 2H, -COCH_2-, J = 7.5 Hz), 3.33 [m, 2H, -COC(CH₂)CH_2 COO-, IA], 4.09, 4.19 (m, 4H, -CH_2CH₂CH₂CH_2-, BD), 4.89 (tt, 1H, HC-O, J_1 = J_2 = 6.3 Hz, 12HS), 5.72, 6.32 (m, 2H, -C=CH_2).

^{13}C NMR (75 MHz, CDCl₃): δ 14.0 (CH_3), 22.5-31.7 (-CH_2-, 12HS), 25.1-25.3 (-CH₂CH_2CH_2CH₂-, BD), 34.7 (-COCH_2-, 12HS), 37.6 [-COC(CH₂)CH₂COO-, IA], 63.6 (-CH_2CH₂CH₂CH₂-, BD), 64.3, 64.4 (-CH₂CH₂CH₂CH_2-, BD), 74.0 (CH-O, 12HS), 127.7, 128.3 (-C=CH₂), 133.8, 134.3 (-C=CH₂), 165.8, 166.0 (-COCCH₂-, IA), 170.4, 170.6 (-CCH₂CH₂COO-, IA), 173.6, 173.8 (-COCH₂-, 12HS).

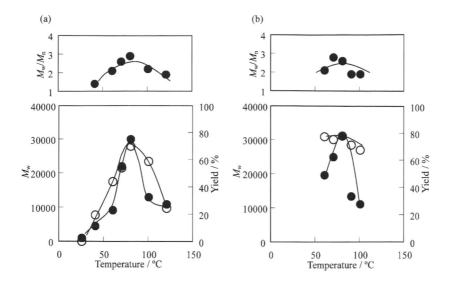

Figure 1. Effects of temperature on M_w, M_w/M_n and polymer yield using lipase CA (a) and lipase PS-D (b). Reaction conditions: 12HS-Me (0.58 mmol), BD (0.072 mmol) and IA-Me (0.072 mmol) were polymerized by lipase CA (50 wt%) or lipase PS-D (100 wt%) in toluene (100 μL) with MS4A placed at the top of the polymerization vessel for 5 d. M_w and M_w/M_n (●); polymer yield (○).

Enzymatic Preparation of Cyclic BD/IA Oligomer

The cyclic BD/IA oligomer was prepared by the reaction of BD and IA-Me using lipase in a dilute toluene solution with MS4A placed at the top of the flask in the vapor phase above the reaction mixture. BD (72 mg, 0.8 mmol) and IA-Me (79 mg, 0.5 mmol) were dissolved in toluene (30.2 mL) in a round-bottomed flask, after which immobilized lipase CA (151 mg, 100 wt% relative to substrate) was added and the reaction mixture was stirred in an oil bath at 80 °C for 48 h. After the reaction, the reaction mixture was diluted with chloroform (20 mL), and insoluble enzyme was removed by filtration. The solvent was then evaporated under reduced pressure to obtain cyclic oligomers consisting mainly of cyclic BD/IA dimer in almost quantitative yield. Purification was carried out by silica gel column chromatography using ethyl acetate-hexane (1:1 v/v, $R_f = 0.40$) as an eluent to obtain the cyclic BD/IA oligomer as white crystals in 40% yield. The molecular structure of the cyclic BD/IA oligmer was confirmed by ^{1}H NMR spectroscopy, MALDI-TOF MS and elemental analysis.

Cyclic BD/IA oligomer

^{1}H NMR (300 MHz, CDCl$_3$): δ 1.70 (m, 4H, -CH_2CH$_2$CH$_2$CH$_2$-), 3.35 [m, 2H, -COC(CH$_2$)CH_2COO-], 4.12, 4.19 (m, 4H, -CH_2CH$_2$CH$_2$CH$_2$-), 5.72, 6.32 (m, 2H –C=CH_2).

The general procedure for the enzymatic ring-opening polymerization of cyclic BD/IA oligomer and 12HS-Me was carried out in a screw-capped vial with MS4A placed at the top of the vial to absorb the condensation byproduct. The preparation of poly(58.8 mol% 12HS/BD/IA) with a M_w of 102000 is described as a typical example. A mixture of cyclic BD/IA oligomer (29.5 mg, 0.16 mmol), 12HS-Me (151 mg, 0.48 mmol) and lipase PS-D (252.7 mg, 140 wt% relative to substrate) was stirred under a nitrogen atmosphere in an oil bath at 80 °C for 4 d. After the polymerization, the reaction mixture was dissolved in chloroform (20 mL), and the insoluble enzyme was removed by filtration. The solvent was then evaporated under reduced pressure to obtain the polymer. The polymer was reprecipitated from chloroform using methanol to remove any unreacted monomers. The molecular weight was determined by SEC. The molecular structure and monomer composition were determined by ^{1}H NMR, ^{13}C NMR and HMBC experiments. The spectral data of poly(58.8 mol% 12HS/BD/IA) obtained by the ring-opening polymerization of cyclic oligomer and 12HS-Me using lipase PS-D was in complete agreement with those obtained by the direct polycondensation of 12HS-Me, BD and IA-Me using lipase PS-D.

Results and Discussion

Synthesis and Characterization

Direct Polycondensation of 12HS-Me, BD and IA-Me

12HS-Me, BD and IA-Me were copolymerized by immobilized lipase CA or lipase PS-D in toluene to produce poly(12HS/BD/IA) as confirmed by ^{1}H and ^{13}C NMR spectroscopy (Scheme 1). It was found that the monomer ratio of the copolymer agreed with the initial monomer feed ratio when lipase PS-D was used. However, when lipase CA was used, 12HS was less reactive compared to BD and IA-Me. Thus, the sequences of the monomers in the copolymer chain differed depending on the lipase origin. Details are discussed in a later section. Figure 1 shows the effects of temperature on M_w, M_w/M_n and polymer yield. Similar tendencies were observed when using lipase CA and lipase PS-D, i.e., M_w and polymer yield increased with increasing temperature from 30 °C to 80 °C. Both the highest M_w and polymer yield were obtained at 80 °C, then both gradually decreased due to thermal crosslinking at the IA moiety and deactivation of the enzyme. Based on these results, further studies were carried out at 80 °C.

Figure 2 shows the time course of the lipase-catalyzed polymerization of 12HS-Me, BD and IA-Me. The M_w of poly(12HS/BD/IA) gradually increased with time and reached a M_w of about 30000 after a 5 d reaction at 80 °C using either lipase CA and lipase PS-D as shown in Figure 2.

Figure 3 shows the effects of enzyme concentration on M_w, M_w/M_n, and polymer yield. When lipase CA was used, the highest M_w of poly(12HS/BD/IA)-1 was produced at an immobilized lipase concentration of 80 wt% as shown in Figure 3a. The molecular weight decreased with a lipase concentration higher than 80 wt%. On the other hand, when using lipase PS-D, the M_w of poly(12HS/BD/IA)-$\underline{2}$ increased with increasing lipase concentration up to 120 wt% and then remained almost constant as shown in Figure 3b.

Lipase-Catalyzed Ring-Opening Polymerization of Cyclic BD/IA Oligomer and 12HS-Me

In order to further increase the M_w of the polymer for better mechanical properties, lipase-catalyzed ring-opening polymerization of the cyclic oligomer may be effective as reported previously (20–22). Therefore, cyclic BD/IA oligomer was first prepared, followed by ring-opening copolymerization with 12HS as shown in Scheme 2. It was found that significantly higher molecular-weight poly(12HS/BD/IA) was produced by the ring-opening copolymerization of cyclic BD/IA oligomer and 12HS-Me using lipase PS-D when compared to direct polycondensation. Figure 4a shows the effects of enzyme concentration on M_w and polymer yield by the copolymerization of cyclic BD/IA oligomer and 12HS-Me. The greatest M_w of poly(12HS/BD/IA)-$\underline{2}$ was produced when 140 wt% lipase PS-D was used after 2 d at 80 °C. Figure 4b shows the time course of the copolymerization of cyclic BD/IA oligomer and 12HS-Me using lipase PS-D. The M_w of poly(12HS/BD/IA)-$\underline{2}$ gradually increased with time and reached the highest M_w of 160000 after a 4 day reaction at 80 °C using 140 wt% immobilized lipase PS-D. On the other hand, the M_w of poly(12HS/BD/IA)-$\underline{1}$ was only 39000 using 70 wt% lipase CA after a 4 d reaction at 80 °C (not shown).

Characterization of Poly(12HS/BD/IA)

Figure 5 shows HMBC spectra of poly(12HS/BD/IA)s produced by the ring-opening polymerization using lipase CA and lipase PS-D. No correlation of 12HS and IA was observed in Figure 5a. This indicates that when the polymerization was carried out using lipase CA, no ester bond formed between the hydroxy group of 12HS and the carboxy group of IA. On the other hand, an ester bond was formed between 12HS and IA by lipase PS-D as confirmed by the HMBC spectrum in Figure 5b (marked x). Based on these results, the proposed molecular structure of poly(12HS/BD/IA) is shown in Scheme 1 and Scheme 2.

In order to further analyze the difference of the two polymer structures produced by lipase CA and lipase PS-D, the reactivities of the two carboxy groups of IA were compared. At first, 12HS-Me having a secondary hydroxy group was reacted with cyclic BD/IA oligomer using lipase CA to produce poly(12HS/BD/IA)-$\underline{1}$ with a M_w of 26000. However, the molar ratio of 12HS and IA of the polymer did not agree with the feed molar ratio. Contrary to this, 12HD having a primary hydroxy group was reacted with cyclic BD/IA oligomer using lipase CA to produce a higher molecular weight poly(12HD/BD/IA) with a M_w of 86000, and the molar ratio of 12HD and IA of the polymer agreed with the feed monomer ratio. This difference might be due to the reactivity of the hydroxy group of 12HD and 12HS towards IA in the enzyme-activated intermediate

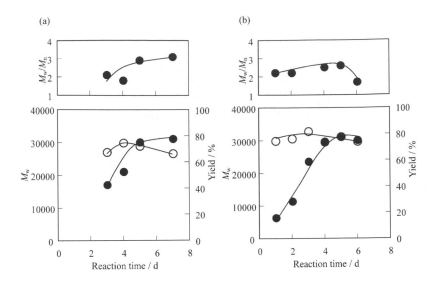

Figure 2. Effects of reaction time on M_w, M_w/M_n and polymer yield using lipase CA (a) and lipase PS-D (b) at 80 °C. Reaction conditions are the same with Figure 1 except reaction time. M_w and M_w/M_n (●); polymer yield (○).

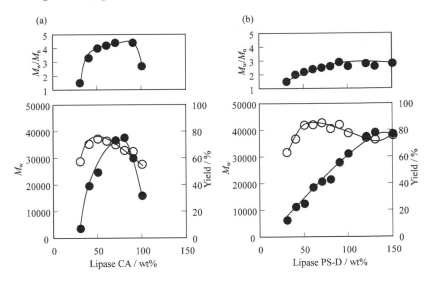

Figure 3. Effects of enzyme concentration on M_w, M_w/M_n and polymer yield using lipase CA (a) and lipase PS-D (b) at 80 °C for 5 d. Reaction conditions are the same with Figure 1 except enzyme concentration. M_w and M_w/M_n (●); polymer yield (○).

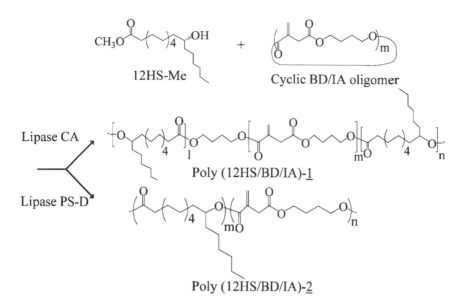

Scheme 2. Ring-opening polymerization of cyclic BD/IA oligomer and 12HS-Me by lipase CA and PS-D.

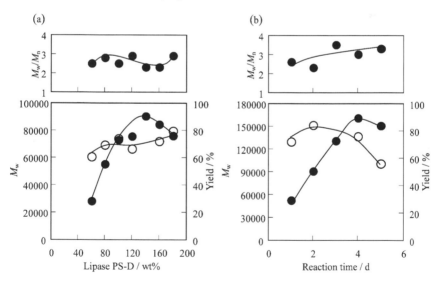

Figure 4. (a) Effects of enzyme concentration on M_w, M_w/M_n and polymer yield. Reaction conditions: Cyclic BD-IA oligomer (0.025 mmol) and 12HS-Me (0.2 mmol) were polymerized by lipase PS-D in the presence of MS4A placed at the top of the polymerization vessel at 80 °C for 2 d. M_w and M_w/M_n (●); polymer yield (○). (b) Time course of the ring-opening copolymerization of cyclic BD/IA oligomer and 12HS-Me. Reaction conditions: Cyclic BD/IA oligomer (0.025 mmol) and 12HS-Me (0.2 mmol) were polymerized by lipase PS-D (140 wt%) at 80 °C. M_w and M_w/M_n (●); polymer yield (○).

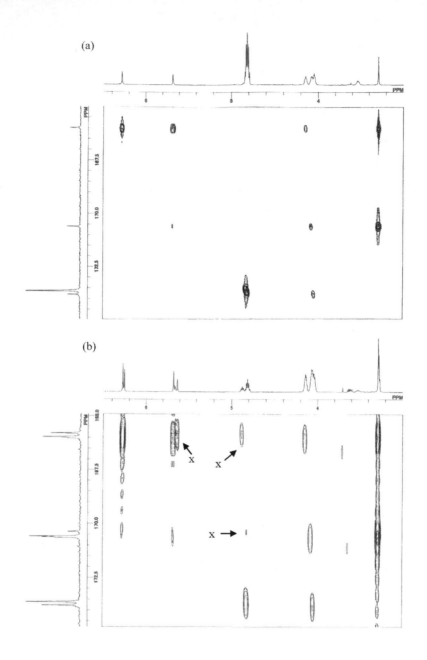

Figure 5. HMBC spectra of poly(12HS/BD/IA) prepared by lipase CA (a) and lipase PS-D (b) (CDCl₃). x indicates the ester bond of 12HS and IA.

(*23–25*). The secondary hydroxy group of 12HS-Me was less reactive with the carboxy group of IA by lipase CA when compared to the primary hydroxy group of 12HD. Also, the reactivity of the two carboxy groups of IA might differ. This difference would be more pronounced in the enzyme-activated intermediate of IA. In order to compare the reactivity of the two carboxy groups of IA, IA-Me was reacted with 1-hexanol (primary alcohol) and 4-decanol (secondary alcohol) using lipase CA and lipase PS-D in toluene. It was found that both carboxy groups of IA-Me (A and B in Table 1) reacted equally with 1-hexanol by both lipases. Also, both carboxy groups of IA-Me reacted equally with 4-decanol by lipase PS-D. However, the carboxy group adjacent to the C=C double bond (A) of IA-Me barely reacted with 4-decanol by lipase CA (Table 1). Based on these results, it was concluded that the carboxy group of IA-Me adjacent to the C=C double bond and the secondary hydroxy group of 12HS-Me were less reactive by lipase CA. Thus, no correlation between 12HS and IA was observed in Figure 5a, because 12HS was exclusively bound to IA via BD when using lipase CA.

Thermal Properties

The thermal properties of poly(12HS/BD/IA)-2, poly(12HS) and poly(BD/IA) were measured using DSC. The results are shown in Table 2. A single crystallization peak at T_c = -40 °C was observed at a cooling rate of -20 °C·min^{-1}. The T_m of the copolymer was measured at a heating rate of 10°C·min^{-1} and a single melting peak at around T_m = -25 °C was observed. Poly(12HS/BD/IA)-2 showed a T_g of around -77 °C, and was a viscous liquid at room temperature. T_m and T_c of the poly(12HS/BD/IA)-2 were similar to those of poly(12HS). On the other hand, T_g values of the poly(12HS/BD/IA)-2 were slightly lower than that of poly(12HS).

Crosslinking Behavior of Poly(12HS/BD/IA)

Poly(12HS/BD/IA)-2 was thermally crosslinked at 180 °C using a hot-press machine and crosslinking was evaluated by a SVNC. The results are summarized in Figure 6. It was found that the viscous poly(12HS/BD/IA)-2 prepared by lipase PS-D was readily crosslinked at 180 °C to produce a crosslinked polymer sheet as shown in Figure 6 (lines 1 - 3). The crosslinking of the polymer occurred between the C=C double bonds of IA. Therefore, the crosslinking was facilitated with increasing IA content. On the other hand, no significant crosslinking of poly(12HS/BD/IA)-1 prepared by lipase CA was observed by heating at 180 °C as shown in Figure 6 (lines 4 and 5). It seems that this difference is caused by the differences in polymer structures. No significant crosslinking was observed for poly(12HS/BD/IA)-1 prepared by lipase CA probably due to the crosslinking site of the polymer, i.e., the C=C double bond of the IA unit might be distributed on the center part of the polymer chain as shown in Scheme 1, with relatively long 12HS chains on both sides. Thus, IA might exist as a core moiety and the 12HS chain might be covering the core moiety like a shell. Therefore, intramolecular crosslinking preferentially occurred rather than intermolecular crosslinking. On the other hand, when poly(12HS/BD/IA)-2 was prepared by lipase PS-D,

Table 1. Reactivity of IA-Me and alcohol by lipase[a]

Substrate	Reaction rate (mol%)			
	Lipase PS-D		Lipase CA	
	A	B	A	B
1-Hexanol	94.2	91.5	97.5	98.0
4-Decanol	50.9	52.6	3.4	21.2

[a] Reaction conditions: IA-Me and alcohol (molar ratio of 1/2) were stirred with 100 wt% lipase in toluene (1.8 g substrate/mL toluene) at 80 °C for 2 d.

Table 2. Thermal properties of poly(12HS/BD/IA)

12HS in polymer (mol%)	T_m (°C)	T_c (°C)	T_g (°C)
100[a]	-24.6	-37.0	-55.7
88.2	-24.2	-39.0	-84.8
82.4	-25.4	-39.7	-76.2
79.7	-25.8	-40.5	-76.6
71.9	-25.7	-41.0	-77.6
67.2	-25.6	-40.0	-78.7
0[b]	-	-	-42.0

[a] poly(12HS),   [b] poly(BD/IA)

the IA unit was randomly distributed in the polymer chain and intermolecular crosslinking occurred.

The crosslinked poly(12HS/BD/IA)-2 sheet with a thickness of 1 mm was soaked in chloroform in order to remove any uncrosslinked soluble fractions. As determined by the weight loss of the sample, the insoluble gel fraction of the crosslinked polymer was 95.8%. The FT-IR absorption peak at 1640 cm[-1], corresponding to the C=C of the polymer, became weak after crosslinking. This indicated that the crosslinking reaction occurred at the C=C group by a free radical crosslinking mechanism (26, 27).

The crosslinked polymer sheet was a soft and transparent elastomer. The light transmission rate of the crosslinked polymer sheet as measured by Haze meter was 92.3%, indicating high transparency similar to acrylate resin and glass. The haze of the crosslinked polymer sheet was 20.3%.

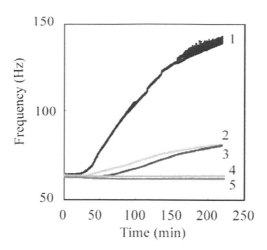

Figure 6. Crosslinking behavior of poly(12HS/BD/IA) prepared by lipase PS-D (lines 1-3) and lipase CA (lines 4 and 5) using a hot-press machine at 180 °C. Crosslinking was evaluated by SVNC. 1: M_w = 25000, IA 19.7 mol%; 2: M_w = 37000, IA 15.9 mol%; 3: M_w = 9000, IA 14.9 mol%; 4: M_w = 33000, IA 7.5 mol%; 5: M_w = 6000, IA 23.7 mol%.

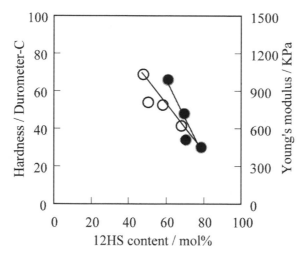

Figure 7. Effects of the 12HS content on the hardness by durometer C (●) and Young's modulus (○) of thermally crosslinked poly(12HS/BD/IA)-2 film.

Hardness and Mechanical Properties of the Crosslinked Poly(12HS/BD/IA)-2

The crosslinked poly(12HS/BD/IA)-2 sheet showed a hardness of 30 - 66 based on the durometer C as shown in Figure 7. This indicated that the crosslinked poly(12HS/BD/IA) is softer than conventional natural and synthetic rubbers. The hardness of the poly(12HS/BD/IA) sheet was dependent on the 12HS content, decreasing with increasing 12HS content of the copolymer as shown in Figure 7.

The hexyl side group of 12HS may endow a flexibility to the elastomer, thus, with the increasing 12HS content, the hardness of the polymer sheet became softer.

A poly(78.4 mol% 12HS/BD/IA)-$\underline{2}$ film with a M_w of 131000 showed a tensile strength at break of 310 KPa and an elongation at break of 130%. On the other hand, poly(47.4 mol% 12HS/BD/IA) film with a M_w of 105000 showed a tensile strength at break of 630 KPa and an elongation at break of 66%. The Young's modulus of poly(12HS/BD/IA) films with varying monomer compositions were measured and the results are shown in Figure 7. The Young's modulus of the copolymer gradually decreased with increasing 12HS content. The mechanical properties of the elastomers were highly dependent on the 12HS content.

Conclusion

Poly(12HS/BD/IA) having a M_w of 30000 was produced by the direct polycondensation of 12-Me, BD and IA-Me using 120 wt% immobilized lipase PS-D and 80 wt% immobilized CA in toluene at 80 °C for 5 d. Significantly higher molecular weight poly(12HS/BD/IA)-$\underline{2}$ having a M_w of 160000 was produced by the ring-opening polymerization of cyclic BD/IA and 12HS-Me using 140 wt% immobilized lipase PS-D in toluene at 80 °C for 4 d. Poly(12HS/BD/IA) was a viscous liquid having low T_c of $-$ 40 °C, T_g of -77 °C and T_m of -25 °C. Poly(12HS/BD/IA)-$\underline{2}$ produced by lipase PS-D was readily crosslinked by hot-press at 180 °C to form a soft and transparent elastomer. Crosslinking was facilitated by increasing the IA content in the copolymer.

Acknowledgments

Immobilized lipase from *Candida antarctica* (Novozym 435) was kindly supplied by Novozymes Japan Ltd. (Chiba, Japan). This work was supported by High-Tech Research Center Project for Private Universities: matching fund subsidy from MEXT, 2006-2011.

References

1. Nayak, P. L. *J. Macromol. Sci., Rev. Macromol. Chem. Phys.* **2000**, *C40*, 1–21.
2. Pryde, E. H.; Princen, L. H.; Mukherje, K. D., Eds. *New sources of fats and oils*; Monograph No. 9; American Oil Chemists' Society: Champaign, IL, 1981.
3. Shabeer, A.; Sundararaman, S.; Chandrashekhara, K.; Dharani, L. R. *J. Appl. Polym. Sci.* **2007**, *105*, 656–663.
4. Can, E.; Wool, R. P. *J. Appl. Polym. Sci.* **2006**, *102*, 1497–1504.
5. Swain, S. N.; Biswal, S. M.; Nanda, P. K.; Nayak, P. L. *J. Polym. Environ.* **2004**, *12*, 35–42.
6. Sperling, L. H.; Manson, J. A.; Quereshi, S.; Fernandez, A. M. *Ind. Eng. Chem. Prod. Res. Dev.* **1981**, *20*, 163–166.
7. Petrovic, Z. S. *Polym. Rev.* **2008**, *48*, 109–155.

8. Gruber, B.; Höfer, R.; Kluth, H.; Meffert, A. *Fat Sci. Technol.* **1987**, *89*, 147–151.

9. Sharma, V.; Kundu, P. P. *Prog. Polym. Sci.* **2008**, *33*, 1199–1215.

10. Zlatanic, A.; Lava, C.; Zhang, W.; Petrovic, Z. S. *J. Polym. Sci., Part B* **2004**, *42*, 809–819.

11. Petrovic, Z. S.; Yang, L.; Zlatanic, A.; Zhang, W.; Javni, I. *J. Appl. Polym. Sci.* **2007**, *105*, 2717–2727.

12. Tsujimoto, T.; Uyama, H.; Kobayashi, S. *Biomacromolecules* **2001**, *2*, 29–31.

13. Tsujimoto, T.; Uyama, H.; Kobayashi, S. *Macromol. Biosci.* **2002**, *7*, 329–335.

14. Tsujimoto, T.; Uyama, H.; Kobayashi, S. *Macromol. Rapid Commun.* **2003**, *24*, 711–714.

15. Ebata, H.; Toshima, K.; Matsumura, S. *Macromol. Biosci.* **2007**, *7*, 798–803.

16. Ebata, H.; Yasuda, M.; Toshima, K.; Matsumura, S. *J. Oleo Sci.* **2008**, *57*, 315–320.

17. Ebata, H.; Toshima, K.; Matsumura, S. *Macromol. Biosci.* **2008**, *8*, 38–45.

18. Tasselli, F.; Donato, L.; Drioli, E. *J. Membr. Sci.* **2008**, *320*, 167–172.

19. Fernández-García, M.; Fernández-Sanz, M.; de la Fuente, J. L.; Madruga, E. L. *Macromol. Chem. Phys.* **2001**, *202*, 1213–1218.

20. Sugihara, S.; Toshima, K.; Matsumura, S. *Macromol. Rapid Commun.* **2006**, *27*, 203–207.

21. Yamamoto, Y.; Kaihara, S.; Toshima, K.; Matsumura, S. *Macromol. Biosci.* **2009**, *9*, 968–978.

22. Yanagishita, Y.; Kato, M.; Toshima, K.; Matsumura, S. *ChemSusChem* **2008**, *1*, 133–142.

23. Uyama, H.; Takeya, K.; Kobayashi, S. *Bull. Chem. Soc. Jpn.* **1995**, *68*, 56–61.

24. Kobayashi, S. *Macromol. Symp.* **2006**, *240*, 178–185.

25. Kobayashi, S. *Macromol. Rapid Commun.* **2009**, *30*, 237–266.

26. Silverman, J.; Zoepfl, F. J.; Randall, J. C.; Markovic, V. *Radiat. Phys. Chem.* **1983**, *22*, 583–585.

27. Isaure, F.; Cormack, P. A. G.; Sherrington, D. C. *J. Mater. Chem.* **2003**, *13*, 2701–2710.

Syntheses of Polyamides and Polypeptides

Chapter 18

Synthesis of Poly(aminoamides) via Enzymatic Means[†]

H. N. Cheng[*,a] and Qu-Ming Gu[b]

[a]Southern Regional Research Center, USDA Agricultural Research Service, 1100 Robert E. Lee Blvd., New Orleans, LA 70124
[b]Ashland Inc., Ashland Research Center, 500 Hercules Road, Wilmington, DE 19808
[*]h.n.cheng@ars.usda.gov
[†]Names are necessary to report factually on available data; however, the USDA neither guarantees nor warrants the standards of the products, and the use of the name USDA implies no approval of the products to the exclusion of others that may also be suitable.

Poly(aminoamides) constitute a subclass of polyamides that are water-soluble and useful for several applications. Commercially they are made via chemical reaction pathways. A review is made in this work of the enzymatic approaches towards their syntheses. Lipases and esterases have been found to be suitable enzymes to produce high-molecular-weight polyamides under relatively mild reaction conditions. A large number of different polymer compositions can be synthesized through enzymatic means. The design of the polymer structure and synthetic considerations are included in this review.

Introduction

Poly(aminoamides) are interesting polymers that have been found to be useful in many different applications. For example, the poly(aminoamide) of adipic acid and diethylene triamine (DETA) is well known as a prepolymer for a cationic resin that is used to improve wet strength and as a creping aid in paper (*1*). A quaternized poly(amidoamine) has been reported as a corrosion control agent (*2*). Modified poly(aminoamides) are claimed to be retention and drainage aids in paper manufacturing (*3*). A poly(aminoamide) dendrimer is used for silica scale control in water technology (4). Poly(aminoamide) resins are used as adhesion promoters of poly(vinyl chloride) plastisols (5). Poly(aminoamides) with UV-absorbing functionalities are used for protection of skin and hair (6, 7). A polyamido-polyethyleneimine has been claimed to be an adhesive coating for polyester films (8); the same polymer is used as a retention aid for paper (9). In biochemical applications, a hybrid siloxane-poly(aminoamide) has been shown to absorb heparin from blood (10).

In the past chemical pathways via condensation polymerization of monomers have been used for the synthesis of poly(aminoamides) (1). Typically a polyamine and a diacid are heated at high temperatures to conduct the polycondensation reaction. Recently there has been a lot of progress to use enzymes to synthesize polyamides. This latest development is reviewed in this work.

Lipase-Catalyzed Synthesis of Polyamides

For many years there has been a lot of interest in using enzymes for polyamide synthesis, but earlier work tended to use protease to produce polypeptides (11–13). It has been found that proteases mostly produce oligopeptides (14), with a few exceptions (15). Earlier, Gu et al (16) used four proteases (chymotrypsin, trypsin, subtilisin, and papain) in attempts to make polyamide from dimethyl adipate and diethylene triamine, but only oligoamides were found. There have also been reports on the use of dipeptidyl transferase (17) and cyanophycin synthetase (18) for peptide synthesis.

An alternative approach is to use lipases (and esterases), some of which are known to catalyze amide formation under suitable reaction conditions. Prior to 2000, there have been several publications on the use of lipases (particularly porcine pancreatic lipase, PPL) to synthesize dipeptides and tripeptides (19–22). In one of these papers, So et al (22) screened 15 different commercial lipases for the synthesis of dipeptides from D-amino acids, and found PPL to be the only effective lipase.

In a U.S. patent application filed in 2000 (and granted in 2004) Cheng et al (23) reported that several commercial lipases could be good catalysts for the synthesis of polyamides from diesters and diamines. The polyamides thus produced have molecular weights in the range of 4,000 to 12,000. For these polymerizations, the reactants were reacted with a lipase either in the absence of solvent, or in the presence of one or more protic solvents such as methanol, ethanol, ethylene glycol, glycerol, t-butanol, isopropanol, or in a water/salt mixture such as water/NaCl. This patent is the first report of the synthesis of high-molecular-weight polyamides using lipase.

In a follow-up work, Gu et al (24) reported the use of lipases to facilitate the synthesis of a family of poly(aminoamides). The polyamides are made by Michael addition reaction of a diamine with an acrylic compound (like methyl acrylate) in a 1:2 molar ratio, respectively, in the first step, and polymerization of additional diamine with the resulting diester or diacid prepolymer at 70-140 °C or in the presence of an enzyme at 60-80 °C.

In 2005, Azim, Sahoo, and Gross (*25*) reported the use of immobilized lipase B from *Candida antarctica* (Novozym® 435) as a catalyst for the formation of amide bonds between diethyl esters and diamines under mild reaction conditions. Oligoamides were produced.

In 2005, Panova et al (*26*) filed a patent application (granted in 2009) where they carried out a detailed study using lipases to produce cyclic amide oligomers from diesters and diamines. The cyclic amide oligomers are useful for the subsequent production of higher molecular weight polyamides.

Also in 2005, Kong et al (*27*) filed a patent application on the preparation of an aqueous polyamide dispersion by lipase-catalyzed polycondensation reaction of a diamine compound and a dicarboxylic compound in aqueous medium. In a separate patent application (*28*), they reported the preparation of an aqueous polyamide dispersion by lipase-catalyzed reaction of an aminocarboxylic acid compound in aqueous medium.

Recently, Loos et al (*29*) reported the synthesis of poly(β-alanine) via lipase-catalyzed ring-opening of 2-azetidinone. After removal of cyclic side products and low molecular weight species pure linear poly(β-alanine) is obtained. The average degree of polymerization of the obtained polymer is limited to DP=8 by its solubility in the reaction medium. A follow-up work has extended the DP to 18 (*30*).

Design of Poly(aminoamide) Structure

From the point of view of applications, it is useful to vary the poly(aminoamide) structure in order to optimize the properties. Certainly molecular weight is an important variable. Another important variable is the amount of amine functionality relative to the number of carbons present in the polymer backbone. With more amine moieties present, the polymer tends to be more water-soluble and can have higher charge density. A cationic polymer with a high level of amine content is strong in alkalinity at high pH and possesses a large amount of positive charges at low pH. Amine groups can be alkylated or acylated with a variety of reactive reagents at alkaline pH. Many applications of poly(aminoamides) require the quaternization of the amine (*1*, *2*), or other derivatizations of the amine functionality (*3*, *6*, *7*). The ability to vary the number of amines versus the number of carbons gives more flexibility in polymer design. For example, there is current interest in creating a comb-like polymer architecture for biomedical applications (*31*). More amine functional groups on the polymer backbone should facilitate the design of such materials.

Some of the poly(aminoamide) structures produced from condensation of DETA and a diacids or diester are shown in Figure 1. In the case of the well-known poly(aminoamide) produced from a condensation of adipic acid and DETA, there is one amine functionality (NH) for every 10 backbone carbons (C) in the repeat unit of the polymer (Structure **1** in Figure 1). Through lipase-catalyzed polymerization, it is now possible to vary this ratio (NH/C) by using different starting materials (*23*) and different chemistry (*24*, *32*).

Structures **2** and **3** shown in Figure 1 are the poly(aminoamides) from dialkyl fumarate and dialkyl malonate, respectively. The composition for structure **2** cannot easily be made via chemical synthesis because Michael addition occurs at the same time as the polycondensation reaction at high temperatures, thereby resulting in a water insoluble material due to crosslinking of the polymer. Structure **3** shown in Figure 1 cannot be synthesized chemically at high temperatures because of other reactions taking place at the polymer chain end that terminate condensation polymerization. Structure **4** in Figure 1 is the copolymer of DETA with dialkyl malonate and dialkyl oxalate; this structure cannot be made chemically as well. Nevertheless, all four poly(aminoamides) **1-4** shown in Figure 1 can be synthesized readily via lipase-catalyzed polycondensation reactions between a polyamine (e.g., DETA) and a diester. Structure **5** in Figure 1 can be made via a two-step synthesis between DETA and methyl acrylate. The synthetic details are given in the next section (Experimental Considerations).

An additional handle in structure design is the use of triethylene tetraamine (TETA) and tetraethylene pentaamine (TEPA). The structures of the polyamides made with dialkyl adipate are shown (structures **6** and **7**) in Figure 2. Enzymatic synthesis of TETA and TEPA with dialkyl malonate, dialkyl oxalate, dialkyl fumarate, or methyl acrylate can potentially produce many more poly(aminoamides) (*23, 24*). These structures can provide poly(aminoamides) with an even larger range of NH/C ratios.

In addition to these compositions, other related polyamides with unique and interesting chemical and physical properties can also be synthesized in a similar fashion (*23, 24*). Two examples are shown in Figure 3.

Thus, lipase catalysis enables many new polyamide structures to be made. The reactions described herein are effective and entail mild reaction conditions and less byproducts. These are good applications of green polymer chemistry.

Experimental Considerations

The above structures can be produced through the following synthetic procedures. For ease of reference, typical procedures are given below. More information is available in the original patents (*23, 24*).

Lipase-Catalyzed Polymerization of Aliphatic Diester and Polyamine

This procedure (*23*) can be used for the synthesis of Structures **1, 2, 3, 4, 6,** and **7**. The polyamine and diester monomers are oligomerized and then reacted at a mild temperature, in the presence of enzyme, to allow polymerization of the oligomers. The reaction product is dissolved in an aqueous solution such as water or alkyl alcohol (e.g., methanol), and the enzyme is removed via filtration. This process allows polymerization of reactants under mild conditions to provide high-molecular-weight reaction products with a relatively narrow molecular weight distribution. In addition, the reaction products are relatively pure due to the use of enzyme and substantial absence of solvents. Further still, the mild conditions

1 NH/C = 1/10

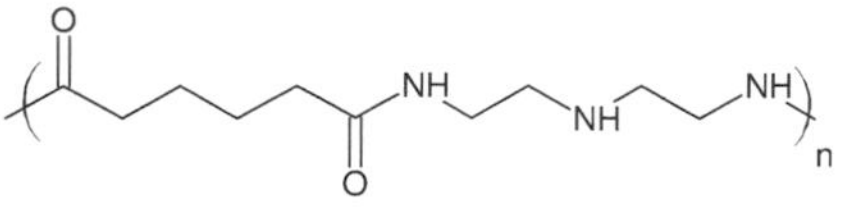

2 NH/C = 1/8

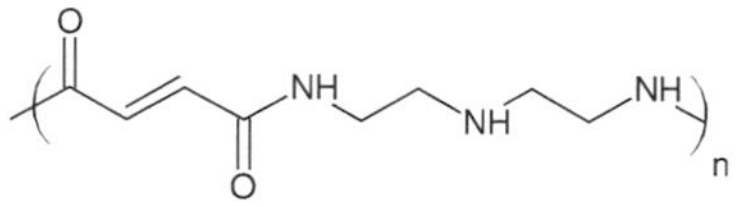

3 NH/C = 1/7

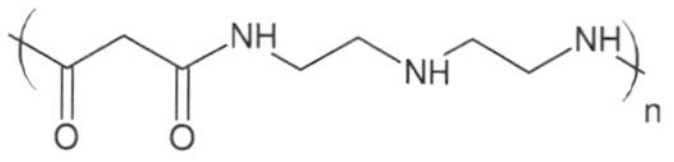

4 NH/C = 1/6.5

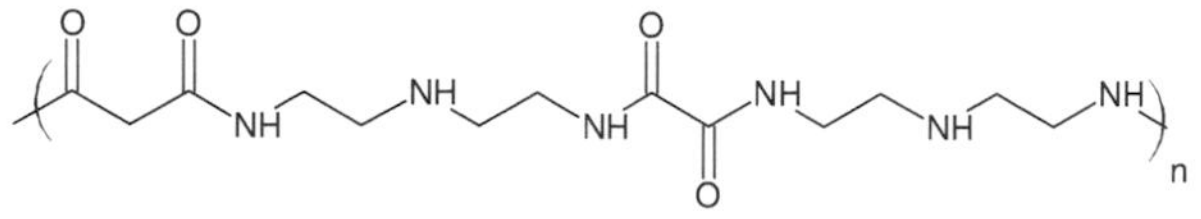

5 NH/C = 1/3.5

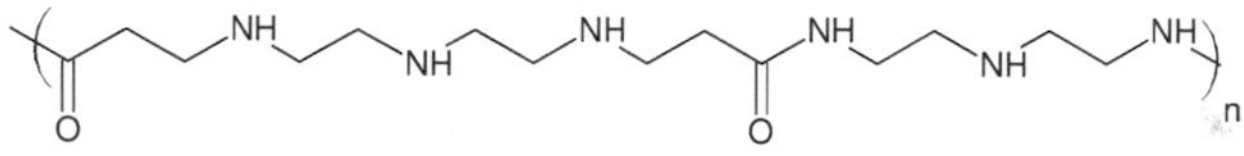

Figure 1. Poly(aminoamide) from DETA and diacid or diester

6 NH/C = 1/6

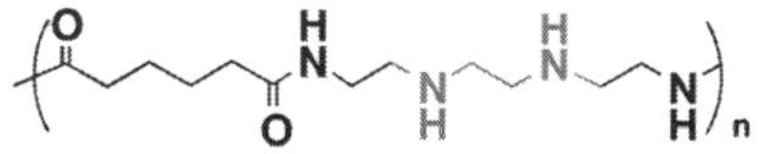

TETA-based

7 NH/C = 1/4.7

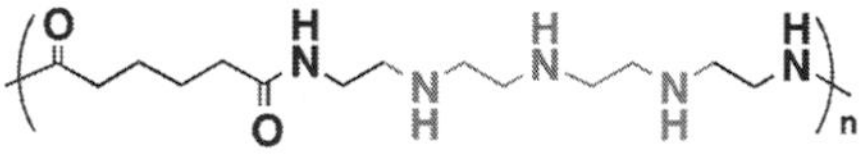

TEPA-based

Figure 2. Poly(aminoamides) from polyamine and diacid and diester

prevent denaturation of the enzyme catalysts, and allow them to be optionally recycled for further use.

In a typical procedure, dimethyl adipate (43.55 g, 0.25 mol), diethylene triamine (28.33 g, 0.275 mol) and Novozym® 435 lipase (2.5 g) are mixed in a 250-ml flask. The reactants are then heated in an oil bath to 90°C in an open vessel with a stream of nitrogen Figure 4. Completion of the reaction is indicated by the appearance of a yellowish solid. Methanol (150 ml) is then added to dissolve the poly(aminoamide) product. The immobilized enzyme is insoluble in the methanol solution and is removed by filtration. Remaining methanol is removed by a rotary evaporator under low pressure. The final product is a yellowish solid with a yield of 48 g, M_w of 8,400 Daltons, and M_w/M_n of 2.73.

a. Aromatic poly(aminoamide)

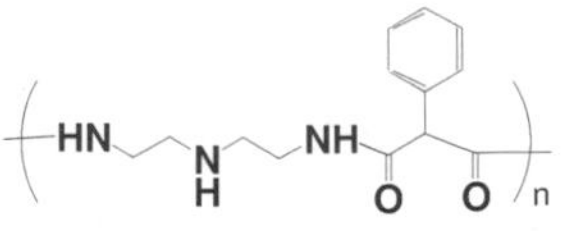

b. PEG-containing polyamide

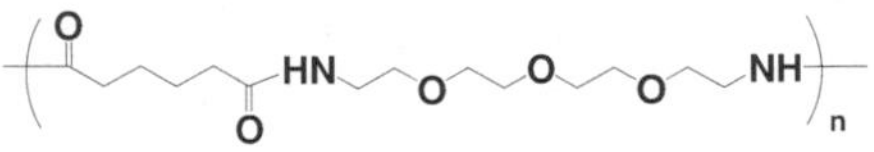

Figure 3. Unconventional polyamide compositions

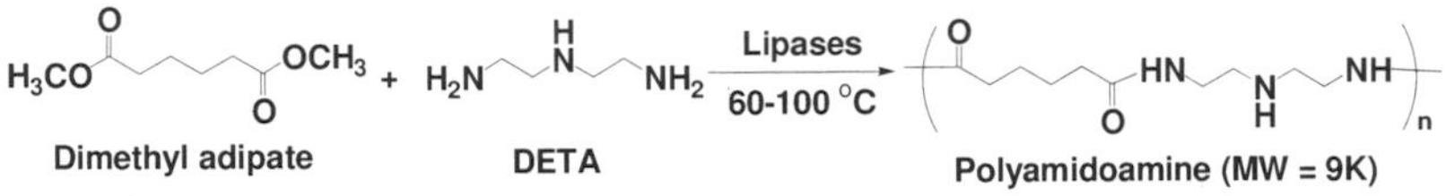

Dimethyl adipate   DETA   Polyamidoamine (MW = 9K)

Figure 4. Enzyme-catalyzed synthesis of poly(aminoamides)

Lipase-Catalyzed Polymerization of Polyamine and Methyl Acrylate

This procedure (*24, 32*) can be used for the synthesis of Structure **5** in Figure 1. As shown in Figure 5, the process used for the current synthesis consists of two discrete steps. In the first process step, exactly one mole of polyamine molecule (such as DETA) is gradually added to two moles of alkyl acrylate (such as methyl acrylate) to form an amine-containing diester in the absence of a solvent (Michael addition). The reaction vessel should be cooled through suitable means because the reaction is exothermic. The reaction temperature for this step can be 10-60°C, preferably 15-40°C, and most preferably 20-30 °C. The addition of water to the reaction mixture enhances the rate of the Michael addition.

In the second process step, amidation of the diester with another mole of either the same polyamine or a different polyamine gives a high-molecular-weight polyamide. This reaction can be achieved at 60-70°C with the assistance of a lipase as the catalyst. Two preferred lipases are those from the yeast *Candida antarctica* (e.g., Novozym® 435) and *Rhizomucor miehei* (e.g., Palatase®), both from Novozymes A/S.

Alternately, the second step of the reaction can be achieved without an enzyme by heating up the reaction mixture to 120-140°C for several hours.

In a typical procedure, methyl acrylate (43.05g, 0.5 mol) is gradually added to DETA (25.84g, 0.25 mol) at 20°C, and the temperature is gradually increased to 40°C with stirring. The addition took about 30 minutes, and the reaction mixture was stirred further at 24°-30°C for about 60 minutes, whereby the intermediate pre-polymer reaction product was formed. Another portion of DETA (25.84g, 0.25 mol) was added, followed by the addition of 4 grams of immobilized lipase *Candida antarctica* (Novozym® 435). The reaction mixture was stirred at 65°C for 16 hours. The viscous product was dissolved in 100mL of methanol at 65°C, and the immobilized enzyme was removed by filtration. The yield was 75 grams.

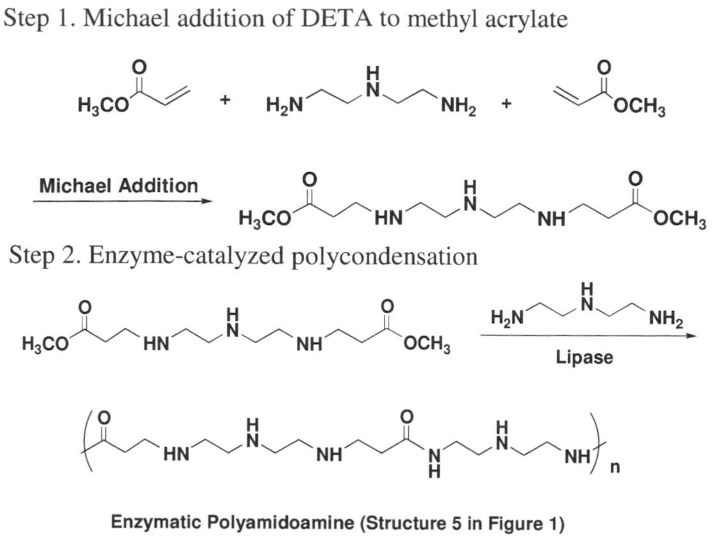

Step 1. Michael addition of DETA to methyl acrylate

Step 2. Enzyme-catalyzed polycondensation

Enzymatic Polyamidoamine (Structure 5 in Figure 1)

Figure 5. Poly(aminoamide) synthesis via a combination of chemical and enzymatic approaches

The molecular weight (M_w) of the final product, based on SEC analysis, was 8,450 Daltons and the polydispersity (M_w / M_n) was 2.75.

Lipase-Catalyzed Polymerization of DETA and Phenylmalonate

The synthetic procedure for Structure **a** in Figure 3 is given here. Diethyl phenylmalonate (23.6 g, 0.10 mol), diethylene triamine (10.3 g, 0.10 mol) and Novozym® 435 lipase (1 g) are mixed in a 500-ml flask and heated in an oil bath to 100 °C. The viscous mixture is stirred at 90-100 °C for 24 hrs in an open vessel with a stream of nitrogen. The mixture solidifies at the end of the reaction. The product is not soluble in most organic solvents and in water at neutral pH. It is soluble in water at pH 3. 150 ml of water are added and the pH is adjusted to 3 by adding concentrated HCl. The immobilized enzyme (being insoluble in water) is removed by filtration. The aqueous solution is lyophilized to give the product as a white solid. The yield is 26.9 grams. The molecular weight (M_w) of the final product, based on SEC analysis, is 3600 Daltons and the polydispersity (M_w / M_n) is 2.70.

Polymerization of Dimethyl Adipate and Triethylene Glycol Diamine

The synthetic procedure for Structure **b** in Figure 3 is described here. Dimethyl adipate (17.42 g, 0.10 mol), triethylene glycol diamine (15.60 g, 0.105 mol) and Novozym® 435 lipase (1.0 g) are mixed in a 250-ml open vessel. The reactants are heated in a stream of nitrogen in an oil bath to 70°C for 24 hours with stirring. The reaction mixture is then cooled and provides a viscous product. Methanol (100 ml) is added to dissolve the viscous product. The immobilized

261

enzyme is insoluble in methanol and is removed by filtration. The remaining methanol in the reaction mixture is removed by a rotary evaporator under low pressure. The final product is a semi-solid with a yield of 28 grams. The molecular weight (M_w) of the final product, based on SEC analysis, is 4,540 Daltons and the polydispersity (M_w /M_n) is 2.71.

Conclusion

Enzyme catalysis has been used to produce many poly(aminoamides). These polymers can be used *as is* as polyeletrolytes, or functionalized further to produce specialty polymers. The advantages of the enzymatic processes (relative to the chemical processes) are: 1) lower process temperature, thereby decreasing energy usage, 2) narrower molecular weight distributions of the products, 3) less branching in the products, 4) enzymatic processes allowing some poly(aminoamides) that cannot be synthesized chemically to be made, e.g., the polyamides derived from dialkyl malonate, malonate/oxalate, phenylmalonate, fumarate, and maleate. A disadvantage is the cost of the enzyme used, which can be partly mitigated if the enzyme is immobilized and recycled (and this is possible in the case of Novozym® 435 lipase).

References

1. For example, (a) Espy, H. H. In *Wet-Strength Resins and Their Applications*; Chan, L. L., Ed.; 1994; pp 13–44. (b) Espy, H. H. *TAPPI J.* **1995**, *78* (4), 90. (c) Maslanka, W. W. U.S. Patent 5,994,449, 11/30/1999.
2. Redmore, D.; Outlaw, B. T. U.S. Patent 4,315,087, 2/9/1982.
3. Hoppe, L.; Behn, R. U.S. Patent 4,052,259, 10/4/1977.
4. Mavredaki, E.; Stathoulopoulou, A.; Neofotistou, E.; Demadis, K. D. *Desalination* **2007**, *210*, 257.
5. Leoni, R.; Rossini, A.; Taccani, M. European Patent Application EP19850201239.
6. Forestier, S.; Lang; G.; Sainte Beuve, E. U.S. Patent 4,866,159, 9/12/1989.
7. Hessefort, Y.; Mei, M.; Carlson, W. U.S. Patent Application 20090297462.
8. Siddiqui, J. A. U.S. Patent 5,453,326, 9/26/1995.
9. BASF - Product Information Polymin® SK at worldaccount.basf.com
10. Cauzzi, D.; Stercoli, A.; Predieri, G. In *Sol-Gel Methods for Materials Processing*; Innocenzi, P., Zub, Y. L., Kessler, V. G., Eds.; 2008; pp 277−282.
11. Guzman, F.; Barberis, S.; Illanes, A. Peptide synthesis: chemical or enzymatic. *Electron. J. Biotechnol.* **2007**, *10* (2), 279–314.
12. Lombard, C.; Saulnier, J.; Wallach, J. M. Recent trends in protease-catalyzed peptide synthesis. *Protein Pept. Lett.* **2005**, *12*, 621–629.
13. Kumar, D.; Bhalla, T. C. *Appl. Microbiol. Biotechnol.* **2005**, *68*, 726–736.
14. Selected protease papers in the past 5 years include: (a) Salam, S. M. A.; Kagawa, K.; Kawashiro, K. *Tetrahedron: Asymmetry* **2006**, *17* (1), 22−29. (b) Morcelle, S. R.; Barberis, S.; Priolo, N.; Caffini, N. O.; Clapes, P. *J. Mol. Catal. B: Enzym.* **2006**, *41* (3–4), 117−124. (c) Quiroga, E.;

Priolo, N.; Obregon, D.; Marchese, J.; Barberis, S. *Biochem. Eng. J.* **2008**, *39* (1), 115–120. (d) Belyaeva, A. V.; Bacheva. A. V.; Oksenoit, E. S.; Lysogorskaya, E. N.; Lozinskii, V. I.; Filippova, I. Y. *Russ. J. Bioorg. Chem.* **2005**, *31* (6), 529–534. (e) Meng, L. P.; Joshi, R.; Eckstein, H. *Chim. Oggi-Chemistry Today* **2006**, *24* (3), 50–53. (f) Joshi, R.; Meng, L. P.; Eckstein, H. *Helv. Chim. Acta* **2008**, *91* (6), 983–992. (g) Li, G.; Kodandaraman, V.; Xie, E. C.; Gross, R. A. *ACS Polym. Prepr.* **2009**, *50* (2), 60–61.

15. Two examples are: (a) Uyama, H.; Fukuoka, T.; Komatsu, I.; Watanabe, T.; Kobayashi, S. *Biomacromolecules* **2002**, *3* (2), 318–323. (b) Wong, C.-H.; Chen, S. T. *J. Am. Chem. Soc.* **1990**, *112*, 945–953.

16. Gu, Q.-M.; Maslanka, W. W.; Cheng, H. N. *ACS Polym. Prepr.* **2006**, *47* (2), 234–235.

17. Heinrich, C. P.; Fruton, J. S. *Biochemistry* **1968**, *7* (10), 3556–3565.

18. Hai, T.; Oppermann-Sanio, F. B.; Steinbuechel, A. *Appl. Environ. Microbiol.* **2002**, *68* (1), 93–101.

19. Margolin, A. L.; Klibanov, A. M. *J. Am. Chem. Soc.* **1987**, *109*, 3802–3804.

20. West, J. B.; Wong, C.-H. *Tetrahedron Lett.* **1987**, *28*, 1629–1632.

21. Kawashiro, K.; Kaiso, K.; Minato, D.; Sugiyama, S.; Hayashi, H. *Tetrahedron* **1993**, *49*, 4541–4548.

22. So, J. E.; Kang, S. H.; Kim, B. G. *Enzyme Microb. Technol.* **1998**, *23* (3–4), 211–215.

23. Cheng, H. N.; Gu, Q.-M.; Maslanka, W. W. U.S. Patent 6,677,427, 11/13/2004.

24. Gu, Q.-M.; Michel, A.; Cheng, H. N.; Maslanka, W. W.; Staib, R. R. U.S. Patent 6,667,384, 11/23/2003.

25. Azim, A.; Azim, H.; Sahoo, B.; Gross, R. A. *ACS PMSE Prepr.* **2005**, *93*, 743–744.

26. Panova, A.; Dicosimo, R.; Brugel, E. G.; Tam, W. U.S. Patent 7,507,560, 3/24/2009.

27. Kong, X.-M.; Yamamoto, M.; Haring, D. U.S. Patent Application 2008/0275182 A1, 11/6/2008.

28. Kong, X.-M.; Yamamoto, M.; Haring, D. U.S. Patent Application 2008/0167418 A1, 7/10/2008.

29. Schwab, L. W.; Kroon, R.; Schouten, A. J.; Loos, K. *Macromol. Rapid Commun.* **2008**, *29*, 794–797.

30. Schwab, L. W.; Kroon, R.; Schouten, A. J.; Loos, K. *ACS Polym. Prepr.* **2009**, *50* (2), 19–20.

31. Davis, N. E.; Karfeld-Sulzer, L. S.; Ding, S.; Barron, A. E. *Biomacromolecules* **2009**, *10*, 1125–1134.

32. Gu, Q.-M.; Michel, A.; Maslanka, W. W.; Staib, R. R.; Cheng, H. N. *ACS Polym. Prepr.* **2009**, *50* (2), 54–55.

Chapter 19

Mechanistic Insight in the Enzymatic Ring-Opening Polymerization of β-Propiolactam

Leendert W. Schwab,[a] Iris Baum,[b] Gregor Fels,[b,*] and Katja Loos[a,*]

[a]Department of Polymer Chemistry and Zernike Institute for Advanced Materials, University of Groningen, Groningen, The Netherlands
[b]Department of Chemistry, Faculty of Science, University of Paderborn, Paderborn, Germany
*fels@uni-paderborn.de; k.u.loos@rug.nl

Here we report the polymerization of β-propiolactam to poly(β-alanine) catalyzed by the immobilized *Candida antarctica* lipase B (N435). The polymer is characterized by ^{1}H-NMR spectroscopy and MALDI-ToF mass spectrometry. The best results were obtained with N435 (dried for 24 hours *in vacuo* at 46 °C) in toluene at 55 °C for a period of 96 hours. Polymerization of β-alanine was not feasible via the N435 catalyst, showing that the β-propiolactam ring is needed for this reaction. This finding ruled out reaction mechanisms that resembled enzymatic polymerization of ε-caprolactone. Instead, a detailed eight-step mechanism was developed by computational simulation that provided an appropriate description of ring-opening polymerization of β-propiolactam.

Introduction

Enzymes can catalyze reactions at ambient temperatures with a high selectivity, and they can do so outside their natural environment in an organic solvent (*1, 2*). Virtually all classes of polymers have been synthesized by enzymatic catalysis (*3*) including some polysaccharides (*3*) that are difficult or impossible to produce by conventional catalysis. The enzymatic ring-opening polymerization of lactones, lactides, cyclic carbonates and depsipeptides by various hydrolases has been studied extensively over the past decade as reviewed by different authors (*3–5*). Gu et al. (*6*) reported polyamide synthesis through enzyme-catalyzed polycondensation of dicarboxylic acid derivatives with diamines. In this work, we present the enzymatic ring-opening polymerization of β-propiolactam as an alternative synthetic route to obtain a polyamide (*7*) and propose a mechanism for the polymerization developed via a synergistic approach combining experimental results with computational simulation. The enzymatic pathways to polyamide synthesis can help reduce energy consumption and potentially decrease the use of toxic chemical compounds in industrial processes.

Enzymatic Ring-Opening Polymerization of Lactones

Enzymatic ring-opening polymerization was first reported in 1993 and involved the ring-opening polymerization of ε-caprolactone and δ-valerolactone by lipases from *Pseudomonas fluorescens*, *Candida cylindracea* and porcine pancreatic lipase (*5*, *8*).

Several unsubstituted lactones with ring sizes from 4 to 16 have since been polymerized by using *Candida antarctica* lipase B (Cal-B) among other lipases from different microorganisms (Figure 1). The generally accepted mechanism for the ring-opening polymerization of lactones by Cal-B is depicted in Figure 2, as proposed by Dong et al. (*9*) It consists of three steps. The first step is called monomer activation; i.e., the monomer is activated by the formation of an acyl-enzyme intermediate by a nucleophilic attack of serine-105 on the lactone carbonyl group. Secondly, the building block for the growing chain (hydroxy acid) is released from the serine residue by hydrolysis of the intermediate (initiation step). In the third (propagating) step a new intermediate is formed and it is attacked by the previously released hydroxy acid. The growing chain of length n+1 is released into the reaction medium. When serine-105 attacks a growing chain it can either react at the carboxyl chain end (called chain-end activation) or at one of the ester carbonyl groups in the chain (called intra-chain activation). Macrocycles can also be formed depending on the type of solvent and the concentration (*10*, *11*). Furthermore, it is a common feature of polycondensation reactions to form macrocycles (Jackobson-Stockmayer theory) (*12*). It is expected that the lactam polymerization will proceed according to a very similar mechanism.

Enzymatic Hydrolysis of β-Lactams

The enzymatic hydrolysis of β-lactams is a convenient method to obtain enantiomerically pure β-amino acids (*13–15*). The number of lipases available for this reaction is limited due to the irreversible binding of β-lactams and β-lactones to the serine in the active site of some lipases and proteases (*16*, *17*). For this reason β-lactams are used in antibiotics. They bind irreversibly to enzymes that are involved in cross-linking reactions in bacterial cell wall synthesis, e.g., the penicilin binding proteins (*18*, *19*). One of the lipases that is not inhibited by these compounds is *Candida antarctica* lipase B. It is capable of performing the enantioselective ring opening of β-lactams without deactivation as was shown by Fülop and coworkers (Figure 3). Substituents on the β-lactam ring can be cyclic or bicyclic, and with or without aromaticity (*20–23*).

266

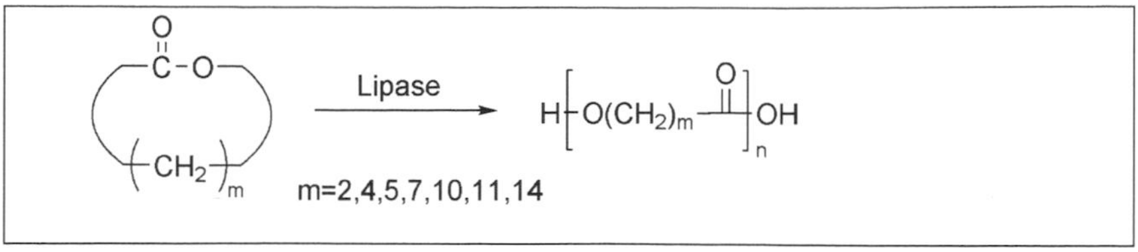

Figure 1. Cal-B catalyzed ring-opening polymerization of lactones.

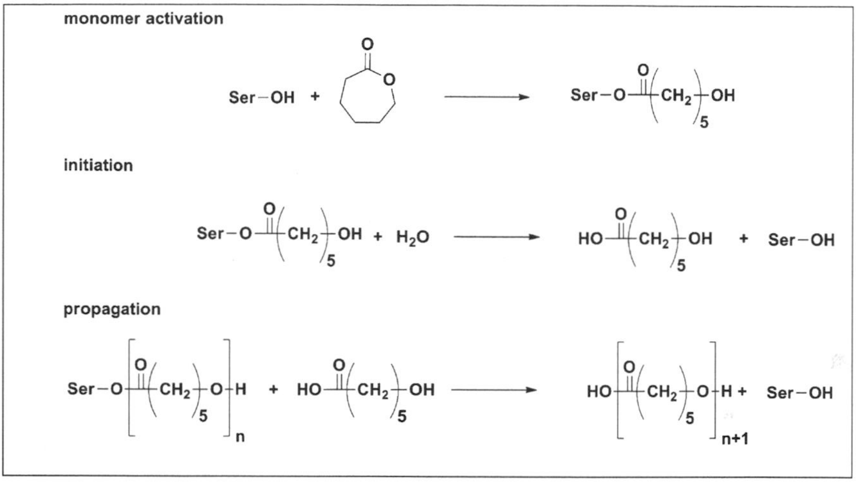

Figure 2. Proposed mechanism for enzymatic ε-caprolactone polymerization.

The reaction was performed in an organic solvent (e.g., toluene and isopropyl ether, or a mixture of these solvents with different alcohols). Reasonable conversions (30% or higher) were found with long chain alcohols and secondary or tertiary alcohols as a co-solvent (*21*), but the hydrolysis was also performed in a solvent-free system (*22*). In addition, the enzymatic formation of lactam rings (size 5-7) by ring closure of β-amino acid esters in organic solvents was reported by Gutman and coworkers (*24*). The reaction was catalyzed by several proteases and porcine pancreatic lipase.

All these examples illustrate that *Candida antarctica* lipase B can bind to a lactam ring and perform reactions on it. In this article, we report the mechanistic insights into the enzymatic ring-opening polymerization of β-propiolactam to synthesize poly(β-alanine) (see Figure 4). The Cal-B used was Novozym® 435 (N435) lipase from Novozymes A/S.

Results and Discussion

Polymer Synthesis

We started our attempts to use Cal-B to polymerize β-propiolactam with diisopropyl ether at 60 °C for 24 hours in order to simulate the system used by Fülöp *et al.* (*21*) A series of experiments was carried out to determine

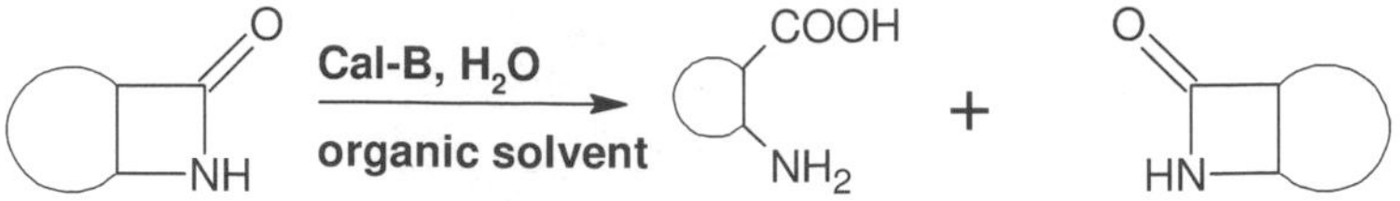

Figure 3. Hydrolysis of substituted β-lactams

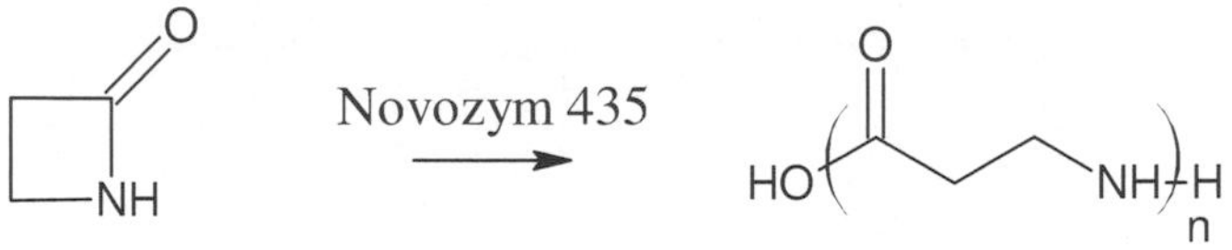

Figure 4. Enzymatic ring-opening polymerization of β-propiolactam

the appropriate reaction conditions. We discovered that polymerization at 90 °C for 96 hours in toluene produced the desired polymer at 30 % yield. The polymer obtained was characterized by MALDI-ToF mass spectrometry and ^{1}H-NMR-spectroscopy. Optimization of the reaction conditions and drying of the enzyme is discussed below.

MALDI-ToF Mass Spectra of Poly(β-alanine)

After purifying the poly(β-alanine) by a washing step with ethanol, we obtained a white solid. The MALDI-ToF mass spectrum (Figure 5) showed a mass increment of 71 m/z corresponding to one monomeric unit. In the ionization process adducts with sodium and potassium were formed. The sodium adducts formed the primary distribution of peaks, while the potassium adducts produced the second distribution of peaks between 300 and 700 m/z. The distribution had a maximum at 467 m/z (DP=6) and extended to 1319.7 m/z (DP=18).

^{1}H-NMR-Spectra of Poly(β-alanine)

The polymer was analyzed by ^{1}H-NMR spectroscopy (Figure 6). The main chain protons (δ= 3.30, 2.29 ppm) and the protons next to the endgroups (amine δ= 3.12 ppm; carboxylic acid δ= 2.52 ppm) could be identified according to the literature (*25, 26*). Although polymerization was carried out successfully, the yield was not high. In our process, low-molecular-weight cyclic products and β-alanine were removed when we washed the product with ethanol.

The average degree of polymerization (DP) determined from the ^{1}H-NMR spectrum is about 7. This value is in good agreement with the maximum of the peak distribution found in the MALDI-ToF data (DP=6). The conversion of β-propiolactam is not complete, as only 30 wt% polymer is obtained after purification. Some of the monomer is hydrolyzed or left unchanged as shown by ^{1}H-NMR spectroscopy. MALDI-ToF mass spectrometry showed that some of the monomer is converted into cyclic products that are also washed away during purification (data not shown). Hydrolysis is caused by the presence of water in the

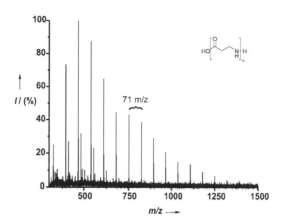

Figure 5. The MALDI-ToF mass spectrum of the poly(β-alanine). At a Δm/z of 71 the [M-Na]⁺ ions of the polymer are found.

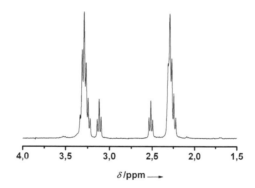

Figure 6. ¹H-NMR spectrum of the poly(β-alanine) with protons of the main chain at δ= 3.30, 2.29 ppm and protons next to end-groups at δ= 3.12, 2.52 ppm.

catalyst. Optimization of the reaction time and temperature as well as changing the drying procedure for the catalyst is needed to improve both the yield and the degree of polymerization of poly(β-alanine).

Optimization

By lowering the reaction temperature to 55 °C and using a reaction time of 96 hours, the polymer was obtained in 80% yield (Table 1). Although the activity assay showed that the enzyme remained active after 96 hours of incubation in toluene, there was apparently some thermal deactivation. According to Novozymes, the N435 lipase shows the highest productivity at temperatures between 40 and 60 °C (27). N435 contained water that was removed by drying over P_2O_5 for 48 hours at 46 °C under reduced pressure. Because it is known that the amount of water in the enzyme is critical for its function (2, 28), the drying conditions need to be optimized.

Table 1. Temperature optimization

T (°C)	t (h)	Yield (%)	DP (^{1}H-NMR)
55	96	81	5
55	144	83	5
90	96	30	5

Table 2. Optimization of drying conditions

time (h)	T (°C)	DP (by ^{1}H-NMR)
2	55	6.7
24	55	7.3
24	46	7.6
48	46	5.0

The degree of polymerization was examined after drying the enzyme for 2, 24 or 48 hours over P_2O_5 *in vacuo* at 55 °C and 46 °C. The results are summarized in Table 2. The best results were obtained when the enzyme was dried for 24 hours, giving a polymer with DP 7. When the enzyme was dried for a shorter period (2h), there was too much water left that hydrolyzed the lactam and reduced the DP to 6.7. After 48 hours of drying, too much water was removed and the enzyme started to deactivate, resulting in a polymer with a lower DP of 5.

Mechanism of the Reaction

Reactions without N435 lipase did not yield any polymeric product, illustrating that N435 was indeed needed as the catalyst. N435 always contains some water that can possibly initiate polymerization or hydrolyze β-propiolactam to form β-alanine which may be the monomer for the polymerization. Both possibilities have been ruled out by performing experiments with deactivated N435 and adding water, but the experiments did not lead to polymer formation. Moreover, N435 was not able to produce any polymer when β-alanine was offered as a monomer for polymerization.

Our result is in accordance with the recent findings of Hollmann et al. (*29*) who showed an inhibition of Cal-B by organic acids with pK_a-values of < 4.8. This inhibition most likely is caused by protonation of His-224 which then no longer can assist in the essential deprotonation of Ser-105 during the catalytic process. The first pK_a of β-alanine is 3.6, which is clearly below the critical value. Hence a release of β-alanine could cause a reduction of the overall enzyme activity by blocking His-224. The solubility of β-alanine in the reaction medium (toluene) is low (10^{-5} mol L^{-1}) and the detrimental effect of the β-alanine will be limited. During the enzymatic polymerization, a competition between elongation of the acyl-enzyme complex and release of the chain by hydrolysis takes place and small amounts of β-alanine will be present during polymerization.

From these results it can be concluded that N435 catalyzes the reaction and neither water, β-alanine nor the carrier material starts the reaction. Because β-alanine is not polymerized by N435, it seems that the ring structure of the β-propiolactam is essential for the formation of polymer; i.e., poly(β-alanine) can only be formed by a ring-opening polymerization utilizing β-lactam as monomer which elongates an acyl-enzyme intermediate. Because of the low nucleophilicity of the lactam nitrogen, this process requires activation of β-lactam by a water molecule. The crystal structures of native Cal-B (1TCA, 1TCB) (30) exhibit a structurally conserved water molecule near His-224-N which is ideally positioned to perform the activation of β-lactam.

On this basis we propose a mechanism for the Cal-B mediated polymerization of β-lactam that consists of two starting steps (I and II), five elongation steps (III-VII) and one termination step (VIII), as depicted in Figure 7, utilizing the catalytic triad Ser-105, His-224 and Asp-187 for the polymerization process, and the oxyanion hole (Gln-106, Thr-40) for stabilization of the negatively charged reaction intermediates. Our mechanism is compatible with experimental data and is supported by in-depth computational calculation of the reactions involved, which was reported elsewhere (31).

The formation of the first acyl-enzyme complex (Figure 7 C) by acylation of the active side serine and ring-opening of the former lactam corresponds to the generally accepted initial step of hydrolysis by serine proteases. The Ser105-O carries out nucleophilic attack at the carbonyl carbon of the β-lactam, while the serine proton is passed onto His224-N (A), and a first tetrahedral intermediate is formed (B). The next step is ring-opening of the former lactam by transfer of a proton from His224 to the lactam-N, which yields the acyl-enzyme complex (C). The subsequent chain elongation is carried out by an activated monomer (E), which is formed by reaction of β-lactam and the catalytic water molecule (D), to yield a dimeric tetrahedral intermediate (F), which rearranges to form a terminally bound tetrahedral intermediate (G). Intermediate G exhibits its hydroxyl group towards His224-NH, whereupon a molecule of water is cleaved off and a new acyl-enzyme complex is formed. The water released corresponds to the water molecule used for the activation of β-lactam.

Molecular Modeling

Computational simulation techniques were employed to assist in the proposed catalytic scheme (Figure 7), starting with a study of the position of the acyl chain of the acyl-enzyme complex inside the enzyme pocket by use of a covalent docking procedure. As expected, the acyl chain occupies the so-called acyl side with its carbonyl oxygen inside the oxyanion hole, whereas the alcohol side of the active side is vacant (Figure 8 C). In order to model step III of the catalysis cycle, a β-lactam monomer was docked into the acylated enzyme in the presence of the crystallographic water molecule in front of His224-N. As can be seen from Figure 8 D the β-lactam in the resulting orientation shows a distance between the carbonyl carbon of the β-lactam and the oxygen of the water of 3.6

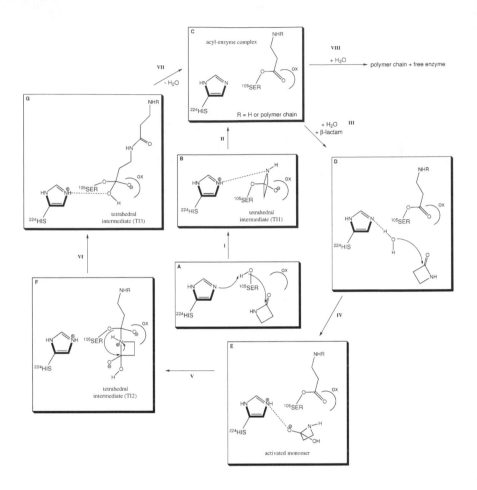

Figure 7. Catalytic cycle of Cal-B mediated polymerization of β-lactam.

Å with an H-O-C angle of 118°. This position is still too far for an OH attack at the carbonyl group, but with the assumption of small movements of this water molecule and the β-lactam, respectively, a nucleophilic attack of the OH group on the carbonyl carbon of the β-lactam takes place while the hydrogen of the water molecule is accepted by His224 (step IV). Docking studies of this activated monomer reveal the structure depicted in Figure 8 E where the activated monomer is stabilized by hydrogen bonds to His224 (2.5 Å) and to Thr40 (3.2 and 2.8 Å, respectively). Owing to the elimination of its double bond character, the activated β-lactam exhibits a higher nucleophilicity compared to the unactivated monomer which should be strong enough to attack the carbonyl carbon of the acyl-enzyme complex. Although the distance between the nitrogen and the acyl-carbonyl carbon still is 3.3 Å the lone pair of the former amide nitrogen is ready to attack the acyl-chain at Ser105. The former β-lactam is no longer planar but slightly puckered (about 16°). The orientation of the nitrogen lone pair towards the acyl-carbonyl carbon is certain because of the rapid inversion of the NH-proton between a *cis* and *trans* configuration with respect to the hydroxyl group.

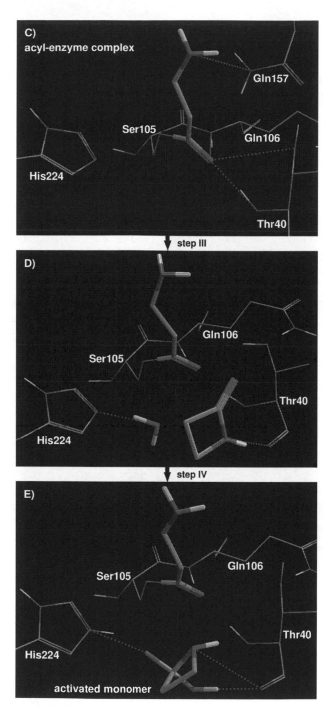

Figure 8. Covalent docking of acyl-enzyme complex (C), docking of second monomer in presence of catalytic water (D) and activation of β-lactam by hydroxyl group of this water molecule (step IV, E) (see color insert)

273

An attack of the activated monomer on the acyl-enzyme intermediate results in a short-lived dimeric tetrahedral intermediate TI2 (9 F). This structure was generated by connecting the nitrogen of the activated β-lactam monomer to the serine bound carbonyl carbon followed by a covalent docking procedure. The negatively charged oxygen of the resulting dimeric tetrahedral intermediate is still stabilized by the oxyanion hole (3.0 Å, 3.3 Å and 3.2 Å, respectively), and the negative charge is compensated by the positive charge of the attacking nitrogen. Further stabilization is achieved from hydrogen bonding between the hydroxy group of the attacking monomer and Ser-O (2.8 Å). The resulting chiral center at the former carbonyl carbon of the acyl-enzyme complex has (S)-configuration established by the orientation of the attacking monomer. The NH proton of the dimeric tetrahedral intermediate is *cis* oriented with respect to the negatively charged carbonyl oxygen.

QM/MM optimizations of the dimeric tetrahedral intermediate (TI2, Figure 9 F) indicate an instantaneous opening of the former lactam ring followed by rearrangement of Ser105-O form the central to the terminal carbon, similar to the insertion of a monomer in transition metal-catalyzed polymerizations. As a result a terminal bound tetrahedral intermediate is formed which again is stabilized by the oxyanion hole (3.4 Å, 3.3 Å and 2.7 Å, respectively) and hydrogen bonding between the OH group of TI3 and His224-NH (2.7 Å). This hydroxyl group can then be cleaved off after protonation by the proton from His224-NH. The released water molecule corresponds with the catalytically used water molecule for activation of the β-lactam monomer (8 D). The newly generated acyl-enzyme intermediate can either be elongated further (Figure 7, step III), where the water balance of the enzyme is neutral, or be liberated by attack of a water molecule (Figure 7, step VIII), which is consumed in this process.

Conclusion

To our knowledge, this investigation is the first study of an enzymatic polymerization process on a molecular level and results in a mechanism in which the growing polymer chain does not have to leave the active site during chain elongation but rather stays bound to serine of the catalytic triad. A liberation of the polymer would imply the risk of the chain leaving the active site and not returning to the binding site in a correct orientation for chain elongation. The limiting factor of the chain length is the necessary presence of a water molecule close to the catalytic triad. The polymerization process can be seen as a permanent competition between chain elongation via activation of a monomer by water and chain termination via attack of the acyl-enzyme intermediate by water. The dried enzyme preparation has to provide one water molecule for every released polymer chain (which should not hinder the polymerization as the polyester formation with lactones as starting materials results in high-molecular-weight polymers). A major obstacle of the Cal-B-catalyzed poly(β-alanine) polymerization is the potential enzyme-catalyzed hydrolysis of β-lactam yielding β-alanine, which itself is not a substrate for Cal-B but rather leads to an overall decrease in enzyme activity and is the major reason for the low degree of polymerization observed.

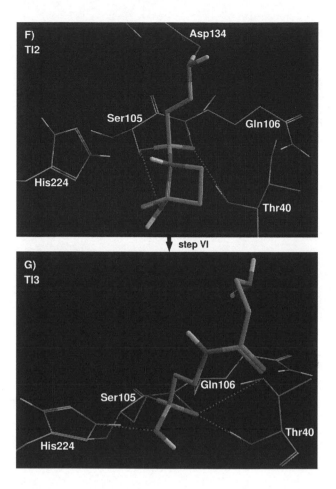

Figure 9. Rearrangement of short-lived intermediate (TI2) in which Ser105-O jumps from central carbon to terminal carbon yielding terminal bound tetrahedral intermediate (TI3). (see color insert)

Experimental

Polymerization of β-Propiolactam

The glassware was flame-dried before polymerization. A mixture of β-propiolactam (100 mg, 1.41 mmol), N435 (100 mg), and dry toluene (5 mL) was stirred for 96 h at 90 °C under a N_2-blanket. After cooling to room temperature, toluene was removed by rotary evaporation. The remaining solids were stirred with ethanol for 15 minutes and filtrated. By extracting the residue with water, pure poly(β-alanine) was obtained. (yield 30 %). ^{1}H-NMR (D_2O): δ = 3.3 (m, 2H; CH_2) 3.12 (t, 2H; CH_2) 2.52 (t, 2H; CH_2) 2.29 (m, 2H; CH_2)

Control Reactions

The following reactions were performed and the lack of polymer observed excluded initiation by water, β-alanine or carrier material. At the same time it shows that β-propiolactam is the sole monomer in this reaction.

- Polymerization without a catalyst present
- Polymerization by deactivated N435 (by heating it to 150 °C for two hours hydrolytic activity was gone after this treatment).
- Polymerization with N435 and additional water
- Polymerization with deactivated N435 and additional water
- Polymerization of β-alanine was attempted
- Polymerization of β-propiolactam in the presence of β-alanine

Computational Methods

The crystal structure of *Candida antarctica* lipase B complexed with a covalently bound phosphonate inhibitor (pdb code 1LBS) (*32*) was supplied by the Protein Data Bank (*33*) and was used as the starting point for the modeling studies. In addition, the pdb files 1TCA and 1TCB of the native enzyme were used to extract specific crystallographic water molecules which were copied to 1LBS at the appropriate position and energy-minimized afterwards. The crystallographic water molecule showing hydrogen bonding to His224 was used for some modeling studies. The protein structure of 1LBS was prepared for the modeling investigation using the program MOE (*34*).

Docking experiments were performed with QXP-Flo (Quick eXPlore) (*35*) using a subset of the protein consisting of a cylindrical 15 Å sphere around the center of the active site. The binding pocket was defined by manual coloration of guided atoms. Amino acids that lie with at least one atom inside this sphere were completely incorporated in the subset and polar hydrogen atoms were added. During the docking process protein atoms were kept fixed except for the hydroxy group of Ser105, for the NH group of His224 and for the functional groups (OH, NH) which formed the oxyanion hole (Thr40, Gln106). The heteroatoms involved were allowed a constrained movement (max. 2 Å) while the polar hydrogens of Ser105, His224 and the amino acids of the oxyanionhole were unconstrained. In docking experiments into the acylated enzyme, the enzyme bound ligand was held fixed except for its functional groups. Docking was usually performed in the absence of crystallographic water molecules if not otherwise stated. In case water molecules were included in the modeling process, the position of the water molecule was taken from the native Cal-B structures as described above (pdb code 1TCA and 1TCB respectively), and energetically minimized in the lipase B - phosphonate complex (pdb code 1LBS). The oxygen atom of these water molecules were also allowed a constrained movement (max. 2 Å) during the docking process, while the corresponding hydrogens were unconstrained.

Docking experiments started with an initial manual placement of the ligand outside the protein and a positioning of the ligand into the active site by the sdock tool of QXP, which merely placed the ligand in an appropriate position without any conformational search. Afterwards a full Monte Carlo search (mcdock) including a conformational search and a local Monte Carlo search with limited degrees of freedom (mcldock) were done. Docking results were scored via rating of the energies (*36*) combined with a principal component analysis. Energy minimization of the protein was done in MOE applying the force field AMBER99, followed by a minimization of the ligand and the surrounding amino acids of a given docking result using the forcefield MMFF94x.

For QM/MM (*37*) calculations the software NWChem (Version 5.1, Pacific Northwest National Laboratory, USA) (*38*) was employed. The enzyme-substrate models were partitioned in a QM region (ligand, His_{224} and Ser_{105}) and an MM subsystem (residual moiety). The active site, which participated in making and breaking bonds, was treated by *DFT* quantum chemical methods at B3LYP level using Ahlrichs-pVDZ basis set. The rest of the enzyme was described by the methods of molecular mechanics using AMBER99 force field. The QM/MM boundary was capped with H atoms and treated by the pseudo bond approach (*39, 40*).

References

1. Klibanov, A. M. *ChemTech* **1986**, 354.
2. Zaks, A.; Klibanov, A. M. *J. Biol. Chem.* **1988**, *263*, 3194–3201.
3. Kobayashi, S.; Uyama, H.; Kimura, S. *Chem. Rev.* **2001**, *101*, 3793–3818.
4. Srivastava, R. K. *Adv. Drug Delivery Rev.* **2008**, *60*, 1077–1093.
5. Uyama, H.; Kobayashi, S. *Chem. Lett.* **1993**, *22*, 1149–1150.
6. Gu, Q.-M.; Maslanka, W. W.; Cheng, H. N. *Polym. Prepr.* **2006**, *47*, 234.
7. Schwab, L. W.; Kroon, R.; Schouten, A. J.; Loos, K. *Macromol. Rapid Commun.* **2008**, *29*, 794–797.
8. Knani, D.; Gutman, A. L.; Kohn, D. H. *J. Polym. Sci., Part A: Polym. Chem.* **1993**, *31*, 1221–1232.
9. Dong, H.; Cao, S.-G.; Li, Z.-Q.; Han, S.-P.; You, D.-L.; Shen, J.-C. *J. Polym. Sci., Part A: Polym. Chem.* **1998**, *37*, 1265–1275.
10. Cordova, A.; Iversen, T.; Hult, K.; Martinelle, M. *Polymer* **1998**, *39*, 6519–6524.
11. Thurecht, K. J.; Heise, A.; deGeus, M.; Villarroya, S.; Zhou, J.; Howdle, S. M. *Macromolecules* **2006**, *39*, 7967–7972.
12. Jacobson, H.; Stockmayer, W. H. *J. Chem. Phys.* **1950**, *18*, 1600–1606.
13. Juaristi, E. *Enantioselective Synthesis of beta-Amino Acids*; Wiley-VCH: New York, 1997.
14. Juaristi, E.; López-Ruiz, H. *Curr. Med. Chem.* **1999**, *6*, 983–1004.
15. Abdel-Magid, A. F.; Cohen, J. H.; Maryanoff, C. A. *Curr. Med. Chem.* **1999**, *6*, 955–970.
16. Knight, W. B.; Green, B. G.; Chabin, R. M.; Gale, P.; Maycock, A. L.; Weston, H.; Kuo, D. W.; Westler, W. M.; Dorn, C. P. *Biochemistry* **1992**, *31*, 8160–8170.
17. Wilmouth, R. C.; Kassamally, S.; Westwood, N. J.; Sheppard, R. J.; Claridge, T. D. W.; Aplin, R. T. *Biochemistry* **1999**, *38*, 7989–7998.

18. Pauline, M.; Carlos, C.; Viviana, J.; Otto, D.; Andréa, D. *FEMS Microbiol. Rev.* **2006**, *30*, 673–691.

19. Walsh, C. *Antibiotics: Actions, Origins, Resistance*; ASM Press: Washington, DC, 2003.

20. Forró, E.; Fülöp, F. *Tetrahedron: Asymmetry* **2003**, *15*, 573–575.

21. Park, S.; Frorró, E.; Grewal, H.; Fülöp, F.; Kazlauskas, R. J. *Adv. Synth. Catal.* **2003**, *345*, 986–995.

22. Forró, E.; Fülop, F. *Tetrahedron: Asymmetry* **2008**, *19*, 1005–1009.

23. Tasnádi, G.; Forro, E.; Fülop, F. *Tetrahedron: Asymmetry* **2007**, *18*, 2841–2844.

24. Gutman, A. L.; Meyer, E.; Yue, X.; Abell, C. *Tetrahedron Lett.* **1992**, *33*, 3943–3946.

25. Jia, L.; Ding, E.; Anderson, W. R. *Chem. Commun.* **2001**, 1436–1437.

26. Applequist, J.; Glickson, J. D. *J. Am. Chem. Soc.* **1971**, *93*, 3276–3281.

27. Novozym AS. Novozym 435 datasheet; 2004; pp 1–3

28. Klibanov, A. M. *J. Am. Chem. Soc.* **1986**, *108*, 2767–2768.

29. Hollmann, F.; Grzebyk, P.; Heinrichs, V.; Doderer, K.; Thum, O. *J. Mol. Catal. B: Enzym.* **2009**, *57*, 257–261.

30. Uppenberg, J.; Hansen, M. T.; Patkar, S.; Jones, T. A. *Structure* **1994**, *15*, 293–308.

31. Baum, I.; Schwab, L. W.; Loos, K.; Fels, G. Manuscript in preparation.

32. Uppenberg, J.; Öhrner, N.; Norin, M.; Hult, K.; Kleywegt, G. J.; Patkar, S.; Waagen, V.; Anthonson, T.; Jones, T. A. *Biochemistry* **1995**, *34*, 16838–16851.

33. RCSB Protein DataBank, http://www.pdb.org.

34. Chemical Computing Group Inc. Montreal, http://www.chemcomp.com.

35. Mcmartin, C.; Bohacek, R. S. *J. Comput.-Aided Mol. Des.* **1997**, *11*, 333–344.

36. Alisaraie, L.; Haller, L. A.; Fels, G. *J. Chem. Inf. Model.* **2006**, *46*, 1174–1187.

37. Valiev, M.; Kawai, R.; Adams, J. A.; Weare, J. H. *J. Am. Chem. Soc.* **2003**, *125*, 9926–9927.

38. Apra, E. Pacific Nortwest National Laboratory. *NWChem, a computational package for parallel computers*, version 4.6; 2004.

39. Valiev, M.; Garrett, B. C.; Tsai, M. K.; Kowalski, K.; Kathmann, S. M.; Schenter, G. K.; Dupuis, M. *J. Chem. Phys.* **2007**, *127*, 051102–051104.

40. Zhang, Y.; Lee, T. S.; Yang, W. *J. Chem. Phys.* **1999**, *110*, 46–54.

Syntheses and Modifications of Polysaccharides

Chapter 20

Production of Natural Polysaccharides and Their Analogues via Biopathway Engineering

Lei Li,[a,b] Wen Yi,[b,c] Wenlan Chen,[b] Robert Woodward,[b] Xianwei Liu,[a] and Peng George Wang[b,*]

[a]National Glycoengineering Research Center and The State Key Laboratory of Microbial Technology, School of Life Sciences, Shandong University, Shandong 250100, China
[b]Departments of Chemistry and Biochemistry, The Ohio State University, Columbus, OH 43210, USA
[c]Present address: Division of Chemistry and Chemical Engineering, Howard Hughes Medical Institute, California Institute of Technology, Pasadena, CA 91125, USA
*pgwang@chemistry.ohio-state.edu

Polysaccharides constitute one of the major classes of bio-macromolecules in living organisms. In the past several decades, tremendous advances in glycobiology and cell biology revealed that polysaccharides play essential roles in a variety of important biological processes. Various natural polysaccharides and their analogues have been synthesized via *in vitro* enzymatic polymerization. This chapter describes the *in vitro* biosynthesis of O-polysaccharides from *Escherichia coli*, as well as *in vivo* production of lipopolysaccharide and its analogues via biopathway engineering.

Introduction

Polysaccharides constitute one of the major classes of bio-macromolecules in living organisms. They are ubiquitous in nature and play essential roles in a variety of important biological processes, such as embryonic development, energy supply, signal transduction, mediation of cell-cell interaction, and stimulation of immune responses (*1–3*). Polysaccharides are divided into two broad classes (Table 1): Homo-polysaccharides, which consist of residues of only one kind of monosaccharide and hetero-polysaccharides, which consist of residues of different monosaccharides. Naturally occurring homo-polysaccharides, (e.g. cellulose, chitin, starch, glycogen) primarily perform structural and energy storage roles in organisms (4, 5). On the other hand, hetero-polysaccharides play much more diverse roles concerning their structural complexities. They are essential components of cell walls and extracellular matrices (peptidoglycan, glycosaminoglycan), major pathogenic factors in bacteria (O-polysaccharides, capsular polysaccharides, exopolysaccharides), and critical in cell development and cell-cell interactions.

The cell surface of bacteria is decorated with remarkable variations of polysaccharide structures. Based on the way in which they are associated with the cell surface, bacterial polysaccharides can be classified into three major groups: O-polysaccharides (O-PS), capsular polysaccharides (CPS), and exopolysaccharides (EPS) (Table 1). These sugars mediate the direct interactions between the bacterium and its host, and have been implicated as an important virulence factor of many plant and animal pathogens.

O-Polysaccharides (O-PS)

O-polysaccharide (also called O-antigen) is a major component of the bacterial cell surface lipopolysaccharide (LPS) (Figure 1). It is composed of multiple copies (as many as 100 copies) of an oligosaccharide repeating unit (O-unit). O-PS are presented on the cell surface via the covalent attachment to the core oligosaccharide and the lipid-A, both of which are essential components of LPS. O-PS is the most structurally varied component among different bacterial species. For example, there are more than 180 O-PS structures found in E. coli (6–8). The variations arise from different sugar compositions, positions, and stereochemistries of the glycosidic linkages within the O-units, as well as the linkages between each other (7, 9). These extensive structural variations of the O-PS allow classifications of the bacterial species into different O-serological groups (6–8) such as E. coli O86. Increasing evidences suggest that the O-PS plays an important role in bacteria-host interactions such as the effective colonization of the host and the resistance to complement-mediated immune responses (10).

Capsular Polysaccharides (CPS)

Capsular polysaccharides are linked to the cell surface through covalent attachments to either phospholipids or lipid-A molecules (22, 23). They are highly hydrated hetero-polysaccharides which can be substituted by non-sugar residues such as acetyl, acyl and sulfate. Similarly, CPS also displays enormous structural variations (23–26). For example, Streptococcus pneumoniae, a gram-positive pathogen that causes pneumonia, meningitis and blood stream infections, can produce over 90 chemically and serologically different CPS (27, 28). Importantly, CPS have been shown to be virulence factors in infectious bacteria. Such a property has been exploited to develop polysaccharide-based vaccines against infectious diseases such as pneumonia and meningitis. Current pneumococcal vaccines containing CPS purified from 23 recognized serotypes are proven to

Table 1. Selected examples of natural polysaccharides

Homopolysaccharides			
Name	**Structure**	**Function**	**Refs**
Glycogen/Starch	—(G c-α-1 4-G c)ₙ	Storage polysaccharides of animals	(4,5)
Chitin	—(GlcNAc-β1,4-GlcNAc)ₙ	Structural components of exoskeletons of invertebrates	(11)
Heteropolysaccharides			
Peptidoglycan	—(GlcNAc-β1,4-MurNAc)ₙ	Important components of bacterial cell wall	(12,13)
Glycosaminoglycan (GAG)	Heparin, Hyaluronan —(GlcNAc-β1,4-GlcA)ₙ Chondroitin —(GalNAc-β1,4-GlcA)ₙ	Structural scaffolds in extracellular matrix, promotes cell adhesion, growth modulation, growth factor concentration, signal transducton	(14,15)
O-polysaccharides	—(4Fuc-α1,2-Gal-β1,3-GalNAc-α1,3-GalNAc1)ₙ 3 ↑ 1 *E. coli* O86:H2 α-Gal	Mediates bacteria-host interaction, effective colonization, resistance to complement-mediated immune response, serotyping	(16,17)
Capsular polysaccharides	—(4Glc-β1,6-GlcNAc-β1,3-Gal-β1)ₙ 4 ↑ 1 *S. pneumoniae* type 14 β-Gal	Important pathogenic factor, promotes bacterial adhesion, resistance to phagocytosis, facilitates biofilm formation	(18,19)
Exo-polysaccharides	—(4Glc-β1,4-Glc-β1)ₙ 6 ↑ 1 *X. campestris* β-Gal1,4-αGlcA-1,2-α-Man	Formation of biofilm, viscosifier, stabilizer, emulsifier, suspending agent	(20,21)

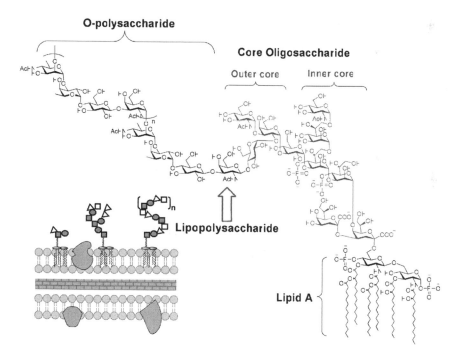

Figure 1. Gram-negative bacterial cell wall and O-polysaccharides (E. coli O157 as an example).

exert serotype specific protection against 90% of the disease causing serotypes (29).

Exopolysaccharides (EPS)

Many microorganisms are capable of producing exopolysaccharides, which are secreted outside the cells (*20, 30, 31*). EPS are released onto the cell with no visible means of attachment and are often sloughed off to form slime. EPSs have a much higher molecular weight compared to both O-PS and CPS, possessing more than 1000 repeating units (*32*). Expression of EPS has important applications in both biomedical research and the food industry (*20, 21, 33–36*). For example, EPS are critical for biofilm formation, and are key components of the biofilm matrix in many biofilms (*37–39*). Furthermore, EPS also play an important role in immune evasion and tolerance toward antibacterial agents. The food industry uses EPS for their unique properties to create viscosifiers, stabilizers, emulsifiers, or gelling agents (*31, 40*).

Biosynthesis of Bacterial Polysaccharides

In spite of the remarkable structural diversity of the O-PS, CPS and EPS, only three major biosynthetic pathways are utilized in nature, as shown by extensive genetic studies (*9, 17*). This implies that nature has come up with only a few solutions to the ubiquitous problem of polysaccharide biosynthesis. In this work, we will mainly focus on the synthesis of O-PS since O-PS have been used as model systems to study polysaccharide biosynthesis in the past decades. Usually, O-PS are synthesized separately before being transferred to the core-Lipid A to form LPS (*41*). The genes involved in the synthesis of O-PS are clustered at the locus historically known as *rfb*. This locus encodes enzymes involved in biosynthesis of sugar nucleotide precursors, including glycosyltransferases, synthesize O-unit or nascent O-PS, and other enzymes required for polysaccharide processing. There are three enzymatic stages in the biosynthesis of O-PS (*41*) (Figure 2).

Initiation

The initiation reaction involves the addition of a sugar phosphate to an undecaprenyl phosphate (Und-*P*) to form a monosaccharide-PP-Und (*42, 43*). This reaction is conserved among all three pathways, and takes place at the cytosolic face of the plasma membrane where the sugar nucleotides are available. Depending on the identity of the sugar residue (usually an *N*-acetyl hexosamine or hexose), the reaction is catalyzed by WecA or WbaP which belong to the PNTP and PHTP protein families, respectively (*44, 45*).

Elongation/Polymerization

The three major biosynthetic pathways are classified depending on the assembly and translocation mechanisms during this step: *Wzy-dependent pathway*: This pathway occurs in the synthesis of the majority of O-PS. It involves the sequential addition of different sugar residues onto the monosaccharide-PP-Und intermediate via catalysis by specific glycosyltransferases. These glycosyltransferases are predicted to be soluble or peripheral membrane bound proteins. Upon completion of the O-units-PP-Und, they are translocated to the periplasmic side of the inner membrane by the flippase **Wzx** (*46, 47*). In the periplasm, the Und-PP-O-units are subsequently polymerized into a sugar polymer by the integral membrane protein **Wzy**. The polymerization reaction involves the transfer of the nascent polymer from its Und-PP carrier to the non-reducing end of the new O-units-PP-Und (*48*). The final component of the *wzy*-dependent pathway is the **Wzz** protein, which functions as a chain length regulator to generate strain specific polysaccharide chain lengths (*42, 49*). *ABC transporter-dependent pathway* and *Synthase-dependent pathway* are detailed reviewed in reference (*17*).

Termination and Ligation

The ligation step involves the transfer of the O-polysaccharide to the nascent core-Lipid A. The ligation reaction is common to all three pathways, whereas its mechanism is still unclear. WaaL, a predicted integral membrane protein with 8 or more transmembrane segments is currently the only known protein required for the ligation. It has been reported that WaaL proteins lack discrimination for the sugar structures of O-units-PP-Und (*50*). The work suggests that they recognize the Und-P carrier rather than the saccharide attached to it, which is consistent with our *in vitro* biochemical characterization of WaaL (Unpublished data). Translocation of complete LPS to the outer membrane involves the protein complex Imp/RlpB (*51*).

Over the years, we have been using *E. coli* O86 as a model system to investigate the *wzy*-dependent pathway, and have gained substantial understanding of the structure, genetics, enzyme functions, and some details of the biosynthetic mechanism.

Sequence of *E. coli* O86 O-Antigen Gene Cluster and Characterization of Glycosyltransferases in the Pathway

E. coli O86 is interesting since it exhibits strong human blood group B activity (*52*). The blood group B activity was further confirmed by an immune response on human, and those individuals with blood group O and A responded with a significant rise of anti-B antibody (*53*). This phenomenon strongly suggested that *E. coli* O86 contains human blood group B antigen trisaccharide moiety (Figure 3a), which was confirmed by the determination of the complete structure (*54*) (Figure 3b).

We sequenced the O-antigen biosynthetic gene cluster of *E.coli* O86:H2 and characterized four glycosyltransferases responsible for the repeating unit biosynthesis (*55*). Random TOPO shotgun sequencing approach (Invitrogen) was used to obtain the gene cluster. A total 13,990 bp sequence (covered a continuous region from galF gene to the start of gnd gene) was obtained. Thirteen open reading frames were identified and annotated based on their sequence similarities to genes in the databases (Figure 4). As expected, three types of biosynthetic genes are found in the gene cluster, including sugar nucleotide (GDP-Fuc)

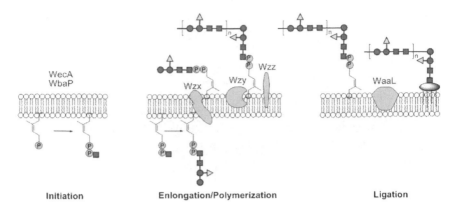

Initiation   Enlongation/Polymerization   Ligation

Figure 2. The wzy-dependent O-antigen biosynthesis pathway (E. coli O86 as an example).

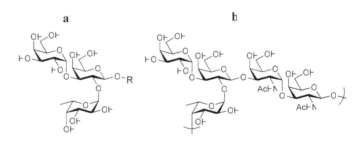

Figure 3. Structure of human blood group B antigen (a) and repeating unit of E coli O86:H2 O-polysaccharide.

biosynthetic genes (*manB, manC, gmd* and *fcl*), glycosyltransferase genes (*wbnH, wbnJ, wbnK* and *wbnI*), and O-antigen processing genes (*wzx and wzy*).

To biochemically characterize the specific function of each glycosyltransferase gene, we cloned the genes individually into expression vectors and expressed them in *E. coli*. After chromatographic purification of each enzyme into near homogeneity, an enzymatic activity assay using either mass spectrometry or a radioactive protocol was carried out. When incubating GalNAc-PP-lipid with WbnH protein in the presence of UDP-GalNAc, we identified in the mass spectrum a prominent peak at m/z = 414.4, consistent with the formation of GalNAc-GalNAc-PP-lipid product. The product was then isolated by gel filtration chromatography, and analyzed by NMR, confirming the predicted structure (*56*). Further biochemical study showed that pyrophosphate lipid moiety is a substrate requirement for WbnH, even though different lipids with various chain lengths can be well tolerated. WbnH belongs to GTB glycosyltransferase superfamily, and its activity is not dependent on divalent cations as cofactors.

Next, WbnJ was expressed with a N-terminal His-tag. Sugar donor assays showed that WbnJ was capable of transferring galactose residue from UDP-Gal to GalNAcα-OMe at high efficiency, indicating wbnJ encodes a β1,3-galactosyltransferase in the gene cluster. Then by using purified WbnJ

286

galF galE wbnH gmd   fcl gmm manC   manB   wzx wbnI wzy   wbnJ wbnK

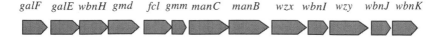

Figure 4. The O-antigen biosynthetic gene cluster of E. coli O86:H2

enzyme, we synthesized disaccharide Galβ1,3GalNAc-OMe. The disaccharides were purified by gel filtration chromatography (57, 58), and the structure was confirmed by NMR spectroscopy.

WbnK was also expressed in *E. coli* as GST fusion at its N-terminus. Its fucosyl transferase function was confirmed by donor specificity assay. Acceptor specificity assay showed that the fucosyltransferase activity was identified only when Galβ1,3GalNAc (T-antigen) was used as a substrate. The change of linkage from 1→3 to 1→4, or the configuration fromβ to α will make very poor substrates for WbnK (Table 2). This was consistent with the structure of O86 O-unit. The trisaccharide (Fucα1,2Galβ1,3GalNAcα-OMe) was then subsequently synthesized by using WbnK-GST fusion protein and the structure was verified by NMR.

WbnI was demonstrated to encode another galactosyl transferase. The acceptor screening showed that WbnI strongly prefers terminal fucosylated sugar structures, and trisaccharide Fucα1,2Galβ1,3GalNAcα–OMe was the best accepter. This indicated that WbnI catalyzed the formation of α1,3-linkage galactose residue at the terminal after the fucose residue was added.

In Vitro Reconstitution of *E. coli* O-Polysaccharide Biosynthetic Pathway

To reconstitute *E. coli* O-polysaccharide biosynthetic pathway *in vitro*, there are two major challenges. First, it is difficult to obtain structurally defined substrates. The repeating unit-diphospho-undecaprenyl (RU-PP-Und) is present in very small quantities in bacterial cells and cannot be isolated in sufficient quantities for biochemical studies. The only solution to overcome this problem is via chemical synthesis. With our expertise in chemo-enzymatic synthesis of complex oligosaccharides we have developed facile routes to obtain substrate analogues on milligram scales. Second, the putative O-polysaccharide polymerase, Wzy, with 11 predicted transmembrane segments, has been hard to overexpress to even a Western Blotting detectable quantity. With several years' efforts, we can now overexpress and purify Wzy in an active form, and reconstitute the O-polysaccharide pathway via utilizing the purified Wzy.

Chemical Synthesis of GlaNAc-PP-lipids

In the biosynthetic pathway of O-PS, WecA protein catalyzes the formation of GalNAc-PP-lipid precursors by coupling UDP-GalNAc with lipid monophosphate. Since WecA is an inner membrane protein with 10 predicted transmembrane segments, it is not a trivial task to obtain sufficient

Table 2. Activity assay of putative glycosyltransferases WbnI and WbnK. (Adapted from reference (55))

WbnI		WbnK	
Acceptor	R/A*	Acceptor	R/A*
Fucα1,2Galβ1,3GalNAcα-OMe	100%	Galβ1,3GalNAcα-OMe	100%
Galβ1,3GalNAcα-OMe	12%	Galα1,3Galβ1,4Glc	N/A
Fucα1,2Galβ-OMe	30%	Galα1,4Galβ1,4Glc	N/A
Galβ1,4Glcβ-OPh	N/A	Galβ1,4Glc	N/A
Galα1,3Galβ1,4Glc	N/A	Galβ-OMe	15%
Galα1,4Galβ1,4Glc	N/A	GalNAcα-OMe	N/A

pure enzymes for enzymatic synthesis. We and other groups have explored the chemical synthetic route for this type of compound. We designed several substrate analogues with truncated lipid moieties as well as with the natural lipid, undecaprenyl (Scheme 1). Towards this direction, we synthesized three GalNAc-PP-lipid analogues which contain a) farnesyl, a sesqui terpeniod lipid; b) pentaprenyl and c) undecaprenyl. These substrates can be easily handled and are also long enough to partition into artificial membranes or micelles, enabling us eventually to investigate the effects of membrane interfaces on enzyme activity.

In Vitro Assembly of *E. coli* O86 O-Polysaccharide Repeating Unit-PP-lipid

The RU-PP-lipids was elaborated from GalNAc-PP-lipids via the sequential addition of sugar residues through use of the following four glycosyltransferases: WbnH, WbnJ, WbnK and WbnI, respectively (Scheme 2). Hereinabove, we have biochemically characterized each glycosyltransferase and determined the respective substrate specificity. We were thus able to directly utilize the purified enzymes to synthesize the authentic repeating unit substrate, RU-PP-lipids, in a step-wise manner (Scheme 2). The intermediates and final products formed in each enzymatic step were analyzed using LC-MS. In addition, RU-PP-lipids were labeled with 3H in the final enzymatic reaction via using of WbnI in the presence of UDP-[³H]Gal, which made it easier to detect Wzy polymerization.

Expression of Wzy and O-Polysaccharide Biosynthesis

Wzy is proposed to function as a sugar polymerase. Even though it lacks the distinctive signature motif, it contains relatively conserved membrane topology. In the *E. coli* O86 gene cluster, *orf11* encodes a membrane protein with 11 predicted transmembrane segments. Topology analysis showed the presence of a relatively large periplasmic domain around 120 amino acids near the C-terminus, a feature shared by a number of Wzy homologs analyzed so far. To confirm the polymerase function *in vivo*, we knocked-out the gene by replacing it with CAT gene using the RED recombination system (*59*). The LPS from the mutant

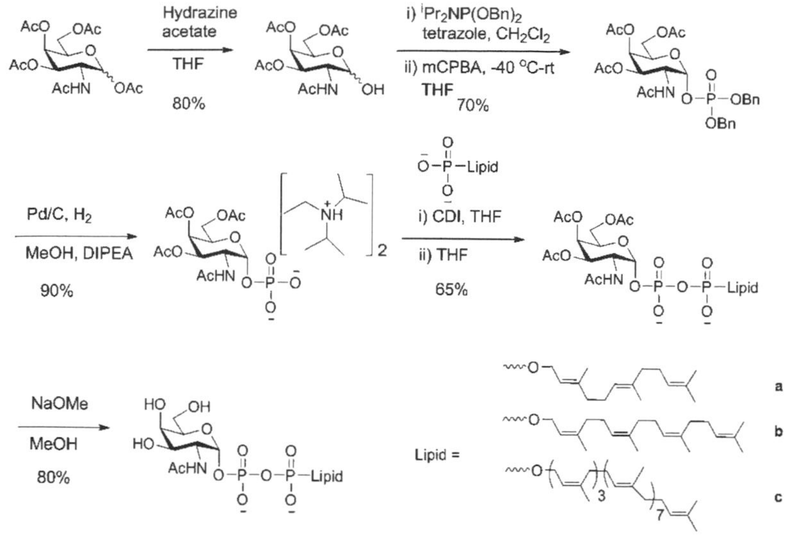

Scheme 1. Chemical synthesis of GalNAc-PP-lipid analogues.

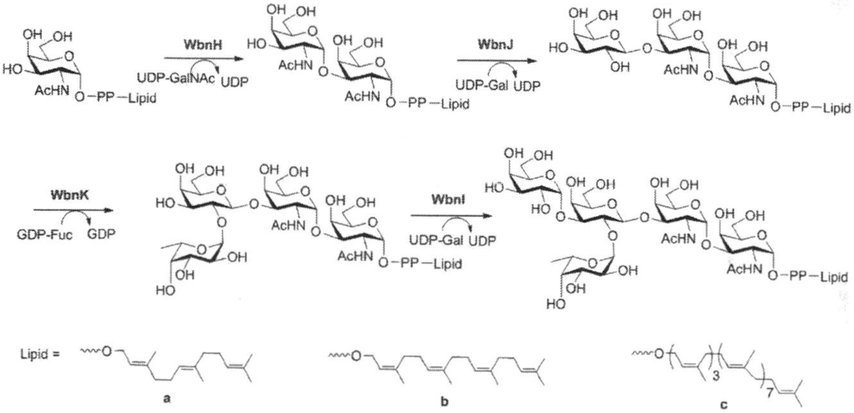

Scheme 2. Enzymatic synthesis of repeating RU-PP-lipid

strain was extracted and visualized on PAGE gel coupled with silver staining, a well-established assay for LPS (Figure 5). The LPS profile of wild-type *E. coli* O86 showed a ladder-like pattern, with the difference of a single O-unit between each band, a characteristic feature of wild-type LPS termed smooth LPS. On the other hand, the ladder-like pattern was abolished in the mutant strain, leaving a major band corresponding to one O-unit linked to the lipid-A-core moiety, a feature termed semi-rough LPS phenotype. Once we reintroduced *orf11* back into the mutant strain via a plasmid, the smooth LPS phenotype was restored. The result indicated that *orf11* encodes a functional Wzy protein that acts as a polymerase *in vivo* to generate polysaccharides.

The difficulty in expressing Wzy in practical amounts is another obstacle that hampers the efforts to study the biosynthesis of O-PS (*48*). Recently, our effort to

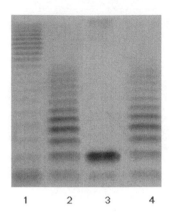

Figure 5. LPS profile of E. coli O86. Lane1: E.coli O86:H2; Lane 2: E. coli O86:B7; Lane 3: E.coli O86:B7 Δwzy; Lane 4: E.coli O86:B7 Δwzy with plasmid carrying E.coli O86:B7 Wzy gene.

express Wzy with a series of expression vectors and expression strains yielded a very promising result. The expression system we identified contains the *wzy* gene constructed in pBAD-myc-His vector (Invitrogen), GroEL & GroES chaperon expression vector. These two vectors were co-transformed into C43(DE3) cells. The expression of *wzy* was induced with L-arabinose, and Wzy was purified using detergent DDM. From the result we can clearly identify a band present in the membrane fraction and elution fraction that is stained positively with anti-myc and has an apparent MW of 48 kD, slightly lower than the theoretical MW (Figure 6), which is common towards membrane proteins. Some bands with higher molecular weight were also stained positively with anti-myc, suggesting the possible oligomerization of Wzy under the experimental conditions. This would be the first time the Wzy protein has been produced in practical amounts.

The function of Wzy has been exclusively studied *in vivo*. However, no definitive biochemical evidence has been presented. With the synthetic substrates (RU-PP-Und) in hand, we were able to demonstrate the polymerization activity of Wzy *in vitro*. RU-PP-lipids were labeled with ^{3}H in the WbnI-catalyzed reaction in the presence of UDP-[^{3}H]Galactose. The radio-labeled substrates were then incubated with a partially purified Wzy membrane fraction for 3 hr at room temperature. The reaction mixture containing the membrane which was isolated from expression of empty vector had the highest radioactivity at an Rf of 0.75 on the chromatography paper. However, in the reaction mixture containing Wzy, the radioactivity appeared at areas with low Rf values (Figure 7). This result suggested the formation of higher molecular weight compounds in the presence of Wzy. To further verify this result, we subjected the reaction mixture to SDS-PAGE and detection by autoradiography (Unpublished data). The radioactive signal on the gel gave a ladder-like pattern, consistent with the formation of multiple repeating units. It is evident from the PAGE assay that Wzy catalyzes the polymerization of repeating units, an essential reaction in PS biosynthesis that had never been unambiguously reconstituted *in vitro* prior to this study. RU-PP-pentaprenyl can also be the substrate of Wzy, but with lower conversion ratio, whereas RU-PP-farnesyl is not a substrate of Wzy reaction.

In summary, we have chemo-enzymatically synthesized *E. coli* O86 RU-PP-lipid, expressed and purified Wzy to an active form. Subsequently, the *in vitro* reconstitution of *E. coli* O86:B7 O-polysaccharide was performed for the first time. This *in vitro* system provides a tool for investigating the detailed molecular mechanism of polysaccharide biosynthesis.

Remodeling Bacterial Polysaccharides by Metabolic Pathway Engineering

Studies have been carried out to engineer several types of bacterial polysaccharides from both *in vitro* and *in vivo* aspects. These efforts have produced structurally modified polysaccharides with potentially significant biomedical applications. One of the most successful examples of *in vitro* modification is the generation of N-propionylated group B meningococcal polysaccharide (GBMP) vaccine conjugate by Jennings and coworkers. The native GBMP is a homopolymer of α(2,8)-linked sialic acids, and has very poor immunogenicity. Chemical modification of the native GBMP via N-deacetylation with strong base and subsequent N-acetylation produced N-propionylated GBMP, which induced high titers of GBMP specific IgG antibody in mice. *In vivo* modification has been conducted on bacterial cell wall and LPS. Bacterial cell wall contains a common polysaccharide structure with repeating GlcNAc-β(1,4)-muramic acid linked with pentapeptide. Nishimura and coworkers have chemically synthesized a series of modified precursors tagged with fluorescein and keto-group, and were able to show that these precursors could be taken up by bacterial biosynthetic pathway and metabolized onto the cell wall. This result provides a novel approach to study bacterial adhesion and to develop vaccines against infectious diseases. Modification of LPS structure via genetic manipulation normally resulted in augmented or novel immunological activity compared with wild-type LPS (*60*). Usually, such modification can be achieved in two ways: 1) cloning the entire O-PS biosynthetic gene cluster from an original strain into a host strain. The resulting LPS contains both components from the original strain and the host strain (*60*); 2) manipulating the gene cluster by either partial deletion or mutation of certain genes, resulting in the change of O-antigen structure in LPS (*61, 62*). The most widely studied example is the modification of *E. coli* O9a O-PS, which contains both α1,2- and α1,3-linked mannose residues in the native structure. Studies showed that elimination of C-terminal domain of mannosyltransferase WbdA resulted in the synthesis of an altered O-polysaccharide structure consisting of only of α1,2-linked mannose (*63, 64*).

Despite the continuing research efforts, bioengineering of bacterial polysaccharides is still a relatively unexplored area. Firstly, wzy-dependent pathway has just started to be exploited or manipulated for the purpose of generating modified polysaccharides. There exists much room for further exploration considering the complexity of the pathway and the fact that this pathway accounts for the synthesis of more than 90% of the existing

Figure 6. Western blotting of Wzy. Lane 1: purified wzy; Lane 2 membrane fraction.

polysaccharide structures. Secondly, except for bacterial cell wall, unnatural sugar substitutions have not been explored in other types of polysaccharides. Much work has been done to metabolically engineer mammalian cell surface glycoconjugates with unnatural sugars, and has represented an important approach to elucidate glycoconjugate functions. On the other hand, in views of the rich structural information in bacterial polysaccharides and their essential roles in bacteria-host interaction and pathogenecity, exploring the modified structure with unnatural sugar substitution will open up an exciting area of dissecting bacterial polysaccharide structure and function relationships.

GDP-Fucose Biosynthesis -- Two Pathways

There are two pathways for GDP-Fuc biosynthesis as shown in Scheme 3. The *de novo* pathway starting from GDP-Man broadly exists in all domains of life. We have identified and characterized the enzymes involved in GDP-Fuc *de novo* pathway from *Helicobacter pylori* (*65, 66*). Understanding the *de novo* pathway may direct metabolism engineering to synthesize GDP-Fuc or fucosylated oligosaccharide *in vivo*. GDP-Fuc can also be synthesized from L-fucose, ATP and GTP from the salvage pathway. This pathway is only found in eukaryotes until a bifunctional enzyme responsible for GDP-Fuc synthesis (FKP) was reported in human symbiont *Bacteroides fragilis* 9343 (*67*). We have cloned and overexpressed this enzyme and constituted a large scale production approach for GDP-Fuc.

Remodeling *E. coli* O86:B7 Polysaccharides

Fucose is a common monosaccharide component in bacterial polysaccharides. It has been implicated as a main determinant of bacterium-induced immune reactivity and is often a virulence factor (*67*). Inspired by the above results and recent research on metabolic incorporation of unnatural sugars into mammalian glycans, we replaced the *E. coli* native GDP-Fuc *de novo* pathway with the salvage pathway from *B. fragilis*, with which we were able to introduce a panel of fucose analogues into polysaccharides with high efficiency and fidelity (*68*).

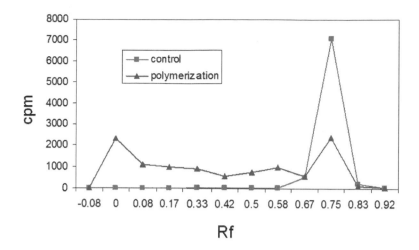

Figure 7. In vitro polymerization assay using RU-PP-Und as substrate. The paper chromatography assay indicates the formation of high molecular products.

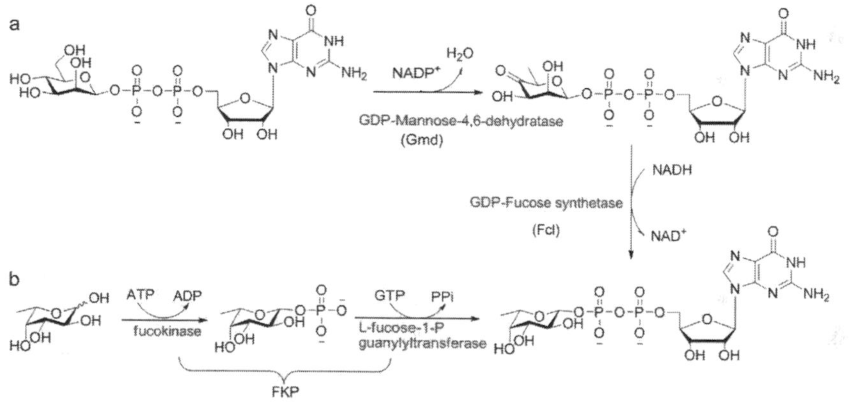

Scheme 3. The biosynthesis pathway for GDP-Fuc. (A) De novo pathway, starts from GDP-Mannose, catalyzed by two enzymes Gmd and Fcl. (B)Salvage pathway, starts from L-fucose, catalyzed by one single enzyme FKP.

Fucose containing bacterial polysaccharide from *E. coli* O86:B7 was chosen as the model, the structure of which has been solved by our group (*69*) (Figure 8a).

Firstly, we disrupted the *de novo* GDP-Fuc pathway by knockout the *gmd* and *fcl* gene in *E. coli* O86:B7. The LPS profile of wild-type strain shows a ladder pattern, whereas the one of disrupted strain shows semi-smooth pattern LPS, with only Lipid-A core and a trisaccharide expressed. However, when we use plasmid carrying either *gmd-fcl* gene or *fkp* gene feeding with fucose, the LPS profile restored to a ladder pattern same as the wild-type strain, see reference (*68*) Figure SI 1. These results suggested that the *de novo* pathway can be functionally replaced by the salvage pathway of GDP-Fuc biosynthesis in *E. coli*.

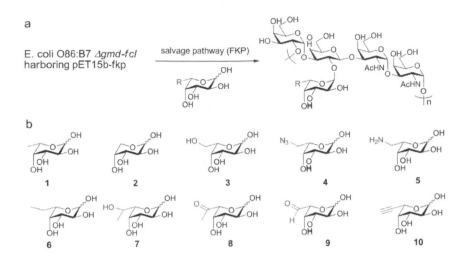

Figure 8. Introducing modifications into polysaccharides. (a) Generation of polysaccharides containing fucose analogues (E. coli O86 as a model system). (b) Fucose analogues used in the study. Modifications are located at C-6 position of fucose. (Adapted from reference (68) Figure 2).

At the same time, we explored the substrate flexibility of FKP via using a panel of fucose analogues, with substitutions on the 6 position (Figure 8b). The results exhibits that FKP can take all the analogues as substrate and produce corresponding GDP-Fuc analogues, although some of them are more preferable than others, see reference (*68*) figure 3.

Subsequently, we examined whether fucose analogues can be incorporated into *E. coli* O-polysaccharides. The *Δgmd-fcl(fkp)* strain was cultured in LB with each of those fucose analogues. The LPS isolated from the strains showed smooth phenotypes, but, with lower repeating units polymerized. This may reflects the specificity of Wzy, the polymerase, or even WbnK, the fucosyltransferase. Nevertheless, the results show that the fucose analogues can be incorporated into O-polysaccharides to form modified LPS. The modification was further confirmed by large scale isolation of LPS from each comparable strain and capillary electrophoresis mass spectrometry (CE-MS) technique, see reference (*68*) figure 4.

Conclusion

The present paper summarized our work in understanding the wzy-pathway in bacterial O-polysaccharide synthesis, from DNA level to protein, carbohydrate level, from *in vivo* investigation to *in vitro* reconstitution. Furthermore, *in vivo* remodeling of O-polysaccharide biosynthesis was performed to generate structurally modified LPS. Using *E. coli* O86 as a model strain, we have identified the glycosyltransferases for the O-polysaccharides biosynthesis, and for the first time, repeating unit-diphospho-lipids (RU-PP-lipids) were chemo-enzymatically synthesized as the substrate of Wzy polymerization. Under the appropriate reaction conditions, the repeating units were successfully polymerized, confirming that Wzy serves as the O-polysaccharide polymerase. Furthermore, we demonstrate a general, facile and effective approach to introduce modifications into polysaccharides *in vivo*, via utilizing the promiscuous GDP-fucose salvage biosynthetic pathway. These works provide a powerful tool for understanding and applying bacterial O-polysaccharide biosynthetic pathway.

Acknowledgments

P.G.W. acknowledges the National Cancer Institute (R01 CA118208), NIH (R01 GM085267), NSF (CHE-0616892) and Bill & Melinda Gates Foundation (51946) for financial support.

References

1. Osborn, M. J. Biosynthesis and assembly of the lipopolysaccharide of the outer membrane. *Bacterial Outer Membranes*; John Wiley and Sons, Inc.: New York, 1979.
2. Westphal, O.; Jann, K.; Himmelspach, K. *Prog. Allergy* **1983**, *33*, 9–39.
3. Aspinall, G. O. *Classification of Polysaccharides. The Polysaccharides*; Academic Press: New York, 1983.
4. Ball, S. G.; Morell, M. K. *Ann. Rev. Plant Biol.* **2003**, *54*, 207–233.
5. Preiss, J.; Sivak, M. *Compr. Nat. Prod. Chem.* **1999**, *3*, 441–495.
6. Caroff, M.; Karibian, D.; Cavaillon, J.-M.; Haeffner-Cavaillon, N. *Microbes Infect.* **2002**, *4*, 915–926.
7. Reeves, P. R. *FEMS Microbial. Lett.* **1992**, *79*, 509–516.
8. Rietschel, E. T.; Brade, L.; Schade, U.; et al. *Adv. Exp. Med. Biol.* **1990**, *256*, 81–99.
9. Erridge, C.; Bennett-Guerrero, E.; Poxton, I. R. *Microbes Infect.* **2002**, *4*, 837–851.
10. Skurnik, M.; Bengoechea, J. A. *Carbohydr. Res.* **2003**, *338*, 2521–2529.
11. Bartlett, D. H.; Azam, F. *Science* **2005**, *310*, 1775–1777.
12. Dziarski, R.; Gupta, D. *J. Endotoxin Res.* **2005**, *11*, 304–310.
13. Welzel, P. *Chem. Rev.* **2005**, *105*, 4610–4660.
14. Habuchi, H.; Habuchi, O.; Kimata, K. *Trends Glycosci. Glycotechnol.* **1998**, *10*, 65–80.
15. DeAngelis, P. L. *Glycobiology* **2002**, *12*, 9R–16R.
16. Wang, L.; Huskic, S.; Cisterne, A.; Rothemund, D.; Reeves, P. R. *J. Bacteriol.* **2002**, *184*, 2620–2625.
17. Raetz Christian, R. H.; Whitfield, C. *Ann. Rev. Biochem.* **2002**, *71*, 635–700.
18. Karlyshev, A. V.; Champion, O. L.; Joshua, G. W. P.; Wren, B. W. *Campylobacter* **2005**, 249–258.
19. Miyake, K.; Iijima, S. *Adv. Biochem. Eng./Biotechnol.* **2004**, *90*, 89–111.
20. Jamrichova, S. *Bull. Potravinarskeho Vyskumu* **2004**, *43*, 127–137.
21. Myszka, K.; Czaczyk, K. *Zywnosc* **2004**, *11*, 18–29.

22. Robbins, J. B.; Schneerson, E.; Egan, W. B.; Vann, W.; Liu, D. T. Virulence properties of bacterial capsule polysaccharides - unanswered questions. *The molecular basis of microbial pathogenicity*; Verlag Chemie: Weinheim, 1980.

23. Jann, K.; Jann, B. *Can. J. Microbiol.* **1992**, *38*, 705–710.

24. Boulnois, G. J.; Jann, K. *Mol. Microbiol.* **1989**, *3*, 1819–1823.

25. Boulnois, G.; Drake, R.; Pearce, R.; Roberts, I. *FEMS Microbiol. Lett.* **1992**, *100*, 121–124.

26. Frosch, M.; Muller, A. *Mol. Microbiol.* **1993**, *8*, 483–493.

27. Kolkman, M. A. B.; Morrison, D. A.; van der Zeijst, B. A. M.; Nuijten, P. J. M. *J. Bacteriol.* **1996**, *178*, 3736–3741.

28. Kolkman, M. A. B.; Wakarchuk, W.; Nuijten, P. J. M.; van der Zeijst, B. A. M. *Mol. Microbiol.* **1997**, *26*, 197–208.

29. Park, Y. S.; Koh, Y. H.; Takahashi, M.; Miyamoto, Y.; Suzuki, K.; Dohmae, N.; Takio, K.; Honke, K.; Taniguchi, N. *Free Radical Res.* **2003**, *37*, 205–211.

30. Ophir, T.; Gutnick, D. L. *Appl. Environ. Microbiol.* **1994**, *60*, 740–745.

31. Stingele, F.; Neeser, J. R.; Mollet, B. *J. Bacteriol.* **1996**, *178*, 1680–1690.

32. Guo, H.; Yi, W.; Song, J. K.; Wang, P. G. *Curr. Top. Med. Chem.* **2008**, *8*, 141–151.

33. Tieking, M.; Gaenzle, M. G. *Trends Food Sci. Technol.* **2005**, *16*, 79–84.

34. Bhaskar, P. V.; Bhosle, N. B. *Curr. Sci.* **2005**, *88*, 45–53.

35. Colliec-Jouault, S.; Zanchetta, P.; Helley, D.; Ratiskol, J.; Sinquin, C.; Fischer, A. M.; Guezennec, J. *Pathol. Biol.* **2004**, *52*, 127–130.

36. Geider, K.; Zhang, Y.; Ullrich, H.; Langlotz, C. *Acta Hortic.* **1999**, *489*, 347–351.

37. Sutherland, I. W. *Microb. Extracell. Polym. Subst.* **1999**, 73–92.

38. Pratt, L. A.; Kolter, R. *Curr. Opin. Microbiol.* **1999**, *2*, 598–603.

39. Sutherland, I. W. *Carbohydr. Polym.* **1999**, *38*, 319–328.

40. Stingele, F.; Vincent, S. J.; Faber, E. J.; Newell, J. W.; Kamerling, J. P.; Neeser, J. R. *Mol. Microbiol.* **1999**, *32*, 1287–1295.

41. Morona, R.; Stroeher, U. H.; Karageorgos, L. E.; Brown, M. H.; Manning, P. A. *Gene* **1995**, *166*, 19–31.

42. Franco, A. V.; Liu, D.; Reeves, P. R. *J. Bacteriol.* **1998**, *180*, 2670–2675.

43. Burrows, L. L.; Lam, J. S. *J. Bacteriol.* **1999**, *181*, 973–980.

44. Amer, A. O.; Valvano, M. A. *J. Bacteriol.* **2000**, *182*, 498–503.

45. Wang, L.; Liu, D.; Reeves, P. R. *J. Bacteriol.* **1996**, *178*, 2598–2604.

46. Liu, D.; Cole, R.; Reeves, P. R. *J. Bacteriol.* **1996**, *178*, 2102–2107.

47. Wacker, M.; Feldman, M. F.; Callewaert, N.; Kowarik, M.; Clarke, B. R.; Pohl, N. L.; Hernandez, M.; Vines, E. D.; Valvano, M. A.; Whitfield, C.; Aebi, M. *Proc. Natl. Acad. Sci. U.S.A.* **2006**, *103*, 7088–7093.

48. Daniels, C.; Vindurampulle, C.; Morona, R. *Mol. Microbiol.* **1998**, *28*, 1211–1222.

49. Franco, A. V.; Liu, D.; Reeves, P. R. *J. Bacteriol.* **1996**, *178*, 1903–1907.

50. Heinrichs, D. E.; Yethon, J. A.; Whitfield, C. *Mol. Microbiol.* **1998**, *30*, 221–232.

51. Wu, T.; McCandish, A. C.; Gronenberg, L. S.; Chng, S.-S.; Silhavy, T. J.; Kahne, D. *Proc. Natl. Acad. Sci. U.S.A.* **2006**, *103*, 11754–11759.
52. Springer, G. F.; Horton, R. E.; Forbes, M. *J. Exp. Med.* **1959**, *110*, 221–244.
53. Kochibe, N.; Iseki, S. *Jpn. J. Microbiol.* **1968**, *12*, 403–411.
54. Andersson, M.; Carlin, N.; Leontein, K.; Lindquist, U.; Slettengren, K. *Carbohydr. Res.* **1989**, *185*, 211–223.
55. Yi, W.; Shao, J.; Zhu, L.; Li, M.; Singh, M.; Lu, Y.; Lin, S.; Li, H.; Ryu, K.; Shen, J.; Guo, H.; Yao, Q.; Bush, C. A.; Wang, P. G. *J. Am. Chem. Soc.* **2005**, *127*, 2040–2041.
56. Yi, W.; Yao, Q.; Zhang, Y.; Motari, E.; Lin, S.; Wang, P. G. *Biochem. Biophys. Res. Commun.* **2006**, *344*, 631–639.
57. Shao, J.; Zhang, J.; Kowal, P.; Lu, Y.; Wang, P. G. *Chem. Commun.* **2003**, 1422–1423.
58. Liu, Z.; Lu, Y.; Zhang, J.; Pardee, K.; Wang, P. G. *Appl. Environ. Microbiol.* **2003**, *69*, 2110–2115.
59. Yi, W.; Zhu, L.; Guo, H.; Li, M.; Li, J.; Wang, P. G. *Carbohydr. Res.* **2006**, *341*, 2254–2260.
60. Paeng, N.; Kido, N.; Schmidt, G.; Sugiyama, T.; Kato, Y.; Koide, N.; Yokochi, T. *Infect. Immun.* **1996**, *64*, 305–309.
61. Kido, N.; Ohta, M.; Iida, K.; Hasegawa, T.; Ito, H.; Arakawa, Y.; Komatsu, T.; Kato, N. *J. Bacteriol.* **1989**, *171*, 3629–3633.
62. Sugiyama, T.; Kido, N.; Kato, Y.; Koide, N.; Yoshida, T.; Yokochi, T. *J. Bacteriol.* **1998**, *180*, 2775–2778.
63. Kido, N.; Suglyama, T.; Yokochi, T.; Kobayashi, H.; Okawa, Y. *Mol. Microbiol.* **1998**, *27*, 1213–1221.
64. Kido, N.; Kobayashi, H. *J. Bacteriol.* **2000**, *182*, 2567–2573.
65. Wu, B.; Zhang, Y.; Zheng, R.; Guo, C.; Wang, P. G. *FEBS Lett.* **2002**, *519*, 87–92.
66. Wu, B.; Zhang, Y.; Wang, P. G. *Biochem. Biophys. Res. Commun.* **2001**, *285*, 364–371.
67. Ma, B.; Simala-Grant, J. L.; Taylor, D. E. *Glycobiology* **2006**, *16*, 158R–184R.
68. Yi, W.; Liu, X.; Li, Y.; Li, J.; Xia, C.; Zhou, G.; Zhang, W.; Zhao, W.; Chen, X.; Wang, P. G. *Proc. Natl. Acad. Sci. U.S.A.* **2009**, *106*, 4207–4212.
69. Yi, W.; Bystricky, P.; Yao, Q.; Guo, H.; Zhu, L.; Li, H.; Shen, J.; Li, M.; Ganguly, S.; Bush, C. A.; Wang, P. G. *Carbohydr. Res.* **2006**, *341*, 100–108.

Chapter 21

Glycosaminoglycan Synthases: Catalysts for Customizing Sugar Polymer Size and Chemistry

Paul L. DeAngelis*

Dept. of Biochemistry and Molecular Biology, University of Oklahoma
Health Sciences Center, Oklahoma Center for Medical Glycobiology,
940 S.L. Young Blvd., Oklahoma City, OK 73126, USA
*paul-deangelis@ouhsc.edu

Synthesis of sugar polymers has always been a challenge. Total organic chemistry approaches are appropriate for smaller oligosaccharides (less than 6 monosaccharides), but as the chain length increases, the efficiency and yields decrease while the production of non-target compounds increases. In addition, typical carbohydrate chemistry results in much waste solvent and spent toxic reagents. To assist the chemist, enzymes, in particular glycosyltransferases and hydrolases, have been employed with success to make natural and artificial structures. Enzyme catalysts have high efficiency, great stereo-selectivity and regio-specificity, and usually operate in aqueous ('green') systems. Here, the production of a variety of short (5 monosaccharides) to long (~12,000 monosaccharides) monodisperse heteropolymers with many potential medical applications using biosynthetic enzymes is described.

Introduction

Pasteurella multocida bacteria produce extracellular capsules composed of the glycosaminoglycans [GAGs] hyaluronan, chondroitin, and heparosan (*1*). These linear polysaccharides with repeating [GlcA-HexNAc] disaccharide subunits also form the backbones of polymers in many vertebrate tissues. The bacterial GAG capsules are virulence factors that allow the microbes to be more successful pathogens by acting as molecular camouflage that render host defenses less effective. We have harnessed several bacterial GAG synthases, the bifunctional enzymes that polymerize the GAG chains using UDP-sugar precursors (Figure 1), for chemoenzymatic synthesis *in vitro* to make novel GAGs with potential for a variety of medical applications. Our biotechnology focus is on drug delivery (both bio-inert stealthy or targeted vehicles) and biomaterial platforms (implantable gels and cell scaffolds).

Key knowledge for developing these new GAG polymer production systems was the identification of the GAG synthases, the dual-action enzymes that polymerize the GAG chains using both UDP-GlcUA and UDP-HexNAc precursors according to the following reaction:

$$n \text{ UDP-GlcUA} + n \text{ UDP-HexNAc} \rightarrow (\text{GlcUA-HexNAc})_n + 2n \text{ UDP}.$$

Sugar polymers, especially molecules with chain lengths longer than five monosaccharides, are difficult to produce by strictly organic synthesis in a monodisperse, defined form as well as usually generate ~1,000:1 waste to target molecules. 'Green' chemoenzymatic synthesis offers the potential to harness enzyme catalysts for rapid, efficient reactions. We have developed methods to construct GAG polysaccharides of any desired size from 10 to 25,000 kDa in synchronized reactions as well as make short GAG oligosaccharides (0.8 to 5 kDa) in step-wise addition reactions. We have also enhanced the potential synthetic repertoire by employing novel UDP-sugars that allow further functionalization of GAGs. For example, a variety of new polymers with unnatural chemical groups in various positions (either single or multiple novel sugar units) have now been made facilitating coupling reactions including 'click' chemistry.

Experimental

Chemoenzymatic Synthesis of Polymers

Reactions containing UDP-sugars (natural or synthetic), an acceptor (either native or biotin-derivatized GAG oligosaccharides or synthetic glycosides), and GAG synthase enzymes (either recombinant maltose-binding protein-PmHS fusions (*2*) or PmHAS (*1–3*) truncations (*3*) were combined in liquid phase reactions in a similar fashion to our previous reports. The products were analyzed by agarose gel electrophoresis with Stains-all detection, polyacrylamide gels with Alcian Blue detection, mass spectroscopy and/or gel filtration chromatography coupled to a light scattering detector.

Results and Discussion

Naturally occurring polysaccharides and oligosaccharides from various organisms are often difficult to prepare in a pure, defined, monodisperse form. Sugar polymers, especially molecules with chain lengths longer than five monosaccharides, are also difficult to synthesize by strictly organic synthesis. In contrast, chemoenzymatic synthesis *in vitro* offers the potential to harness enzyme catalysts for rapid, efficient reactions.

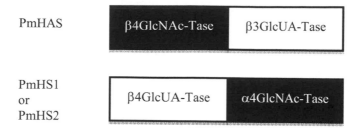

PmHAS   β4GlcNAc-Tase   β3GlcUA-Tase

PmHS1
or   β4GlcUA-Tase   α4GlcNAc-Tase
PmHS2

*Figure 1. Schematic of Recombinant PmHAS and PmHS Enzymes. Some
useful GAG synthases are the Pasteurella enzymes that make HA, PmHAS, and
heparosan, PmHS1 or PmHS2. Even though the same monosaccharides are
transferred during HA and heparosan biosynthesis, the glycosidic linkages
are different and the PmHAS or PmHS protein sequences are <u>not</u> very similar.
Each polypeptide chain contains two relatively independent glycosyltransferase
(Tase) activities (each with an acceptor site and a donor site). Mutagenesis
of one active site often leaves the remaining active site unperturbed thus
creating new catalysts for step-wise reactions rather than polymerization. The
synthase proteins will catalyze GAG synthesis in vivo or in vitro as long as the
UDP-sugars are supplied.*

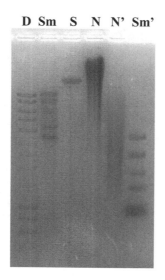

D Sm S N N' Sm'

*Figure 2. Agarose gel analysis of monodisperse HA Polymers. D, DNA
standards; Sm, mixture of five synthetic HAs ranging from 1,500 to 495 kDa ;
2.4 MDa synthetic HA, N, natural HA from rooster comb; N', natural HA from
Streptococcus bacteria; Sm', mixture of five synthetic HAs ranging from 495 to
27 kDa.*

We have developed methods to construct large (0.01 to 8,000 kDa) GAG
polysaccharides of a desired size by controlling stoichiometry in synchronized
reactions (Figure 2) (*3*). The use of an acceptor (a short GAG chain that mimics

the nascent polymer terminus) to prime the GAG synthase circumvents the random, slow initiation of the *de novo* synthesis thus all GAG chains are elongated in parallel and achieve the same size (*i.e.*, nearly monodisperse; polydispersity = 1.02-1.2 depending on final polymer size). In contrast, when the acceptor is not employed, a wide variety of chain sizes (*i.e.* polydisperse) are formed from asynchronous elongation events.

We have also created GAG oligosaccharide synthesis systems employing immobilized mutant enzyme reactors in a step-wise sugar addition strategy; pentamers to 22-mers have been prepared (*4*). A normally bifunctional enzyme is mutated to inactivate one of the transferase activities, but the other transferase remains functional. In an example of the oligosaccharide synthesis strategy, a mutant catalyst (*e.g.*, a GlcNAc-tase) is used to transfer a single sugar to an acceptor (*e.g*, a chain terminating with a GlcUA unit), and then the reaction mixture is removed and allowed to react with the next mutant catalyst (*e.g.*, a GlcUA-tase). The strategy may be repeated to build a variety of GAG polymers. Certain GAG synthases exhibit relaxed acceptor specificity allowing non-cognate molecules to be elongated. For example, the creation of hybrid or chimeric sugar molecules containing both HA-like and chondroitin-like disaccharide repeats have been prepared.

In our most recent work, we have expanded the GAG chemical functionality repertoire. Our main approach is to make synthetic UDP-sugars containing unnatural substitutions (including azido, alkyne, alkyl, fluoro, or protected amine groups) that the GAG synthases can recognize and incorporate into sugar polymers (Figure 3). Most analogs do not work as well as the authentic natural precursors, but a few are even better substrates. By manipulating the sugar addition strategy and the reaction conditions, we have made a variety of GAG-like polymers from 5 to ~10,000 sugars where either one or multiple artificial sugar units are added to a single chain. The new groups in the GAG chain promise to enhance their potential for use in chemical reactions and/or possess new biological activities.

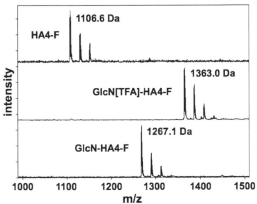

Figure 3. Mass spectrometric analyses of a HA tetramer tagged with a fluorescein (top), the protected amine (GlcN[TFA]) addition pentamer product made using UGP-GlcN[TFA] with PmHAS enzyme (middle) and the de-protected pentamer with a free amine (bottom).

Conclusion

The study of the GAG synthases is passing its infancy, but as more knowledge is gained, the production of new generation GAG-like polymers with better biological and/or chemical properties is expected. Some of the expected novel GAG therapeutics include safer anti-coagulants, biomaterials that gel after injected into the body, and improved non-toxic drug delivery systems.

Acknowledgments

The author of this paper would like to thank the various researchers who contributed to this work including: Dixy E. Green, Nigel J. Otto, F. Michael Haller, Wei Jing, Alison E. Sismey, Regina C. Visser, Robert J. Linhardt, Michel Weiwer, Martin E. Tanner and Gert-Jan Boons. The National Institutes of Health, the Oklahoma Center for the Advancement of Science and Technology, and NSERC of Canada, provided financial support of this research.

References

1. DeAngelis, P. L. *Glycobiology* **2002**, *12*, 9R.
2. Sismey-Ragatz, A. E.; Green, D. E.; Otto, N. J.; Rejzek, M.; Field, R. A.; DeAngelis, P. L. *J. Biol. Chem.* **2007**, *282*, 28321.
3. Jing, W.; DeAngelis, P. L. *J. Biol. Chem.* **2004**, *279*, 42345.
4. DeAngelis, P. L.; Oatman, L. C.; Gay, D. F. *J. Biol. Chem.* **2003**, *278*, 35199.

Chapter 22

Development and Applications of a Novel, First-in-Class Hyaluronic Acid from *Bacillus*

Khadija Schwach-Abdellaoui,* Birgit Lundskov Fuhlendorff, Fanny Longin, and Jens Lichtenberg

Novozymes Biopolymer A/S, Kroghshoejvej 36, DK-2880 Bagsvaerd, Denmark
*khsa@novozymes.com

Since 1934, hyaluronic acid (HA) has been extracted from rooster combs and employed for various applications such as ophthalmic surgery, orthopedic and wound healing. A breakthrough in HA production occurred in 1985, when the more sophisticated streptococcal fermentation process was implemented and considered the 'gold standard' since then. Novozymes Biopolymer radically shifted the paradigm by employing a unique and non-pathogenic strain, *Bacillus subtilis*, as an "industry factory" to produce a new biosynthetic HA product: HyaCare®. HyaCare® powder is composed of microparticles with enhanced surface properties that improve solubilization time compared to conventional processes. This new technology combined with the molecular and physico-chemical properties of HA provides improved safety, purity, consistency, stability and ease of filtering. HyaCare® is produced using an advanced fermentation process without ingredients derived from animal sources or organic solvents. It is an extremely pure source of HA in which there are no exotoxins and low level of proteins. These properties make it suitable for use in biomedical and pharmaceutical applications.

Introduction

Hyaluronic acid (HA) is a natural linear polysaccharide consisting of D-glucuronic acid and *N*-acetyl-D-glucosamine linked through β-1,3 glycosidic bonds while consecutive disaccharide repeating units are linked through β-1,4 bonds (Figure 1). In vertebrates HA is ubiquitous in all organs and fluids and in the extracellular matrix of soft connective tissues where it provides a backbone for the distribution and organization of proteoglycans, fibrin, fibronectin and collagen. Since HA was discovered in 1934, it has been extracted from rooster combs and employed for various applications such as ophthalmic surgery, orthopedic and wound healing. Recovery of HA from rooster combs necessitates extensive purification using harsh organic solvents to remove antigenic avian proteins. A breakthrough in HA production occurred in 1985, when the more sophisticated streptococcal fermentation process was implemented and since then considered as the 'gold standard.' Streptococci are fastidious organisms to grow, are natively pathogenic, and have the potential to produce exotoxins. Moreover it is difficult to control HA molecular weight from streptococcal fermentation. Novozymes Biopolymer radically shifted the paradigm by employing a unique and non-pathogenic strain, *Bacillus subtilis*, as an 'industry factory' to produce a new biosynthetic HA product: HyaCare® (*1*).

B. *subtilis* is one of the most well-characterized gram-positive micro-organisms that can be genetically manipulated using a wide array of tools available. It does not produce, nor does the genome sequence encode, a hyaluronidase which could degrade HA. Thus, these organisms offer several advantages as possible hosts for producing HA of well defined molecular weight and narrow polydispersity (*2*). Recombinant *Bacillus* species have been used for several decades to produce industrial enzymes and small molecules such as riboflavin and amino acids and many of these products have achieved GRAS (Generally Recognized As Safe) status. The *has*A gene from *Streptococcus equisimilis*, which encodes the enzyme hyaluronan synthase, has been introduced into *Bacillus subtilis* and expressed, resulting in the production of authentic HA. HyaCare® is produced using an advanced fermentation process during which no ingredients derived from animal sources or organic solvents are used. It is an extremely pure source of HA containing no exotoxins and very low level of proteins.

Here we report the physicochemical properties, thermal stability and filterability of this new source of HA compared to streptococcal HA (sHA) as well as its interaction with ingredients and drugs. Preliminary toxicity evaluation is also reported here.

Experimental

HA powder was dissolved into PBS buffer to mimic the most common formulations used in Eye Care at concentrations ranging from 0.1 to 2%. Samples were autoclaved under various temperatures and times (110°C/20 min, 121°C/16 min and 134°C/3 min). Molecular weights (MW) were determined using SEC/MALLS/RI (Size-Exclusion-Chromatography combined with Multi-Angle-Laser-Light-Scattering and Refractive Index detector).

HA solutions were prepared in PBS buffer at concentrations ranging from 0.025% to 0.5% (w/v), preserved with Polyhexamethylene Biguanide (PHMB,

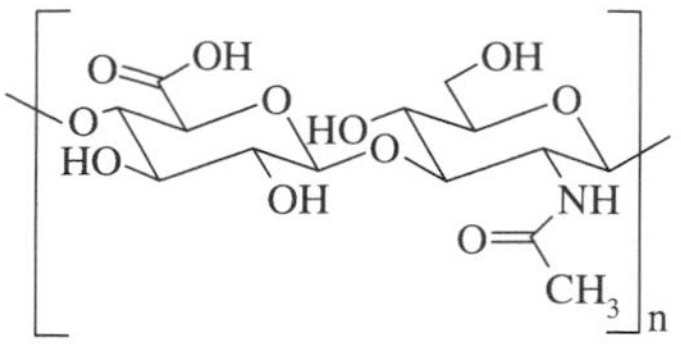

Figure 1. Molecular structure of the HA disaccharide repeating unit.

Cosmocil® CQ) and were filtered under house vacuum through a Durapore® membrane filter (0.22 μm GV) adapted on a Büchner funnel. The filtration time was recorded with t=0 when the solution had just been loaded on the filter.

PBS solutions were prepared with HyaCare concentrations ranging from 0.1 to 0.5% and containing 0.3 or 0.5% gentamicin. Clearance was evaluated by visual inspection and dynamic viscosity was measured using Brookfield Viscometer LVDV-II +Pro, (E-246-99) equipped with a small sample adapter.

HyaCare® was evaluated in two separate *in vivo* toxicological studies where the animals were treated either by the intravenous (IV) or the subcutaneous (SC) route of administration. In each of the two studies twenty-four male rats of the Wistar strain were randomly divided into four groups each comprising of 6 rats per group. The animals were treated with HyaCare® at the dose levels of 25 (low dose), 50 (mid dose) and 100 (high dose) mg/kg b.wt/day, respectively for a period of 14 consecutive days. All animals were closely monitored for body weight changes, clinical signs and at termination they were subjected to a full gross necropsy.

Results and Discussions

HyaCare® MW was investigated for several batches using SEC-MALLS and showed very good consistency in both MW and polydispersities (I) (Figure 2). This is due to the fact that bacillus-derived HA is excreted extracellularly and extracted using a gentle aqueous process mainly comprising ultrafiltration and ions exchange steps and that has minor effect on HA degradation. Control of the molecular weight and polydispersity of HA is very important for the functionality of the HA-based products, especially in the biomedical and pharmaceutical applications.

In case of HyaCare®, detailed study of the MW decrease upon autoclavation was allowed by plotting the relative retained MW as a function of the HA concentration (Figure 3). Results showed that the higher the HA concentration, the higher the retained HyaCare® molecular weight after autoclavation, regardless of the autoclavation conditions. For most concentrations, the autoclavation at 134°C for 3 min was milder or significantly milder than the other selected autoclavation conditions. It is noteworthy that the purity of HA is of outmost importance when carrying out an autoclavation step. It is well-established that cations such as Cu^+, Fe^{2+} and Sn^{2+} enhance the depolymerization of HA as they are involved in redox reactions that may lead to the formation of radicals, which in turn degrade the HA chains.

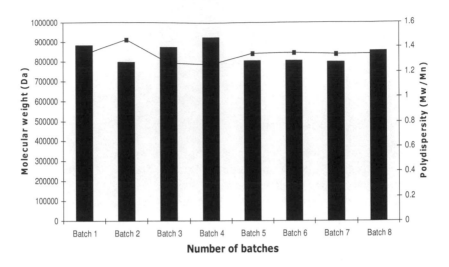

*Figure 2. Consistency in MW (■) and polydispersities (—) for several HyaCare®
batches*

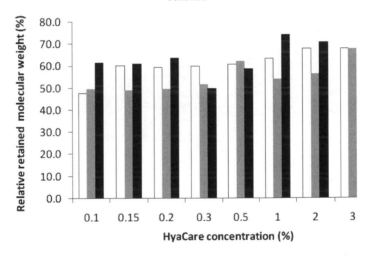

*Figure 3. HA retained molecular weight after autoclavation of HyaCare®
solutions at 110°C/20 min (white), 121°C/16 min (gray) and 134°C/3 min (black)*

Results show that high MW HA is more sensitive to autoclavation than
medium MW HA at 0.1% and 1% concentration (Figure 4). A possible explanation
could be the increased number of available sites to thermal degradation in high
MW HA compared to medium MW HA. HyaCare® was significantly more stable
than sHA high MW and sHA medium MW.

Topical ophthalmic formulations containing thermolabile drugs are typically
sterilized by filtration at production scale. However, those containing HA may
be difficult to filter due to the relatively high viscosity of this ingredient even at
low concentrations. The molecular weight does not affect the filtration time up to
0.1%. However, at 0.2% concentration, it takes up to three times longer to filter

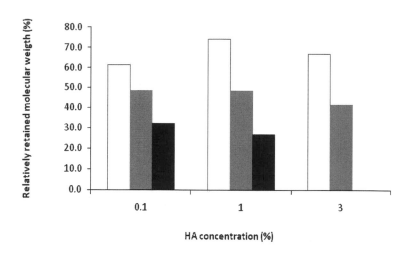

Figure 4. HA retained molecular weight after autoclavation at 134°C/3 min comparing HyaCare® (white) with medium MW sHA (HA2, gray) and high MW sHA (HA1, black). The MW was normalized

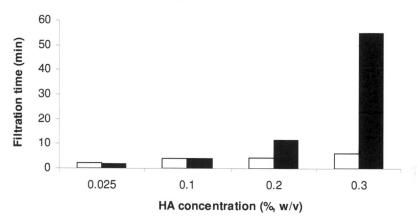

Figure 5. Filtration time of HA solutions at various concentrations; (white) medium MW HyaCare®, (0.89 MDa); (black) high MW s-HA (2.25 MDa).

a high MW HA solution than a medium MW HA solution (Figure 5). The effect is even more pronounced when the concentration is increased to 0.3%. Today, commercial eye drops recommended for the treatment of dry eye contain from 0.1% to 0.3% HA. Sterile filtration of these ophthalmic solutions is a crucial step in the manufacturing process in that its optimization can not only reduce the number of adverse events linked to the handling of highly viscous solutions, but also lower the production costs (*3*).

When stored with the commonly used preservatives in pharmaceutical formulations namely, benzalkonium chloride and polyhexamethylene biguanide, HyaCare® did not show any degradation at 25 and 40°C or interaction with the preservatives (results not shown).

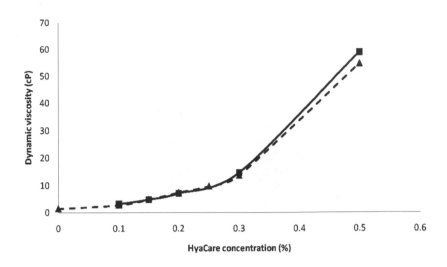

Figure 6. Dynamic viscosity of HyaCare formulations with (---) and without gentamicin (—)

Gentamicin is one of the most commonly used antibiotics in ophthalmology and dermatology and other biomedical applications due to its wide antibacterial spectrum and because, unlike many antibiotics, it is stable at the high temperatures used during sterilization. Sustained high local concentrations of antibacterial compounds such as gentamicin are required to minimize and even to cure some diseases such as chronic osteomyelitis. Results of the combination of Hyacare® and gentamicin show that the drug did not interfere with HA at any concentration studied. The clearance of the solution was maintained even at high concentrations of HA and the dynamic viscosity was unchanged with or without gentamicin (Figure 6). We believe that HyaCare® will represent the ideal carrier for sustained released of gentamicin as HA is already used and well tolerated for ophthalmic, dermatological and other biomedical applications.

The toxicological findings following IV administration to rats comprised of clinical signs such as mild lethargy observed immediately after dosing in some of the rats from high dose treated group, which however recovered within 10 to 20 minutes after dosing. No significant difference was observed in the body weight gain in any of the treatment groups compared to the control. The gross pathological examination of the HyaCare® treated groups did not reveal any treatment related changes compared to the control group.

Rats treated by SC administration for 14 days showed no clinical signs of toxicity or body weight changes in any groups during the treatment period. The gross pathological examination of the HyaCare® treated groups did not reveal any treatment related changes except marginal reddening and/or hardening at the injection site. The present findings are supported by several *in vitro* studies previously conducted including cytotoxicity, dermal and eye irritancy tests (**data not shown**). The low toxicity of HyaCare® is mainly due to its high purity and to the fact that HA is a natural component of the extracellular matrix (*4*).

In conclusion, based on the present *in vivo* toxicological studies and the previously conducted *in vitro* tests HyaCare® is considered exhibiting no or a very limited toxicological potential at all end points.

Conclusion

Several HA-based products in the biomedical space are currently using high molecular weight HA. We believe that in the majority of the cases this is due to the fact that mainly high MW was available from streptococcal fermentation when these products were developed. Our studies demonstrate that medium molecular weight from *Bacillus* fermentation has superior properties in term of consistency in molecular weight and polydispersity and the ease of sterile filtration. Moreover the higher purity of HyaCare® compared to the available sources of HA offers the possibility of heat sterilization with minor degradation under given conditions and allows its use with various ingredients, preservatives and drugs without degradation, precipitation or decrease in viscosity. Last but not least the encouraging preliminary toxicity studies conducted on rats are very promising and show that even IV injection of HA will now be possible when safe and high purity source is used.

We believe that high MW is still needed for applications such as ophthalmic surgery where strong cohesive properties are needed and could not be obtained by using medium MW even at very high concentrations. High MW HyaCare from *Bacillus subtilis* fermentation is thus under current development.

Finally, these unique properties will undoubtedly make HyaCare® the product of choice for biomedical and pharmaceutical applications, especially for drug delivery.

References

1. Widner, W.; Behr, R.; Von Dollen, S.; Tang, M.; Heu, T.; Sloma, A.; Sternberg, D.; DeAngelis, P.; Weigel, P.; Brown, S. *Appl. Environ. Microbiol.* **2005**, *71*, 3747.
2. Tang, M. R.; Sternberg, D.; Behr, R. K.; Sloma, A.; Berka, R. M. *Ind. Biotechnol.* **2006**, *2*, 66.
3. Guillaumie, F.; Furrer, P.; Felt-Baeyens, O.; Fuhlendorff, B. L.; Nymand, S.; Westh, P.; Gurny, R.; Schwach-Abdellaoui, K. *J. Biomed. Mater. Res.* **2009**, in press.
4. Hori, K.; Esumi, Y.; Takaichi, M. *Jpn. Pharmacol. Ther.* **1994**, *22* (3), 325.

Biocatalytic Redox Polymerizations

Chapter 23

Enzymatic Synthesis of Electrically Conducting Polymers

Ryan Bouldin,[1] Akshay Kokil,[2] Sethumadhavan Ravichandran,[2] Subhalakshmi Nagarajan,[3] Jayant Kumar,[3] Lynne A. Samuelson,[5] Ferdinando F. Bruno,[5] and Ramaswamy Nagarajan[4,*]

[1]Department of Chemical Engineering, [2]Department of Chemistry, [3]Department of Physics, [4]Department of Plastics Engineering, University of Massachusetts, Lowell, MA 01854, USA
[5]U.S. Army Natick Soldier Research Development and Engineering Center, Natick, MA 01760, USA
*Current Address: Department of Plastics Engineering, Ball Hall, One University Ave., Lowell, MA, 01854, email: Ramaswamy_Nargarajan@uml.edu, Ph: 978-934-3454

The field of 'green chemistry' is rapidly expanding to include synthesis of polymers and advanced functional materials. Using enzymes as catalysts has been advocated as a viable environmentally friendly alternative to strong oxidants in polymer synthesis. Enzymes permit the use of milder reaction conditions while providing higher efficiency in terms of product yield and ease of product separation. In this review, we summarize the use of oxidoreductases as catalysts for the synthesis of electrically conducting polymers that are based on aniline, pyrrole, and thiophene. The mechanism of enzyme catalysis in the context of oxidative polymerization is presented. The roles of redox potential in determining reaction feasibility and strategies to influence the initiation of polymerization are also discussed. Reaction parameters such as pH, temperature, dopant type, dopant concentration, and their influence on the polymerization and product properties are presented. While significant progress has been made in the synthesis of conducting polymers using enzymes, many challenges still remain in understanding the interactions between monomer, dopant, enzyme, and the initiator. Further investigations on these aspects will expand this field and provide opportunities for the creation of commercially viable products.

1. Introduction

With the world demanding increasingly safe, clean, and environmental friendly products, enzymatic and biocatalytic synthetic strategies have enormous growth potential within the polymer industry. An area that could stand to benefit from the new-found environmental awareness is the field of conducting polymers (CPs). With their increasing applications in organic electronics and photovoltaics, photochromics, chemical and biological sensing, and biomaterials (1), there is increasing interest in environmentally friendly routes to the synthesis of conducting polymers. Enzymatically-produced conducting polymers have the opportunity to be at the forefront of transforming the production methods for new and evolving polymeric materials towards more sustainable and environmentally friendly routes.

Herein we present a review on enzymatic strategies for the synthesis of conducting polymers derived from aniline, pyrrole, and thiophene. It is written to serve as a comprehensive summary of the current state-of-the-art in enzymatic synthesis of electrically conducting polymers. This article is expected to benefit researchers interested in developing these advanced functional materials using environmentally friendly methods. It discusses the synthetic approaches currently utilized for the enzymatic synthesis of electrically conducting polymers and summarize the influence of reaction parameters such as pH, temperature, and the role of the charge balancing dopant on polymer properties. Very recently, there have been several reports of cleaner/"greener" efforts to synthesize conducting polymers. Some of these include using iron (2, 3) and copper (4) salts, enzyme mimics (5–7) or just hydrogen peroxide and high ionic strength solutions (8), as catalyst; however while novel and interesting to the conducting polymer community, they are outside of the scope of this review.

1.1. Introduction to Conducting Polymers

Since the time of their discovery in 1977 (9), conjugated polymers have become a field of great scientific and technological importance leading to a plethora of new generation materials with attractive opto-electronic properties. The synthesis of conducting polymers offers promising new, high value, materials that retain many of the attractive mechanical properties and processing advantages of traditional plastics. The realization of the importance of these materials resulted in the Nobel Prize in Chemistry (2000) for Alan J. Heeger, Alan MacDiarmid and Hideki Shirakawa (10).

Typically, conducting polymers are produced by direct oxidation of a monomer with potent oxidants, or via electrochemical means in corrosive solutions or organic solvents. In both chemical and electrochemical synthesis the first step of the reaction involves the oxidation of the monomer leading to the formation of a monomer radical cation. This initiates the growth of the polymer chain. The initial oxidation step requires either strong oxidants or the use of organic solvents. Alternatively, the initial oxidation can also be accomplished using electrochemical methods. The lack of scalability of the reaction to produce polymers in higher yields is the biggest limitation of electrochemical methods. While the harsh conditions used for large-scale chemical synthesis have not initially supplanted the investigation of conducting polymers, increasing regulatory mandates and the general public's growing environmental consciousness are likely to impede their commercial utility unless more benign synthetic schemes are developed.

1.2. Biocatalytic Synthesis – A 'Greener' Option

The pursuit of a 'greener' approach that uses predominantly aqueous reaction media and milder, degradable oxidants has lead to the exploration of biocatalysis as a viable alternative. Hydrogen peroxide, a naturally occurring mild oxidant, is generally recognized as safe (GRAS) according to FDA. Hydrogen peroxide (H_2O_2) is often produced in biological systems as a byproduct of oxygen metabolism and enzymes like peroxidases harmlessly decompose low concentrations of H_2O_2 to water and oxygen. From a purely thermodynamics perspective, H_2O_2 should be able to directly oxidize monomers such as aniline and pyrrole to initiate the polymerization. However, these reactions are extremely slow and have been shown to cause structural defects in the polymer (11), and thus there is a need for a catalyst to increase the rate and yield of these reactions.

An effort to utilize natural processes and a desire to explore biomimicry of in vivo processes has lead to the use of enzymes for a variety of applications (12). While the use of natural catalysts to accomplish chemical transformations predates most recorded history, the systematic study of biocatalysis has only been started in the last century. The number of documented enzymes has significantly risen from around 700 in the 1960s to over 4100 today (13). Fortunately, several of these enzymes have been shown to exhibit many of their inherent catalytic properties in vitro, and thus have been extensively used as organic catalysis for the synthesis of materials, like polymers. Enzymes have been shown to be catalysts for the synthesis of polysaccharides (12), polyamides (14), polyphenol (15), and some vinyl monomers (16). The use of enzymes as catalysts offers enhanced regulation of chemoselectivity and stereochemistry, while being non-toxic. Enzyme catalysis is typically more energy efficient and produces higher yields of product when compared to traditional polymerization methods. Most enzyme catalyzed reactions are typically simple: one pot, readily scalable, and require little purification steps.

Among the large variety of enzymes available, oxidoreductases are the most relevant class biocatalysts for the synthesis of electrically conducting polymers. Oxidoreductases classified as EC 1 (Enzyme commission number) are a class of enzymes that catalyze oxidation-reduction reactions. Peroxidases (EC 1.11.1) are a sub-class of oxidoreductases that act on peroxides as substrates.

1.3. Importance of Redox Potential in Oxidoreductase Catalyzed Polymerization

Biocatalysts do not operate by different scientific principles than conventional chemical catalysts. Their mechanism of action although complicated are governed by the laws of thermodynamics. Understanding the role of redox potential is important for selection of the appropriate type of enzyme and monomer combinations that can result in successful enzymatic polymerization. The thermodynamics of the system insist that the overall change in Gibbs free energy ($\Delta G = -nFE°$) for a chemical reaction to spontaneously occur should be equal to zero or a negative value. In other words, the standard electrode potential (net E°) should be positive. Every step in the reaction should be thermodynamically feasible; i.e. the difference between the reduction potentials of the reduced and oxidized species must be greater than zero. Therefore, the reduction potential of the initiator or enzyme substrate must be greater than the reduction potential of the enzyme to generate the oxidized form of the enzyme. Once oxidized or "activated", the enzyme can oxidize a monomer to its radical cation form if its reduction potential is greater than that of the monomer. The radical cation of the monomer can then propagate toward forming a polymer by reacting with other radical cations or monomer units. As mentioned earlier, in theory, the substrate molecule (typically H_2O_2 or oxygen), may oxidize the monomer directly. However, these reactions have been shown to be very slow, and thus there is a need for the enzyme to catalyze the reaction.

To simplify this somewhat confusing argument, the standard electrode potential equations are given below for two reactions. In example one, horseradish peroxidase (HRP) is used to oxidize phenol, while the second example highlights that HRP is unable to oxidize the thiophene monomer. The first step in both examples is the oxidation of HRP by the substrate molecule, H2O2. The reported reduction potentials of H_2O_2, HRP, activated HRP, phenol, and thiophene are approximately 1.76 V, -0.90 V (*17*), 0.90 V (*17*), - 0.60 V (*18*), and - 2.0 V (*19*) versus a normal hydrogen electrode (NHE), respectively.

Overall Reaction

$$E°_{cell} = E°_{ox} + E°_{red} > 0$$

1st step : Enzyme Oxidation

$$- E°_{red} \ HRP + E°_{red} \ substrate > 0$$

$$- 0.90 \ V + 1.76 \ V > 0$$

$$Result = HRP \xrightarrow{\text{oxidation}} activated \ HRP$$

Example 1 :   Phenol Oxidation

$$- E^\circ_{red} \text{ phenol} + E^\circ_{red} \text{ activated HRP} > 0$$

$$- 0.60 \text{ V} + 0.90 \text{ V} > 0$$

$$\text{Result} = \text{phenol} \xrightarrow{\text{oxidation}} \text{phenol radical cation}$$

Example 2 : Thiophene Oxidation

$$- E^\circ_{red} \text{ thiophene} + E^\circ_{red} \text{ activated HRP} < 0$$

$$- 2.00 \text{ V} + 0.90 \text{ V} < 0$$

$$\text{Result} = \text{No Reaction}$$

2. Proposed Mechanisms for Enzymatic Polymerization Reactions

2.1. Peroxidases

The mechanism for peroxidase-type enzyme-catalyzed polymerization reactions is well known (20). The most commonly used peroxidases used for the polymerization of CPs include HRP, soybean peroxidase (SBP), and palm tree peroxidase (PTP). The reaction involves an initial two-electron oxidation of the native enzyme (ex. HRP) to form an oxidized intermediate (ex. HRP-I) by hydrogen peroxide. The monomer is then oxidized (by HRP-I) to produce monomeric radical species, which can undergo coupling to form a dimer. Upon oxidation of the monomer, the enzyme is reduced to another intermediate state (ex. HRP-II) that can oxidize an additional monomer unit. Therefore during the peroxidase cycle, two moles of radical cations are generated per mole of H_2O_2 reacted with the enzyme. The cycle can repeat itself with the dimer molecule being oxidized to its radical cation and theoretically repeat until eventually the polymer is formed. However, there is still some debate about whether there is a molecular weight cutoff for the growing polymers' interaction with the enzymes active site, since enzymes are known to be substrate specific. It is postulated that after appreciable growth the polymer chain can not be directly oxidized by the enzyme due to its size. Therefore, any further growth of the polymer chain must proceed through another mechanism. Figure 1 presents the heme active site commonly found in all peroxidases and a schematic of its catalytic cycle.

2.2. Laccase

In contrast to peroxidases with an iron-based heme active center, the active site of a laccase contains four copper atoms that can be oxidized from the ground state Cu^I to Cu^{II} in the presence of oxygen (O_2), Figure 2. The activated enzyme oxidizes a monomer to a radical cation. In the process, the activated enzyme is reduced back to the Cu^I ground state. The catalytic cycle functions similarly to the peroxidase cycle; however for each mole of O$_2$ that oxidizes a Cu active center, only one mole of a monomer radical cation is produced.

2.3. Use of Redox Mediators

In order to expand the catalytic utility of oxidoreductases, the possibility of using redox mediators have been explored (22). Especially in some cases where the reduction potential of an activated enzyme is insufficient to oxidize a monomer, a redox mediator can be used to initiate the polymerization reaction. With the use of a redox mediator, the substrate still oxidizes the enzyme in the same manner as before; however, in the next step, the redox mediator (not the monomer) is oxidized to its radical cation form by the activated enzyme. Subsequently, the radical cation of the redox mediator can be transferred to the monomer to initiate polymerization.

One of the earliest reports on the use of redox mediators for the electrochemical synthesis of conducting polymers utilized thiophene oligomers to initiate the polymerization of the parent polymer (23). An increase in the rate of electrochemical polymerization of thiophene was observed in the presence of small amounts of bithiophene or terthiophene. The use of bithiophene/terthiophene reduced the amount of applied potential needed for the electrochemical polymerization. Greater rate enhancement was observed with terthiophene compared to bithiophene. This was attributed to the lower oxidation potential of the terthiophene. A mechanism for the electrochemical polymerization of thiophene in the presence of bithiophene/terthiophene was proposed, as depicted in Figure 3. When the electrochemical polymerization was initiated in the presence of small amounts of bithiophene, the bithiophene was oxidized to the radical cation in the first step to initiate growth of the polymer chain. The bithiophene radical cation then reacted with a neutral thiophene monomer to yield a trimer. This process was suggested to continue iteratively to form the CP.

Redox mediators have been used to polymerize substituted phenols using peroxidases (24, 25) and laccases (26). The most commonly used redox mediators for the enzymatic synthesis of conducting polymers include 2,2'-azino-bis (3-ethylbenzthiazoline-6-sulfonic acid) (ABTS), and terthiophene. These mediators have reported reduction potentials in the range of 0.7 - 1.2 V (vs. NHE), which closely matches those found in HRP, SBP, and several laccases. The mechanism of radical transfer between the redox mediator and the monomer is not well understood. There are inconsistencies in explanation provided for this process since the mediator's reduction potential is typically insufficient to oxidize the monomer. However, it has also been reported that the redox mediator may be incorporated as part of the backbone of the growing polymer. This would indicate that at least some degree of radical chain transfer does occur when a redox mediator is used.

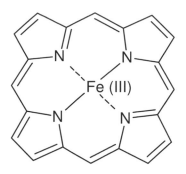

Peroxidase active centre

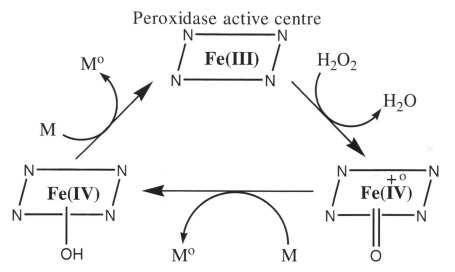

Figure 1. Active Site and Catalytic Cycle of a heme cofactor conatining Peroxidase, M – Monomer, Mo – Radical Cation of Monomer (20).

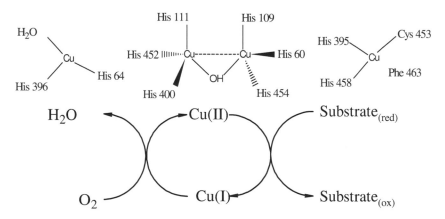

Figure 2. Active Site and Catalytic Cycle of Laccase (21).

3. Enzymatic Synthesis of Polyaniline

The first report of any enzyme-catalyzed polymerization of a conducting polymer used the oxidoreductase, billirun oxidase, to polymerize aniline (27). However, since this initial report, the overwhelming majority of publications on the synthesis of conducting polymers have utilized the oxidoreductase horseradish peroxidase (HRP). The reported redox potentials for HRP and aniline are both in the range of 0.9 V (vs. NHE) (17). Therefore, it is expected that the polymerization proceeds direct by the mechanism highlighted above for peroxidases. To date, the authors are unaware of any reports of conducting polymers of aniline being made through the redox mediator mechanism.

HRP has been shown to be a very effective catalyst for the synthesis of highly conductive form of polyaniline (PANI) with conductivity reported as high as 10 S/cm (28). Figure 4 presents a schematic of the structural differences found in the many doped forms of conducting PANI.

In addition to producing high conductivity, the enzymatic reaction conditions are much more environmentally friendly than those used for traditional chemical and electrochemical polymerizations. In the reports of HRP catalyzed PANI synthesis, the pH of the reaction, ratio of aniline to dopant/template, and concentration of hydrogen peroxide were some of the key factors that influenced the properties. The reactions were predominantly carried out at room temperature; however when lower temperature synthesis was applied, the conductivity of the obtained polymer increased (29). Generally, the reactions were performed within the pH range of 3-6 and the molar ratio of monomer to dopant were maintained at 1:1.

3.1. Use of a Polyelectrolyte Template

3.1.1. Poly(styrene-4-sulfonate)

The most commonly used template for the polymerization of aniline with HRP has been poly(styrene-4-sulfonate, sodium salt) (PSS). In the reaction mixture, PSS serves three distinct functions. First, the strong anionic charge of PSS ($\sim$ pKa = 0.7) (30) couples with the protonated form of aniline (anilinium ion) (pKa = 4.63) to electrostatically align the monomer along the polymer backbone. Secondly, the sulfonate group serves as charge-balancing dopants along the electroactive PANI backbone. Finally, the residual sulfonate groups solubilize the PANI-PSS complex. This renders the complex dispersible in aqueous and some organic solvents, making it valuable for practical applications. The use of PSS has been shown to produce predominantly para-coupled benzenoid-quinoid (head-to-tail) chains of the green colored conductive emeraldine salt form of PANI (31). It has been shown through solid state NMR studies that when PANI was synthesized with HRP without a template, the resulting polymer was a highly branched and insulating form of PANI (32). Figure 5 presents a schematic of the resulting PANI produced with and without the presence of PSS.

The PANI-PSS complex was shown to be electroactive using spectroscopy to monitor shifts in the absorption spectrum at pHs between 3.5-11 (34). Electrical

D.C conductivity measurements of PANI-PSS were highly dependent on the ratio of aniline monomer to PSS present in the reaction mixture. The conductivity for samples made at pH 4.3 (optimum pH) ranged from 6×10^{-5} S/cm to 5×10^{-3} S/cm when molar ratios of PANI:PSS of 0.6:1 to 2.2:1 were used, respectively. All samples could be protonically doped via exposure to hydrochloric acid (HCl) vapor to increase the conductivity by roughly an order of magnitude. Furthermore, a molar ratio of 2.0:1.0 PANI:PSS was shown to be the minimum ratio required to maintain solubility/disperablility of the final complex (33).

The growth of the polymer complex could be monitored over time via gel-permeation chromatography (GPC). A UV detector, set to the absorption max of PSS, was used to monitor changes in the molecular weight over time. As the reaction progressed, a bimodal signal developed. The elugram consisted of a growing peak assigned to the newly formed PANI-PSS complex, and a second peak remaining at the initial retention time, signifying uncomplexed PSS (34).

In addition to HRP, a peroxidase enzyme isolated from the Royal palm tree was shown to be an efficient catalysis for PANI-PSS synthesis (35). Optimum reaction conditions were observed at a pH of 3.5 and a PANI:PSS molar ratio of 1:1. No conductivity values were reported, however UV-Vis spectroscopy revealed a highly doped PANI that could exhibit conductivities in the range of those produced by HRP or higher.

3.1.2. Biologically Derived Templates

Once it was discovered that PSS could be used as a template for the synthesis of PANI researchers began investigating the possibility of substituting biological molecules in the place of PSS. The gentle reaction conditions of enzymatic synthesis allow for more biologically delicate molecules, RNA and DNA, to be used as templates. It was believed that if conducting polymers could be electrostatically bound to the hydrophilic areas of oligonucleotides, the resulting complex could be a viable biosensor. These new complexes could provide some sensitivity to distinguish between perfectly matched and mismatched "target" RNA/DNA complements during hybridization. The use of these complexes should provide a route to faster, less tedious, and more portable detection systems aimed at detection of complementary DNA strands of specific pathogens.

Molecular complexes of PANI and genomic DNA (calf thymus DNA) were synthesized using HRP as the catalyst (36–38). The PANI:DNA complex was electroactive, and the PANI exhibited a distinct handedness that mimicked the secondary structure of the DNA, Figure 6. Similarly, controlled molecular weight oligoanilines were polymerized with HRP along programmed sequences of oligonucleotides. Sequential cytosine nucleotides within larger DNA backbones served as templates for PANI oligomer formation (39). The polymerization was shown to be restricted to single strands of DNA.

In a double layer template approach, bovine serum albumin (BSA) was coupled with SDS for the polymerization of aniline. The positively charged BSA was shown to first align sodium dodecylsulfate (SDS) along its backbone. The SDS could serve as a template for the polymerization of aniline (40). The BSA's positive charge, however, was said to repeal the anilinium ion, and thus only oligoanilines were formed with this approach.

3.1.3. Other Polymeric Templates

A variety of other charged and neutral polymeric templates have been used during the enzymatic synthesis of PANI to varying degrees of success. These templates include poly(diallydimethylammonium chloride), poly(vinylamine), poly(ethylene oxide), poly(vinyl alcohol), poly(acrylic acid), poly(maleic acid co-olefin), and others (31). The UV-Visible absorption spectra of a few of these complexes are presented in Figure 7. The conducting form of polyaniline is known to have strong absorption between 800-1200 nm due to polaron transitions. It was observed that when cationic or neutrally charged templates were used no polaron absorption bands were observed. Slight polaron absorption was observed with weak acid templates compared to those observed with the strong acid template, PSS.

In contrast to the cationic, neutral, and weak acid templates mentioned above, poly(vinylphosphonic acid) (PVP) (41) and poly(vinylsulfonic acid, sodium salt) (42) have been shown to produce an emeraldine salt of PANI with a conductivity as high as 5.5×10^{-2} S/cm and 4.78×10^{-1} S/cm, respectively. The PANI-PVP complexes was especially sensitive to the ratio of aniline monomer to PVP used and the amount of hydrogen peroxide present in the reaction mixture. When the molar ratio of PANI:PVP exceeded 1:5, the free phosphonate groups could no longer solubilize the growing PANI chain and the complex precipitated out of solution. In a broader perspective, the work highlighted here validates the need for strong interactions between the positive (cationic) charge of the CP backbone and the negative (anionic) charge of a template or dopant to produce a conductive form of the polymer.

Furthermore, it was shown that even small molecule strong acid dopants when combined with weak acid polymeric templates could produce a conducting PANI. Chiral PANI nanocomposites were synthesized with HRP in the presence of poly(acrylic acid) (PAA) and 10-camphorsulfonic acid (CSA) (43). Conductivity values were in the range of 10^{-2} S/cm. Surprisingly, the PANI/PAA/CSA composite was produced in an enantiospecific manner by HRP regardless of the chirality of the CSA used. This suggests that the enzyme was selectively controlling the secondary structure of the newly formed polymer.

3.1.4. Micellar-Based Template

Anionic micelles have been shown to be effective templates for PANI synthesis. Sodium dodecylsulfate (SDS) (40), sodium dodecylbenzenesulfonic acid (DBSA) (44), and sodium dodecyl diphenyloxide disulphonate (DODD) (45) produced conductive samples of PANI when catalyzed by HRP. DBSA has also been used along with a laccase derived from the fungi *Trametes* hirsute to produce conducting PANI (46, 47). Additionally, a chiral PANI was produced with palm tree peroxidase (PTP), DBSA, and low concentrations of CSA (*48*). PTP was shown to have a higher enzymatic activity compared to HRP both with and without the presence of DBSA. It was found, however, that the conditions that favored higher conductivity did not favor the formation of chiral PANI and vice versa.

It has been proposed that anionic surfactants, like DBSA could serve as nanoreactors when organized as micelles. The localized anionic charge of the sulfonate groups was shown to sufficiently recreate the unique local environment found with PSS. However, it was noted that small percentages of branched products were formed with micellar templates. Unsurprisingly, when sodium benzene sulfonate (SBS), hexadecyltrimethylammonium bromide (CTAB), and polyoxyethylene(10) isooctylphenyl ether (Triton X-100) were used as templates and with HRP as the catalyst, the resulting polymers were branched and insulating (*31*). These results directly mirrored those produced from polymeric templates containing cationic or neutral charges along the backbone. SBS, despite containing an anionic sulfonate group was unable to self-assemble into any organized template and produce highly conducting PANI. Even DBSA, when used at concentrations below its critical micelle concentration, was shown to produce an insulating PANI (*31*). These results clearly indicated the requirement for a well organized anionic template to facilitate HRP catalyzed polymerization of aniline.

3.1.5. Template Free Synthesis

Conductive PANI was successfully synthesized in the presence of toluenesulfonic acid by the enzyme, SBP (*49*). This was the first report of using an non-self-assembled small molecule dopant to produce a conductive form of PANI. The template-free polymerization was run at pH 3.5 and 1°C. The resulting conductivity (2.4 S/cm) was shown to be nearly identical to chemically produced PANI; however, x-ray photoelectron spectroscopy (XPS) did reveal a higher proportion of chain defects in the samples catalyzed by SBP compared to traditional chemical oxidation with ammonium peroxydisulfate (*50*). Unfortunately, the authors did not give specific reasons to why the reaction was able to proceed with SBP without the use of a organized template. SBP has over 50% of the amino acid sequence as the HRP, but exhibits a higher thermal stability (*51*), higher reduction potential (*52*), and greater activity (*51*) at lower pH. It is unclear whether any of these properties played a role in enabling the polymerization to occur. However, it should also be noted that a lower temperature was used for the synthesis.

Aniline was also polymerized with the enzyme glucose oxidase (GOx) without a template at a wide variety of pHs and temperatures (*53*). GOx has been shown to produce hydrogen peroxide when exposed to oxygen in the presence of glucose. During the reaction, glucose is converted into gluconic acid, which may serve as a weak dopant for the PANI produced. In polymerization reactions run at pH conditions 4-7, the obtained PANI exhibited absorption spectra characteristic of branched and insulating form.

3.1.6. Synthesis of PANI Catalyzed by Immobilized Enzyme

In an effort to improve the economics of enzyme catalyzed polymerization, several reports have been published on immobilization strategies for HRP. One example that tested the reuseability of the immobilized HRP utilized ionic liquids as the encapsulating media. HRP was dissolved in the ionic liquid (IL), 1-butyl-3-methylimidazolium, to generate a biphasic system with water. Biphasic/interfacial systems have been shown in the past to be excellent media for the chemical synthesis of nanofibular PANI (54, 55). This strategy was shown to effectively catalyze the polymerization of aniline to the emeraldine form of PANI with HRP. Figure 8 presents a schematic of how IL were recycled during the synthesis of PANI. The IL/HRP could be recycled up to five times to produce PANI with conductivity on the order of 10^{-3} S/cm (56). In another immobilization approach, a laccase was immobilized on carboxymethylcellulose using Woodhead's reagent, 2-ethyl-5-phenylisoxazolium-3'-sulfonate (57). This was then used along with poly(2-acrylamido-2-methyl-1-propanosulfonic acid) at a variety of pH below 4.5 to produce PANI with a 2-probe conductivity of 1.3×10^{-2} S/cm. The immobilized laccase was recycled only two times to show reusability.

Additionally, HRP entrapped in crosslinked matrices of chitosan (58) or covalently attached to polyethylene (PE) surfaces (59) have been used to polymerize aniline. The PANI produced in both cases appear spectroscopically similar to those produced via free enzyme; however, the recyclability of these immobilized enzymes were not tested.

3.2. Polyaniline Derivative Syntheses

In efforts to expand the potential applications of enzymatically made PANI, many derivatives of PANI have been polymerized with HRP and various laccases. Poly(2-ethylaniline) (60), poly(o-toluidine) (61), and several poly(alkoxyanilines) (62) have each been polymerized under nearly identical conditions as the parent polymer with HRP as the catalyst and PSS as the dopant/solubilizer. All polymers had polaron absorptions between 730-750 nm, with each exhibiting reversible electroactivity in the UV absorption spectra. N-substituted derivatives (N-methyl, N-ethyl, N-butyl, and N-phenyl) have also been enzymatically polymerized with HRP in the presence of PSS (63). All these derivatives were water soluble, electroactive, and exhibited polaron (benzenoid form) absorption maximum at wavelengths greater than 1000 nm.

Polyaniline derivatives have also been polymerized without the use of a template with HRP. In a systematic study of methoxy-derivatives of aniline, it was found that the reaction was optimum at up to 15% water miscible polar co-solvents while in the presence of camphorsulfonic acid (64). It was also noted that peroxyacetic acid was a suitable replacement for hydrogen peroxide as an oxidant for HRP (64).

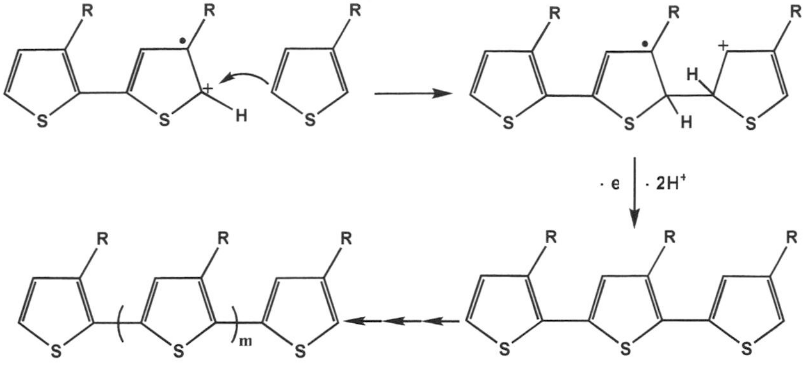
R
R
R
R
R
R
S
S
H
S
S
S
H
H
S
· e
· 2H⁺
R
R
R
R
R
R
S
S
m
S
S
S
S

Figure 3. Proposed mechanism for the electrochemical polymerization of thiophene in presence of bithiophene (23).

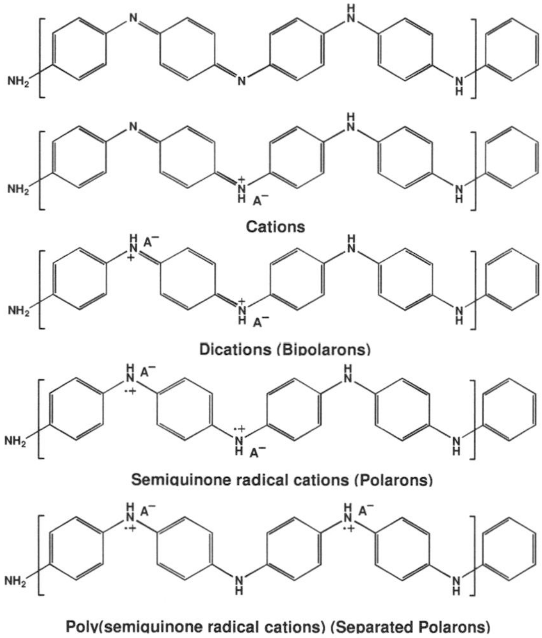
NH2
Cations
Dications (Bipolarons)
Semiquinone radical cations (Polarons)
Poly(semiquinone radical cations) (Separated Polarons)

Figure 4. Different Doped Forms of Conducting PANI (28).

3.3. Self-Doped Polyaniline

To improve the processability and eliminate the need for external dopants, several researchers have reported on enzymatically synthesized self-doped PANI derivatives. 3-amino-4-methoxybenzenesulfonic acid (SA) was polymerized with HRP and in the presence of tetradecyltrimethylammonium bromide (MTAB) and a polycationic template, poly(vinylbenzydimethylhydroxyethylammonium chloride) (PVAC) (*65*). The inherently negatively charged monomer self-assembled along the cationic micelle and template. The organized SA was then polymerized on the surface of the cationic templates with HRP. The products

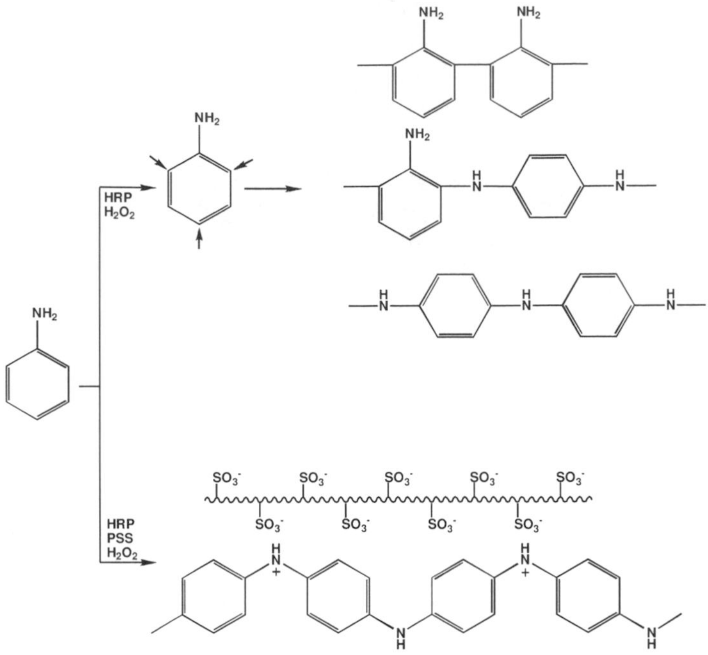

Figure 5. Schematic of the PANI produced by HRP with and without PSS (33).

exhibited changes in conductivity (10^{-4} S/cm to 10^{-8} S/cm) that could be explained by transformation between expanded and collapsed coil orientation, with the expanded coil yielding higher conductivity values.

3.4. Applications of Enzymatically Made Polyaniline

Two reports have been published recently that directly utilized enzymatically made PANI for surface patterning applications. 4-aminothiophenol was polymerized with HRP for direct dip-pen nanolithography (DPN) (66). Nanoscale patterns were produced by covalently bonding thiol groups of the monomer onto a gold substrate, followed by polymerization with HRP. In a second application, environmentally benign photo patterns were generated via the polymerization of PANI on the surface of a photo-crosslinked thymine based polymer with pendent phenylsulfonate groups (67). The PANIs produced were electroactive and had conductivities greater than 10^{-2} S/cm. Figure 9 presents a schematic of the photocrosslinking reaction and subsequent polymerization of PANI with HRP.

A variety of nanostructured complexes have been produced from enzymatically made PANI. Conductive fibers of PANI-PSS were enzymatically made with HRP and dry spun into fibers of approximately 100 μm diameter (68). The oriented fibers were shown to have a conductivity of an order of magnitude higher than thin films made of the same material. Nanostructured colloids of conducting PANI were polymerized with SBP in the presence of chitosan (69). The morphology and stability of colloids produced using SBP was shown to be sensitive to both temperature and pH (70).

328

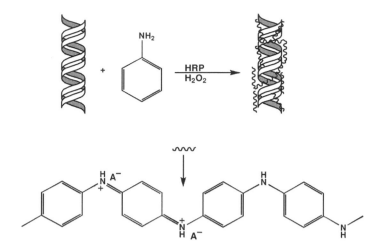

Figure 6. Schematic representing the formation of PANI on DNA (36).

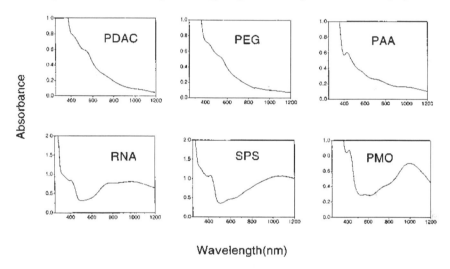

Wavelength(nm)

Figure 7. UV-Visible Absorption Spectra of PANI's made with a variety of
polymeric templates (31).

Optical and electrochemical sensors for the detection of saccharides were
produced by enzymatically synthesizing the self doped copolymer, poly(aniline-
co-3-aminobenzeneboronic acid), with HRP (71). It was shown that detection
of various saccharides was greatly improved by using the enzymatically made
copolymer compared to one produced chemically.

4. Polypyrrole and Poly(3,4-ethylenedioxythiophene)

The oxidative polymerization of pyrrole and 3,4-ethylenedioxythiophene
(EDOT) monomers yields two important classes of intrinsically conducting
polymers, Polypyrrole (Ppy) and Poly(3,4-ethylenedioxythiophene) (PEDOT)

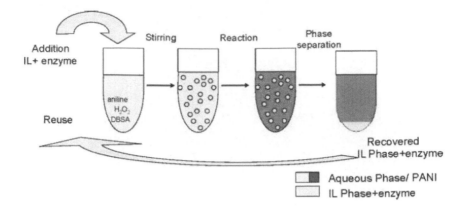

Figure 8. Use of ionic liquid immobilized HRP for the synthesis of conducting PANI (56).

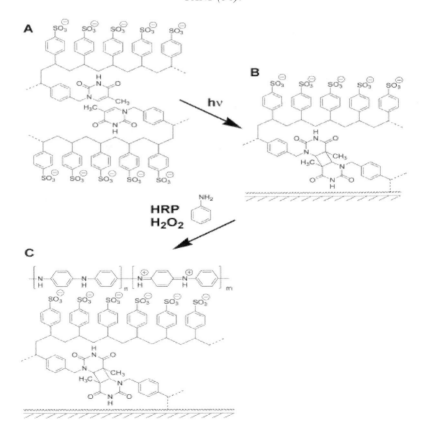

Figure 9. Schematic of photopatterning of conductive PANI with immobilized polyelectrolytes (67).

respectively (*72*), other than Polyaniline (discussed in the previous sections). Although the enzyme catalyzed oxidative polymerizations of these two monomers have received less attention as compared to the enzymatic synthesis of PANI, there are recent research efforts for the development of benign "greener" routes for the synthesis of Ppy and PEDOT synthesis. We believe that the initial delay on extending enzyme catalysis to the synthesis of Ppy and PEDOT was partially due to the fact that the reduction potentials of pyrrole (*73*) and EDOT (*74*) are both higher than the reduction potential of HRP (*17*) (~0.9V Vs NHE). We believe that this mismatch in potentials results in an inability to polymerize pyrrole and EDOT directly with HRP, and therefore requires either a redox mediator or the use of a different enzyme with an appropriate redox potential. In this section, we will focus on the recent literature reports about enzyme catalyzed synthesis of Ppy and PEDOT. We expect the fundamental research and the applications generated by these polymers will rapidly increase due to recent breakthroughs in the direct synthesis of these polymers catalyzed by the commercially available enzyme, Soybean Peroxidase (SBP).

4.1. Enzymatic Synthesis of Polypyrrole (Ppy)

Ppy is classically synthesized using chemical (*75–77*), and/or electrochemical (*78, 79*) oxidation of the pyrrole monomer. Ppy has found applications in diverse areas including electromechanical systems (*80, 81*), biosensors (*82, 83*), and drug delivery systems (*84*) owing to its excellent stability and more importantly, better biocompatibility compared to other conducting polymers. Therefore, there are tremendous benefits to develop methods to synthesize Ppy under benign and non-toxic conditions.

It was reported that chemical oxidation of pyrrole occurred directly with H_2O_2, to obtain Ppy. However, the rate of the reaction was slow. The direct oxidation of pyrrole with hydrogen peroxide also yields considerable derivatives of 2-pyrrolidinone (Figure 10) (*11*). Such defects in the Ppy backbone, due to over oxidation, has an effect of lowering the electrical conductivity of the resulting polymer.

In the first report on the synthesis of water soluble Ppy utilizing enzymatic oxidative polymerization (*85*), HRP and H_2O_2 were employed as enzyme catalyst and oxidant, respectively (Figure 11). Poly(sodium 4-styrene sulfonate) (PSS) was used as the charge balancing dopant as well, to impart solubility to the synthesized Ppy-PSS complex. In a typical reaction setup, the monomer, enzyme catalyst, and dopant were first dissolved in a buffer solution at a predetermined pH and the polymerization was initiated with addition of an aliquot of the oxidant. The oxidant was then incrementally added and the reaction was stirred for a predetermined time to complete the polymerization. The authors monitored the reaction using UV-Visible spectroscopy and in the absence of H_2O_2 the characteristic bipolar transition peak at 450 nm and the free charge carrier peak at 800 nm (due to doping) were not observed, suggesting that the enzyme alone did not initiate pyrrole polymerization. The authors also reported that the reaction was completed after two days at pH 2. The reaction was performed at varying pH values and it was reported that the polymerization did not occur in reaction medium that had a pH higher than 4. Fourier transform infrared spectroscopy (FTIR) was utilized for confirming the structure of the resulting polymers and an absorption at 1710 cm⁻¹ was observed. This was attributed to a carbonyl stretch and suggests over-oxidation of the resulting polymers. The authors did not report any electrical conductivity data of the synthesized polymers.

Kupriyanovich and co-authors recently reported the enzymatic polymerization of pyrrole utilizing HRP and H_2O_2 at pH 4.5 (86). The authors observed that the polymerization did not proceed in the absence of HRP, suggesting that the enzyme also played a key role in the reaction. The soluble low molecular weight reaction products were isolated from the insoluble resulting polymer and gel permeation chromatography (GPC) was performed on the soluble fraction. GPC revealed that the soluble fraction contained heptamers, octamers, and some higher molecular weight species. It is interesting to note that the synthesized Ppy polymers and low molecular weight compounds displayed the presence of a carbonyl group suggesting over-oxidation. The presence of oligomers that contain carbonyl groups suggests that 2-pyrrolidone derivatives similar to those displayed in Figure 9 might be present in the soluble fraction. To improve the polymerization rate the authors utilized a redox mediator 2,2′-azino-bis(3-ethylbenzthiazoline-6-sulfonic acid)diammonium salt (ABTS) (86). Since the rate of enzymatic oxidation of the ABTS salt is several orders of magnitude higher than pyrrole, ABTS radical cations form in the initial step of the polymerization, which then oxidize the pyrrole monomer while getting reduced (Figure 12). Due to the presence of the sulfonate groups ABTS also acts as a dopant for the Ppy. A conductivity value of 1.2×10^{-6} S/cm was reported for the Ppy synthesized with ABTS. Upon doping with PSS along with ABTS, the conductivity increased to 2×10^{-4} S/cm. The conductivity increased by an order of magnitude with secondary doping with I_2. Similarly, Cruz-Silva and co-workers employed ABTS as a redox mediator for the enzymatic oxidative polymerization of pyrrole with HRP (87). Using poly(vinyl alcohol) in the reaction mixture the authors obtained sterically stabilized highly water-dispersible Ppy colloids.

In a different approach, Ppy was synthesized at pH 4 using laccase as the enzyme catalyst and oxygen as the oxidant (26). As displayed in Figure 13, the laccase catalyzes the oxidation of pyrrole to its radical cation with simultaneous reduction of oxygen to water. Utilizing UV-Visible absorbance spectroscopy, the authors were able to monitor pyrrole concentration during the reaction. The rate of polymerization of pyrrole with laccase enzyme catalyst was determined to be 0.11 µM/min. The polymerization of pyrrole in acidic reaction medium and in the absence of the laccase was also reported and the rate of polymerization for the acid catalyzed mechanism was determined to be 0.068 µM/min. The conductivity of the Ppy synthesized using ABTS as the redox mediator and dopant was 0.2 S/cm.

In another report, the enzyme lactate oxidase (LOD) was used as a catalysis for the polymerization of pyrrole. LOD catalyzes the conversion of lactate to pyruvic acid and H_2O_2 in the presence of oxygen. The generated H_2O_2 then serves as a direct oxidant for the polymerization of pyrrole (88). Long reaction times were reported for the polymerization. The authors additionally noted the product's UV-Vis spectrum was typical of reduced forms of Ppy obtained by electrochemical dedoping. The LOD molecules are negatively charged and are

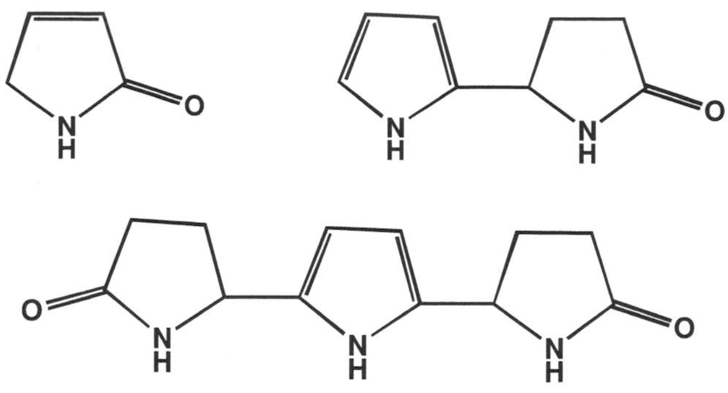

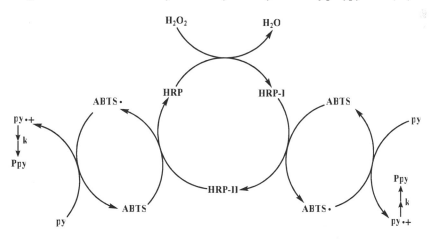

Figure 10. Structure of 2-pyrrolidinone derivatives upon direct oxidation of pyrrole with hydrogen peroxide (11).

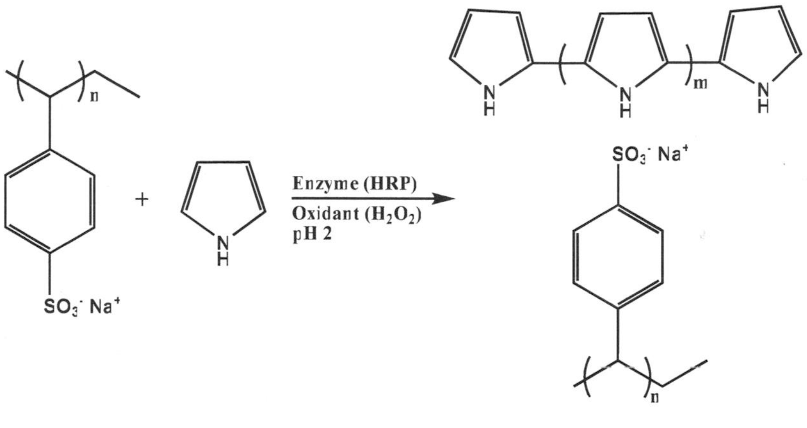

Figure 11. Reaction scheme for the enzymatic synthesis of polypyrrole (85).

Figure 12. Reaction schematic for polymerization of pyrrole with HRP (enzyme), H202 (oxidant) and ABTS (redox mediator) (87).

said to act as seeds for the polymerization. At the end of the reaction the authors obtained Ppy-LOD nanoparticles which further assembled to form nanometer and micron sized architectures. Additionally when carbon nanotubes (CNT) were added to the reaction mixture, the CNTs functioned as seeds for Ppy formation and a CNT-Ppy composite nanofilm was obtained.

In our laboratories, Ppy was successfully synthesized in high yields using soybean peroxidase (SBP) as catalyst, H_2O_2 as oxidant, and PSS as the dopant at pH 3.5 (89). The influence of reaction time, temperature, and concentration of the dopant on the reaction was systematically studied. For all the synthesized polymers no absorption peaks were observed between 1600 cm^{-1} and 1800 cm^{-1} suggesting that over-oxidation of Ppy did not occur. The reaction was monitored using UV-Visible spectroscopy and the absorption peaks at 460 nm (bipolaron) and 900 nm (free carrier) reached the maximum value after 2h indicating reaction completion. Interestingly it was observed that the yield of the reaction and the conductivity of Ppy-PSS obtained was highly dependant on the reaction temperature, Figure 14. It was observed that the reaction yields and the conductivity of the product increased with decreasing reaction temperatures. Maximum conductivity in the order of 10^{-1} S/cm was observed for Ppy-PSS synthesized at 2 °C. To further understand the reaction kinetics studies are underway in our laboratories.

4.2. Enzymatic Synthesis of Poly(3,4-ethylenedioxythiophene)

PEDOT is one of the more commercially important and well known conducting polymer, which can be synthesized either chemically or electrochemically (74). PEDOT-PSS has found applications in varying areas including electrochromic devices (90), biosensing (91), photovoltaic devices (92) etc.

The first reported enzymatic synthesis of water soluble PEDOT was catalyzed using HRP (Figure 15) (93). The reaction was performed at 4 °C and pH 2, with H_2O_2 as the oxidant and PSS as the dopant. After a 16 hour reaction a blue colored polymer solution was obtained. The authors observed that pH 2 was the optimal condition of the reaction medium and the reaction did not proceed at pH 4 and pH 6. Since the activity of HRP is greatly reduced in acidic medium, the authors suggested that the HRP is preferentially localized in the hydrophobic monomer droplets. The conductivity of the synthesized PEDOT was 2 x 10^{-3} S/cm. Recently the authors reported that the HRP solution in EDOT can be separated from the aqueous dispersion of the synthesized PEDOT-PSS (94). This isolated HRP solution was reused for further PEDOT polymerization. This process could be repeated up to 10 times with the HRP solution separated out each time and the conductivity obtained for all the obtained polymers was in the order of 10^{-3} S/cm.

In our laboratories, SBP was utilized as the enzyme catalyst to synthesize PEDOT/PSS using milder conditions (pH > 2) in the presence of terthiophene as the redox mediator and PSS as the dopant (95) as shown in Figure 16. SBP was not observed to catalyze EDOT polymerization in the absence of the terthiophene redox mediator. The utilized terthiophene (<1 wt%) gets oxidized and facilitates the polymerization of EDOT in the presence of PSS. It was found that at least 0.5

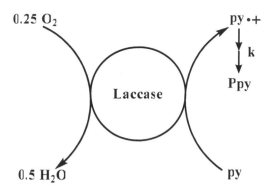

Figure 13. Reaction schematic for polymerization of pyrrole with laccase as enzyme catalyst and O2 as the oxidant (26).

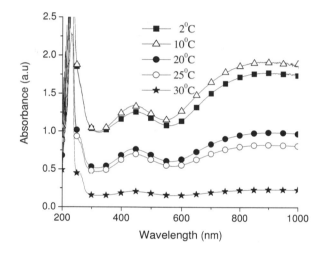

Figure 14. Synthesis temperature effects on the absorption spectra of Ppy-PSS made with SBP (89).

wt % of terthiophene was required in the reaction mixture for the polymerization to proceed. The polymerization was performed at varying pH and utilizing UV-Visible spectroscopy it was shown that even at pH 4 the reaction proceeded to afford PEDOT-PSS (Figure 17). The UV-Visible absorption spectrum and the FTIR spectrum of the synthesized PEDOT-PSS closely resembled the spectrum for the commercially available polymer. The synthesized polymers could be reversibly dedoped (increasing pH) and re-doped (decreasing pH). The conductivities of the synthesized polymers were in the range of $10^{-3} - 10^{-4}$ S/cm. Unlike the cases where ABTS is used as the mediator, here the mediator is actually a trimer of parent thiophene this ensures that if the conjugated mediator becomes incorporated into the growing polymer chain, the electrical properties of the resulting polymers will remain unaffected.

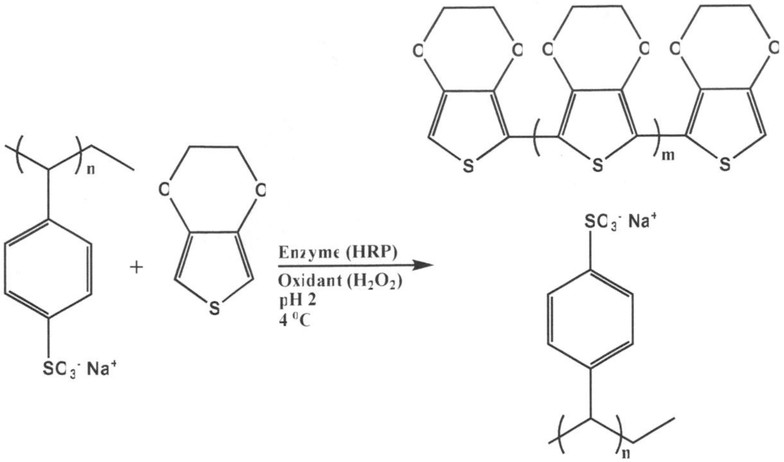

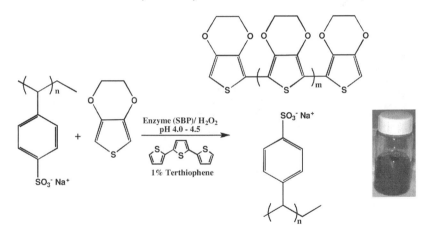

Figure 15. Reaction scheme for the enzymatic synthesis of PEDOT-PSS using HRP as the enzyme catalyst and H_2O_2 as the oxidant (93).

Figure 16. Reaction Scheme for the SBP catalyzed polymerization of EDOT using terthiophene as a redox mediator (95).

5. Conclusions

The use of enzymes to produce advanced functional materials such as electrically conducting polymers has recently been highlighted as a new field of "green polymer chemistry". Enzymatic synthesis offers a viable alternative to the traditional oxidative polymerization methods. Enzymatic catalysis is environmentally friendly, energy efficient, non-toxic, readily scalable, and produces high yields of products. The oxidoreductase enzymes, HRP, SBP, PTP, and laccase have been shown to be effective catalysts for producing highly conductive forms of PANI, Ppy, and PEDOT. The reactions are sensitive to the pH, ratio of monomer to dopant/template, and temperature. Judicious choice of these parameters ensures the production of an electrically conductive species

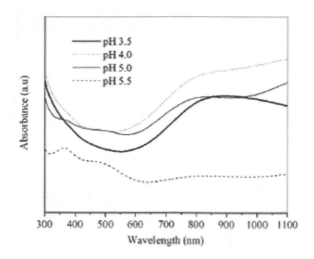

Figure 17. UV-Visible absorption spectra for polymerization of EDOT with SBP (enzyme catalyst), H₂O₂ (oxidant), and terthiophene (redox mediator) at varying pH (95).

with properties comparable to those produced using traditional chemical or electrochemical methods.

Even though progress has been made in development of "greener" enzymatic routes for the polymerization of aniline, pyrrole and EDOT further investigations are required to explore other aspects of these reactions. The research efforts to date have only answered some questions regarding reaction feasibility for a given combination of enzyme and monomer. The effect of temperature and pH on the kinetics of the reaction needs to be better understood. In many cases the enzyme activity is determined/calibrated utilizing an assay involving a phenolic monomer (pyrogallol (96) or guaiacol (97)). However the activity of the enzyme towards a chosen conjugated monomer in the presence of ionic species (polyelectrolytes) might be very different. Understanding the influence of the ionic environment on enzyme activity can provide new enzyme selection criteria for specific polymerization reactions.

Balancing redox potentials provides a good starting point for selection of a monomer and enzyme that can catalyze a successful polymerization reaction. In some cases small molecules such as easily oxidizable aromatics (ABTS for example), and conjugated oligomers can act as redox mediators for the enzyme to initiate the polymerization. However this mechanism as well as the interaction of oligomers and polymers with the enzyme's active site needs to be better understood for the advancement of this field. With the increased commercial availability of a greater variety of enzymes, new dopants (oligomers and polyelectrolytes) combined with possibilities for expanding redox mediators' usage, innumerable exciting opportunities would become available for developing efficient and sustainable routes to conducting polymers.

References

1. Richardson-Burns, S. M.; Hendricks, J. L.; Foster, B.; Povlich, L. K.; Kim, D.-H.; Martin, D. C. *Biomaterials* **2007**, *28*, 1539–1552.
2. Dias, H. V. R.; Fianchini, M.; Rajapakse, R. M. G. *Polymer* **2006**, *47*, 7349–7354.
3. Wang, Y.; Jing, X.; Kong, J. *Synth. Met.* **2007**, *157*, 269–275.
4. Dias, H. V. R.; Wang, X.; Rajapakse, R. M. G.; Elsenbaumer, R. L. *Chem. Commun.* **2006**, 976–978.
5. Bruno, F. F; Fossey, S. A.; Nagarajan, S.; Nagarajan, R.; Kumar, J.; Samuelson, L. A. *Biomacromolecules* **2006**, *7*, 586–589.
6. Roy, S.; Fortier, J. M.; Nagarajan, R.; Tripathy, S.; Kumar, J.; Samuelson, L. A.; Bruno, F.F. *Biomacromolecules* **2002**, *3*, 937–941.
7. Nagarajan, S.; Nagarajan, R.; Bruno, F. F.; Samuelson, L. A.; Kumar, J. *Green Chem.* **2009**, *11*, 334–338.
8. Surwade, S. P.; Agnihotra, S. R.; Dua, V.; Manohar, N.; Jain, S.; Ammu, S.; Manohar, S. K. *J. Am. Chem. Soc.* **2009**, *131*, 12528–12529.
9. Shirakawa, H; MacDiarmid, A.; Heeger, A. *J. Chem. Soc., Chem. Commun.* **1977**, *16*, 578–580.
10. Shirakawa, H.; MacDiarmid, A.; Heeger, A. *Chem. Commun.* **2003**, *1*, 1–4.
11. Bocchi, V.; Chierici, L.; Gardini, G. P.; Mondelli, R. *Tetrahedron* **1970**, *26*, 4073–4082.
12. Kobayashi, S.; Makino, A. *Chem. Rev.* **2009**, *109*, 5288–5353.
13. Numbers obtained from active entries in enzyme nomenclature database 2010.
14. Gu, Q. M.; Maslanka, W. W.; Cheng, H. N. *ACS Symp. Ser.* **2008**, *999*, 309–319.
15. Ayyagari, M.; Akkara, J. A.; Kaplan, D. L. *Acta Polym.* **1996**, *47*, 193–203.
16. Kalra, B.; Gross, R. A. *Green Chem.* **2002**, *4*, 174–178.
17. Jantschko, W.; Furtmüller, P. G.; Allegra, M.; Livrea, M. A.; Jakopitsch, C.; Regelsberger, G.; Obinger, C. *Arch. Biochem. Biophys.* **2002**, *398*, 12–22.
18. Steenken, S.; Neta, P. *J. Phys. Chem.* **1982**, *86*, 3661–3667.
19. Audebert, P.; Guyard, L.; Nguyen Dinh An, M.; Hapiot, P.; Chahma, M.; Combelas, C.; Thiebault, A. *J. Electroanal. Chem.* **1996**, *407*, 169–173.
20. Dunford, H. B. . In *Peroxidases in Chemistry and Biology*; Everse, J., Grisham, M. B., Eds.; CRC Press: 1991; Vol II, pp 1–24.
21. Riva, S. *Trends Biotechnol.* **2006**, *24*, 219–226.
22. Husain, M.; Husain, Q. *Crit. Rev. Environ. Sci. Technol.* **2008**, *38*, 1–42.
23. Wei, Y.; Chan, C.-C.; Tian, J.; Jang, G.-W.; Hsueh, K. F. *Chem. Mater.* **1991**, *3*, 888–897.
24. Won, K.; Kim, Y. H.; An, E. S.; Lee, Y. S.; Song, B. K. *Biomacromolecules* **2004**, *5*, 1–4.
25. Chelikani, R.; Hwan Kim, Y.; Yoon, D.Y.; Kim, D.S. *Appl. Biochem. Biotechnol.* **2009**, *157*, 263.
26. , H.-K.Song; Palmore, G.T.R. *J. Phys. Chem. B.* **2005**, *109*, 19278–19287.
27. Aizawa, M.; Wang, L.L.; Shinohara, H.; Ikariyama, Y. *J. Biotechnol.* **1990**, *14*, 301–310.

28. Sahoo, S. K.; Nagarajan, R.; Roy, S.; Samuelson, L. A.; Kumar, J.; Cholli, A. L. *Macromolecules* **2004**, *37*, 4130–4138.

29. Cruz-Silva, R.; Romero-García, J.; Angulo-Sánchez, J. L.; Ledezma-Pérez, A.; Arias-Marín, E.; Moggio, I.; Flores-Loyola, E. *Eur. Polym. J.* **2005**, *41*, 1129–1135.

30. Lide, D. R. *Handbook of Chemistry and Physics*; 68th ed.; CRC Press: Boca Raton, FL, 1993; pp D159−161.

31. Liu, W.; Cholli, A. L.; Nagarajan, R.; Kumar, J.; Tripathy, S.; Bruno, F. F.; Samuelson, L. *J. Am. Chem. Soc.* **1999**, *121*, 11345–11355.

32. Sahoo, S. K.; Nagarajan, R.; Chakraborty, S.; Samuelson, L. A.; Kumar, J.; Cholli, A. L. *J. Macromol. Sci., Part A: Pure Appl. Chem.* **2002**, *39*, 1223–1240.

33. Samuelson, L.; Anagnostopoulous, A.; Alva, K. S.; Kumar, J.; Tripathy, S. K. *Macromolecules* **1998**, *31*, 4376–4378.

34. Lui, W.; Kumar, J.; Tripathy, S.; Senecal, K. J.; Samuelson, L. *J. Am. Chem. Soc.* **1999**, *121*, 71–78.

35. Sakharov, I. Y.; Vorobiev, A. C.; Castillo Leon, J. *J. Enzyme Microb. Technol.* **2003**, *33*, 661–667.

36. Nagarajan, R.; Liu, W.; Kumar, J.; Tripathy, S.; Bruno, F. F.; Samuelson, L. A. *Macromolecules* **2001**, *34*, 3921–3927.

37. Nagarajan, R.; Roy, S.; Kumar, J.; Tripathy, S. K.; Dolukhanyan, T.; Sung, C.; Bruno, F. F.; Samuelson, L. A. *J. Macromol. Sci., Part A: Pure Appl. Chem.* **2001**, *38*, 1519–1537.

38. Roy, S.; Nagarajan, R.; Bruno, F. F.; Tripathy, S. K.; Kumar, J.; Samuelson, L. A. *Proc. ACS, PMSE* **2001**, *85*, 202–203.

39. Datta, B.; Schuster, G. B. *J. Am. Chem. Soc.* **2008**, *130*, 2965–2973.

40. Gu, Y.; Chen, C. C.; Ruan, Z. W. *Synth. Met.* **2009**, *159*, 2091–2096.

41. Nagarajan, R.; Tripathy, S.; Kumar, J. *Macromolecules* **2000**, *33*, 9542–9547.

42. Shen, Y.; Sun, J.; Wu, J.; Zhou, Q. *J. Appl. Polym. Sci.* **2004**, *96*, 814–817.

43. Thiyagarajan, M.; Samuelson, L. A.; Kumar, J.; Cholli, A. *J. Am. Chem. Soc.* **2003**, *125*, 11502–11503.

44. Liu, W.; Kumar, J.; Tripathy, S.; Samuelson, L. A. *Langmuir* **2002**, *18*, 9696–9704.

45. Rumbau, V.; Pomposo, J. A.; Alduncin, J. A.; Grande, H.; Mecerreyes, D.; Ochoteco, E. *Enzyme Microb. Technol.* **2007**, *40*, 1412–1421.

46. Steltsov, A. V.; Morozova, O. V.; Arkharova, N. A.; Klechkovskaya, V. V.; Staroverova, I. N.; Shumakovich, G. P.; Yaropolov, A. I. *J. Appl. Polym. Sci.* **2009**, *114*, 928–934.

47. Streltsov, A. V.; Shumakovich, G. P.; Morozova, O. V.; Gorbacheva, M. A.; Yaropolov, A. I. *Appl. Biochem. Microbiol.* **2008**, *44*, 264–270.

48. Caramyshev, A. V.; Lobachov, V. M.; Selivanov, D. V.; Sheval, E. V.; Vorobiev, A. K.; Katasova, O. N.; Polyakov, V. Y.; Makarov, A. A.; Sakharov, I. Y. *Biomacromolecules* **2007**, *8*, 2549–2555.

49. Cruz-Silva, R.; Romero-Garcia, J.; Angulo-Sanchez, J. L.; Ledezma-Perez, A.; Arias-Marin, E.; Moggio, I.; Flores-Loyola, E. *Eur. Polym. J.* **2005**, *41*, 1129–1135.

50. Cruz-Silva, R.; Romero-García, J.; Angulo-Sánchez, J. L.; Flores-Loyola, E.; Farías, M. H.; Castillón, F. F.; Días, J. A. *Polymer* **2004**, *45*, 4711–4717.

51. Geng, Z.; Rao, K. J.; Bassi, A. S.; Gijzen, M.; Krishnamoorthy, N. *Catal. Today* **2001**, *64*, 233–238.

52. McEldoon, J. P.; Pokora, A. R.; Dordick, J. S. *Enzyme Microb. Technol.* **1995**, *17*, 359–365.

53. Kausaite, A.; Ramanaviciene, A.; Ramanavicius, A. *Polymer* **2009**, *50*, 1846–1851.

54. Huang, J.; Virji, S.; Weiller, B. H.; Kaner, R. B. *J. Am. Chem. Soc.* **2003**, *125*, 314–315.

55. Huang, J.; Kaner, R. B. *J. Am. Chem. Soc.* **2004**, *126*, 851–855.

56. Rumbau, V.; Marcilla, R.; Ochoteco, E.; Pomposo, J. A.; Mecerreyes, D. *Macromolecules* **2006**, *39*, 8547–8549.

57. Vasil'eva, I. S.; Morozova, O. V.; Shumakovich, G. P.; Yaropolov, A. I. *Appl. Biochem. Microbiol.* **2009**, *45*, 27–30.

58. Jin, Z.; Su, Y.; Duan, Y. *Synth. Met.* **2001**, *122*, 237–242.

59. Alvarez, S.; Manolache, S.; Denes, F. *J. Appl. Polym. Sci.* **2003**, *88*, 369–379.

60. Nabid, M. R.; Entezami, A. A. *Iran. Polym. J.* **2003**, *12*, 401–406.

61. Nabid, M. R.; Entezami, A. A. *Eur. Polym. J.* **2003**, *39*, 1169–1175.

62. Nabid, M. R.; Sedghi, R.; Entezami, A. A. *J. Appl. Polym. Sci.* **2007**, *103*, 3724–3729.

63. Nabid, M. R.; Entezami, A. A. *Polym. Adv. Technol.* **2005**, *16*, 305–309.

64. Kim, S. C.; Huh, P.; Kumar, J.; Kim, B.; Lee, J. O.; Bruno, F. F.; Samuelson, L. A. *Green Chem.* **2007**, *9*, 44–48.

65. Kim, S. C.; Kim, D.; Lee, J.; Wang, Y.; Yang, K.; Kumar, J.; Bruno, F. F.; Samuelson, L. A. *Macromol. Rapid Commun.* **2007**, *28*, 1356–1360.

66. Xu, P.; Kaplan, D. L. *Adv. Mater.* **2004**, *16*, 628–633.

67. Trakhtenberg, S.; Hangun-Balkir, Y.; Warner, J. C.; Bruno, F. F.; Kumar, J.; Nagarajan, R.; Samuelson, L. *J. Am. Chem. Soc.* **2005**, *127*, 9100–9104.

68. Wang, X.; Schreuder-Gibson, H.; Downey, M.; Tripathy, S.; Samuelson, L. *Synth. Met.* **1999**, *107*, 117–121.

69. Cruz-Silva, R.; Escamilla, A.; Nicho, M. E.; Padron, G.; Ledezma-Perez, A.; Arias-Marin, E.; Moggio, I.; Romero-García, J. *Eur. Polym. J.* **2007**, *43*, 3471–3479.

70. Cruz-Silva, R.; Arizmendi, L.; Del-Angel, M.; Romero-García, J. *Langmuir* **2007**, *23*, 8–12.

71. Huh, P.; Kim, S. C.; Kim, Y.; Wang, Y.; Singh, J.; Kumar, J.; Samuelson, L. A.; Kim, B. S.; Jo, N. J.; Lee, J. O. *Biomacromolecules* **2007**, *8*, 3602–3607.

72. *Handbook of Conducting Polymers*; Skotheim, T. A., Elsenbaumer, R. L., Reynolds, J. R., Eds.; Marcel Dekker: New York, 1998.

73. Arjomandi, J.; Holze, R. *J. Solid State Electrochem.* **2007**, *11*, 1093–1100.

74. Schweiss, R.; Lubben, J. F.; Johannsmann, D.; Knoll, W. *Electrochim. Acta* **2005**, *50*, 2849–2856.

75. Oh, E. J.; Jang, K. S.; MacDiarmid, A. J. *Synth. Met.* **2001**, *125*, 267–272.

76. MacDiarmid, A. J. *Synth. Met.* **1997**, *84*, 27–34.

77. Lee, J.Y.; Song, K. T.; Kim, S. Y.; Kim, Y.C.; Kim, D. Y.; Kim, C. Y. *Synth. Met.* **1997**, *84*, 137–140.
78. Zhou, M.; Pagels, M.; Geschke, B.; Heinze, J. *J. Phys. Chem. B* **2002**, *106*, 10065–10073.
79. MacDiarmid, A. J.; Epstein, A. J. *Synth. Met.* **1994**, *65*, 103–116.
80. Jager, E. W. H.; Smela, O.; Inganas, O. *Science* **2000**, *290*, 1540–1546.
81. Spinks, G. M.; Xi, B.; Zhou, D.; Truong, V.-T.; Wallace, G. G. *Synth. Met.* **2004**, *140*, 273–280.
82. Geetha, S.; Rao, C. R. K.; Vijayan, M.; Trivedi, D. C. *Anal. Chim. Acta* **2006**, *568*, 119–125.
83. dos Santos Riccardi, C.; Yamanaka, H.; Josowicz, M.; Kowalik, J.; Mizaikoff, B.; Kranz, C. *Anal. Chem.* **2006**, *78*, 1139–1145.
84. Szumerits, S.; Bouffier, L.; Calemczuk, R.; Corso, B.; Demeunynck, M.; Descamps, E.; Defontaine, Y.; Fiche, J. B.; Fortin, E.; Livache, T.; Mailley, P.; Roget, A.; Vieil, E. *Electroanalysis* **2005**, *17*, 2001–2017.
85. Nabid, M. R.; Entezami, A. A. *J. Appl. Polym. Sci.* **2004**, *94*, 254–258.
86. Kupriyanovich, Y. N.; Sukhov, B. G.; Medvedeva, S. A.; Mikhaleva, A. I.; Vakul'skaya, T. I.; Myachina, G. F.; Trofimov, B. A. *Mendeleev Commun.* **2008**, *18*, 56–58.
87. Cruz-Silva, R.; Amaro, E.; Escamilla, A.; Nicho, M. E.; Sepulveda-Guzman, S.; Arizmendi, L.; Romero-Garcia, J.; Castillon-Barraza, F. F.; Farias, M. H. *J. Colloid Interface Sci.* **2008**, *328*, 263–269.
88. Cui, X.; Li, C. M.; Zang, J.; Zhou, Q.; Gan, Y.; Bao, H.; Guo, J.; Lee, V. S.; Moochhala, S. M. *J. Phys. Chem. C* **2007**, *111*, 2025–2031.
89. Bouldin, R.; Ravichandran, S.; Garhwal, R.; Nagarajan, S.; Kumar, J.; Bruno, F. F.; Samuelson, L. A.; Nagarajan, R. *Polym. Prepr.* **2009**, 50.
90. Nicoletta, F. P.; Chidichimo, G.; Cupelli, D.; DeFilpo, G.; DeBenedittis, M.; Gabriele, B.; Salerno, G.; Fazio, A. *Adv. Funct. Mater.* **2005**, *6*, 995–999.
91. Krishnamoorthy, K.; Gokhale, R. S.; Contractor, A. Q.; Kumar, A. *Chem. Commun.* **2004**, *7*, 820–821.
92. Coakley, K.; Mcgehee, M. D. *Chem. Mater.* **2004**, *16*, 4533–4542.
93. Rumbau, V.; Pomposo, J. A.; Eleta, A.; Rodriguez, J.; Grande, H.; Mecerreyes, D.; Ochoteco, E. *Biomacromolecules* **2007**, *8*, 315–317.
94. Sikora, T.; Marcilla, R.; Mecerreyes, D.; Rodriguez, J.; Pomposo, J. A.; Ochoteco, E. *J. Polym. Sci., Part A: Polym. Chem.* **2009**, *47*, 306–309.
95. Nagarajan, S.; Kumar, J.; Bruno, F. F.; Samuelson, L. A.; Nagarajan, R. *Macromolecules* **2008**, *41*, 3049–3052.
96. Chance, B.; Maehly, A. C. *Methods Enzymol.* **1955**, *11*, 773–775.
97. Sessa, D. J.; Anderson, R. L. *J. Agric. Food Chem.* **1981**, *29*, 960–965.

Chapter 24

Sustained Development in Baeyer-Villiger Biooxidation Technology

Peter C. K. Lau,[a],* Hannes Leisch,[a] Brahm J. Yachnin,[b] I. Ahmad Mirza,[b] Albert M. Berghuis,[b] Hiroaki Iwaki,[c] and Yoshie Hasegawa[c]

[a]Biotechnology Research Institute, National Research Council Canada, 6100 Royalmount Avenue, Montreal, QC, H4P 2R2, Canada
[b]Departments of Biochemistry and Microbiology & Immunology, McGill University, 3655 Prom Sir William Osler, Montreal, QC, H3G 1Y6, Canada
[c]Department of Life Science & Biotechnology and ORDIST, Kansai University, Suita, Osaka, 564-8680, Japan
*peter.lau@cnrc-nrc.gc.ca

From the first crystal structures of a prototypical cyclohexanone monooxygenase (CHMO) complexed with both FAD and NADP$^+$ cofactors to whole genome mining of new microbial Baeyer-Villiger monooxygenases (BVMOs), this review highlights the recent progress in Baeyer-Villiger biooxidation technology. Protein engineering of BVMOs as enantioselective enzymes in asymmetric catalysis to accommodate new substrates is an active pursuit. Cofactor recycling appears to no longer be an issue, and we have gained a greater understanding of whole-cell biotransformation dynamics and limitations, leading us towards industrial scale realization. In terms of substrate profiling, new naturally-occurring substrates have been found in addition to a seeming rebirth of interest in BVMO-mediated oxidation of steroids.

Introduction

The Baeyer-Villiger (BV) chemical oxidation of ketones into esters or cyclic ketones to lactones using peracids is a more than century–old heritage reaction. It began in 1899 with the oxidation of menthone to the corresponding ring-expanded lactone (7-isopropyl-4-methyl-2-oxo-oxepanone) using a mixture of sodium persulfate and concentrated sulfuric acid (aka Caro's acid) (*1*). While it has been proven to be a powerful reaction in organic synthesis, it has its shortcomings that include usage of toxic or highly reactive oxidants (typically organic peroxyacids), generation of more waste than product, and in the case of enantioselective reactions, the products are often not of sufficiently high optical purity (Figure 1).

Baeyer-Villiger monooxygenases (BVMOs) are nature's mimicry of the abiotic BV reaction. They are flavoproteins and redox enzymes capable of nucleophilic oxygenation of a wide range of linear or cyclic ketones, yielding optically pure esters or lactones. Molecular oxygen is used as the oxidant, vis-à-vis peracids (e.g., *m*-chloroperbenzoic acid) in the chemical BV oxidation, and water is the only byproduct. Besides the "green" attributes, the chemo-, regio- and enantio-selectivities of BVMOs make them highly valuable bioreagents in asymmetric catalysis.

Cloned entities of BVMOs are needed, not only because they can be abundantly produced in a heterologous host, but more importantly to address the need to decouple the BVMO-encoding gene from the ring-opening hydrolytic step that normally goes hand-in-hand in the natural host as it metabolizes molecules such as cyclohexanol or cyclohexanone as the sole carbon source for energy and growth. Before 1999, the centennial mark of the BV chemical reaction, only one cloned entity existed - the cyclohexanone monooxygenase (CHMO) derived from *Acinetobacter* sp. NCIMB 9871 (*2*). This cloned biocatalyst has served as a model and benchmark for whole cell biotransformation in a number of notable organic syntheses. Numerous reviews covering the BVMO reactions have been published, including the latest comprehensive review by Margaret Kayser in early 2009 (*3*). This chapter will discuss the most recent development of the BV biooxidation technology that spans new BVMO discovery, the first crystal structures of a CHMO and its comparison to two other available BVMO structures, protein engineering, bioprocess development, and bioproduction of large ring-size lactones. Of special relevance to a fledgling bioeconomy, we will introduce the biooxidation of ketones derived from natural sources such as camphor, a naturally occurring and abundant terpenone.

Dawn of a New Century of Cloned BVMOs

Table 1 is a list of currently available cloned BVMOs, largely derived from bacterial sources, and mostly of Type I that have been classified as dependent on flavin adenine dinucleotide (FAD) and NADPH for activity and bearing the fingerprint or sequence signature of FxGxxxHxxxW(D/P) (*4*). Type II BVMOs are flavin mononucleotide (FMN) and NADH dependent and thus far appear to be rather limited, not only in the number of clones but in characterized examples (*3l,3n*). These are two-component proteins typified by the 2,5-diketocamphane monooxygenase and 3,6-diketocamphane monooxygenase in the metabolism of (+/-) camphor by *Pseudomonas putida* NCIMB 10007 or ATCC 17465 (*5*). In each case, a reductase component is required to reduce FMN to $FMNH_2$, which is necessary for the oxygenase component to perform the BV oxidation.

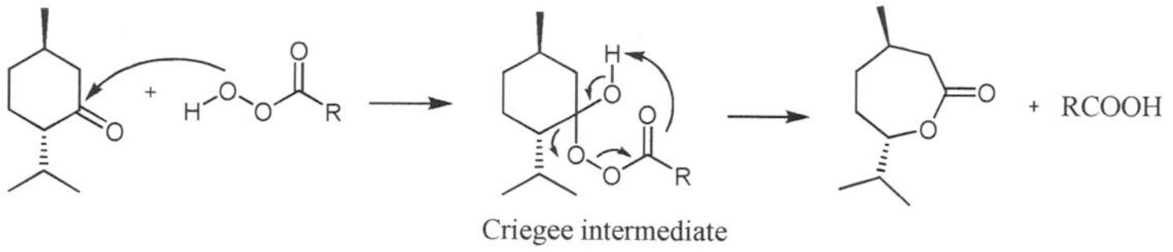

Criegee intermediate

Figure 1. Mechanism of the chemical oxidation of (-)-menthone with peracids.

Table 1. Growing list of heterologously expressed BVMOs

Type I BVMO (Class B FMO)

Strain origin	Common name	E. coli expression plasmid/ strain or gene locus	Reference
Acinetobacter sp. NCIMB 9871	CHMO	JM105pKK223-3	(2)
		BL21(DE3)pMM4;	(14)
		JM109(pCM100);	(15)
		JM109(DE3)pET-22b;	
		TOP10[pQR239]	(16)
Acinetobacter sp. SE19	CHMO	Cosmid clone	(17)
Arthrobacter sp. BP2	CHMO	pTrc-His-topo	(18)
Arthrobacter sp. L661	CHMO	BL21(DE3)pETCHMO-His	(19)
Brachymonas petroleovorans	CHMO	-	(20)
Brevibacterium epidermis HCU	CHMO	DH10BpPCB3	(21)
B. oxydans IH-35A	BVMO	pSD80	Unpublished
Comamonas sp. NCIMB 9872	CPMO	DH5α(pCMP201)	(22)
		JM109(DE3)pET-22b	(23,24)
Gordonia sp. TY-5	ACMO	Rosetta(DE3)pEACMA	(25)
Mycobacterium tuberculosis H37Rv	BVMO	pDB1 (Rv0892)	(11)
	EtaA	pDB2 (Rv0565c)	
		pDB3 (rv3854c)	(12)
		pDB4 (Rv1393c	
		pDB5 (Rv3049c)	(26)
		pDB6 (Rv3083)	
Nocardia sp. NCIMB 11399	CHMO	pSD80	Unpublished
Pseudomonas aeruginosa PAO1	CHMO	Rosetta Gami (DE3)pGEX-KG	(27)
P. fluorescens ACB	HAPMO	BL21(DE3)pLysS; pET-5a	(28)
P. fluorescens DSM 50106	BmoF1 (alkane)	JM109(pABE)	(29)
P. putida JD1	HAPMO	RosettapET22b(+)PpJD14HAPMO	(30)

Continued on next page.

345

Table 1. (Continued). Growing list of heterologously expressed BVMOs

P. putida KT2440	BVMO	JM109pJOE-KT2440pGro7	*(31)*
P. putida NCIMB 10007	MO2 : OTEMO	MO2 in pSD80	Unpublished
Pseudomonas sp. HI-70	CPDMO	BL21(pCD201)	*(32)*
P. veronii MEK700	MEKMO	BL21pAM262	*(33,34)*
Rhodococcus jostii RHA1	BVMO (MO 1-23)	Rosetta 2 (DE3)pLysS pET-YSBLIC-3C (MO 1-23)	*(13)*
R. rhodocrous IFO3338	SMO	BL21(DE3)pSMO-EX	*(35)*
R. ruber SC1	CDMO	pDCQ7 or pDCQ8	*(36)*
Rhodococcus sp. HI-31	CHMO	pSDRmChnB1	*(37)*
Rhodococcus sp. Phi 1	CHMO	pTrc-His-topo	*(18)*
Rhodococcus sp. Phi2	CHMO	pTrc-His-topo	*(18)*
Rhocococcus sp. TK6	CHMO	BL21(DE3) pET21a(+)	*(38)*
Streptomyces aculeolatus NRRL18422	BVMO (ORF8HoxC)	Gene cluster	*(39)*
S. avermitilis	PtlE	BL21(DE3)pET31bMox	*(40)*
Streptomyces coelicolor A3(2)	CHMO (MO 96) CHMO (MO 103)	Rosetta Gami (DE3)pGEX-KG)	*(27)*
Streptomyces sp. Eco86	BVMO (ORF13HoxC)	Gene cluster	*(39)*
Thermobifida fusca YX	PAMO	TOP10pPAMO	*(41,42)*
Xanthobacter sp. ZL5	CHMO	BL21(DE3)p11X5.1(pET11a)	*(43,44)*

Type II BVMO (class C FMO)

Pseudomonas putida NCIMB 10007	MO1 : 2,5-1-DKCMO 2,5-2-DKCMO 3,6-DKCMO	2,5-1 in pSD80 2,5-2 in pSD80 3,6 in pSD80	Unpublished

Atypical BVMO

Streptomyces argillaceus	MtmOIV	BL21(DE3)pLysS PRSETb Histag	*(45,46)*

ACMO: acetone monooxygenase; CDMO: cyclodecanone monooxygenase; CHMO: cyclohexanone monooxygenase; CPDMO: cyclopentadecanone monooxygenase; CPMO: cyclopentanone monooxygenase; DKCMO: diketocamphane monooxygenase; EtaA: ethionamide monooxygenase; HAPMO: 4-hydroxyacetophenone monooxygenase; MEKMO: methylethylketone monooxygenase; MtmOIV: mithramycin monooxygenase; OTEMO: 2-oxo-Δ3-4,5,5-trimethylcyclopentynyl-acetylCoA monooxygenase; PAMO: phenylacetone monooxygenase; PtlE: 1-deoxy-11-oxopentalenic acid monooxygenase; SMO: steroid monooxygenase.

In the superfamily of flavin monooxygenases (FMO), Type I and II BVMOs have been referred to as class B and C, respectively (*6*).

Notable examples of Type I BVMOs that have been characterized to some extent, but not yet cloned, are monocyclic monoterpene ketone monooxygenase

(MMKMO) of *Rhodococcus erythropolis* DCL14 (*7*); a 74-kDa CHMO of a black fungus, *Exophiala jeanselmei* strain KUFI-6N (*8*); and those involved in steroid metabolism in *Penicillium lilacinum* AM83 and AM111 (*9, 10*). The microbial genome database is a rich ground for potential BVMOs, particularly those of Type I. The multiplicity of BVMOs in *Mycobacterium tuberculosis* H37Rv (*11, 12*) and *Rhodococcus jostii* strain RHA1 (*13*), 6 in the former and 23 in the latter, provides two of the first opportunities for researchers to explore the possible novelty of the biocatalysts in organic synthesis (*11, 13*).

Thus far, all the identified Type I BVMOs including those that are predicted from the sequenced genomes (>400) seem to belong to bacteria and fungi only, being absent in archaebacteria, plants, animal and the human genome (*3n*). Actinomycetes such as the nocardiaform *Rhodococcus* appear to harbor the greatest number in a given genome, perhaps a reflection of the presence of linear plasmids which encode various catabolic traits or production of secondary metabolites.

Prototypical *Acinetobacter* CHMO and Its Correct Sequence

The first DNA sequence and predicted amino acid sequence of CHMO was obtained from *Acinetobacter* sp. strain NCIMB 9871 (*2*). This sequence, however, was found to be incorrect by mass spectrometry protein sequencing of both the recombinant and native forms of CHMO from strain 9871 (*47*). Instead, the sequence of the CHMO-encoding gene (*chnB*) (543 amino acids) deposited by Iwaki and coworkers in 1999 (*15*) under GenBank accession number AB006902 was certified to be the correct one. Care should be taken when sequence comparison of BVMOs is concerned and proper referencing to the *bone fide* CHMO sequence is encouraged. The CHMO-encoding gene (*chnB*) of strain 9871 was established as part of the cyclohexanol degradation pathway. A comparison of this full gene complement to the same pathway in the related SE19 strain (*17*) has been carried out by Iwaki and coworkers (*48*).

First Crystal Structures of BVMOs

For many years, despite intense interest, structural studies of BVMOs were not forthcoming; however, a breakthrough came in 2004 with the publication of the crystal structure of a thermostable and monomeric BVMO, phenylacetone monooxygenase (PAMO) (Figure 2A) (*41*). This structure of the FAD-bound, basal state of the enzyme revealed for the first time its overall fold, and made it necessary to propose that domain rotations and conformational changes must be occurring during catalysis. More recently, two distinct NADP+-bound structures of a new CHMO from *Rhodococcus* sp. HI-31, representing the prototypical CHMO, were solved (Figure 2B) (*37*). This monomeric enzyme is a close homologue to both the canonical CHMO from *Acinetobacter* sp. NCIMB 9871 (55% sequence identity) and PAMO (42% identity). The structures provided crucial insights into the complex domain motions that are taking place during the BVMO catalytic cycle. At about the same time, the structure of the atypical BVMO, MtmOIV, was also elucidated (Figure 2C) (*46*). Even though it functions as a BVMO, its structure is completely different from PAMO or CHMO. MtmOIV does not contain the fingerprint sequence of Type I BVMOs, and is also unrelated to Type II BVMOs. Sequence identity with PAMO, for example, is merely 8%. This enzyme is in fact more closely related by structure and sequence to the FAD-dependent hydroxylases, PgaE and CabE of the glucocorticoid receptor (GR$_2$) subfamily, involved in the biosynthetic pathways of aromatic polyketides (angucycline) in *Streptomyces* spp. (*46*).

Overall Structure of BVMOs

The initial structure of PAMO (*41*) (Figure 2A) revealed a two-domain structure, where both domains have a dinucleotide binding, or Rossmann, fold. The larger domain contains the FAD binding site which is positioned such that the isoalloxazine ring is located at the interface of the two domains (*41*). The crystal structures of CHMO (*37*) revealed a very similar fold, and demonstrated that the second domain is the location of the NADP$^+$ cofactor binding site. In fact, careful analysis of domain motions indicated that this second domain should actually be considered to be composed of two domains: the NADPH binding domain, and a helical domain (Figure 3A). The NADPH cofactor binds to the NADPH binding domain, but sits in the cavity between the NADPH and FAD binding domains (Figure 3A).

In contrast, the FAD-binding domain in the MtmOIV structure is a large central domain sandwiched by a middle domain, which is expected to be involved in NADPH-binding, and a C-terminal domain (Figure 2C) (*46*). This represents a markedly different fold for an enzyme that is mechanistically quite similar to the traditional BVMOs. Another difference lies in the binding of FAD. While in the standard BVMOs, the FAD is buried deep in the FAD-binding domain (*37, 41, 49*), MtmOIV has FAD bound to the surface of the protein. This is likely responsible for the much lower FAD content observed with MtmOIV than with other BVMOs (*46*).

Structural Aspects of Standard BVMO-Mediated Catalysis

While the catalytic mechanism of BVMOs has been well-characterized from a kinetic standpoint (*50, 51*), the crystal structures of BVMOs have been instrumental in understanding how these enzymes are able to carry out such a complex mechanism. In addition to learning about the overall fold of the enzyme, the structure of PAMO revealed the binding site of FAD, and by extension, the putative active site of the enzyme (*41*). This also revealed the key role of Arg 337 (PAMO numbering) in catalysis, as this residue was observed in two conformations. One conformation, called the "in" position, would allow this residue to stabilize the peroxyanion intermediate (*41*); however, Arg 337 would have to move into the "out" position to allow NADPH to come in and reduce FAD (*41*). This was corroborated in one of the two structures of CHMO, which shows Arg 329 (*Rhodococcus* sp. HI-31 CHMO numbering – henceforth referred to as CHMO numbering) in a position similar to the "out" position, thereby allowing

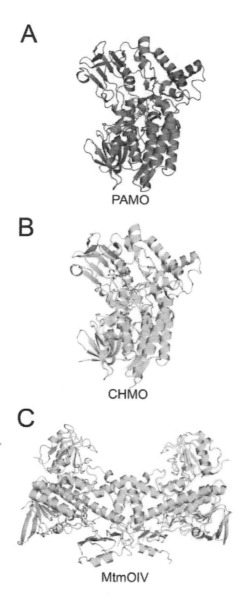

A

PAMO

B

CHMO

C

MtmOIV

Figure 2. Comparison of the three BVMO crystal structures. (A) Overall structure of PAMO, shown in blue. (B) Overall structure of the closed conformation of CHMO, shown in green. (C) Overall structure of MtmOIV, with one protomer shown in yellow and the other shown in wheat. (see color insert)

NADP$^+$ to bind (*37*). In the second structure of CHMO, Arg 329 occupies a third, "push" position where the nicotinamide head is pushed deeper into the enzyme, allowing Arg 329 to stabilize either the peroxyanion intermediate or the Criegee intermediate (*37*).

The CHMO structures revealed that the fold for standard BVMOs should be divided into three domains (Figure 3A), as indicated by the domain motions

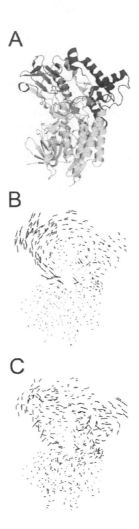

Figure 3. Domain rotations in Type I BVMOs. (A) Overall structure of the closed conformation of CHMO, colored by domain. The FAD-binding domain is shown in wheat, the NADPH-binding domain in blue, the helical domain in brown, the linker regions in green, the key mobile, catalytic loop in orange, and the BVMO signature sequence in yellow. (B) Equivalent atom representation of the open and closed conformations of CHMO. After aligning the structures using the FAD-binding domain, vectors were drawn between equivalent C-α atoms. (C) Same as in (B), but comparing the open conformation of CHMO with PAMO (37). (see color insert)

observed when comparing the various crystal structures. Two distinct enzyme conformations of CHMO were observed: the open and closed conformations (*37*). When comparing the open and closed conformations of CHMO, a subtle but significant rotation of the NADPH-binding domain relative to the helical and FAD-binding domains can be observed (Figure 3B). A comparison of the PAMO structure with the open conformation of CHMO, however, shows the

NADPH-binding and helical domains moving as a single unit (Figure 3C) (*37*). These structures reveal that the binding of substrates/cofactors induces significant domain movements. It had already been speculated, based on the PAMO structure, that such domain motions must be necessary to effect the complex reaction mechanism of BVMOs (*41*).

The two structures of CHMO also revealed a putative substrate binding site that differed in each structure. In the closed conformation (Figure 4A), a well-defined substrate binding pocket is observed close to the isoalloxazine ring of FAD, but is blocked off from the bulk solvent by the NADP$^+$ (*37*). In contrast, the open conformation has a less well-defined substrate binding site that is open to the solvent (Figure 4B) (*37*). In addition to the overall domain motions, this re-organization of the substrate binding site is likely controlled by a shift in the position of NADP$^+$, as well as the ordering of the loop composed of residues 487-504 (CHMO numbering) (*37*). In PAMO, this region sticks out into solvent and is held there through crystal packing interactions, suggesting that these residues form a disordered, solvent-exposed loop in solution (*41*). This is corroborated in the open conformation of CHMO, which shows no clear density corresponding to these residues (*37*). In the closed conformation, however, this loop folds back in towards the NADP$^+$ co-factor, causing the previously mentioned shift in the position of NADP$^+$. Residue W492 (CHMO numbering) contacts the 2' hydroxyl of the nicotinamide ribose, and mutation of this residue reduces the activity to only 14% of the wild type enzyme (*37*). As this loop forms part of the putative substrate binding site, this strongly suggests that the position of this loop plays a critical role in both catalysis and the domain organization of the enzyme (*37*).

Using the substrate binding site observed in the closed conformation of CHMO, it was possible to model the Criegee intermediate into the active site to provide some clues regarding the key steps in the catalytic cycle (Figure 4C) (*37*). The mechanism is discussed below.

A New Role of the BVMO Signature Sequence

Mutation of the central histidine in the signature sequence of HAPMO had a dramatic effect on its enzyme activity (*4*). Interestingly, in the structures of CHMO and PAMO, this signature sequence lies in part of the NADPH-binding domain that does not directly contact either cofactor or the substrate binding site. As this precludes the direct involvement of the signature sequence in catalysis, it suggests that this sequence is involved in stabilizing the NADPH-binding domain (*37*), particularly through the aromatic residues at either end of the sequence. Furthermore, the invariant histidine 166 (CHMO numbering) (Figure 4D) may play a key role in the domain movements of the enzyme. In the PAMO structure, this histidine is pointing out into space (*41*). In the NADP$^+$-bound structures of CHMO, however, it forms a hydrogen bond with a glycine in a loop region that links the NADPH-binding and FAD-binding domains. It also appears to be involved in positioning NADP$^+$, as that linker contacts the cofactor directly (*37*, *49*). This could mean that the histidine acts as a switch, causing the enzyme to move in response to the presence of the NADPH cofactor (*37*).

Comparison to the Structures of FMOs

In addition to being able to catalyze the BV oxidation reaction via the Criegee intermediate, BVMOs are also able to oxidize non-ketone, nucleophilic substrates (*52*). The catalysis of this alternative reaction, which does not employ the Criegee intermediate, resembles that performed by FMOs (*53*). It is not surprising, therefore, that the structures of FMOs are strikingly similar to those of CHMO and PAMO. For example, an FMO from the methylotrophic bacterium *Methylophaga* sp. SK1 (mFMO) also displays a similar domain architecture (Figure 5), but with the NADPH-binding and FAD-binding domains somewhat rotated in comparison to CHMO and PAMO (*49*). As mFMO is a dimer in solution, it is possible that this slightly altered architecture could be seen in the multimeric BVMOs, such as HAPMO (dimeric) (*54*), acetone monooxygenase (tetrameric) (*25*), or CPMO (tetrameric) (*55*). Intriguingly, the $NADP^+$ retains its role of blocking access to the isoalloxazine ring of FAD in mFMO, as in the case of BVMOs (*49*). The authors also speculated on the ability of the enzyme to undergo domain movements, representing another similarity to the BVMOs (*49*).

Even with the limited number of available BVMO structures, it is clear that our understanding of the structure-function relationship of BVMOs has increased dramatically. This data has given us a greater understanding of how these enzymes can perform the complex enzyme mechanism that has been otherwise characterized using enzyme kinetics; however, many key questions still remain, including how the domain movements change throughout the catalytic cycle, as well as how to reconcile the substrate specificity, regiospecificity, and enantiospecificity with the structural data.

Mechanistic Studies of PAMO

Up to 2008, two kinetic studies of CHMO were conducted, one by Walsh and coworkers in 1982 (*50*) and a more comprehensive study nearly 20 years later (*51*). Recently, the group of Fraaije investigated the reaction mechanism of the thermostable PAMO from *T. fusca* by performing a detailed steady-state and pre-steady-state kinetic analysis (*56*). PAMO has been subject of several studies including elucidation of its crystal structure as discussed above (*41*), substrate profiling (*57*), and enzyme engineering (*58*).

The postulated mechanism for the oxidation of phenylacetone by PAMO is shown in Figure 6, and at a first glance is very similar to the proposed catalytic cycle of CHMO. First NADPH binds to PAMO and FAD is reduced by the (*R*)-hydrogen of the nicotinamide, which was established by the observation of a kinetic isotope effect using deuterated NADPH. In contrast, the available crystal structures of CHMO suggest a pro *S*-hydride transfer (*37*). Similar to CHMO and liver microsomal FMO (*59*), PAMO was found to stabilize the reactive C4a-peroxyflavin intermediate, produced upon reaction of reduced FAD with molecular oxygen. The peroxy species reacts with the substrate to form an intermediate which could not be unequivocally assigned in this study, representing either the Criegee intermediate or a C4a-hydroxyflavin form (Figure 6). The decay of this unassigned intermediate to fully oxidized PAMO was found

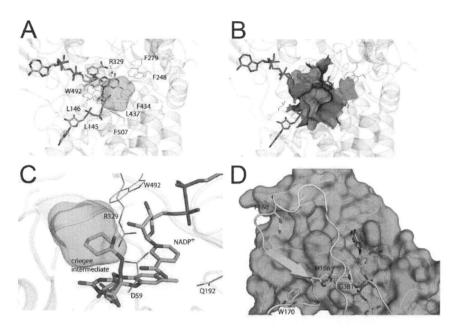

Figure 4. (A) The putative substrate binding cavity as seen in the closed conformation of CHMO, formed primarily by hydrophobic residues. (B) In comparison, the larger binding cavity in the open conformation caused by the movement of the loop formed by residues 487-504 can be seen. (C) Model of the Criegee intermediate based on the structures of CHMO. (D) The BVMO signature sequence (yellow) is shown here to be anchored by F160 and W170, which interacts with hydrophobic pockets in the NADPH-binding domain. H166 is shown to be interacting with G381, part of a key linker (cyan) that connects the FAD- and NADPH-binding domains (37). (see color insert)

to be the rate limiting step of the catalytic cycle. The release of NADP$^+$ from the oxidized coenzyme was relatively fast in the case of PAMO, which differs considerably from the results with CHMO, where the dissociation of NADP$^+$ step was found to be rate-determining in catalysis. In addition, the group of Fraaije prepared two mutant enzymes, R337A and R337K, which were still able to stabilize the peroxyflavin intermediate, but were unable to oxidize phenylacetone or benzyl methyl sulfide. This demonstrates the crucial role of R337 in PAMO for biocatalysis, a conserved residue present in other BVMOs sequences as well (e. g. R329 in CHMO).

Diversity of New BVMOs

To increase the diversity of the BVMO "toolkit", new or novel BVMOs with varying substrate specificities, high regioselectivities, or excellent enantio- and diastereoselectivities were sought.

Bornscheuer and coworkers reported the cloning, expression, characterization, and biocatalytic investigation of a 4-hydroxyacetophenone monooxygenase (HAPMO) from *Pseudomonas putida* JD1 *(30)*. Although this is the third described BVMO (after HAPMO of *P. fluorescens* ACB and PAMO from *T. fusca*) that preferentially oxidizes arylaliphatic ketones, the newly described HAPMO was able to oxidize a wide range of *para*-substituted aromatic ketones including 2-acetyl pyrrole, aliphatic ketones, methyl-4-tolylsulfide, and β-hydroxy ketones. Noteworthy is the kinetic resolution of 3-phenyl-2-butanone, which was achieved in excellent enantioselectivity (*E* value >200) as shown in Figure 7.

Altenbuchner and coworkers further investigated the methylethylketone (MEK) degradation pathway of *Pseudomonas veronii* MEK700 (*33*), left off by Onaca and coworkers in 2007 (*34*), by cloning the BVMO-encoding gene designated *mekA*. Expressed as a C-terminal strep-tag fusion protein in *E. coli* under the control of a L-rhamnose inducible promoter, MekA was purified. Substrate profiling revealed high activity towards linear aliphatic ketones, whereas cyclic and aromatic ketones showed lower conversion rates. Purified MekA was also used for the kinetic resolution of aliphatic β-hydroxy ketones with 4-hydroxy-2-octanone as the most suitable substrate (*E* value 22). A possible curiosity item of MekA is its apparent dual usage of NADPH and NADH as cofactors, the latter still imparting 45% activity with MEK as substrate.

Studies on the biosynthesis of pentalenolactone (**3**), a sesquiterpenoid antibiotic with activity against both Gram-positive and Gram-negative strains of bacteria, led to the discovery of a new BVMO, named PtlE, from *Streptomyces avermitilis* (*40*). With reference to characterized BVMOs, the 594-amino acid sequence of PtlE is 68% similar (52% identical) to the cyclopentadecanone monooxygenase (CpdB; CPDMO) of *Pseudomonas* sp. HI-70 (*32*). PtlE expressed in *E. coli* BL21(DE3) was found to oxidize 1-deoxy-11-oxopentalenic acid (**1**) to neopentalenolactone D (**4**), and not, as expected, to pentalenolactone D (**2**) as shown in Figure 8. A new branch of the pentalenolactone family tree that consisted of 9 members, including CpdB, was proposed. PtlE from *S. avermitilis* is one of very few Type I BVMOs which has been assigned a specific biosynthetic role and a defined natural substrate, the primary example being that of MtmOIV (*45, 46*).

Genome Mining: A Good Business

The fact that every sequenced microbial genome leaves at least 50% of its coding capacity uncharacterized has prompted the prospect of genome mining as an efficient, if not economical, way for new biocatalyst discovery. A prime example of this approach in new BVMO discovery is the characterization of the thermostable PAMO that led to its 3D structure (*41*). Grogan and coworkers took advantage of two bacterial genomes, viz. *Mycobacterium tuberculosis* H34Rv and *Rhodococcus jostii* RHA1, that contain a multitude of potential BVMO sequences (*11, 13*). The latest effort was the cloning of the entire putative BVMO gene set, 23 of them that share 12 to 43% sequence identity with the *Acinetobacter* CHMO. It is noteworthy that a previously developed "ligation independent cloning" (LIC) strategy was used to generate the multiple clones in a high throughput mode. The end result was the successful expression as soluble proteins of 13 candidate genes that enabled substrate profiling using both chiral and achiral substrates

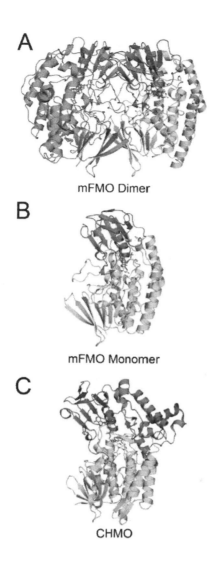

A

mFMO Dimer

B

mFMO Monomer

C

CHMO

Figure 5. Comparison of the crystal structure of mFMO with the open conformation of CHMO. (A) The overall structure of mFMO. The FAD- and NADPH-binding domains of one protomer are shown in wheat and blue, respectively. The second protomer in the dimeric mFMO is shown in grey. (B) One protomer of mFMO, using the same colour scheme as in (A). (C) The overall structure of the open conformation of CHMO, using the same colour scheme as in (A). (see color insert)

(13). Diversity in activity, regio-, and enantioselectivty among the BVMOs for the transformation of chiral substrates was observed. Furthermore, a correlation between the biocatalysts' sequences and their selectivities was established. This also allowed the identification of sequence motifs that are specific to subgroups of the BVMOs from *R. jostii* and other organisms.

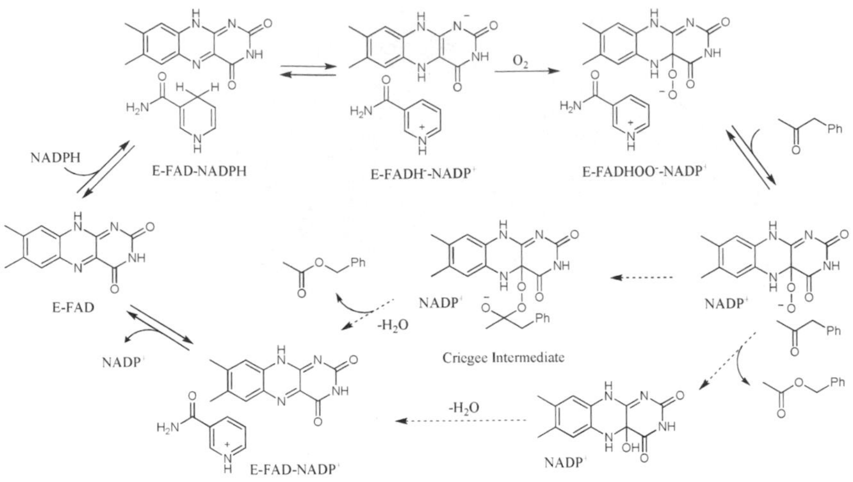

Figure 6. Proposed mechanism of PAMO (56).

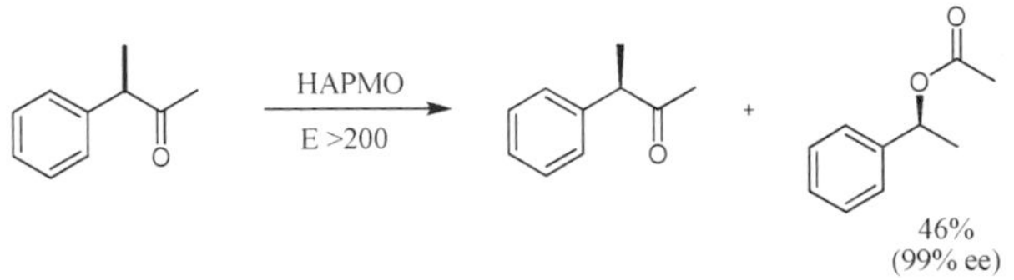

Figure 7. Kinetic resolution of 3-phenyl-2-butanone (30).

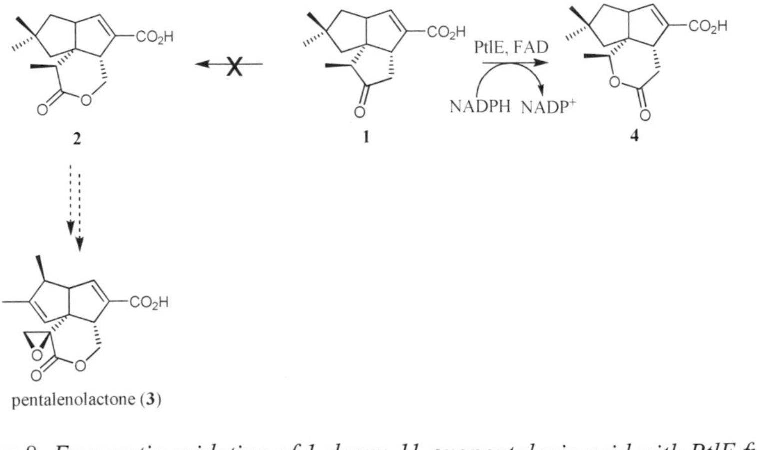

Figure 8. Enzymatic oxidation of 1-deoxy-11-oxopentalenic acid with PtlE from Streptomyces avermitilis to neopentalenolactone but not pentalenolactone (40).

Protein Engineering

Biocatalysts suitable for industrial applications are characterized by a number of features, including high process stability, high enantio-, regio-, and chemoselectivity, and high catalytic turnover numbers for the specific reaction being applied to. As most of the natural occurring biocatalysts including BVMOs lack one or more of those characteristics, improvements are needed and can be relatively easily achieved by protein engineering, either by rational design or variations of directed evolution.

In this domain, Reetz is the *man* one should not *fret* about. Readers are referred to his inspiring Perspective on the occasion of the 2009 Arthur C. Cope Award (*60*). One of his group's latest contribution in method development is making use of a reduced "amino acid alphabet" for simultaneous randomization of four amino acids (residues 441 to 444), found in a previously identified loop in the PAMO structure (aka the bulge) that influenced substrate acceptance and enantioselectivity. Judicious sequence alignment of homologous enzymes was the key to reducing the codon choice (alphabet) thereby minimizing the otherwise laborious screening effort. The research was aimed at expanding the narrow substrate spectrum of PAMO that is limited to phenylacetone and its derivatives (*61*). As a result, the much reduced library of mutants was screened for the effectiveness in the kinetic resolution of 2-phenyl cyclohexanone (very slow reaction with an E-value of 1.2 (*S*) for wild type PAMO). A significant improvement in both the activity (reaction rate) as well as enantioselectivity (E-value (R) of up to 70) was observed for a number of mutants.

Although the new mutagenesis approach led to a significant improvement of the biocatalytic activity of PAMO, the substrate scope was still not considerably broadened. The "battle of the bulge" continued with a new strategy (*62*). This new strategy combined information obtained from sequence alignment of eight BVMOs, the known structure of PAMO, and induced-fit docking experiments. In addition, highly conserved prolines (position 440 and 437) adjacent to the loop were targeted for saturation mutagenesis. Two focused libraries were produced and were evaluated by screening about 200 mutants each for the ability to oxidize *rac*-2-ethylcyclohexanone (Table 2). Saturation mutagenesis at position 437 resulted in only inactive mutants whereas mutagenesis at position 440 led to the identification of numerous active and highly selective mutants (E-value >200). Seven mutants showed particularly high activity and those mutant enzymes were further evaluated for their potential in the resolution of 2-substituted cyclohexanones, showing promising E-values from 3.4 to >200 depending on the substrate as well as on the mutant applied. The success of the "proline hypothesis" led to the suggestion of "proline scanning" a term after alanine scanning that is commonly employed as a strategy in saturation mutagenesis.

Kirschner and Bornscheuer applied a directed evolution strategy to the improvement of BmoF1 from *Pseudomonas fluorescens* DSM 50106 (*63*). Previously this biocatalyst was successfully applied to the kinetic resolution of 4-hydroxy-2-ketones. Ketones were oxidized to the corresponding hydroxyalkyl acetates with an E value ~ 55. Error-prone polymerase chain reaction (epPCR) was used to introduce random mutations and the first round of mutants (>3500 clones) were screened for improved activity using a modified adrenaline assay as shown in Figure 9. Second round mutants were created by combining beneficial mutations from the first round using consecutive cycles of QuikChange® site-directed mutagenesis resulting in a double mutant (H51L/S136L) with improved enantioselectivity (E ~ 86).

Latest Trends in Cofactor Recycling

Although whole-cell mediated BVMO oxidations have been shown to be extremely effective for the production of lactones and esters, the use of isolated enzymes would be advantageous in some cases. The use of isolated enzymes for BVMO catalyzed reactions is quite costly, since stoichiometric amounts of NAD(P)H are required and therefore the application of isolated BVMOs has been limited to small scale. The most applied solution to this problem is the coupling of the main reaction with a second ancillary reaction, capable of recycling the cofactor. Typically, NAD(P)$^+$-dependent dehydrogenases are applied to regenerate the cofactor at the expense of a cheaper sacrificial substrate.

An innovative way to effect the reduction of NADP$^+$ was introduced by the groups of Mihovilovic and Fraaije (64). They successfully prepared fusion proteins of a number of representative BVMOs (CHMO, CPMO, and PAMO) covalently linked to a NADPH-regenerating phosphite dehydrogenase (PTDH). The oxidation of phosphite to phosphate with PTDH was chosen since this reaction is basically irreversible and the enzyme is highly selective for phosphate, which eliminates the possibility of any unwanted reactions with the main substrate. The constructed fusion proteins were applied to the oxidation of a wide range of substrates and showed similar results (regioselectivities, enatioselectivities) compared to those obtained with individual enzymes. Interestingly, neither NADP$^+$ nor NADPH had to be added to crude extracts of cells when used for biotransformations, since sufficient cofactors were released during the lyses of the *E. coli* hosts. The fusion proteins were referred to as self-sufficient BVMOs, analogous to the self-sufficient cytochromes P450 in which the electron supply unit (reductase) is integrated with the hydroxylating component in one polypeptide.

Subsequently, an improved generation of the self-sufficient BVMO was constructed in order to address the instability of the PTDH over time (65). To achieve this, a thermostable variant of PTDH was used, and its codon use was optimized for expression in *E. coli*. A His-tag was also added to the N-terminus of the BVMO. This improvement led to biocatalytic profiles that were similar to the individual enzymes, and cell free extracts were successfully applied to the conversion of several ketones without the loss of activity.

A very different strategy for the cofactor regeneration was introduced by Reetz and coworkers in 2007 (66). The strategy was to reduce flavin monooxygenases by visible light using ethylenediaminetetraacetate (EDTA) as a sacrificial electron donor (Figure 10). Using a mutant of PAMO (PAMO-P3) as the BVMO, the enzyme was shown to be stable for several hours and its selectivity was not affected by the reaction conditions. However, the observed catalytic activity of the system was one to two magnitudes lower than of whole cells or *in-vitro* systems with conventional cofactor recycling systems. In a subsequent publication, the group investigated the scope and limitations of the light driven biocatalytic oxidation (67). The major limitations of the regenerating system were identified to be an uncoupling of the light-driven flavin reduction reaction from the regeneration of the prosthetic group of the monooxygenases and slow electron transfer kinetics

Table 2. Kinetic resolution of 2-ethylcyclohexanone with generated mutants of PAMO (*62*)

Mutant	E-value	K_m (mM)	k_{cat} (s^{-1})	k_{cat}/K_m $(M^{-1} s^{-1})$
P440F	26	0.89	1.2	1300
P440L	>200	1.6	0.72	450
P440I	>200	2.7	0.66	240
P440N	>200	2.2	1.5	680
P440H	34	1.0	0.83	830
P440W	>200	1.3	1.3	1000
P440Y	95	1.9	1.1	580

E-value: enantioselectivity value; K_m: Michaelis constant; k_{cat}:turnover number

presumably due to a hindered steric access of the mediator to the PAMO-bound FAD.

Substrate Profiling

This section of the review will discuss the latest results of substrate profiling using clones of some naturally occurring BVMOs. In most cases, these old and new BVMOs have not been genetically modified.

The biocatalytic profile of a BVMO from *Xanthobacter* sp. ZL5 was evaluated by Mihovilovic and coworkers (*68*). Whole-cell biotransformations were carried out on various 2-substituted cyclohexanones, fused cyclobutanones, and terpenone derivatives. The observed regio- and enantioselctivities were nearly identical to results obtained with the prototypical *Acinetobacter* CHMO. This was interpreted as a reflection of the 58% sequence similarity between the two proteins. In a subsequent publication, the same BVMO was applied to the stereoselective desymmetrization of *meso* substrates (*69*). In addition to the conversion of a wide range of prochiral ketones, with high enantioselectivities similar to previous results obtained with the *Acinetobacter* CHMO, the BVMO from *Xanthobacter* sp. ZL5 was able to oxidize structurally demanding ketones with high enantioselectivites. It is noteworthy to mention the epoxidation reaction of a non-activated C=C bond catalyzed by this BVMO as shown in Figure 11.

Prochiral 2,6-substituted perhydropyran type ketones as suitable substrates for BVMOs were intensively studied by the same group. Eight recombinant *E. coli* cells expressing different bacterial BVMOs were applied to the enzymatic oxidation of eight heterocyclic ketones (*70*). CHMO from *Acinetobacter* proved to be the most effective catalyst, being able to convert four of the eight substrates in moderate to excellent yield and in the case of prochiral substrates, products with high enantiomeric excesses (98 to 99.5%) were obtained. Similar to the trend of previous publications, the authors observed two distinct groups

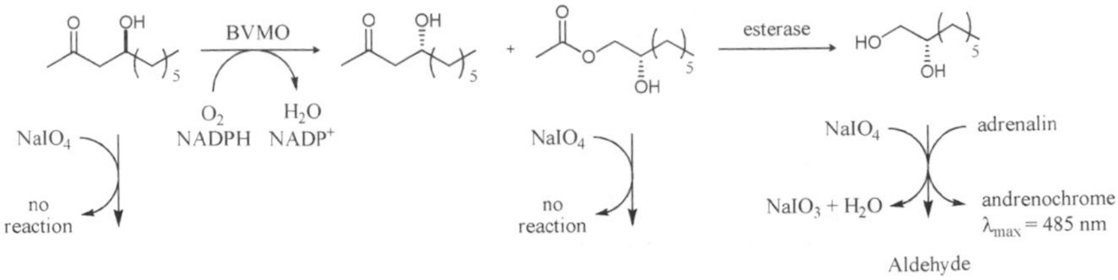

Figure 9. Adrenaline assay for the detection of BVMO activity in the conversion of 4-hydroxy-2-ketones (63).

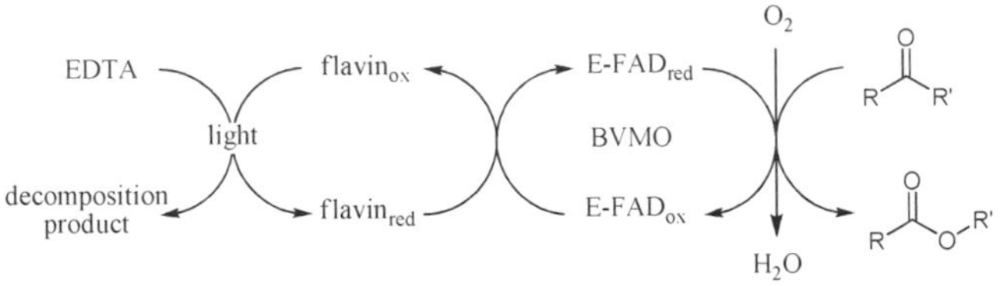

Figure 10. Simplified light driven regeneration approach for BVMO (66).

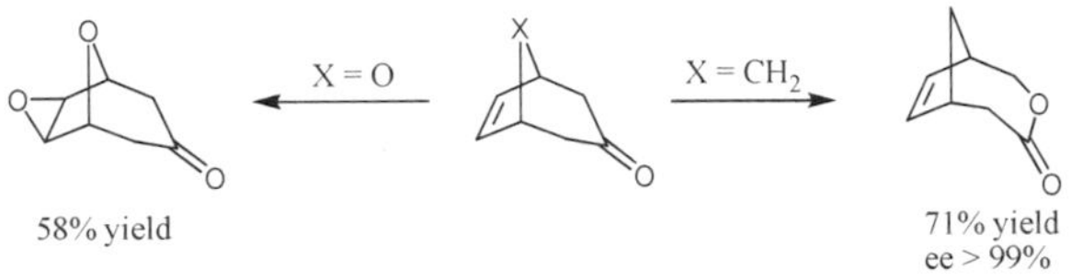

58% yield

71% yield
ee > 99%

Figure 11. Xanthobacter sp. ZL5 BVMO mediated oxidations of bridged ketones (69).

of BVMOs, "CHMO-type" and "CPMO-type", with respect to the substrate acceptance. This complementary behavior of BVMOs was also observed in the regiodivergent oxidation of fused cyclic butanones (71). "CHMO-type" BVMOs gave approximately a 1 : 1 ratio of both normal and abnormal lactones in excellent yield and good to excellent enantiomeric excesses, whereas "CPMO-type" BVMOs oxidized the fused ketones nearly exclusively to the "normal" lactones in racemic form as shown in Figure 12.

Gotor and coworkers studied the enantioselective kinetic resolution of four different substituted 3-phenylbutan-2-ones and α-acetylphenylacetonitril in non-conventional media employing purified PAMO as a biocatalyst (72). It was shown that the addition of different amounts of cosolvents to the conventional buffer media could improve the biocatalytic properties of PAMO. Higher conversion rates in the BVMO-catalyzed oxidations were achieved by the addition of 10% ethyl acetate, presumably due to an increase of the substrate solubility. Furthermore, the addition of up to 50% (v/v) methanol led to an increase in the enantioselectivity for the resolution of 3-phenylbutan-2-ones as shown in Table 3.

The oxidation of benzo-fused ketones (tetralones, indanones, and benzocyclobutanones) using three purified BVMOs (HAPMO, PAMO, and M-PAMO, a M446G mutant) was studied by the same group (73). Tetralones proved to be poor substrates for all BVMOs, whereas lactones derived from

360

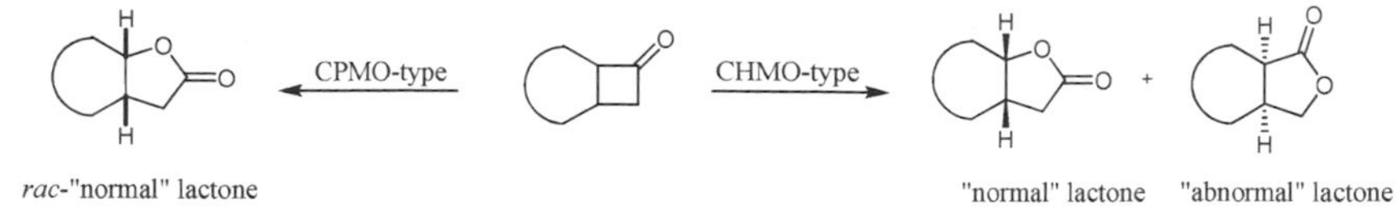

Figure 12. Regiodivergent oxidations of fused cyclobutanones (71).

indanones and benzocyclobutanones were produced in good yield. In the case of 1-indanone, the regiocomplementary formation of lactones was observed, the expected 3,4-dihydrocoumarin was produced by the action of HAPMO, whereas M-PAMO produced the unexpected product, 1-isochromanone as shown in Figure 13. Interestingly, the performance of the isolated BVMO could be significantly improved by the addition of organic solvents (up to 5%) to the aqueous media.

The same three isolated enzymes (HAPMO, PAMO, and M-PAMO) were applied to the kinetic resolution of α-alkyl benzyl ketones (74). High enantioselectivities were achieved by matching the substrate with the right biocatalyst to access (R)-α-alkyl benzyl ketones and (S)-α-alkyl benzyl esters. For example, all three enzymes showed an excellent E value in the resolution of 3-(3-methylphenyl)butan-2-one, whereas ketones with electron withdrawing substituents (nitro or trifluoromethyl) on the aromatic ring were resolved with excellent enantioselectivities by HAPMO and PAMO, for example.

The kinetic resolution of aliphatic acyclic β-hydroxy ketones was achieved using whole-cells overexpressing BVMOs by joint efforts of the groups of Mihovilovic and Bornscheuer (75). 12 different BVMOs from bacterial origin were used and it was found that β-hydroxy 2- and 3-ketones were good substrates for most of the BMVOs, especially for those previously described as cyclic ketone converting enzymes (Table 4). The oxidation of 4-hydroxy-2-ketones with excellent enantioselectivities ($E > 100$) was observed with CHMO from *Acinetobacter*, and protected (acyl, formyl) derivatives were converted with CPMO in excellent enantioselectivities. Interestingly, the resolution of 5-hydroxy-3-ketones led to the predominant formation of the "abnormal" esters when a BVMO from *P. fluorescens* DSM 50106 was applied.

BVMO-Mediated Oxidations of Steroids

In addition to the oxidation of cyclic and aryl aliphatic ketones using BVMOs, there has been a recent revival in the interest of the enzymatic Baeyer-Villiger oxidation of steroids using either isolated enzymes, recombinant whole cells, or fungal strains. Steroids were the first organic molecules which were identified as substrates for an enzymatic BV reaction (76). After a lively interest in the oxidation of steroids in the 1950s, only a few papers have been published covering this area of BVMO technology.

Swizdor and coworkers reported BMVO activity of the fungal strain *Penicillium lilacinum* AM111 on 3β-hydroxy-5-ene-steroids (9). Biotransformation of dehydroepiandrosterone (DHEA, **6**) gave 3β-hydroxy-17a-oxa-D-homo-androst-5-en-17-one (**7**) as the only product in quantitative yield, whereas androstenedione (**9**) yielded testololactone (**10**). The

Table 3. Kinetic resolution of 3-phenyl-2-butanone in the presence of organic solvents (*72*)

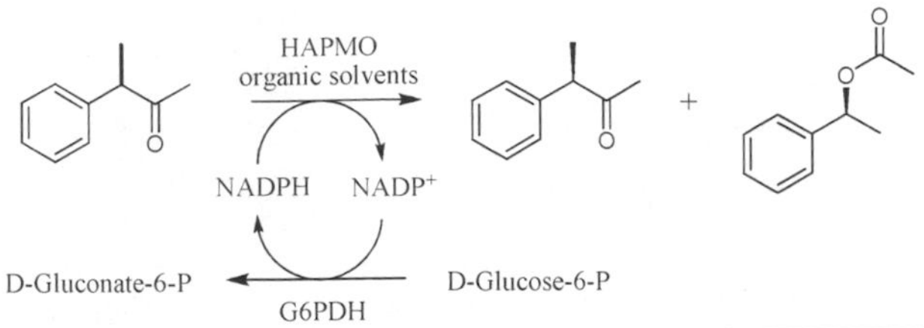

co-solvent	time (h)	ee (%) (R)-substrate	ee (%) (R)-product	conversion (%)	E-value
None	1	44	91	31	32
30% DMSO	3	23	90	20	24
30% DMF	3	7	93	7	30
10% MeOH	1	21	94	18	43
30% MeOH	3	14	95	13	49
50% MeOH	8	15	95	13	46
30% Dioxane	3	60	50	54	5.5
30% EtOH	3	6	13	31	1.5
30% i-PrOH	3	39	54	42	5
30% AcOEt	4	20	86	19	16
30% i-Pr$_2$O	4	17	61	22	5
30% Hexane	4	54	79	40	14

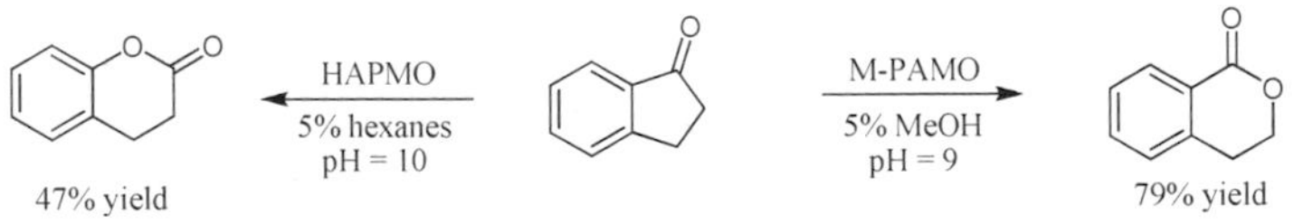

Figure 13. Regioselective oxidation of 2-indanone (73).

oxidation of pregnenolone (**5**) using the fungal strain *Penicillium lilacinum* AM111 gave a mixture of 5 steroids after 30 hours of fermentation time (Figure 14).

Subsequent studies by the same group led to the identification of high BVMO activity in fungus *Penicillium camemberti* AM83 (*10*). This strain was able to convert DHEA (**6**), pregnenolone (**5**), androstenedione, and progesterone (**8**) to testolactone (**10**) in excellent yields in each case. Detailed time-course experiments indicated that the 17β-side chain cleavage and oxidation of the ketones at C-17 were catalyzed by two different substrate induced BVMOs.

The group of Hunter showed that incubation of a series of steroids (progesterone, testosterone acetate, 17β-acetoxy-5α-androstan-3-one, testosterone, and androst-4-en-3,17-dione) with the thermophilic fungus *Myceliophthora thermophilia* gave modifications at all four rings of the steroid nucleus and the C-17β side-chain (*77*). A novel BV oxidation of the A-ring

Table 4. BVMO mediated oxidations of β-hydroxy-2-ketones (75)

Substrate	BVMO	time (h)	conv. (%)	ee_S (%)	ee_P (%)	E-value
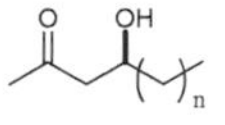 n = 3	CHMO$_{Acineto}$	8	49	91	96	156
	CPMO	3	53	64	56	7
	BVMO$_{Pfl}$	24	46	85	>99	>200
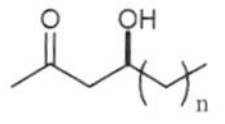 n = 4	CHMO$_{Acineto}$	20	49	96	>99	>200
	CPMO	2	44	48	62	7
	BVMO$_{Pfl}$	8	48	89	98	>200
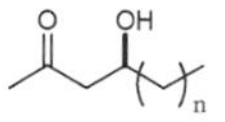 n = 5	CHMO$_{Acineto}$	20	50	92	93	90
	CPMO	2	44	39	50	4
	BVMO$_{Pfl}$	6	41	68	96	100
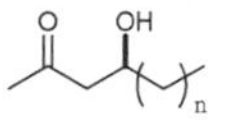 n = 6	CHMO$_{Acineto}$	30	20	13	52	4
	CHMO$_{Rhodo2}$	10	18	17	80	10
	CPMO	2	65	37	20	2

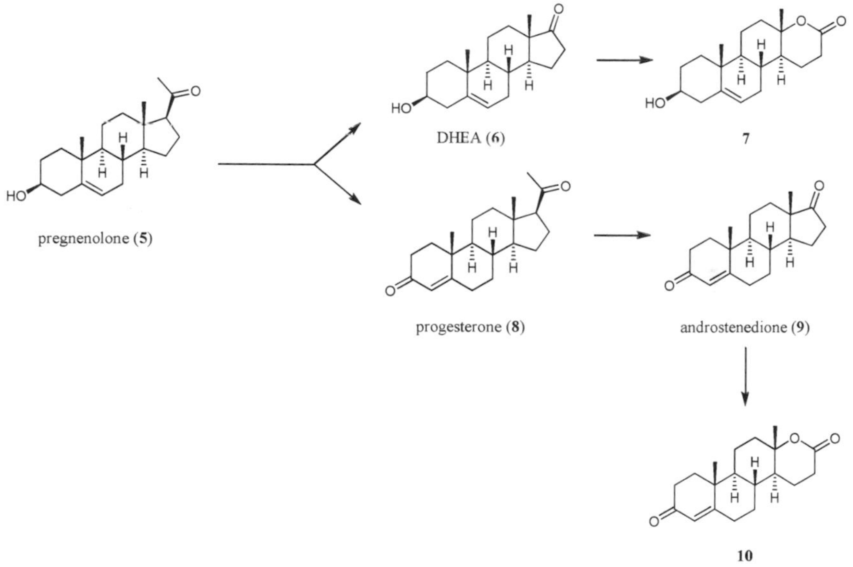

Figure 14. Penicillium lilacinum AM111 mediated BVMO oxidations (9).

was observed when 17β-acetoxy-5α-androstan-3-one was incubated with the thermophilic fungus, in addition to the BV oxidation of the 17β-methylketone moiety of progesterone. Immediate saponification of the produced 7-membered lactone by a presumptive lactone hydrolase in the fermentation media yielded 4-hydroxy-3,4-seco-pregn-20-one-3-oic acid, whose structure was confirmed by X-ray crystallography.

363

Since the cloning of a steroid monooxygenase-encoding gene from *R. rhodocrous* that catalyzed the BV oxidation of progesterone to testosterone acetate reported in 1999 (*35*), the most recent report of steroid oxidations using a cloned BVMO was published by Ottolina and coworkers (*78*). Cyclopentadecanone monooxygenase (CPDMO) from *Pseudomonas* sp. strain HI-70 was chosen as biocatalyst, since it was able to oxidize a broad range of compounds, including large cyclic and bicyclic ketones. Among 33 steroidal substrates screened, CPDMO did not oxidize exocyclic ketones of pregnene steroids and steroids with a keto group in the C-ring. However, numerous A and D ring ketosteroids were accepted as substrates and oxidized to the corresponding lactones with full control of regiochemistry. A 42% product yield was achieved in the case of 3β-hydroxy 17α-oxa-17α-homo-androst-5-ene-17-one. In addition, it was shown that both the isolated CPDMO with *in situ* NADPH regeneration as well as whole cells overexpressing CPDMO were suitable biocatalysts for the various transformations.

Bioprocess Development

Despite the fact that the BVMO technology offers numerous biocatalysts with varying substrate spectra, it has not yet been applied to industrial processes. Most BVMO reactions have been performed on mg–scale and only few publications have addressed the scale up and/or the process optimization of the reactions. Key process bottlenecks in a preparative BVMO oxidation have been identified as substrate and product inhibition, cell toxicity, oxygen supply limitations, and substrate solubility. The following overview will discuss the latest trends in bioprocess development.

Flow cytometry was applied in an elegant way by Woodley and coworkers to monitor the effects of substrate as well as product concentration on whole cells overexpressing the *Acinetobacter* CHMO (*79*). Both the physiological and metabolic status of the whole-cell biocatalyst with respect to the concentration of organic substances, bicyclo[3.2.0]hept-2-en-6-one (**11**) as substrate and (-)-(1*R*,5*S*)-3-oxabicyclo[3.3.0]oct-6-en-2-one (**13**) as product, were studied. The results demonstrated that high substrate concentrations (above 3 g/L) was associated with cell damage that occurred rapidly (within 30 minutes, 73% cell viability) and was considerably more harmful than high product concentrations (cell viability 30 min post-bioconversion initiation remained at 85% at up to 10 g/L product concentration). In contrast, the product concentration affected substantially the cell viability over time and the ability of the whole-cells to carry out the BVMO reaction.

A very effective method to reduce both the substrate and product concentration in the BVMO fermentation media was introduced by Furstoss and Alphand back in 2001 (*80*). The use of an adsorbent resin (e.g., Dowex Optipore L-493) in the fermentation medium allows for *in situ* substrate feeding and product removal (SFPR), and therefore the oxidation of organic substrates at high concentrations without affecting the activity of the biocatalyst. Recently, the same group further elaborated on this concept and published a detailed protocol for the preparative scale production of (-)-(1R,5S)-3-oxa-bicyclo[3.3.0]oct-6-en-2-one (13) from enantiopure (-)-bicyclo[3.2.0]hept-2-en-6-one (11), including detailed information on troubleshooting for potential difficulties during the fermentation (81). A 25g/L starting material was used as the proof-of-concept, and the lactone yield was 75-80%. Enantiomeric purity (ee) of the product is >99%.

A different approach to limit substrate and product concentration during fermentations is the use of a biphasic fermentation medium. Lau and coworkers described the first bioproduction of lauryl lactone of C_{12}-ring size from cyclododecanone (14) using the recombinant CPDMO as shown in Figure 15 (82). First, the optimum organic solvent was selected based on resting cell-activity assays in 100 mL scale. Using hexadecane as organic phase, 10 to 16 g of lauryl lactone (15) were produced in a 3-L bioreactor that was operated in a semi-continuous mode of the two-phase system, compared to 2.4 g of lactone in batch mode. Disappearance of the substrate concomitant with product formation was monitored by ReactIR 4000, a FTIR-based technology (83) (Figure 16).

One factor that has not received much attention in the optimization of the BVMO bioprocess is host engineering. To the best of our knowledge, BVMOs have been expressed in either *E. coli* or *Saccharomyces cerevisae* until Park and coworkers reported the expression of the *Acinetobacter* CHMO in *Corynebacterium glutamicum* (84). The Gram-positive bacterium is best known for its production of amino acids, its high generation of cofactors and stability in the presence of organic molecules. The recombinant *C. glutamicum* was applied to the oxidation of cyclohexanone in a fed-batch culture. ε-Caprolactone could be produced in up to 16 g/L culture media under growing conditions, which compared favorably to oxidations carried out with recombinant *E. coli* carrying the same gene (11 g/L for growing (85) and 7.9 g/L for non-growing cells (86)).

The first pilot-plant 200-L scale biotransformation employing recombinant *E. coli* overexpressing CHMO as whole cell biocatalyst was reported in 2008 (87). Although multiphase solutions, such as the use of adsorbent resins or water immiscible organic solvents have been employed successfully for limiting product as well as substrate concentration in the fermentation media, Woodley and coworkers investigated the oxidation of racemic bicyclo[3.3.0]hept-2-en-6-one (11), employing a controlled substrate addition concept. The ketone was fed to resuspended whole cells in a dilute sodium chloride solution (10 g of NaCl / L H_2O) at two feed rates: 0 - 2 h at 0.6 g/L h and 2 - 4.5 h at 1.1 g/L h. A simple downstream process, saturation of the aqueous layer with sodium chloride followed by ethyl acetate extraction, allowed for the recovery of 495 g of a 1 : 1 regioisomeric mixture of lactones, both in excellent enantiomeric excess. Comparison of previous data from bioconversion of the bicyclic ketone at 1.5, 50 and 200 L scale showed that although the bioprocess was easily scalable, very similar yields were obtained at the three scales (3.5, 3.9 and 4.5 g/L, respectively). The 50-L conversion was actually more efficient on a volumetric basis. It is worth noting that the arabinose-induced CHMO activities in *E. coli* TOP10[pQR239] cells were in the range of 400-900 U/g dry weight depending on the fermentation conditions.

In a subsequent publication, Wohlgemuth and coworkers addressed the problem of the preparative-scale separation of the optically pure lactones derived

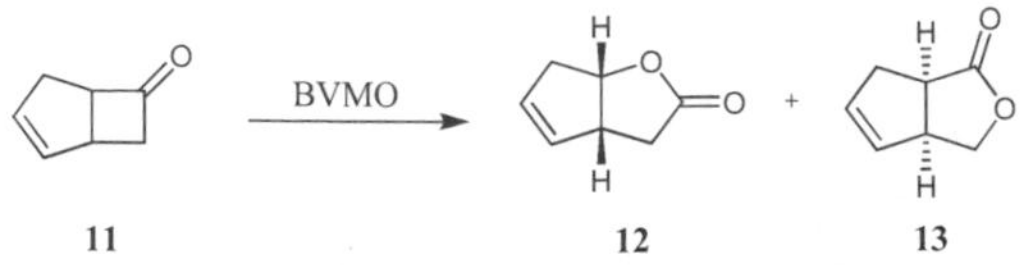

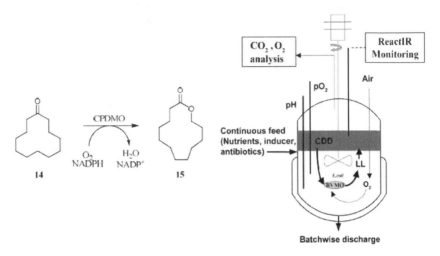

Figure 15. BVMO-mediated oxidation of bicyclo[3.2.0]hept-2-en-6-one.

Figure 16. Bioproduction of lauryl lactone (LL) from cyclododecane (CDD) in a two-phase semi-continuous bioreactor with at-line IR-monitoring (82).

from the oxidation racemic bicyclo[3.3.0]hept-2-en-6-one (**11**) (*88*). By using simulated moving bed (SMB) chromatography technology, the regioisomers were separated with a productivity of 0.026 g/g stationary phase per day with a solvent consumption of 0.363 L/g feed. The SMB chromatography was employed to reduce solvent consumption as well as providing an efficient and scalable separation technology.

Conclusions and Outlook

The prospect for BVMO technology is a bright one. Although the structures of a prototype CHMO that is in complex with its prosthetic FAD and cofactor NADPH have been solved for the first time, much more research is required to fully understand the mechanistic process. The complexity and intricate nature of the domain movements or rotations in this representative BVMO, together with lessons learned from the structures of PAMO and that of the atypical MtmOIV, both in the absence of NADP+, have just started to be unveiled. In this regard, the three structures are a "company" rather than a "crowd". Additional structures of BVMOs with bound cofactors and substrates would help to consolidate the "true face" of cofactor orientation (if not multiple personality on a case-by-case basis), and formation of Criegee intermediate, etc. during the catalytic circle. This would provide a scenario of BVMO dynamics at its best.

Inevitably, more whole genome-mined BVMO candidates will be cloned and characterized in addition to what a metagenomics approach would bear. However, what target and why will be the most challenging questions insofar as new biocatalysts discovery and the characterization thereof, or even protein engineering pursuits, are concerned. The green stamp of the BVMO reaction as a single criterion may not be enough.

Substantial progress has been made in the past years on the process engineering or bioprocessing side. The first 200-L scale of a CHMO-catalyzed biotransformation of a key chiral compound has been realized. The challenges in scale-up regarding oxygenation, stability of enzymes, substrate and product inhibition, for example, are still immense, though not necessarily insurmountable. In the bioprocess design, simplicity appears to be the key as the economics in the flowsheet of the bioprocess can quickly add up. Bigger is not necessarily better in product yield, as the 200-L biotransformation (*87*) has shown in comparison to a 50-L run. However, a case specific evaluation is recommended, as the particular characteristics of a given substrate and product pair may not be universal. The same applies to separation methods of the chemicals.

Besides the many examples of Type I BVMO members, we have seen the best of what appears to be in a class of its own, viz. the MtmOIV, an atypical BVMO catalyzing a key step in the biosynthesis of the anticancer agent mithramycin. However, the world of BVMOs will not be complete without the Type II enzymes. Exemplified by the 2,5-diketocamphane and 3,6-diketocamphane monooxygenases (2,5-DKCMO and 3,6-DKCMO) of the camphor degradation pathway in *P. putida* NCIMB 10007, these are two-component systems, each consists of a reductase that utilizes NADH to reduce FMN and an oxygenase that uses the reduced flavin to perform the BV oxidation. Seminal work on the biochemistry of these enzymes were carried out by Trudgill and coworkers (*5a,5b*). As a matter of fact, the structure solution of the oxygenating component of 3,6-DKCMO has reportedly been solved (*89*) after the much awaited crystallization of the protein (*90*). In the context of a biobased economy that promotes a greater use of renewable feedstocks from biomass, we are interested in the generation of possible green polymers from camphor derived lactones. This will be the story of another day.

Note Added in Revision

Since the submission of this chapter, Type I BVMOs have been the subject of another review by Fraiije's group (*91*), an attestation of interest and fast moving pace in this technology. On the research side, encapsulation of the prototypical CPMO in polyelectrolyte complex as an immobilization matrix was attempted and met with reasonable success in terms of substrate conversion and enantiomeric excess compared to the performance of free cells (*92*). In steroid biotransformation, a series of 3α-substituted steroidal substrates were screened for hydroxylation activity by the fungus *Aspergillus tamarii* (*93*). Interestingly, lactone formation from substrates, e.g., 3α-hydroxy-5α-androstan-17-one and 3α–acetoxy-5α-androstan-17-one, confirmed the *in vitro* bioconversion results obtained previously by recombinant biocatalyst CPDMO (*78*). Last but not least, Reetz's group in their directed evolution endeavour continues to amaze by in(tro)ducing allostery to PAMO, a phenomenon that is not known for any BVMO (*94*). The significance of this relates to protein engineering of different shaped binding pockets, aimed at an induced fit and facilitated by domain movements and allosteric effects.

References

1. Baeyer, A.; Villiger, V. *Ber. Dtsch. Chem. Ges.* **1899**, *32*, 3625–3633.
2. Chen, Y. C.; Peoples, O. P.; Walsh, C. T. *J. Bacteriol.* **1988**, *170*, 781–789.
3. For reviews on BVMOs see: (a) Kayser, M. M. *Tetrahedron* **2009**, *65*, 947–974. (b) Mihovilovic, M. D.; Rudroff, F.; Grötzl, B. *Curr. Org. Chem.* **2004**, *8*, 1057–1069. (c) Kamerbeek, N. M.; Janssen, D. B.; van Berkel, W. J. H.; Fraaije, M. W. *Adv. Synth. Catal.* **2003**, *345*, 667–678. (d) Mihovilovic, M. D.; Müller, B.; Stanetty, P. *Eur. J. Org. Chem.* **2002**, 3711–3730. (e) Flitsch, S.; Grogan, G. In *Enzyme Catalysis in Organic Synthesis*, 2nd ed.;Drauz, K., Waldmann, H., Eds.; Wiley-VCH: Weinheim, 2002. (f) Roberts, S. M.; Wan, P. W. H. *J. Mol. Catal. B: Enzym.* **1998**, *4*, 111–136. (g) Colonna, S.; Gaggero, N.; Pasta, P.; Ottolina, G. *Chem. Commun.* **1996**, *20*, 2303–2307. (h) Willetts, A. *Trends Biotechnol.* **1997**, *15*, 55–62. (i) Walsh, C. T.; Chen, Y.-C. *Angew. Chem., Int. Ed.* **1988**, *27*, 333–343. (j) Kelly, D. R. *Chimica Oggi* **2000**, *18*, 33–39 and 52–56; (k) Stewart, J. D. *Curr. Org. Chem.* **1998**, *2*, 195–216. (l) Fraaije, M. W.; Janssen, D. B. In *Modern Biooxidation. Enzymes, Reactions and Applications*; Schmid, R. D., Urlacher, V. B., Eds.; Wiley-VCH: Weinheim, 2007. (m) Kirschner, A.; Bornscheuer, U. T. In *Handbook of Green Chemistry, Volume 3: Biocatalysis*; Crabtree, R. H., Ed.; Wiley-VCH, Weinheim, 2009. (n) Pazmino, D. E. T; Fraaije, M. In *Future Directions in Biocatalysis*; Matsuda, T., Ed.; Elsevier: 2007.
4. Fraaije, M. W.; Kamerbeek, N. M.; van Berkel, W. J. H.; Janssen, D. B. *FEBS Lett.* **2002**, *518*, 43–47.
5. (a) Taylor, D. G.; Trudgill, P. W. *J. Bacteriol.* **1986**, *165*, 489–497. (b) Jones, K. H.; Smith, R. T.; Trudgill, P. W. *J. Gen. Microbiol.* **1993**, *139*, 797–805. (c) Beecher, J.; Willetts, A. *Tetrahedron: Asymmetry* **1998**, *9*, 1899–1916.
6. van Berkel, W. J. H.; Kamerbeek, N. M.; Fraaije, M. W. *J. Biotechnol.* **2006**, *124*, 670–694.
7. van Der Werf, M. J. *Biochem. J.* **2000**, *347*, 693–701.
8. Hasegawa, Y.; Nakai, Y.; Tokuyama, T.; Iwaki, H. *Biosci. Biotechnol. Biochem.* **2000**, *64*, 2696–2698.
9. Kolek, T.; Szpineter, A.; Swizdor, A. *Steroids* **2008**, *73*, 1441–1445.
10. Kolek, T.; Szpineter, A.; Swizdor, A. *Steroids* **2009**, *74*, 859–862.
11. Bonsor, D.; Butz, S. F.; Solomons, J.; Grant, S.; Fairlamb, I. J. S.; Fogg, M. J.; Grogan, G. *Org. Biomol. Chem.* **2006**, *4*, 1252–1260.
12. Fraaije, M. W.; Kamerbeek, N. M.; Heidekamp, A. J.; Fortin, R.; Janssen, D. B. *J. Biol. Chem.* **2004**, *279*, 3354–3360.
13. Szolkowy, C.; Eltis, L. D.; Bruce, N. C.; Grogan, G. *ChemBioChem* **2009**, *10*, 1208–1217.

14. Chen, G.; Kayser, M. M.; Mihovilovic, M. D.; Mrstik, M. E.; Martinez, C. A.; Stewart, J. D. *New J. Chem.* **1999**, *23*, 827–832.
15. Iwaki, H.; Hasegawa, Y.; Teraoka, M.; Tokuyama, T.; Bergeron, H.; Lau, P. C. K. *Appl. Environ. Microbiol.* **1999**, *65*, 5158–5162.
16. Doig, S. D.; O'Sullivan, L. M.; Patel, S.; Ward, J. M.; Woodley, J. M. *Enzyme Microb. Technol.* **2001**, *28*, 265–274.
17. Cheng, Q.; Thomas, S. M.; Kostichka, K.; Valentine, J. R.; Nagarajan, V. *J. Bacteriol.* **2000**, *182*, 4744–4751.
18. Brzostowicz, P. C.; Walters, D. M.; Thomas, S. M.; Nagarajan, V.; Rouvière, P. E. *Appl. Environ. Microbiol.* **2003**, *69*, 334–342.
19. Kim, Y.-M.; Jung, S.-H.; Chung, Y.-H.; Yu, C.-B.; Rhee, I.-K. *Biotechnol. Bioproc. Eng.* **2008**, *13*, 40–47.
20. (a) Brzostowicz, P. C.; Walters, D. M.; Jackson, R. E.; Halsey, K. H.; Ni, H.; Rouviere, P. E. *Environ. Microbiol.* **2005**, *7*, 179–190. (b) Bramucci, M. G.; Brzostowicz, P. C.; Kostichka, K. N.; Nagarajan, V.; Rouvière, P. E. Thomas, S. M. WO 2003/020890 A2.
21. Brzostowicz, P. C.; Blasko, M. S.; Rouvière, P. E. *Appl. Microbiol. Biotechnol.* **2002**, *58*, 781–789.
22. Iwaki, H.; Hasegawa, Y.; Wang, S.; Kayser, M. M.; Lau, P. C. K. *Appl. Environ. Microbiol.* **2002**, *68*, 5671–5684.
23. Clouthier, C. M.; Kayser, M. M.; Reetz, M. T. *J. Org. Chem.* **2006**, *71*, 8431–8437.
24. van Beilen, J. B.; Mourlane, F.; Seeger, M. A.; Kovac, J.; Li, Z.; Smits, T. H. M.; Fritsche, U.; Witholt, B. *Environ. Microbiol.* **2003**, *5*, 174–182.
25. Kotani, T.; Yurimoto, H.; Kato, N.; Sakai, Y. *J. Bacteriol.* **2007**, *189*, 886–893.
26. Snajdrova, R.; Grogan, G.; Mihovilovic, M. D. *Bioorg. Med. Chem. Lett.* **2006**, *16*, 4813–4817.
27. Park, J.; Kim, D.; Kim, S.; Kim, J.; Bae, K.; Lee, C. *J. Microbiol. Biotechnol.* **2007**, *17*, 1083–1089.
28. Kamerbeek, N. M.; Moonen, M. J. H.; Van Der Ven, J. G. M.; Van Berkel, W. J. H.; Fraaije, M. W.; Janssen, D. B. *Eur. J. Biochem.* **2001**, *268*, 2547–2557.
29. Kirschner, A.; Altenbuchner, J.; Bornscheuer, U. T. *Appl. Microbiol. Biotechnol.* **2007**, *75*, 1095–1101.
30. Rehdorf, J.; Zimmer, C. L.; Bornscheuer, U. T. *Appl. Environ. Microbiol.* **2009**, *75*, 3106–3114.
31. Rehdorf, J.; Kirschner, A.; Bornscheuer, U. T. *Biotechnol. Lett.* **2007**, *29*, 1393–1398.
32. Iwaki, H.; Wang, S.; Grosse, S.; Bergeron, H.; Nagahashi, A.; Lertvorachon, J.; Yang, J.; Konishi, Y.; Hasegawa, Y.; Lau, P. C. K. *Appl. Environ. Microbiol.* **2006**, *72*, 2707–2720.
33. Völker, A.; Kirschner, A.; Bornscheuer, U. T.; Altenbuchner, J. *Appl. Microbiol. Biotechnol.* **2008**, *77*, 1251–1260.
34. Onaca, C.; Kieninger, M.; Engesser, K.-H.; Altenbuchner, J. *J. Bacteriol.* **2007**, *189*, 3759–3767.
35. Morii, S.; Sawamoto, S.; Yamauchi, Y.; Miyamoto, M.; Iwami, M.; Itagaki, E. *J. Biochem.* **1999**, *126*, 624–631.

36. Kostichka, K.; Thomas, S. M.; Gibson, K. J.; Nagarajan, V.; Cheng, Q. *J. Bacteriol.* **2001**, *183*, 6478–6486.

37. Mirza, I. A.; Yachnin, B. J.; Wang, S.; Grosse, S.; Bergeron, H.; Imura, A.; Iwaki, H.; Hasegawa, Y.; Lau, P. C. K.; Berghuis, A. M. *J. Am. Chem. Soc.* **2009**, *131*, 8848–8854.

38. Choi, J.-H.; Kim, T.-K.; Kim, Y.-M.; Kim, W.-C.; Park, K.; Rhee, I.-K. *J. Microb. Biotechnol.* **2006**, *16*, 511–518.

39. Banskota, A. H.; McAlpine, J. B.; Sørensen, D.; Aouidate, M.; Piraee, M.; Alarco, A.-M.; Omura, S.; Shiomi, K.; Farnet, C. M.; Zazopoulos, E. *J. Antibiot.* **2006**, *59*, 168–176.

40. Jiang, J.; Tetzlaff, C. N.; Takamatsu, S.; Iwatsuki, M.; Komatsu, M.; Ikeda, H.; Cane, D. E. *Biochemistry* **2009**, *48*, 6431–6440.

41. Malito, E.; Alfieri, A.; Fraaije, M. W.; Mattevi, A. *Proc. Natl. Acad. Sci. U.S.A.* **2004**, *101*, 13157–13162.

42. Fraaije, M. W.; Wu, J.; Heuts, D. P. H. M.; Van Hellemond, E. W.; Spelberg, J. H. L.; Janssen, D. B. *Appl. Microb. Biotechnol.* **2005**, *66*, 393–400.

43. Rial, D. V.; Cernuchova, P.; van Beilen, J. B.; Mihovilovic, M. D. *J. Mol. Catal. B: Enzym.* **2008**, *50*, 61–68.

44. van Beilen, J. B.; Mourlane, F.; Seeger, M. A.; Kovac, J.; Li, Z.; Smits, T. H. M.; Fritsche, U.; Witholt, B. *Environ. Microbiol.* **2003**, *5*, 174–182.

45. Gibson, M.; Nur-E-Alam, M.; Lipata, F.; Oliveira, M. A.; Rohr, J. *J. Am. Chem. Soc.* **2005**, *127*, 17594–17595.

46. Beam, M. P.; Bosserman, M. A.; Noinaj, N.; Wehenkel, M.; Rohr, J. *Biochemistry* **2009**, *48*, 4476–4487.

47. Kneller, M. B.; Cheesman, M. J.; Rettie, A. E. *Biochem. Biophys. Res. Commun.* **2001**, *282*, 899–903.

48. Iwaki, H. ; Hasegawa, Y. ; Teraoka, M. ; Tokuyama, T.; Bernard, L.; Lau, P. C. K. In *Biocatalysis in Polymer Science*; Gross, R. A., Cheng, H. N., Eds.; ACS Symposium Series 840; American Chemical Society: Washington, DC, 2003; Chapter 6, pp 80–92.

49. Alfieri, A.; Malito, E.; Orru, R.; Fraaije, M. W.; Mattevi, A. *Proc. Natl. Acad. Sci. U.S.A.* **2008**, *105*, 6572–6577.

50. Ryerson, C. C.; Ballou, D. P.; Walsh, C. T. *Biochemistry* **1982**, *21*, 2644–2655.

51. Sheng, D.; Ballou, D. P.; Massey, V. *Biochemistry* **2001**, *40*, 11156–11167.

52. Branchaud, B. P.; Walsh, C. T. *J. Am. Chem. Soc.* **1985**, *107*, 2153–2161.

53. Ziegler, D. M. *Annu. Rev. Pharmacol. Toxicol.* **1993**, *33*, 179–199.

54. Kamerbeek, N. M.; Fraaije, M. W.; Janssen, D. B. *Eur. J. Biochem.* **2004**, *271*, 2107–2116.

55. Trudgill, P. W. In *Methods in Enzymology*, Ed.; Mary, E. L.; Academic Press: 1990; Vol. 188, pp 77–81.

56. Torres Pazmiño, D. E.; Baas, B.-J.; Janssen, D. B.; Fraaije, M. W. *Biochemistry* **2008**, *47*, 4082–4093.

57. De Gonzalo, G.; Torres Pazmiño, D. E.; Ottolina, G.; Fraaije, M. W.; Carrea, G. *Tetrahedron: Asymmetry* **2005**, *16*, 3077–3083.

58. (a) Torres Pamino, D. E.; Snajdrova, R.; Rial, D. V.; Mihovilovic, M. D.; Fraaije, M. *Adv. Synth. Catal.* **2007**, *349*, 1361–1368. (b) Bocola, M.;

Schulz, F.; Leca, F.; Vogel, A.; Fraaije, M. W.; Reetz, M. T. *Adv. Synth. Catal.* **2005**, *347*, 979–986.

59. Beaty, N. B.; Ballou, D. P. *J. Biol. Chem.* **1981**, *256*, 4619–4625.

60. Reetz, M. T. *J. Org. Chem.* **2009**, *74*, 5767–5778.

61. Reetz, M. T.; Wu, S. *Chem. Commun.* **2008**, 5499–5501.

62. Reetz, M. T.; Wu, S. *J. Am. Chem. Soc.* **2009**, *131*, 15424–15432.

63. Kirschner, A.; Bornscheuer, U. T. *Appl. Microbiol. Biotechnol.* **2008**, *81*, 465–472.

64. Torres Pamino, D. E.; Snajdrova, R.; Baas, B.-J.; Ghobrial, M.; Mihovilovic, M. D.; Fraaije, M. W. *Angew. Chem., Int. Ed.* **2008**, *47*, 2275–2278.

65. Torres Pamino, D. E.; Riebel, A.; de Lange, J.; Rudroff, F.; Mihovilovic, M. D.; Fraaije, M. W. *ChemBioChem* **2009**, *10*, 2595–2598.

66. Hollmann, F.; Taglieber, A.; Schulz, F.; Reetz, M. T. *Angew. Chem., Int. Ed.* **2007**, *46*, 2903–2906.

67. Taglieber, A.; Schulz, F.; Hollmann, F.; Rusek, M.; Reetz, M. T. *ChemBioChem* **2008**, *9*, 565–572.

68. Rial, D. V.; Cernuchova, P.; van Beilen, J. B.; Mihovilovic, M. D. *J. Mol. Catal. B: Enzym.* **2008**, *50*, 61–68.

69. Rial, D. V.; Bianchi, D. A.; Kapitanova, P.; Lengar, A.; van Beilen, J. B.; Mihovilovic, M. D. *Eur. J. Org. Chem.* **2008**, 1203–1208.

70. Mihovilovic, M. D.; Grötzl, B.; Kandioller, W.; Muskotal, A.; Snajdrova, R.; Rudroff, F.; Spreitzer, H. *Chem. Biodiversity* **2008**, *5*, 490–498.

71. Mihovilovic, M. D.; Kapitan, P.; Kapitanova, P. *ChemSusChem* **2008**, *1*, 143–148.

72. Rodriguez, C.; de Gonzalo, G.; Torres Pazmino, D. E.; Fraaije, M. W.; Gotor, V. *Tetrahedron: Asymmetry* **2008**, *19*, 197–203.

73. Rioz-Martinez, A.; de Gonzalo, G.; Torres Pazmino, D. E.; Fraaije, M. W.; Gotor, V. *Eur. J. Org. Chem.* **2009**, 2526–2532.

74. Rodriguez, C.; de Gonzalo, G.; Torres Pazmino, D. E.; Fraaije, M. W.; Gotor, V. *Tetrahedron: Asymmetry* **2009**, *20*, 1168–1173.

75. Rehdorf, J.; Lengar, A.; Bornscheuer, U. T.; Mihovilovic, M .D. *Bioorg. Med. Chem. Lett.* **2009**, *19*, 3739–3743.

76. Turfitt, G. E. *Biochem. J.* **1948**, *42*, 376–383.

77. Hunter, A. C.; Watts, K. R.; Dedi, C.; Dodd, H. T. *J. Steroid Biochem. Mol. Biol.* **2009**, *116*, 171–177.

78. Beneventi, E.; Ottolina, G.; Carrea, G.; Panzeri, W.; Fronza, G.; Lau, P. C. K. *J. Mol. Catal. B: Enzym.* **2009**, *58*, 164–168.

79. Shitu, J. O.; Chartrain, M.; Woodley, J. M. *Biocatal. Biotransform.* **2009**, *27*, 107–117.

80. (a) Simpson, H. D.; Alphand, V.; Furstoss, R. *J. Mol. Catal. B: Enzym.* **2001**, *16*, 101–108. (b) Hilker, I.; Alphand, V.; Wohlgemuth, R.; Furstoss, R. *Adv. Synth. Catal.* **2004**, *346*, 203–214. (c) Hilker, I.; Gutierrez, M. C.; Alphand, V.; Wohlgemuth, R.; Furstoss, R. *Org. Lett.* **2004**, *6*, 1955–1958.

81. Hilker, I.; Gutierrez, M. C.; Furstoss, R.; Ward, J.; Wohlgemuth, R.; Alphand, V. *Nat. Protocol.* **2008**, *3*, 546–554.

82. Yang, J.; Wang, S.; Lorrain, M. J.; Rho, D.; Abokitse, K.; Lau, P. C. K. *Appl. Microbiol. Biotechnol.* **2009**, *84*, 867–876.

83. Yang, J.; Lorrain, M.-J.; Rho, D.; Lau, P. C. K. *Ind. Biotechnol.* **2006**, *2*, 138–142.

84. Doo, E.-H.; Lee, W.-H.; Seo, H.-S.; Seo, J.-H.; Park, J.-B. *J. Biotechnol.* **2009**, *142*, 164–169.

85. Lee, W. H.; Park, J.-B.; Park, K.; Kim, M. D.; Seo, J. H. *Appl. Microbiol. Biotechnol.* **2007**, *76*, 329–338.

86. Walton, A. Z.; Stewart, J. D. *Biotechnol. Prog.* **2002**, *18*, 262–268.

87. Baldwin, C. V. F.; Wohlgemuth, R.; Woodley, J. M. *Org. Process Res. Dev.* **2008**, *12*, 660–665.

88. Kaiser, P.; Ottolina, G.; Carrea, G.; Wohlgemuth, R. *New Biotechnol.* **2009**, *25*, 220–225.

89. Isupov, M. N.; Lebedev, A. A. *Acta Crystallogr., Sect. D* **2008**, *D64*, 90–98.

90. Mcghie, E. J.; Isupov, M. N.; Schroder, E.; Littlechild, J. A. *Acta Crystallogr., Sect. D* **1998**, *54*, 1035–1038.

91. Pazmino, D. E. T.; Dudek, H. M.; Fraaije, M .W. *Curr. Opin. Chem. Biol.* **2009**, *14*, 1–7.

92. Hucik, M.; Bucko, M.; Gemeiner, P.; Stefuca, V.; Vikartovsla, A.; Mihovilovic, M. D.; Rudroff, F.; Iqbal, N.; Chorvat, D., Jr.; Lacik, I. *Biotechnol. Lett.* **2010**, *32*, 675–680.

93. Hunter, A. C.; Khuenl-Brady, H.; Barrett, P.; Dodd, H. T.; Dedi, C. *J. Steroid Biochem. Mol. Biol.* **2010**, *118*, 171–176.

94. Wu, S.; Acevedo, J. P.; Reetz, M. T. *Proc. Natl. Acad. Sci. U.S.A.* **2010**, *107*, 2775–2780.

Enzymatic Hydrolyses and Degradations

Chapter 25

Embedding Enzymes To Control Biomaterial Lifetime

Manoj Ganesh* and Richard Gross

NSF I/UCRC for Biocatalysis & Bioprocessing of Macromolecules, Polytechnic Institute of NYU, Six Metrotech Center, Brooklyn, New York 11201
*maddymanoj@gmail.com

Bioresorbable materials with 'tunable' lifetimes are needed to meet a continuously expanding range of biomedical applications. Control of polymer lifetime has relied on altering intrinsic properties of synthesized polymeric materials. However, development of new biopolymer compositions is hindered by time and cost constraints. Herein we show that, biomaterials bioresorption rate can be regulated by embedding an enzyme within a bioresorbable polymer matrix when that enzyme is active for matrix hydrolysis. Poly(ε-caprolactone) was selected as a model system since its use has been limited due to its slow hydrolytic degradation (years). *Candida antarctica* Lipase B (CALB) was surfactant paired to enable its dispersion throughout PCL films. *Embedded enzyme matrix hydrolysis* occurs by formation of multiple enzyme-surface interfaces throughout films. Films become highly porous and crystalline as degradation proceeds. The concepts developed herein are applicable to a wide range of accepted biomaterials and can greatly expand the range of applications for which they are useful.

Introduction

Conventional thinking when designing bioresorbable materials is to use the intrinsic properties of synthesized materials to regulate their degradation rate. For example, considerable effort has been dedicated to introducing comonomers to manipulate the bioresorption rate of PCL (*1*). While this is a well-established strategy, each copolymer composition represents a new medical material that must be assessed by an array of expensive tests requiring years. Herein, we describe a simple but powerful alternative approach to control biomaterial lifetime using PCL as the model system. The hypothesis tested is that, biomaterials bioresorption rate can be regulated by embedding an enzyme within a bioresorbable polymer matrix when that enzyme is active for matrix hydrolysis. The PCL-degrading enzyme, *Candida antarctica* Lipase B (CALB), was selected and surfactant paired to render CALB soluble for dispersion within a PCL matrix. This principle termed *embedded enzymatic degradation* can be used for any biomaterial for which an enzyme is known or can be developed for matrix hydrolysis. Therefore, it represents a general approach to be deployed to control biomaterial lifetime.

PCL is an excellent candidate for controlled release and other biomaterial applications due to its availability, biodegradability, ease by which it can be melt-processed into various shaped substances, biocompatibility and miscibility with many drugs (*2*). PCL is a soft material that has a peak melting temperature, tensile stress at break, elongation at break and tensile modulus of 60 °C, 4 Mpa, 800-1000 % and 386 Mpa, respectively (*3*). However, numerous studies have shown that PCL degrades slowly when implanted in animals. For example, PCL capsules, implanted in rats, with initial molecular weight (M_w) 66 000, retained their shape for over 2-years[2]. By 30 months, PCL capsules broke into low molecular weight pieces (M_w 8000) (*2*). The slow degradability of PCL limits its use for systems that require shorter lifetimes (*1*). Nevertheless, PCL and its copolymers continue to be intensely studied for a range of medical applications (*4*).

Herein, surfactant-paired CALB was mixed in toluene with PCL and films were prepared by solution casting. Sodium bis(2-ethylhexyl)sulfosuccinate (AOT), the surfactant used to solubilize CALB, is approved for administration to both humans and animals (*5*) and was found to be safe for encapsulation and delivery of lipophilic drugs in aqueous based delivery systems (*6, 7*). The biocompatibility experiments of CALB are currently under progress, but, if a problem arises, well-known methods are available to identify and remove segments in protein that trigger an immune response (*8*). Furthermore, if needed, other enzymes could be substituted for CALB that are known to catalyze PCL hydrolysis (*9*).

CALB was surfactant paired with AOT to enable enzyme transfer from the aqueous to the organic phase. The method of ion-pairing is described in the following section. Ion-paired CALB was embedded in PCL matrices with enzyme contents of 6.5 and 1.65% (w/w). Weight loss was determined gravimetrically and protein content was quantified using the micro BCA protein assay. Molecular weights were determined at room temperature by gel permeation chromatography (GPC). Scanning electron micrographs were generated with an Amray 1910 SEM and Schottkey field emission gun at 5 kV. DSC measurements were performed using a Perkin-Elmer DSC system with a scan rate of 10 °C/min. CALB was conjugated with fluorescein isothiocyanate (FITC) and this conjugate was embedded in PCL matrices. Imaging of CALB distribution within PCL was performed by confocal microscopy (Leica Microsystems, Bannockburn, IL) using a 488 nm excitation laser.

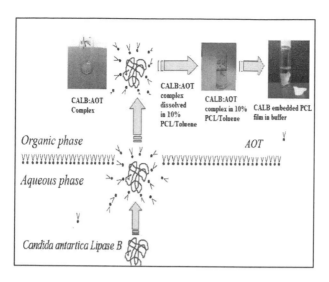

Scheme 1. Surfactant pairing of Candida antartica Lipase B (CALB) with AOT.

Concept: CALB Extraction to Isooctane

Protein solubilization is reportedly achieved by using much lower concentration of surfactant avoiding the formation of reverse micelles, thereby forming ion-paired complex with the protein (*10*). Theconcept behind solubilization of protein molecule in ion-pairing techniqueis based on the interactions that occur between the surface charges of proteins and the charged headgroup of the surfactant. For example, using a negatively charged surfactant such as AOT, proteins that exhibit a net positive charge can be extracted into the organic phase. The mechanism for formation ofion-paired complex involves direct interaction between the surfactant molecules found in the organic phase and the protein molecule in the aqueous phase, henceforth is completely depended on the electrostatic forces of attraction. The process can be visualized (Scheme 1) as occurring at the interface where direct interaction between surfactant and protein moleculeNotes results in extraction of the protein molecule along with some water and ions. To transfer CALB from this aqueous media to an organic phase, the following method was developed that adapts other protocols described in the literature[2]. CALB (0.5 mg/mL) was dissolved in a media consisting of sodium acetate buffer (20 mM, pH: 4.5), 9 mM calcium chloride solution, and 0.25% v/v isopropyl alcohol (IPA). This solution was mixed with an equal volume of 2mM AOT/anhydrous iso-octane that was placed in a bench top shaker at 30 °C for 15 min at 100 rpm. Centrifugation at 30 °C, 3000 rpm for 10 minutes was necessary to achieve adequate separation between organic and aqueous layers. The organic phase in which the protein was extracted was assayed by UV-spectroscopy to determine the absorbance at 280 nm. The extinction coefficient (ε) of CALB, calculated empirically, is 63789 $M^{-1}cm^{-1}$). Then, ion-paired CALB was isolated from the organic layer by evaporating the organic solvent (iso-octane) by rotary-evaporation.

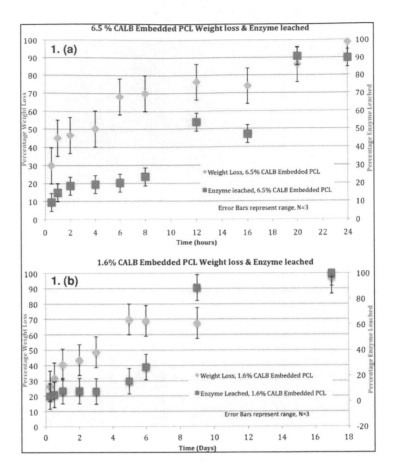

Figure 1. Weight loss and Enzyme leached: (a) 6.5% CALB embedded PCL & (b) 1.6% CALB embedded PCL

Embedding Ion-Paired CALB in PCL

PCL films were solution cast by adding the desired quantity of ion-paired CALB to anhydrous toluene containing 10% w/v PCL. After vortexing the mixture for five minutes the solution was immediately poured onto a glass plate and cast using a Gardco adjustable micrometer "MICROM II" film applicator. The film was then dried in a vacuum chamber (25 °C, 10 torrs) overnight to remove toluene. Films with average thickness of 80 ± 5 μm were then cut into dimensions of 2 by 2 cm resulting in films with average weights of 37 ± 2 mg.

Degradation Profile of Embedded Enzymes

Films with defined quantities of AOT-paired CALB dispersed within a PCL matrix were prepared by solution casting from toluene. Initial surface area, thickness and average weight of films were 4 cm², 80 ± 5 μm and 37 ± 2 mg,

Table 1. Variation of number average molecular weight, M_n & polydispersity index, PDI with respect to percentage degradation

Sample	Weight loss (%)	$M_n{}^c$	$M_w/M_n{}^c$
6.5%[a]	0	117000	1.8
6.5% [a]	30	36400	3.7
6.5% [a]	60	31800	3.6
6.5% [a]	80	24700	5.2
1.6% [a]	0	126000	1.7
1.6% [a]	30	40000	4.1
1.6% [a]	60	34000	4.0
1.6% [a]	80	33200	4.3
external addition[b]	0	124000	1.7
external addition[b]	30	128000	1.7
external addition[b]	60	123000	1.7
external addition[b]	80	117000	1.8

[a] CALB embedded within PCL film [b] CALB added to medium instead of embedding within films [c] M_n and M_w/M_n, determined by GPC, are number average molecular weight and polydispersity (PDI), respectively (M_n, error - ± 5% & M_w/M_n, error - ± 5%, represents range n=3)

respectively. Concentrations of CALB-AOT investigated herein are 19.4 and 5% (w/wto PCL), which corresponds to CALB contents of 6.5 and 1.6%, respectively. Films were incubated in phosphate buffer solution (20 mM, pH 7.10) with shaking (200 rpm) at 37 °C. Media were replaced with fresh buffer at regular time intervals described below. Figure 1a shows PCL films with 6.5% CALB degraded rapidly, so that by 24 h, complete film weight loss was achieved. CALB diffusion from films into the media was determined by micro-BCA analysis. By 2 h, film weight loss and CALB diffusion into the media reached 48% ± 10 and 18% ± 5, respectively. Thereafter, while film weight loss increased to 70% by 6 h, CALB release remained unchanged. Thus, by 8 h, while total film weight loss was 70% ± 10, 80% ± 5 of CALB still remained in films. Surprisingly, from 6 to 16 h, no further film weight loss occurred but protein diffusion into the media increased to 50% ± 5.

By reducing CALB content from 6.5 to 1.6%, the time required for complete film degradation was increased from 1 day to about 17 days (Figure 1b). While a large fraction (40%) of film weight loss occurred during the first day, further film weight loss occurred slowly over the following 16 days. Similar to experiments performed with 6.5% CALB, 1.6% CALB also experienced the phenomena where film weight loss remained constant while a large fraction of CALB diffused from films into the media. For 1.6% CALB, this occurred between 5 and 9 days while CALB depletion from films increased from 15 to 90% ± 10.

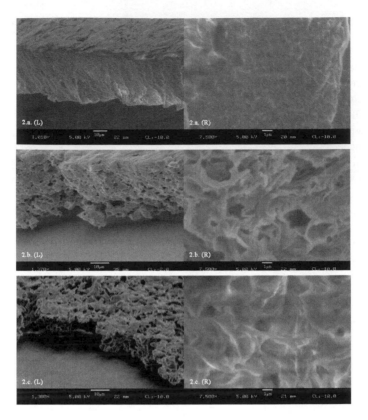

Figure 2. SEM cross-sectional views at different magnifications of PCL films for: (a) CALB externally added to films resulting in 43% weight loss after a 2 h incubation; (b) CALB embedded (6.5%) within films resulting in 30% weight loss after a 2 h incubation; (c) CALB embedded (1.6%) within films resulting in 44% weight loss after a 48 h incubation. L and R designate lower and higher magnification views, respectively.

Molecular Weight Changes: Embedded System vs. External System

The progression of PCL molecular weight change as a function of incubation time was determined for two systems: *i*) enzyme added externally to the polymer matrix and *ii*) 6.5 and 1.6% CALB embedded in the PCL matrix. Plots of M_n versus %-weight loss and polydispersity (PDI, M_w/M_n) are shown in Table 1. Referring to Fig 1, film weight loss reached 30 and 40% early during incubations with 6.5 and 1.6% embedded CALB, respectively. These weight losses were accompanied by large decreases in M_n from 120 000 to about 40 000. Furthermore, PDI increased from 1.7 to about 4.0. For the 1.6%-CALB-PCL system, M_n and PDI remained at 25 000-35 000 and 3.5 to 4.3 till 80% weight loss. These changes in M_n and PDI are explained by CALB degradation of PCL in CALB-embedded films from many surfaces created at numerous interfaces between CALB-AOT aggregates (see below) and the PCL matrix. CALB degradation of PCL occurs by endo-attack (3) creating a fraction of lower molecular weight chains that are trapped within a PCL matrix for which a substantial fraction of the material has not yet been accessed by enzyme and, therefore, remains at its original molecular weight. As time progresses during incubations, lower M_n PCL chains are degraded to water-soluble products that diffuse from the matrix and cause weight loss while new oligomers are generated thereby causing M_n to remain low and PDI to remain high. In contrast, when CALB is added externally to PCL films without embedded enzyme, M_n and PDI remain constant as film weight loss proceeds (4) (Table 1). This remarkable contrast in molecular weight change between embedded and externally added CALB is due to that, when CALB is added externally, CALB can only access PCL at film surfaces. Without access to PCL throughout films, externally added CALB does not affect the bulk molecular weight and, therefore, M_n and PDI do not change. Indeed, from review of the literature we conclude that *embedded enzymatic degradation* occurs by a unique mechanism distinguished by a distinctive progression of events that occur during the degradation process. This is further illustrated by comparison of *embedded enzymatic degradation* to bulk chemical hydrolysis of PLA films incubated in buffer. As PLA film hydrolysis proceeds PLA molecular weight average decreases and PDI moves toward a poisson distribution due to random events of cleavage that occur throughout films (5).

Morphological Studies

Cross sectional images recorded by SEM of partially degraded embedded CALB-PCL films, and PCL films to which CALB was added externally, are displayed in Figure 2. SEM images in Fig 2a recorded for externally added CALB incubation with PCL to 43% film weight loss, shows a roughened surface with no evidence of degradation below the surface region. This is consistent with the model that, when CALB is added externally, CALB can only access PCL at film surfaces. SEM photographs displayed in Figs 2b and 2c show cross-sectional views of films with initial embedded CALB loadings of 6.5 and 1.6% that reached weight loss values of 30 and 44%, respectively. In contrast to externally added CALB, incubations of embedded CALB films results in pore formation throughout the material. This is explained by, as pores increase in size, they conjoin creating networks. Furthermore, water continues to reach new film regions where CALB aggregates reside. Once CALB aggregates obtain sufficient water supply they begin actively hydrolyzing adjoining PCL surfaces.

DSC measurements showed significant differences in PCL crystallinity for residual PCL films that had undergone *embedded enzymatic degradation* versus PCL films whose degradation was a consequence of external enzyme addition to media. For example, the 6.5% CALB-PCL embedded system, which underwent a weight loss of 80%, gave a residual film that was 90% crystalline. Furthermore, the 1.6% CALB-PCL system, that underwent a weight loss of 80%, gave residual films that were 80% crystalline. In contrast, when CALB was added externally and film weight loss reached 80%, film crystallinity was 50%, which is close in value to % crystallinity of untreated films. Furthermore, analysis by DSC of control films retrieved after incubation without enzyme at 37 °C over 17 days showed no change in film crystallinity. Thus, preferential degradation of PCL amorphous regions by *embedded enzymatic degradation* results in a highly crystalline porous matrix. When CALB is added externally, CALB's access is limited to the films surface leaving the bulk material unaffected by the enzyme. Consequently, based on the same reasoning, both the crystallinity and molecular weight of residual films from external enzyme addition remain largely unchanged from control films untreated by enzyme.

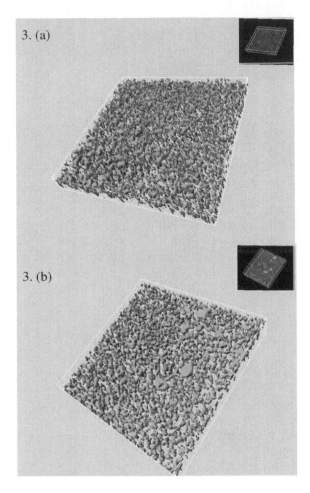

Figure 3. Laser scanning confocal microscopy images: (a) distribution of FITC tagged CALB in 6.5% CALB embedded PCL and (b) distribution of FITC tagged CALB in 1.6% CALB embedded PCL.

Enzyme Distribution in Polymer

In order to determine embedded CALB distribution and the average size of CALB-AOT aggregates in PCL matrices, CALB was labeled with fluorescein isothiocyanate (FITC) and the resulting FITC-CALB embedded PCL films were analyzed by laser scanning confocal microscopy. Resulting images of non-incubated films with 6.5 and 1.6% embedded FITC-labeled CALB are displayed in Figures 3a and 3b, respectively. The entire film volume was scanned to determine average aggregate size and standard deviation (σ). Values were 6.5 ± 2.4 and 4.0 ± 1.6 µm for PCL-CALB films with 6.5 and 1.6% embedded CALB, respectively.

Conclusion

The present paper describes that, by surfactant pairing CALB and dispersing CALB throughout PCL films, wide variations in PCL lifetime was achieved. Thus, without copolymerization, changing initial PCL molecular weight or film crystallinity, but by embedding an enzyme within the PCL matrix that is active for PCL hydrolysis, biomaterials were created that degrade by a novel mechanism due to formation of multiple enzyme-surface interfaces throughout films. By specific manipulation of CALB-surfactant-PCL interfaces we expect that far greater manipulation of film degradation rates can be achieved. Furthermore, identification of different material-hydrolytic enzyme systems will further expand the impact of *embedded enzymatic degradation* on the field of biomaterials science and engineering.

References

1. Little, U.; Buchanan, F.; Harkin-Jones, E.; McCaigue, M.; Farrar, D.; Dickson, G. Accelerated degradation behaviour of poly(3-caprolactone) via melt blending with poly(aspartic acid-co-lactide) (PAL). *Polym. Degrad. Stab.* **2009**, *94*, 213–220.
2. Sun, H.; Mei, L.; Song, C.; Cui, X.; Wang, P. The in vivo degradation, absorption and excretion of PCL-based implant. *Biomaterials* **2005**, *27*, 1735–1740.
3. Gross, R. A.; Kalra, B. Biodegradable polymers for the environment. *Science* **2002**, *297*, 803–807.
4. Bolgen, N.; Vargel, P.; Korkusuz, Z. M.; Menceloglu, Y. Z.; Piskin, E. In vivo performance of antibiotic embedded electrospun PCL membranes for prevention of abdominal adhesions. *J. Biomed. Mater. Res., Part B: Appl. Biomater.* **2006**, *81B*, 530–543.
5. *Physician's Desk Reference*, 49th ed.; Edward R. Barnhart Publishing: New York, 1989.
6. Liu, J.-G.; Xing, J.-M.; Chang, T.-S.; Liu, H.-Z. Purification of nattokinase by reverse micelles extraction from fermentation broth: Effect of temperature and phase volume ratio. *Bioprocess Biosyst. Eng.* **2006**, *28*, 267–273.
7. Gupta, S.; Moulik, S. P. Biocompatible Microemulsions and their prospective uses in drug delivery. *J. Pharm. Sci.* **2008**, *97*, 22–45.
8. Shikano, S.; Coblitz, B.; Sun, H.; Li, M. Genetic isolation of transport signals directing cell surface expression. *Nat. Cell Biol.* **2005**, *7*, 985–992.

9. Nisida, H.; Tokiwa, Y. Distribution of poly(β-hydroxybutyrate) and poly(ε-caprolacton) aerobic degrading microorganisms in different environments. *J. Environ. Polym. Degrad.* **1993**, *1*, 227–233.

10. Paradkar, V. M.; Dordick, J. S. Mechanism of extraction of chymotrypsin into isooctane at very low concentrations of aerosol OT in the absence of reversed micelles. *Biotechnol. Bioeng.* **1993**, *43*, 529–540.

Chapter 26

Surprisingly Rapid Enzymatic Hydrolysis of Poly(ethylene terephthalate)

Åsa M. Ronkvist, Wenchun Xie, Wenhua Lu, and Richard A. Gross*

NSF I/URC for Biocatalysis and Bioprocessing of Macromolecules,
Department of Chemical and Biololgical Sciences, Polytechnic University,
Six Metrotech Center, Brooklyn, New York 11201
*rgross@poly.edu

A detailed study and comparison was made on the catalytic activities of cutinases from *Humilica insolens* (HiC), *Pseudomonas mendocina* (PmC) and *Fusarium solani* (FsC) using low-crystallinity (*lc*) and biaxially oriented (*bo*) poly(ethylene terephthalate) (PET) films as model substrates. Cutinase activity for PET hydrolysis was assayed using a pH-stat to measure NaOH consumption versus time, where initial activity was expressed as units of μmole NaOH added per hour and per mL reaction volume. HiC was found to have good thermostability with maximum initial activity from 70 to 80 °C whereas PmC and FsC performed best at 50 °C. Assays by pH-stat showed that the cutinases had about 10 fold higher activity for the *lc*PET (7% crystallinity) than for the *bo*PET (35 % crystallinity). Under optimal reaction conditions, initial activities of cutinases were successfully fit by a heterogeneous kinetic model. The hydrolysis rate constant k_2 was seven fold higher for HiC at 70 °C (0.62 μmole/cm^2/h) relative to PmC and FsC at 50 and 40 °C, respectively. In a 96 h degradation study using *lc*PET films, incubation with PmC and FsC both resulted in 5 % film weight loss at 50 and 40 °C, respectively. In contrast, HiC catalyzed *lc*PET film hydrolysis at 70°C resulted in 97±3% weight loss in 96 h, corresponding to a loss in film thickness of 30 μm per day. For all three cutinases, analysis of aqueous soluble degradation products showed they consist exclusively of terephthalic acid and ethylene glycol.

Introduction

Poly(ethylene terephthalate) (PET) is the highest-volume polyester produced and is used in numerous applications such as films, fibers, and packaging. It is the most commonly used synthetic fiber (25 million tones produced annually world wide) and is forecasted to account for almost 50% of all fiber materials in 2008 (*1*). A disadvantage of PET is its hydrophobic nature resulting in poorly wettable surfaces. As a consequence, difficulties with PET fibers are encountered when applying finishing compounds and coloring agents. Furthermore, PET fibers with hydrophobic surfaces build up electrostatic charges and are prone to bacterial adhesions. Therefore, methods for PET surface modification are important to those who manufacture PET for textiles, biomedical, microelectronics, and packaging industries. Common means for surface hydrophilization range from chemical (e.g. alkali, etching) to plasma treatments (*2*). An interesting alternative is the use of enzymes for polyester surface modification (*3–9*). Advantages of enzymatic methods include their action under mild conditions with low energy input without the need for expensive machinery. Furthermore, enzyme-catalyzed hydrolysis is restricted to material surfaces due to the large size and incompatibility of enzymes with polymeric substrates. Hence, enzymatic hydrolysis has the potential to provide large changes in surface functionality while not affecting bulk properties (*2, 7*). Thus far, PET hydrolyzing activity has been reported for members of cutinase (*3, 5–8, 10*), lipase (*3, 7, 10*) and esterase (*3, 4*) families. Relative to lipases and esterase, cutinases have shown greatest promise for PET hydrolysis (*3, 7, 10*). It was demonstrated that treatment of PET with cutinases from Pseudomonas mendocina (*5*), Fusarium solani (*3*), and Thermobifida fusca (*3*) result in enhanced surface hydrophilicity and increased cationic dye binding.

Apart from surface modification, enzymes are also a potential tool for PET degradation (*10, 11*). The predicted life time of PET ranges from 25-50 years (*12*). PET products are the premier recyclable plastic with a recycle value only second to aluminum. However, for certain PET product forms, such as thin films, textiles or composite structures, recycling of PET is uneconomical. In these cases, degrading PET would provide an alternative strategy to recover value (*12*). Until recently, PET was regarded as non-degradable with excellent hydrolytic stability (*10–12*). PET can be chemically hydrolyzed but harsh chemical conditions such as using sulfuric acid at 150 °C is required (*11*). Attempts to improve its susceptibility to hydrolysis has included increasing the content of aliphatic repeat units along chains (*10–12*). The potential of using enzymatic catalysis for PET recycling was recently demonstrated by Müller et al (*10*) who treated commercial PET of low-crystallinity (9%) from soft drink bottles with a cutinase from T. fusca at 55°C. They reported achieving a weight loss of 50 % after 3 weeks and a decrease in thickness per week of 17 μm.

Herein, we report a detailed study and comparison of the catalytic activities of cutinase from three different organisms, using low-crystallinity (7%) PET films (lcPET) as substrates. The affect of crystallinity on cutinase-catalyzed degradation was determined using biaxially oriented PET films (boPET) with 35% crystallinity. Three different cutinases were included in this study, specifically those from *Humicola insolens* (HiC), *Pseudomonas mendocina* (PmC) and *Fusarium solani* (FsC). These cutinases were first purified (>95%) and are quantified in nmole/mL. This is in contrast to previous publications on enzyme-catalyzed PET degradation where enzyme quantification relied on activity units (*4, 7, 8, 13*). Studies were conducted to elucidate effects of pH and temperature on initial rates of cutinase-catalyzed lcPET degradation. Kinetics of cutinase-catalyzed lcPET hydrolysis was performed and results were fit to kinetic models. The only other literature report on PET-like substrates where kinetic data was reported was by Figueroa et al (*14*). However, they used cutinase from Novozyme with unknown origin and the substrate was cyclo-tris-ethylene terephthalate. To determine kinetic parameters, rates of cutinase-catalyzed PET hydrolysis were measured with a pH-stat, which directly monitors released acid during ester cleavage. Cutinase-catalyzed PET hydrolysis was also studied by measuring PET film weight loss. Results obtained from film weight loss were successfully correlated to determinations of film degradation by HPLC analysis of formed degradation products. Furthermore, results of weight loss allowed comparisons to extents of enzyme-catalyzed PET degradation reported elsewhere (*3, 5, 7*).

Materials and Methods

Materials

HiC and PmC were kind gifts from Novozymes (Bagsvaerd, Denmark) and Genencor (Danisco US Inc., Genencor Division, Palo Alto, CA), respectively. Wild type FsC cloned into *Pichia pastoris* was purchased from DNA 2.0 (Menlo Park, CA). Methods for fermentative synthesis and purification of FsC are given in our paper (*15*). Low crystallinity PET films (*lc*PET, product number 029-198-54) and biaxially oriented PET films (*bo*PTE, product number 543-716-95), both with a thickness of 250 μm, were purchased from Goodfellow Co. Terephtalic acid (TPA) and bis(2-hydroxylethyl) terephthalate (BHET), used as standard samples for HPLC analysis, were purchased from Fluka and Sigma, respectively. Mono(2-hydroxyl ethyl) terephthalate (MHET), also used for HPLC analysis, was obtained as a mixture with water by hydrolyzing BHET with KOH at 75° C for 60 min. All other chemicals were purchased from Sigma-Aldrich Co in the highest available purity and were used without further purification.

Cutinase Catalyzed PET Hydrolysis Monitored by pH-stat

Cutinase activity for PET hydrolysis was assayed using a pH-stat apparatus (Titrando 842, Metrohm) equipped with Tiamo 1.1 software according to the method given in the paper (*15*). In summary, PET films with dimensions 5 x 5 mm^2 were used in order to achieve good stirring and PET concentration was expressed as cm^2/mL. Initial PET hydrolysis rates were expressed as units of μmole of added NaOH per hour and mL of reaction volume. Studies of enzyme activity on PET in pH-stat experiments were performed over temperatures from 30 to 90 °C, pH values 6.5 to 9.5, cutinase concentrations 0 to 12 nmole/mL and PET concentrations 0 to 40 cm²/mL. All reactions were performed in duplicates. Control samples (without enzyme) were run for each parameter in order to determine background chemical hydrolysis that was subtracted from total hydrolysis to measure enzyme-catalyzed PET hydrolysis.

Degradation Studies

A PET film (15 x 15 mm²) with thickness 250 μm was placed in a vial containing 3 mL of Tris-HCl buffer (pH 8.0) with 10 % glycerol and 10 μM of one of the cutinases. The effect of Tris buffer concentration (0 – 1 M) and time course studies were performed by placing vials in a rotary shaker-incubator at 100 rpm, from 0 to 96 h, at optimal temperature values determined from pH-stat experiments (see above). At sampling times, three replicate vials for each cutinase were removed from the shaker-incubator, films were washed extensively with distilled water, dried in vacuo (30 mmHg) for 24 h at 40° C and then weighed. Weight loss was calculated by subtracting the final weight from the initial weight. Control samples (without enzyme) were also carried out in triplicate in order to account for potential contribution to weight loss by chemical hydrolysis. Final weight loss values due to cutinase-catalyzed hydrolysis were calculated by subtracting background weight loss attributed to chemical hydrolysis.

Scanning Electron Microscopy Imaging (SEM)

Morphology of PET films before and after enzyme exposure was examined by Scanning Electron Microscopy (SEM) on a Hitachi S-570 model at an accelerating voltage of 20 kV. Samples were coated with gold in an argon field using a Pelco sputter coater 91000 for 140 s at 18 mA. Digitized images were brought into Epax genesis software for assembly.

Differential Scanning Calorimetry (DSC)

PET films from degradation studies and original low-crystallinity and biaxially oriented PET film (as received) were analyzed by DSC using a TA instruments Q100 DSC, equipped with an LNCS low-temperature accessory. DSC scans were run in the temperature range from 0 – 300 °C under dry nitrogen atmosphere. All runs were carried out at a heating rate and cooling rate of 10°C/min and PET sample weights were about 10 ± 2 mg.

To establish the degree of crystallinity, the heat fusion, ΔH_m, and cold crystallization, ΔH_c, were determined by integrating areas ($J \cdot g^{-1}$) under peaks. Percent crystallinity was calculated using the following equation:

$$\text{Crystallinity (\%)} = \frac{|\Delta H_m| - |\Delta H_c|}{\Delta H^\circ_m} \times 100 \tag{1}$$

where ΔH°_m is the heat of fusion for a 100 % crystalline polymer, which is estimated to be 140.1 J g⁻¹ (7, 16).

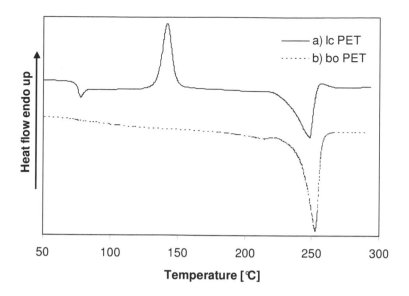

Figure 1. DSC thermograms of a) lcPET, and b) boPET, recorded during first heating scans with a heating rate of 10°C/min.

Reversed-Phase High Performance Liquid Chromatography (HPLC)

The incubation media collected from degradation studies was analyzed by HPLC using a similar method to that developed by Vertommen et al (7). The HPLC system consisted of a Waters 2795 separation module and a Waters 2996 photo diode array detector (Milford, MA, USA). The separation was carried out on a phenomenex Synergi Hydro-RP column (4um, 250mmx4.6mm I.D.) run under isocratic conditions. The mobile phase was composed of 70% water, 20% acetonitrile, 10% formic acid solution (v/v) and the flow rate was 1 mL/min. Standard solutions were mixtures of TPA and BHET at concentrations of 0.5, 1, 5, 10, and 25 µg/mL, respectively. Calibration curves of TPA and BHET were constructed by plotting the peak area of TPA and BHET at wavelength 244 nm versus concentrations of standard solution.

Results and Discussion

DSC Studies of lc- and boPET

Low crystallinity and biaxially oriented PET films (*lc*PET and *bo*PET, respectively) were characterized by DSC. The first heating scan of *lc*PET Figure 1a, showed a clear glass transition temperature (T_g) at 75 °C, a cold crystallization peak at 142 °C and a melting temperature (T_m) at 247 °C. The Cold crystallization peak originates form crystallization of amorphous regions (*17*). In contrast, the thermogram of *bo*PET (Figure 1b) displayed at 253 °C but transitions associated with crystallization and T_g were not seen. Degree of crystallinity values for *lc*PET and *bo*PET, calculated using equation 1 above, are 7.0±0.5% and 35.0±0.5%, respectively. The DSC peaks of *lc*PET and *bo*PET corresponds well with other published work (*16*, *18*, *19*). For example, Karagiannidis et al (*16*) reported that, for PET with degree of crystallinity values higher than 30 % (i.e. similar to *bo*PET), peaks corresponding to T_g and cold crystallization in DSC first heating scans were not observed.

Effect of Temperature and pH on Cutinase-Catalyzed PET Hydrolysis

Details of methods for purification and characterization of the three cutinases used herein are given in the paper (*15*). This showed that HiC, PmC and FsC used herein were highly pure (97±0.4%, 99±0.2% and 96±2%, respectively). Common to previous studies where cutinases were used for polyester hydrolysis reactions, cutinase concentration was not given. Instead, amount of cutinases used in reactions was characterized by their activity for release of *p*-nitrophenol (*p*NP) from *p*NP esters of short chain fatty acids. Subsequently, based on activity of cutinases on these model substrates, their activities were compared for polymer hydrolysis reactions (*4*, *7*, *8*, *13*). Not surprisingly, Heumann et al (*3*) showed there was no correlation between cutinase activity on *p*NP esters and their activity for hydrolysis of PET. Thus, in this study, rather than quantifying enzymes by activity units on a surrogate substrate, concentration of purified cutinases are given in mol/mL.

Cutinase activity for PET hydrolysis was assayed using a pH-stat to measure NaOH consumption versus time. Titration of NaOH by pH stat keeps the pH constant as acid is liberated due to cutinase-catalyzed PET hydrolysis. Initial slopes in μmole NaOH titrated per h were recorded by pH-stat during the first hour of incubations, and these slope values were used in subsequent plots to assess cutinase activity as a function of incubation parameters (e.g. temperature, pH). Figure 2a and 2b shows the effect of temperature on cutinase initial activity on *lc*PET at pH 8.0. As the temperature increases from 30 to 80 °C, HiC activity increases almost 30 fold to 10.6 μmole/mL/h . Thereafter, HiC activity drops steeply to 0 at 90°C, likely due to thermal induced denaturation. Figure 2b shows that, at 50 °C, both PmC and FsC reach their maximum activities of 1.3 and 1.8 μmole mL/h respectively. PmC and FsC lose all activity at about 60 °C, hence, their thermal stabilities are substantially lower than HiC.

Further inspection of temperature dependent HiC-catalyzed *lc*PET hydrolysis results (Fig. 2a) shows that hydrolysis activity dramatically increased at temperatures above 65 °C. This is explained by HiC's high thermal stability that allows incubations of HiC with *lc*PET to be conducted at temperatures nearby *lc*PET's T_g (75 °C). Welzel's Ph.D. thesis (*20*) demonstrated that amorphous phase mobility influences polyester biodegradability. Furthermore, Welzel (*20*) showed that, in order to enable degradation of aromatic polyesters such as PET, the degradation temperature should be close to its T_g. Indeed, Müller et al (*10*) attributed reaching 50 % PET film (surface area of 2.25 cm²) weight loss during incubations with *T. fusca* cutinase with its low crystallinity (10 %). This enabled degradations at 55 °C, a temperature above that used in previous cutinase-catalyzed PET hydrolysis studies (*3*, *5*, *7*, *8*).

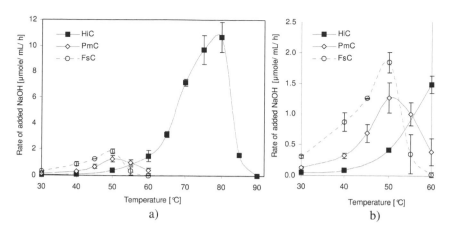

Figure 2. Temperature dependence of cutinase activity for lcPET hydrolysis at pH 8.0 determined by pH-stat, using 10 mM NaOH as titrant, with 13 cm2/mL lcPET and 6 nmole/mL of cutinase in 3 mL of 0.5 mM Tris buffer containing 10% glycerol. Graph b) is an expansion of the 30 -60 °C temperature region in 1 a). Error bars represent standard deviation method based on duplicate repeats.

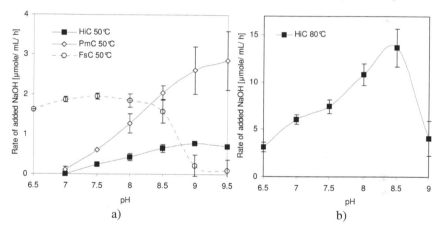

Figure 3. Activity of cutinase for lcPET hydrolysis determined by pH-stat, using 10 mM NaOH as titrant, with 13 cm2/mL lcPET and 6 nmole/mL cutinase in 3 mL of 0.5 mM Tris buffer containing 10% glycerol. Fig. 3a displays pH dependence at 50 °C for HiC, PmC and FsC whereas Fig. 3b displays pH dependence at 80 °C for HiC. Error bars represent standard deviation method based on duplicate repeats.

The effect of pH on initial HiC, PmC and FsC activities at 50 °C, and HiC activity at 80 °C, are illustrated in Figures 3a and 3b, respectively. The activity of FsC is largely unchanged between pH 6.5 and 8.5 (1.9 to 1.6 μmole/mL/h) but drops sharply to 0.2 at pH 9.0. Hence, FsC shows poor stability at alkaline pH values of ≥ 9.0. In contrast, PmC activity increases from 0.2 to 2.8 μmole/mL/h as the pH is increased from 7.0 to 9.5. Hence, PmC shows exceptional stability under alkaline conditions. HiC shows a similar pH activity trend as PmC at 50 °C

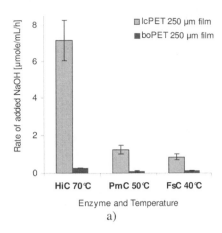 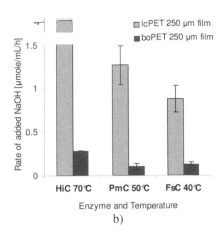

a)   b)

Figure 4. Effect of crystallinity on cutinase-catalyzed hydrolysis of PET at pH 8.0, determined by pH-stat, using 10 mM NaOH as titrant, with 13 cm²/mL of either lcPET (7% crystallinity) or boPET (35 crystallinity) and 0.13 mg/mL of either HiC, PmC or FsC at 70, 50 and 40 °C, respectively. Figure b) is an expansion of the y-axis in graph a) from 0 to 1.5 µmole/mL/h. Error bars represent standard deviation method based on duplicate repeats.

since its activity increases from 0.1 to 0.8 µmole/mL/h as pH increased from 7.0 to 9.5. Study of Figure 2b along with comparison between Figures 3a and 3b further emphasize the relatively low activity of HiC at 50 °C relative to its activity at 70 and 80 °C. Figure 3b shows that at 80 °C and pH 8.5, HiC reaches its maximum initial activity of 13.6 µmole/mL/h. However, unlike at 50 °C, increase in medium pH from 8.5 to 9.0 results in a large decrease in initial activity to 4.1 µmole/mL/h. Hence, HiC's tolerance to pH values of $\geq$ 9.0 is decreased by increases in medium temperature.

Influence of Crystallinity

To investigate the influence of PET crystallinity on HiC, PmC and FsC activity, comparisons were made using *lc*PET and *bo*PET (see above). Observation of Figure 4 shows that, initial activities of HiC, PmC and FsC decreased by 25, 10 and 6 fold when exposed to PET films of higher crystallinity (35 vs. 7%). This is consistent with results reported by many other investigators who studied how alteration in polyester crystallinity influences hydrolytic enzyme activity (*21–26*) Vertommen et al (*7*) showed that FsC displayed higher hydrolytic activity toward amorphous PET than crystalline PET (4% versus 48 % crystallinity). Marten et al (*27*) showed that, in order to improve the biodegradation rate for crystalline or semi-crystalline polyesters, enzyme-polymer incubations should be performed close to the polymers melting point. However, since PET's melting point (250 °C) is well above temperatures at which enzymes are stable, this strategy cannot currently be employed for PET with high crystalline content.

Further study of Fig. 4 shows that, when *lc*PET is the substrate, HiC exhibits a 6 fold higher initial rate relative to PmC and FsC. When *bo*PET is the substrate, HiC has about two-fold higher hydrolytic activity relative to PmC and FsC. As discussed above, the importance of being able to conduct enzyme-PET incubations near or above its T_g cannot be overemphasized. Interestingly, comparison of activities of HiC, PmC and FsC at 50 °C (pH 8.0, see Fig. 2b) on *lc*PET show that their initial rates are 0.42, 1.27 and 1.84 mMh^{-1}, respectively. In other words, if restricted to lower temperature conditions, HiC would not be the preferred enzyme for PET hydrolysis. By remaining active at PET's T_g, HiC has the advantage of access to more mobile chains in the amorphous phase. Thus, the observed higher activity of HiC than FsC and PmC under optimal temperature-pH conditions appears to be largely a function of physical changes occurring in PET at its T_g.

In kinetic studies described below with HiC, PmC and FsC, cutinase-*lc*PET incubations were conducted at pH 8.0 and at 70, 50 and 40 °C, respectively. Hence, although FsC showed maximum initial activity at 50 °C (Fig. 2), its poor thermal stability demonstrated in the paper (*15*) suggested that working at a lower temperature is advised. Furthermore, the decision to carry out kinetic studies with HiC at 70 instead of 80 °C was to reduce the magnitude of uncertainty in measurements (see error bars in Fig. 1a at 70, 80 and 90 °C) as well as to reduce water evaporation during long-term incubations. Moreover, instead of maintaining the pH at 8.5 for PmC and HiC, all three cutinases were studied at pH 8.0 to reduce background chemical hydrolysis while supporting high cutinase activity.

Kinetic Studies

The classical Michaelis-Menten enzyme kinetic model, derived for homogeneous reactions, is based on enzyme-limited conditions. However, in the case of enzymatic polymer degradation where the substrate is usually insoluble and the reaction is limited to the surface of the substrate (substrate-limited), the Michaelis-Menten model will generally not apply. Alternative heterogeneous kinetics models based on substrate-limited conditions have therefore been developed for: *i*) PHB-depolymerase-catalyzed poly(3-hydroxybutyrate) hydrolysis (*28–31*), *ii*) lipase-catalyzed PCL hydrolysis, *iii*) lipase-catalyzed hydrolysis of polyester nanoparticles (*20, 32, 33*) and *iv*) cellulase-catalyzed hydrolysis of cellulose (*34–36*). Mukai et al (*29*) proposed a kinetic model for PHB-depolymerase-catalyzed hydrolysis of poly(β-hydroxybuterate), PHB, to account for enzyme-substrate heterogeneity as well as different roles played by enzyme binding and active sites. This model was later modified by Timmins and Lenz (*28*) in order to account for substrate concentration. Subsequently, Scandola et al (*31*) demonstrated that the same enzymatic degradation could be expressed by a two-step kinetic model with the following initial reaction rate (V_0) equation:

$$V_0 = k_2[S]_0 \, \frac{K[E]_0}{K + [E]_0} \qquad (2)$$

Or as a linearized expression:

$$\frac{[S]_0}{V_0} = \frac{1}{Kk_2[E]_0} + \frac{1}{k_2} \quad (3)$$

where $[S]_0$ is the initial surface concentration, $[E]_0$ is the initial enzyme concentration, and k_2 is the hydrolysis rate constant. Also, K is the adsorption equilibrium constant defined as the ratio of rate constants for adsorption to desorption (k_1/k_{-1}). Interestingly, the rate equations from the kinetic models referred to above for lipases (20, 32, 33) and cellulases (34–36) are expressed similarly as equation (2), but where the constants are defined differently. Herzog et al (33) developed a rate equation for lipase-catalyzed hydrolysis of polyester nanoparticles. Like equation (2), they assumed that the degradation rate is proportional to the surface area, but they also introduced that the molar density of ester bonds of the polymer, and enzyme loading at the polymer surface, are proportional to the degradation rate. The rate equation they proposed to describe enzymatic polyester hydrolysis takes into account important variables that undoubtedly influence the degradation rate. However, by introducing two additional variables, the equation becomes more complicated to solve and requires further information about the system under study. Herzog et al (33) solved for these two variables by introducing an equation that describes the conversion of cleaved ester bonds as a function of time until the reaction reaches a plateau value. While evaluation of how experimental data obtained herein would fit rate expressions developed by Herzog et al (33) is of great interest and was strongly considered, further study led us to conclude this evaluation could not be performed with the current system. That is, the Herzog rate equations require rapid polyester hydrolysis which was achieved by using an aliphatic polyester substrate with high available surface area resulting from nanoparticle formation. In contrast, lcPET degrades relatively slowly and its surface area is limited since it is used in film form. Therefore, the kinetic study presented below was performed by applying equations (2) and (3).

At fixed PET surface concentration (13 cm^2/mL). Incubations were performed for 1 h in 0.5 mM Tris-HCl buffer (3 mL, pH 8) with 10% glycerol, at 70, 50 and 40 °C for HiC, PmC and FsC, respectively. Plots of experimental enzymatic hydrolysis rates vs. enzyme concentration are displayed in Figure 5 (left side). Experimental data points are shown as filled symbols, whereas, theoretical results from equation fitting are shown as solid lines. The right side of Figure 5 presents reciprocal plots based on Eq. 3. It is observed that the hydrolysis rate dependence of $[E]_0$ changes from first to zero-order kinetics. The maximum rates reached by HiC, PmC and FsC are 6.1, 0.8 and 0.9 µmole/mL/h, respectively. Using solver utility in Excel for windows, values for parameters K and k_2 were found by fitting Eq. (2) to experimental data by least mean square regression fit. The results are shown as solid lines on the left side of Figure 5. Similarly, Eq. (3) was fit to the experimental data by applying the calculated K and k_2 from Eq. 2 as input. The reciprocal form of this data and the corresponding values of R^2 in Figure 5 (right side) verify the linearity of experimental data. Furthermore, agreement between experimental data plotted in the reciprocal form suggested by Eq. (3) and

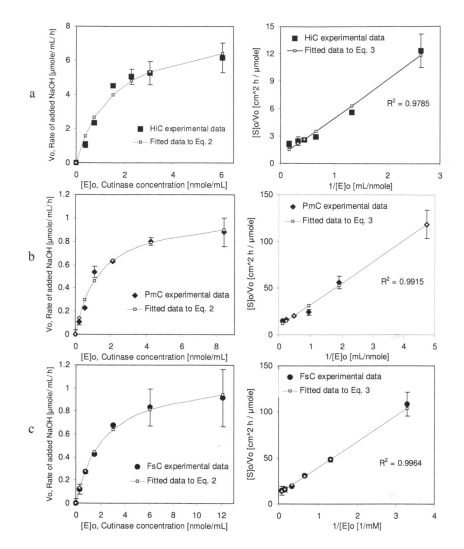

Figure 5. Initial rate of NaOH consumption as a function of cutinase concentration and reciprocal plots based on Eq. 3 are shown on the left and right sides, respetively. Experiments were performed with HiC, PmC and FsC at 70, 50 and 40 °C, respectively, with lcPET concentration fixed at 13 cm²/mL. Experimental values are plotted as filled symbols whereas data fitted to Eq. (2) and Eq. (3) are shown as solid lines on the left and right side, respectively. Error bars represent standard deviation method based on duplicate repeats.

corresponding fitted plots using Eq. (2) show that, hydrolysis of lcPET, using HiC, PmC and FsC, obeys the kinetic model proposed by Scandola (*31*).

The pH-stat assay was used to determine the dependence of lcPET hydrolysis rate on cutinase concentration. Concentrations of the three cutinases were varied The kinetic parameters determined from Eq. (2) for HiC, PmC and FsC are shown Table 1. HiC at 70°C has the highest hydrolysis rate constant k_2 (0.62 µmole/

cm^2·/h), which is 7 fold higher than rates of PmC and FsC at 50°C and 40 °C, respectively. Interestingly, this corresponds to observations in the temperature study (Figure 2) where HiC at 70 °C showed 6 fold higher activity than PmC at 50 °C and FsC at 40 °C. Further study of Figure 2 shows that PmC and FsC have similar hydrolytic activities at 50 and 40 °C, respectively. Also, k_2 results in Table 1 for PmC and FsC are close in value (0.08 and 0.09 μmole/cm^2·/h, respectively). The adsorption constant K shown in Table 1 is a measure of the enzymes affinity to *lc*PET. Though the three cutinases show similar K values, PmC has the highest affinity to PET while FsC possesses the lowest affinity. Interestingly, the same observation was made in our paper (*15*) where PmC showed the highest affinity to PVAc while FsC showed the lowest. However, in PVAc hydrolysis, differences in cutinase affinities were larger in magnitude.

Figure 6 shows the initial hydrolysis rate data measured by NaOH consumption as function of *lc*PET concentration. Concentrations of HiC, PmC and FsC were held constant at 2, 2 and 3 nmole/mL, respectively. The cutinase concentrations selected were at the higher end of those within the linear region of V_0 versus E_0 plots, so that reactions were still under substrate saturated conditions. Experimental hydrolysis rates of *lc*PET are shown as filled symbols and solid lines in Figure 6.They reveal a linear dependency on $[S]_0$ up to 15, 20 and 20 cm^2/mL for HiC, PmC and FsC, respectively. Taking values of K and k_2 from Table 1, Eq. 2 was used to generate a plot of V_0 as a function of $[S]_0$, shown as dotted lines in Figure 6. By comparing experimental and fitted data from Eq 2, it is concluded that Eq. (2) is applicable only under substrate-limiting conditions. This confirms that, under conditions where cutinase activity is determined using *lc*PET in film form as substrate, the kinetic data for all three cutinases, obtained both as a function of enzyme and substrate concentration, fit the kinetic model proposed by Scandola (*31*) (Eq. 2). Of interest would be to perform a similar kinetic analysis using *bo*PET as substrate. Unfortunately, using the pH-stat method described herein and in the paper (*15*), the hydrolysis rate was too slow. Thus, for kinetic studies of *bo*PET with the enzymes studied herein, a more sensitive analytic method will be required.

Table 1. Kinetic parameters from equation (2) determined for PET hydrolysis using HiC, PmC and FsC at 70, 50 and 40 °C, respectively

	$[S]_0$ (μM)	K (μM^{-1})	k_2 $(\mu mole/cm^2·/h)$
HiC	13	0.64	0.62
PmC	13	0.76	0.08
FsC	13	0.41	0.09

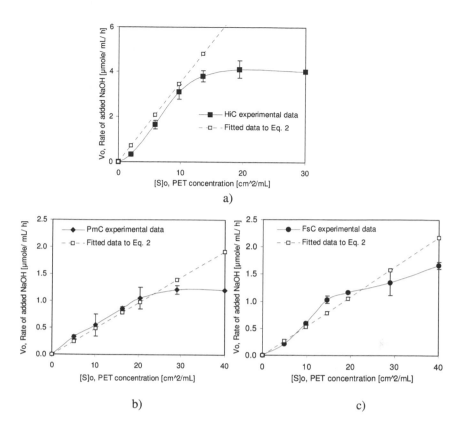

Figure 6. Initial rate of NaOH consumption as a function of lcPET concentration: a) 2 nmole/mL HiC at 70 °C, b) 2 nmole/mL PmC at 50 °C, c) 3 nmole/mL FsC at 40 °C. Experimental values are plotted as filled symbols and solid curves whereas plots generated by fitting Eq. 2 are shown as dotted lines with open symbols. Error bars represent standard deviation method based on duplicate repeats.

Long-Term Incubations of Cutinases with PET

Kinetic and temperature studies were based on results from measured initial rates, determined during the first hour of cutinase-PET incubations. Also of interest is to determine to what degree the three cutinases are able to degrade lcPET film during extended incubation times. Figure 7 shows weight loss of lcPET film (80 mg with dimensions: 15 mm^2 x 15 mm^2 x 250 μm) as a function of incubation time and cutinase used in 1 M Tris-HCl. Consistent with expectations from kinetic studies above, HiC was most active for catalysis of lcPET hydrolysis. In fact, HiC completely degrades lcPET film (97 ± 3 % weight loss) to water soluble products within 96 h. Taking into account that PET films are about 250 μm thick, this represent a degradation rate of 30 μm per day or 210 μm per week (per film side). Regarding enzyme coverage, 0.09 mg of enzyme is available per cm^2. In control studies conducted without HiC, no significant (<1%) lcPET weight loss was observed. In a related study, Müller et al (*10*) exposed a low-crystallinity

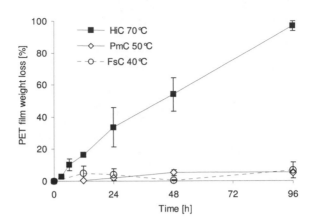

Figure 7. Degradation study of lcPET (2.25 cm²/mL) as a function of incubation time in 1 M Tris-HCl with 10% glycerol, at pH 7.5 and 10 nmole/mL of either HiC, PmC or FsC at 70, 50 and 40 °C, respectively. Error bars represent standard deviation method based on triplicate repeats.

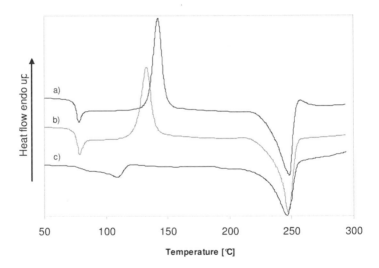

Figure 8. DSC thermograms recorded during first heating scans of: a) as received lcPET, b) lcPET incubated without enzyme addition for 96 h at 70 °C in Tris-HCl (pH 8), and c) recovered film after incubation of lcPET for 96 h with HiC in Tris-HCl (pH 8.0) at 70 °C.

PET film to *T. fusca* cutinase at 55°C. Available *T. fusca* cutinase was 0.22 mg per cm² PET surface. The result was a film weight loss of 50% within 3-weeks and a thickness decrease rate of 17 µm per week. Furthermore, PET used by Müller et al (*10*) had a crystallinity of 9%, T_g at 75 °C and a melting temperature at 249 °C. Similarly, *lc*PET used herein has a degree of crystallinity of 7 ±1 %, T_g at 75 °C and a melting temperature at 247°C. Thus, the PET films used in these two studies have similar physical properties. It is therefore concluded that HiC has remarkably higher activity than *T. fusca* cutinase for *lc*PET hydrolysis. This is largely attributable to the higher thermal stability of HiC. Indeed, incubations of HiC with lcPET at 70°C are close to the T_g of PET. Inspection of Fig. 2 shows that by increasing the incubation temperature of HiC with lcPET from 55 to 70 °C, the initial rate of hydrolysis increased by 7-fold. Thus, HiC more affectively hydrolyzes PET when its amorphous phase is near, at or above PET's T_g where chains at or nearby the film surface have greater mobility.

Figure 7 also shows the results of lcPET incubation with PmC at 50°C and FsC at 40°C. For FsC, 5 ± 1 % weight loss was reached within the first 24 h and, thereafter, little change in film weight loss was observed. For PmC, film weight loss slowly increased so that by 48 h, weight loss reached 5% and no further increase in weight loss was observed to 96 h. The liquid media from both FsC and PmC film incubation studies was separated from films after 96 h. These liquid media were then used to perform weight loss studies with virgin lcPET films. Similar weight loss values were reached with the new films (~5% within 96 h) suggesting that the attainment of a plateau after which further weight loss was not observed is not attributable to deactivation of cutinase remaining in media (not on film surfaces). Vertommen et al (7) reported strong adsorption by FsC to PET surfaces occurs. One hypothesis that explains these results is that PmC and FsC saturate available lcPET surfaces and, over the course of 24 to 48 h, become deactivated. Soluble cutinase in the media may not have access to enzyme saturated surfaces and no further degradation occurs. Further work is planned to investigate the validity of this hypothesis. PmC and FsC, like T. fusca cutinase, lack sufficient thermal stability to allow incubations at or above 70 °C. Being 25 to 35 °C below the Tg of PET is a serious disadvantage as is discussed above.

It is noteworthy that boPET was similarly exposed for 96 h to HiC, PmC and FsC at 70, 50°C and 40 °C, respectively. However no measurable weight loss was detected with all three cutinases.

Of interest is the extent that lcPET crystallinity changes both during exposure to HiC and under control conditions without enzyme at 70 °C. Figure 8 shows DSC thermograms of lcPET recorded during first heating scans: *i*) prior to incubations (scan a), *ii*) a control, incubated for 96 at 70°C (scan *b*), and *iii*) recovered PET after exposure to HiC at 70 °C for 96 h (scan c). Analysis of thermogram *a* shows that non-treated lcPET film has a T_g at 75 °C, a cold crystallization peak at 142 °C, and a peak melting temperature (T_m) at 247 °C. Study of thermogram c, recorded of PET recovered after exposure to HiC for 96 h (~3% of the starting film weight), shows the T_g increases to 80 °C and the cold crystallization transition is not found. Instead of a crystallization transition, a small pre-melting endothermic peak (*16*) appears at about the same temperature region. Furthermore, from analysis of the melting heat of fusion, the crystallinity increased to 27%. Therefore, consistent with results of HiC on *bo*PET (see above), HiC preferentially degrades amorphous regions of the polymeric material. The preferential attack by enzymes of polymer amorphous regions is expected and has been reported by others using various enzyme-polymer systems (7, 10, 21–26, 37, 38). Of six replicate experiments where lcPET was exposed to HiC for 96 h at 70 °C, two of these runs resulted in complete conversion of film to water-soluble products (e.g. 100% film weight loss). Thus, although HiC is more active on amorphous regions of PET, it has sufficient activity on crystalline

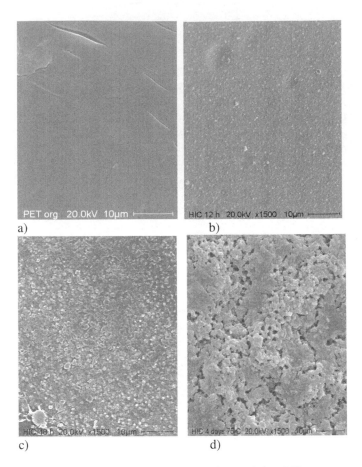

a)

b)

c)

d)

Figure 9. SEM images of lcPET films exposed to HiC for different time periods resulting in varying extents of film weight loss. The following describes incubation times and %-film weight loss values: a) lcPET as received b) 12 h exposure with 18% weight-loss, c) 48 h exposure with 54% weight-loss, d) 96 h exposure with 95% weight-loss.

domains to completely degrade *lc*PET. Another important question was whether *lc*PET crystallinity increases under control conditions when heated without enzyme at 70 °C for 96 h. Recovery of film after this control experiment and analysis of crystallinity by DSC shows only a small increase to 9%. This minor rise in crystallinity is also evident by a shift of the cold crystallization peak to lower temperature compared to the initial sample, a phenomenum known to occur for even small changes in crystallinity (*17*).

Effect on surface roughness caused by incubation of HiC with *lc*PET film was studied by recording SEM photographs displayed in Figure 9. Inspection of SEM images shows that, as incubation time was increased from 12 to 48 and 96 h corresponding to weight loss values of 18, 54 and 95%, surface roughness increases so that, at 96 h, the recovered film surface has holes of up to 5 μm penetrating below the surface.

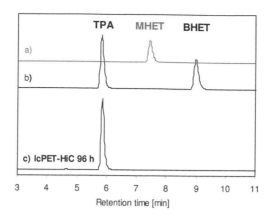

Figure 10. In HPLC chromatogram c, the hydrolysis product (TPA) generated by HiC catalyzed hydrolysis of lcPET (2.25 cm²/mL) by 10 nmole/mL of HiC in 1 M Tris-HCl with 10% glycerol (pH 8.0, 70 °C). Chromatograms a and b show retention times of TPA, MHET and BHET standards prepared for this analysis.

Analysis of Degradation Products

Separation of terephthalic acid (TPA), bis(2-hydroxylethyl) terephthalate (BHET), and mono(2-hydroxyl ethyl) terephthalate (MHET) was achieved by HPLC using a reversed phase column run under isocratic conditions with U.V. detection (see Materials and Methods Section and chromatograms *a* and *b* in Fig. 10). This allowed construction of calibration curves for these compounds from 0.5 to 25 µg/mL with R^2 values of 0.997.

Figure 10c shows the HPLC chromatogram obtained by analyzing the medium after a 96 h incubation of *lc*PET with HiC in 1 M Tris-HCl (pH 8) at 70 °C. The only compound found is TPA. Similarly, only TPA was found when the same HPLC analysis was performed on media from incubations of *lc*PET with PmC and FsC for 96 h corresponding to film weight loss values of 5 % for both enzymes. In contrast, Yoon et al (*5*) reported that, in addition to TPA, treatment of PET fibers with PmC produced non-identified terephtalate containing species whereas, NaOH treated PET fibers yielded only TPA. Vertommen et al (*7*) found that FsC-catalyzed PET hydrolysis predominantly produced MHET, some TPA and small traces of BHET. They also reported that the ratio of MHET to TPA depended on the ratio of cutinase to PET. The smaller the ratio of enzyme to substrate, the more predominant MHET became. Relative to studies carried out herein, previous work discussed in Refs. (*5*) and (*7*) were conducted with differences in reaction temperature, buffer and ratio of cutinase to substrate. Such changes in incubation conditions could lead to incomplete conversion of water-soluble PET degradation products to TPA. However, our study shows definitively that, under the appropriate conditions, HiC, FsC and PmC are all active for complete conversion of *lc*PET to TPA and ethylene glycol.

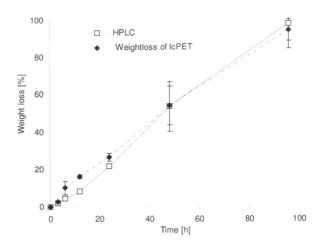

Figure 11. Comparison of lcPET film weight loss determined by measuring TPA concentration accumulated in media and by direct gravimetric measurements to determine film weight loss for incubations with 10 nmole/mL HiC in 1 M Tris-HCl with 10% glycerol at 70 °C and pH 8.0. Error bars represent standard deviation method based on triplicate repeats

Based on the above discovery that TPA is the only degradation product that accumulates in media during HiC-catalyzed degradation of *lc*PET, TPA concentration in media determined by HPLC analysis as a function of incubation time was used to predict the extent of *lc*PET film weight loss that occurred during incubations with HiC at 70 °C and pH 8. Values of film weight loss determined by measuring mass change of films and calculated based on HPLC analysis of TPA in incubation media were in excellent agreement (see Fig. 11). This comparison validates the HPLC method used, along with the conclusion that TPA is the predominant degradation product formed from cutinase-catalyzed hydrolysis of *lc*PET. Furthermore, the results herein show the potential of using cutinases for conversion of PET to high purity TPA and ethylene glycol.

Conclusion

This paper compares the activity of three different cutinases in purified form, notably HiC, PmC and FsC, using PET with low (*lc*, 7%) and higher (*bo*, 35%) crystallinity. Consistent with previous reports by others using different enzyme-polymer systems (*21–26, 37, 39*) increased PET crystallinity negatively affects cutinase ability to degrade PET. Initial activities of cutinase on *lc*PET were successfully fit to a heterogeneous kinetic model. This provides an opportunity to quantitatively compare HiC, PmC and FsC while gaining a better understanding of fundamental differences in their kinetic parameters. For example, results herein revealed that PmC and FsC had the highest and lowest affinities, respectively, for *lc*PET. Furthermore, the initial hydrolysis rate of HiC is 7 fold higher than PmC and FsC. The reason for HiC's remarkable activity for *lc*PET hydrolysis was largely attributed to its ability to remain active at 70 °C, just below PET's T_g. At this temperature, HiC benefits from higher mobility of chains in the amorphous phase thus increasing the accessibility of HiC to PET ester groups. Results from initial degradation rates were consistent with extended time incubations of cutinases with *lc*PET. Thus, incubations of 96 h duration at 70 °C of HiC with *lc*PET resulted in 97±3% film weight loss. In contrast, 96 h incubations of *lc*PET with PmC and FsC at 50°C and 40°C, respectively, resulted in 5 % weight loss. Nevertheless, PmC and FsC remain of great interest for performing surface modification of PET used in textile and biomedical applications where the goal is to increase surface hydrophilicity that requires little or no PET weight loss. Indeed, previous studies by others have demonstrated the utility of PmC and FsC for increasing PET surface hyrophilicity (*3–8*). Results herein show that HiC's activity for *lc*PET hydrolysis is much higher than was previously reported (*10*) and HiC was found to remain active over extended times to completely degrade films that are 250 μm thick. Furthermore, all three cutinases degrade PET to TPA. Thus, the potential to complete convert commercial PET materials of low crystallinity to TPA and ethylene glycol under mild conditions was demonstrated. Indeed, the majority of PET recycling is focused on the bottle manufacturing industry that utilizes PET of low crystallinity to achieve high bottle transparency (*10*).

References

1. Guebitz, G.; Cavaco-Paulo, A. *Trends Biotechnol.* **2007**, *26* (1), 32–38.
2. Fischer-Colbrie, G.; Heumann, S.; Guebitz, G. Enzyme for Polymer Surface Modification. In *Modified Fibers with Medical and Specialty Applications*; Edwards, J. V., Buschle-Diller, G., Goheen, S. C., Eds.; Springer: 2006; pp 125–143.
3. Heumann, S.; et al. *J. Biochem. Biophys. Methods* **2006**, *39*, 89–99.
4. Alisch, M.; et al. *Biocatal. Biotransform.* **2004**, *22* (5/6), 347–351.
5. Yoon, M. Y.; Kellis, J. T.; Poulouse, A. J. *AATCC Rev.* **2002**, *2* (6), 33–36.
6. Kellis, J. T.; Poulouse, A. J.; Yoon, M. Y. Enzymatic Modification of the Surface of a Polyester Fiber or Article. U.S. Patent 6,254,645 B1, 2001, Genencor, Inc.: Rochester, NY.
7. Vertommen, M. A. M. E.; et al. *J. Biotechnol.* **2005**, *120*, 376–386.
8. Araujo, R.; et al. *J. Biotechnol.* **2007**, *128*, 849–857.
9. Eberl, A.; et al. *J. Biotechnol.* **2008**, *135*, 45–51.
10. Mueller, R.-J.; et al. *Macromol. Rapid Commun.* **2005**, *26*, 1400–1405.
11. Mueller, R.-J.; Kleeberg, I.; Deckwer, W.-D. *J. Biotechnol.* **2001**, *86*, 87–95.
12. Gallagher, F. G. Controlled Degradation Polyesters. In *Modern Polyesters: Chemistry and Technology of Polyesters and Copolyesters*; Scheirs, J., Long, T. E., Eds.; John Wiley & Sons: 2003; pp 591–608.
13. Nimchua, T.; Punnapayak, H.; Zimmermann, W. *Biotechnol. J.* **2007**, *2*, 361–364.
14. Figueroa, Y.; Hinks, D.; Montero, G. *Biotechnol. Prog.* **2006**, *22* (4), 1209–1214.
15. Ronkvist, A. M.; et al. *Macromolecules* **2009**, *42* (16), 6086–6097.

16. Karagiannidis, P. G.; Stergiou, A. C.; Karayannidis, G. P. *Eur. Polym. J.* **2008**, *44*, 1475–1486.

17. Pingping, Z.; Dezhu, M. *Eur. Polym. J.* **1997**, *33* (10–12), 1817–1818.

18. Xu, T.; et al. *Macromolecules* **2004**, *37*, 6985–6993.

19. Lee, B.; et al. *Macromolecules* **2004**, *37*, 4174–4184.

20. Welzel. *Einfluss der chemischen Struktur auf die enzymatische Hydrolyse von Polyester-Nanopartikeln*; TU: Braunschweig, 2003; p 188.

21. Huang, S. J. *Compr. Polym. Sci.* **1989**, *6*, 597–606.

22. Nishida, H.; Tokiwa, Y. *J. Environ. Polym. Degrad.* **1993**, *1* (1), 65–80.

23. Mochizuki, M.; et al. *J. Appl. Polym. Sci.* **1995**, *55*, 289–296.

24. Abe, H.; Doi, Y. *Int. J. Biol. Macromol.* **1999**, *25*, 185–192.

25. Yoo, E. S.; Im, S. S. *J. Environ. Polym. Degrad.* **1999**, *7* (1), 19–26.

26. Seretoudi, G.; Bikiaris, D.; Panayiotou, C. *Polymer* **2002**, *43*, 5405–5415.

27. Marten, E.; Muller, R.-J.; Deckwer, W.-D. *Polym. Degrad. Stab.* **2003**, *80*, 485–501.

28. Timmins, M. R.; Lenz, R. W. *Polymer* **1997**, *38* (3), 551–562.

29. Mukai, K.; Yamada, K.; Doi, Y. *Int. J. Biol. Macromol.* **1993**, *15* (December), 361–366.

30. Kasuya, K.; Inoue, Y. *Polym. Degrad. Stability* **1995**, *48*, 167–174.

31. Scandola, M.; Focarete, M. L.; Frisoni, G. *Macromolecules* **1998**, *31*, 3849–3851.

32. Wu, C.; et al. *Polymer* **2000**, *41*, 3593–3597.

33. Herzog, K.; Mull, R. J.; Deckwer, W.-D. *Polym. Degrad. Stab.* **2006**, *91*, 2486–2498.

34. Bailey, J. *Biochem. J.* **1989**, *262*, 1001.

35. Sattler, W.; Esterbauer, H. *Biotechnol. Bioeng.* **1989**, *33*, 11221–1234.

36. Nidetzky, B.; et al. *Enzyme Microb. Technol.* **1994**, *16*, 43–52.

37. Canetti, M.; Urso, M.; Sadacco, P. *Polymer* **1999**, *40*, 2587–2587, 2594.

38. Mueller, R.-J. *Process Biochem.* **2006**, *41*, 2124–2128.

39. Parikh, M.; Gross, R. A.; McCarthy, S. P. The Effect of Crystalline Morphology on Enzymatic Degradation Kinetics. In *Biodegradable Polymers and Packaging*; Ching, C., Kaplan, D. L., Thomas, E. L., Eds.; Technomic: Lancaster Basel, 1993; p 407.

Chapter 27

Polylactic Acid (PLA)-Degrading Microorganisms and PLA Depolymerases

Fusako Kawai*

R & D Center for Bio-based Materials, Kyoto Institute of Technology,
Matsugasaki, Sakyo-ku, Kyoto 606-8585, Japan
*fkawai@kit.ac.jp

This paper summarizes topics on microorganisms able to degrade polylactic acid (PLA) and PLA depolymerases. Although the glass transition temperature for PLA is high (approximately 60 °C), PLA-degrading microbes can degrade solid PLA at far lower temperatures such as 30 or 37 °C. Such degraders are Actinomycetes belonging to family *Nocardiaceae*: PLA depolymerases were purified and cloned as serine proteases from genus *Amycolatopsis*. Thermophilic lipases were obtained from thermophilic *Bacillus* strains able to grow on PLA at 60 °C, although their contribution to degradation of PLA is skeptical at high temperature as PLA is easily hydrolyzable. Commercially available proteases and lipases are known to act as PLA depolymerases. We found that enantioselectivity of protease-type depolymerases is specific to poly(L-lactic acid), but that of lipase-type depolymerases is preferential to poly(D-lactic acid). Thus, proteases and lipases are categorized into two different classes of PLA depolymerases.

Introduction

The start of chemical synthesis of poly(lactic acid) (PLA) dates back to 1932 when Carothers first synthesized PLA of approximately 3,000 Da. In the 1960s, PLA found its use in medical fields as a bio-absorbable material. Since the latter half of the 1980s, plastic waste attracted public concern as an environmental issue. Very recently, rising cost and limited resources of crude oil has turned more attention toward alternative sources. This trend shed light on PLA again as a bio-based material capable of replacing oil-based materials. PLA is chemically synthesized from lactic acid, a representative fermentation product from plant resources, and is thereby defined as a biomass plastic. The biodegradability of biomass plastics is not a main focus, since the CO_2 released from PLA by combustion is thought to equal to the CO_2 absorbed by plants, thereby yielding zero-emission of CO_2. Therefore, these plastics are designated "carbon-neutral." However, the biodegradability of PLA has already been established since the first report on enzymatic hydrolysis of PLA by William that described the feasibility for proteases and unfeasibility for esterases (1). Later, some lipases and esterases were reported to be able to hydrolyze PLA (2, 3), but they seem to be active only for low molecular weights (4) or poly(DL-lactide) (5). Among acid, neutral and alkaline proteases, only alkaline proteases showed appreciable activity (14 positive results among a total of 22) for high molecular weight PLA (6), but all the commercial lipases were negative (7). Lim et al. showed that all the six mammalian and microbial serine proteases hydrolyzed PLA well (8). There are also many reports on the microbial assimilation of PLA, since Pranamuda et al. first isolated the PLA-assimilating *Amycolatopsis* sp. strain HT-32 (9). Tokiwa and Calabia concluded that most of the PLA-degrading microorganisms phylogenetically belong to the family of *Pseudonocardiaceae* and related genera such as *Amycolatopsis, Lentzea*, etc. in which proteinous materials promote the production of the PLA-degrading enzyme (10). PLA-degrading enzymes from PLA-assimilating microorganisms were purified from different strains of *Amycolatopsis* approximately at the same time by two groups (11, 12) and were characterized as proteases. Later, both groups cloned the genes (13, 14). It is notable that almost all the degradation tests have been carried out using poly(L-lactic acid) (PLLA) and that no information regarding the biodegradability of poly(D-lactic acid) (PDLA) is available, except that proteinase K hydrolyzed PLLA but not PDLA (15) and that Tomita et al. isolated a thermophilic *Bacillus stearothermophilus* able to grow at 60 °C on PDLA as a sole carbon source, although they did not mention enzyme activity contributing to the degradation of PDLA (16). As PLA is hydrolyzed at a relatively high rate at high temperatures (>50 °C), whether the strain excretes PDLA-degrading enzyme or utilizes hydrolyzed products depends on future characterization of PDLA-degrading enzyme. Thus, no information has been available on the enzymatic hydrolysis of PDLA until now.

This paper summarizes research on PLA-degrading microorganisms and PLA-degrading enzymes. In addition, I will introduce research of our group on enantioselectivity of PLA-degrading enzymes, based on which PLA depolymerases are categorized into two different classes.

PLA-Degrading Microorganisms and Their PLA Depolymerases

Since Pranamuda et al. first isolated the PLA-degrading *Amycolatopsis* sp. strain HT32 from soil in 1997 (9), many groups isolated various PLA-degrading microorganisms, as shown in Table 1. PLA-degrading Actinomycetes belong to the family of *Pseudonocardiaceae* and related genera, the representative strain of which is *Amycolatopsis* sp. (*10*). Stock culture of *Amycolatopsis* species (15 strains among total 25 strains) made clear zones on PLA agar plates (*11, 17*), suggesting that this genus contributes a lot for PLA degradation in nature. As approximately 95% of all the soil Actinomycetes are *Streptomyces*, limited distribution of PLA-degrading Actinomycetes is in accordance with that PLA film is hardly degraded when buried in soil. Except Actinomycetes, *Bacillus, Brevibacillus*, and *Geobacillus* have been reported as thermophilic degraders (*18–20*). *Paenibacillus amylolyticus* strain TB-13 could decrease turbidity of DL-PLA emulsion at 37 °C, but its degradability toward PLA film is unknown (*21*). As PLA film is degraded quickly in compost, thermophilic degraders are probable. However, degradation rates by thermophilic degraders are still under discussion. Since PLA is hydrolyzed non-enzymatically and remarkably at 60 °C, the enzymatic contribution to hydrolysis of PLA at 60 °C has to be evaluated carefully, compared with non-enzymatic hydrolysis (with regards to change in film weight, molecular mass and products), but this kind of comparison is insufficient in reports. Although we cannot deny the contribution of thermophilic Actinomycetes to PLA degradation in compost, this should be evaluated carefully in the future. Mayumi et al. recently cloned three genes encoding PLA depolymerases, based on metagenome from compost (*22*), one of which coded fo a thermostable esterase homologous to *Bacillus* lipase and showed the binding ability to DL-PLA powders with molecular masses of lower than 20,000. As the expressed enzyme had no activity on L-PLA with molecular masses of approximately 130,000, the enzyme might be able to degrade depolymerized PLA products. We could guess the same mechanism to other thermophilic enzymes. As DL-PLA has lower melting point and higher hydrolytic rate, compared with those of PLLA or PDLA, we should be careful for evaluating biodegradability of PLA (DL-PLA, PLLA and PDLA) in compost. A fungus, *Tritirachium album* ATCC22563, also showed degradation ability of PLLA, silk fibroin and elastin, which was inducible with gelatin, suggesting the induction of a protease (*23*). However, the role of the fungus in nature for degradation of PLA is skeptical, since the enzyme was not induced at all in the absence of gelatin.

Table 1. PLA-degrading microorganisms and their degrading enzymes[a]

Microorganisms	PLA-degrading enzyme	Substrate and evaluation of degradability	°C	Ref.
Actinomycetes				
Amycolatopsis sp. strain HT32	protease	L-PLA, film weight	30	(*9*)
Amycolatopsis sp. strain 3118	"	L-PLA, film weight	30	(*26*)

Continued on next page.

Table 1. (Continued). PLA-degrading microorganisms and their degrading enzymes[a]

Microorganisms	PLA-degrading enzyme	Substrate and evaluation of degradability	°C	Ref.
Amycolatopsis sp. strain KT-s-9	"	L-PLA, silk fibroin, halo on PLA-agar plate	30	*(27)*
Amycolatopsis sp. strain 41	"	L-PLA, silk powder, casein, film weight	30	*(11)*
Amycolatopsis sp. strain K104-1	"	L-PLA, casein, fibroin, turbidity of emusion and film weight	37	*(12)*
Amycolatopsis orientalis IFO12362	"	L-PLA powder, TOC	30, 40	*(24)*
Lentzea waywayandensis	"	L-PLA, film weight	30	*(28)*
Kibdelosporangium aridum	"	L-PLA, film weight	30	*(29)*
Bacteria				
Brevibacillus sp.	unknown	L-PLA, film (TOC, GPC, viscosity)	60	*(19)*
Bacillus stearothermophilus	unknown	D-PLA, film (TOC, GPC, viscosity)	60	*(16)*
Geobacillus thermocatenulatus	unknown	L-PLA, film (TOC, GPC,viscosity)	60	*(20)*
Bacillua sinithii strain PL21	esterase	L-PLA, pellet or powder, GPC	60	*(18)*
Paenibacillus amylolyicus strain TB-13	lipase	DL-PLA, turbidity of emulsion, TOC	37	*(21)*
Metagenome from compost[b]	Lipase (esterase)	DL-PLA powder, TOC, absorption of enzymes	60	*(22)*
Fungi				
Tritirachiium album ATCC 22563	protease	L-PLA, film weight	30	*(23)*
Cryptococcus sp. strain S-1	Cutinase AB102945	L-PLA, turbidity of emulsion	30	*(25)*
Aspergillus oryzae RIB40	cutinase	DL-PLA, turbidity of emulsion	37	*(30)*

[a] TOC: total organic carbon, GPC: gel permeation chromatography, Ref.; reference [b] homologous to a*Bacillus* lipase.

The inside temperature of compost at the secondary fermentation stage is expected to be approximately 60 °C, where degradation of organic materials is most promoted. As vegetative bacterial cells are killed at 55 °C, mature compost is free of general pathogenic bacteria and can be used as safe fertilizers for plants. The reason why PLA is not degraded in soil, but well degraded in compost is most probably due to the low density of *Pseudonocardiaceae* actinomycetes in soil and the high hydrolytic rate of PLA over 50 °C, yielding hydrolyzed products, which are more feasible for microbial degradation than the original high molecular weight of PLA. Enzymatic degradation is probable when enzyme potential and rigidity of molecular chain of a substrate were in accordance with each other, but impossible when binding energy of molecular chain of a substrate is over catalytic energy of an enzyme. However, as shown above, the degradation of PLA was possible at 30 °C far lower than glass transition temperature (Tg: approximately 55 °C), which is not understandable by the flexibility of the substrate molecular chain, but understandable by the reason that PLA absorbs water to cause the collapse of a polymer block, which would produce hydrophilic parts either on the polymer surface or inside of the polymer and become feasible with attack by microbes or enzymes. The fact that hydrolysis of PLA is possible at lower temperatures than Tg seems to support this assumption. Low degradation of PLA film in soil is probably due to low water activity in soil in addition to low density of *Pseudonocardiaceae*. Thus biodegradation of PLA at 30-37 °C is understandable. On the other hand, thermophilic microbes/enzymes are necessary for recycling of PLA, because thermophilic microbes/enzymes are tougher than mesophilic ones and hydrolysis is more efficient at higher temperatures. Actually PLA depolymerase from *A. orientalis* IFO12362 showed twice activity at 40 °C as much as that of proteinase K at 37 °C (*24*).

Since the first report of Pranamuda et al on PLA-degrading activity by culture supernatant of *Amycolatopsis* sp. strain HT32 (*9*), several enzymes from *Pseudonocardiacear* were considered to be proteases, as shown in Table 1. Nakamura et al. purified PLA depolymerase from *Amycolatopsis* sp. strain K104-1 and its N-teminal amino acids had homology with serine protease from earthworm able to degrade fibrin and with serine protease from crab able to degrade collagen (*12*). The same group also cloned the gene for the enzyme, which had homology with serine proteases from eukaryotic cells belonging to chymotrypsin family (endopeptidase) and produced oligolactide and lactic acid from PLA (*13*). The enzyme was feasible to degrade PLA film at 30-37 °C. On the other hand, the enzyme from *Bacillus sinithii* strain PL21 was a thermotolerant esterase (*18*) and that from *Paenibacillus amylolyticus* strain TB-13 showed 45-50% homology with mesophilic *Bacillus* family I-4 lipases (*21*). These enzymes are all from PLA degraders.

On the other hand, Masaki et al. isolated *Cryptococcus* sp. S-2 for use in wastewater treatment and found that the strain excreated the strong lipase activity (*25*). They cloned the gene for a lipase. The gene had higher homology with cutinase (EC 3.1.1.74) than with lipase, which showed stronger degradation ability toward PLA than proteinase K. Therefore we could say that the enzyme is the strongest PLA depolymerase so far known. This is astonishing, since the strain was not intended to show PLA degradation. Another cutinase-type enzyme from *Aspergillus oryzae* RIB40 also had weak activity toward PLA, compared with those toward PBS and PBSA (*30*).

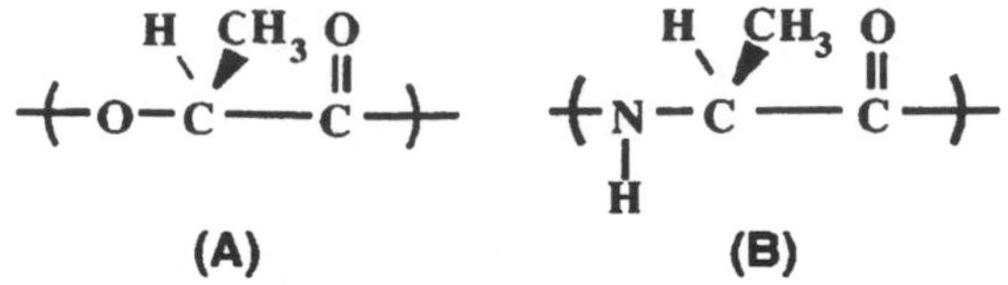

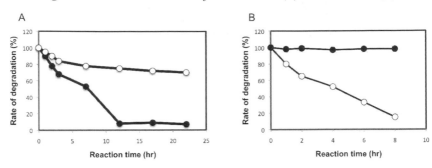

Figure 1. Chemical structure of L-lactic acid (A) and L-alanine (B)

Figure 2. Reaction of lipase-type (A) and protease-type (B) PLA depolymerases with PLLA and PDLA. (A) CLE from Cryyptococcus sp. S-1 (20 μg/ml, pH 7.0); (B) purified recombinant PLA depolymerase from Amycolatopsis sp. K104-1 (20 μg/ml, pH 9,0). Reactions were performed at 37 °C with shaking using PLLA or PDLA (a final concentration of approximately 0.1 wt%). Degradation rate was measured by change of absorbance at 600 nm. Open circle, PLLA; closed circle, PDLA

All the PLA-degraders and their enzymes acted on PLLA or DL-PLA except *Bacillus stearothermophilus* that was reported to degrade PDLA at 60 °C (*16*), but degradation of PLLA and the enzyme activity was not mentioned. Unavailability of PDLA through regular commercial routes and extremely high price of PDLA seemed to have restricted research on degradation of PDLA.

Degradation of PLA by Hydrolytic Enzymes

The biodegradation of PLA was reported first by Williams in 1981 using proteases including proteinase K (*1*). Degradation of PLA by lipases was unfeasible in his report. Later the feasibility of lipases was reported (*2, 3*), but they were active on low molecular weights (*4*) or poly(DL-lactide) (*5*). However, active proteases are widely distributed in *Pseudonocardiaceae* and their activities are in general higher than those of lipases (esterases) (*26–28*). Most of commercially available proteases can catalyze hydrolysis of PLA (*8*): Microbial proteases (proteinase K and subtilisin) and mammalian proteases (α-chymotrypsin, trypsin and elastase) could degrade PLA, but plant proteases could not. This result suggested that PLA as bioabsorbable material used in human body must partly be hydrolyzed by proteases *in vivo*. As proteinase K has the best activity among commercially available enzymes, it is often used for enzymatic degradation of PLA. As the chemical structure of PLA is a polyester or PHA, lipases (esterases) are predicted to be most probable degrading enzymes, but actually proteases are the most potential degrading enzymes, which is in accordance with that PLA depolymerases purified or cloned from PLA-degrading *Amycolatopsis* species are proteases (Table 1 (*11–14*)). It is already well known that the nature does not discriminate natural or synthetic compounds, but the most importance is chemical structures themselves recognizable by enzymes. In general, polyesters are degraded by lipases (esterases) with increased rates in turn of aromatic ones, aliphatic-aromatic ones and aliphatic ones. Tokiwa et al. suggested degradation rates of polyesters as polyhydroxybutyrate (PHB=plycaprolactone (PCL)>polybutylene succinate (PBS) >PLA (*4*). Biodegradation test in field also showed low biodegradability of PLA compared with PHB/V (valerate), PCL and PBS. Biodegadation of polyesters by lipases is dependent on either chemical structure or melting point (mp) (*4*). Accordingly, low biodegradability of PLA is possibly due to its short carbon length of monomer and high mp (PLLA and PDLA: approximately 175 °C).

Catalytic Aspect of PLA Depolymerase

Both of proteases and lipases belong to a serine hydrolase family, but differ in the substrate specificities. All the PLA depolymerases from *Pseudonocardiaceae* are proteases active on PLLA, as shown in Table 1. Proteinase K and a protease-type enzyme from *Tritirachium album* ATCC22563 were PLLA-spccific (*15, 23*). PLA depolymerases from thermotolerant *Bacillus sinithii* strain PL21 and *Paenibacillus amylolyticus* strain TB-13 were an esterase active on PLLA and a lipase active on DL-PLA, respectively (*18, 21*), although their exact biodegradation ability are ambiguous. Commercially available lipases (esterases) act on DL-PLA, but not on optically active PLLA or D-PHB (*4*). Protease-type PLA depolymerases can hydrolyze PLLA, but not D-PHB. PHB depolymerases do not act on PLLA. Cutinase from *Cryptococcus* sp. S-2 showed higher activity toward PLLA than that of proteinase K, suggesting that this acts exceptionally as an esterase on PLLA (*25*). These results suggested that PLA-degrading enzymes are distinctly different from PHB depolymerases.

Specificities of hydrolytic enzymes are in general wide, but recognition of the polyester (PLA) by proteases is biochemically interesting to understand what specificity means. Tokiwa et al. explained that the origin of a protease-type PLA depolymerase from Actinomycetes is a protease working on silk fibroin, based on homology of L-lactic acid with L-alanine as the major component of silk fibroin (Fig. 1) and induction of the enzyme by proteinous compounds such as gelatin (*4*). Reeve et al disclosed that proteinase K was PLLA-specific and did not act on PDLA at all (*15*). Matsuda et al. confirmed that a recombinant PLA depolymerase from *Amycolatopsis* did not work on PCL and PHB (*13*). PHB depolymerase does not act on PLLA (a kind of hydroxyalkanoate), due to differences in optical activities of both substrates and in carbon chain length of 2-hydroxyalkanoate and 3-hydroxyalkanoate. Thus PLA depolymerases have to be categorized differently from PHB depolymerases and PLA is considered as the third type-polyester following synthetic polyesters and polyhydroxyalkanoate (PHA) including PHB.

Using the recombinant purified PLA-degrading enzyme from *Amycolatopsis* sp. K104-1 (*13*) and the recombinant purified cutinase-like enzyme (CLE) from *Cryptococcus* sp. S-2 (*25*), we examined their enantioselectivity toward PLLA and PDLA, as shown in Fig. 2. PLA-degrading enzyme was absolutely PLLA-specific. Together with the report on the enatioselectivity of proteinase K (*15*) and the fact that proteases originally recognize polymer of L-amino acids, we could conclude that protease-type PLA depolymerases are PLLA-specific. On the other hand, CLE acted on both PLLA and PDLA, but the activity was higher on PDLA than on PLLA (PDLA-preferential). The enantioselectivity of two types of PLA depolymerases will be useful for the biological recycling of PLA and the recovery of lactic acid (especially expensive D-lactic acid). We found that some lipases are also PDLA-preferential (unpublished data). The enantioselectivity of crude enzymes could be a good indicator for prediction of the type of enzymes, either protease or cutinase, which would lead to the successful cloning of enzyme genes, based on the conserved regions of each group.

Commercially available true lipases did not act on PLLA and PDLA. True lipases possess a lid covering an active site to lead to the interfacial activation (*31*), but some lipases, esterases and cutinases have neither lid nor interfacial activation. To cover an active site by a lid leading to an interfacial activation, the size of an active site inlet cannot be too big. On the other hand, the inlet of PLA depolymerase has to be big enough to accommodate a macromolecular PLA. Accordingly, lipase-type PLA depolymerases are probably not a typical true lipase, but an esterase (cutinase) without a lid useful for interfacial activation and with an active cavity big enough to accommodate a polymer substrate (cutin is a rather big molecule of a complex structure).

Conclusion

From the aforementioned information, the following conclusions can be obtained for the enzymatic degradation of PLLA and PDLA by PLA depolymerases.

1) Polyester-degrading enzymes are categorized into three groups; synthetic (aliphatic) polyester-degrading group, PHA-degrading groups and PLA-degrading groups.
2) PLA depolymerases are categorized into two enzyme groups; protease-type (type I) and lipase (cutinase)-type (type II).
3) Type I-PLA depolymerases are PLLA-specific, but lipase (cutinase)-type PLA depolymerases are PDLA-preferencial.
4) The presence of a lid in true lipases (causing the interfacial activation) interferes the access of PLA to the catalytic amino acid in the active cavity.
5) Absence of a lid and the geometry of the active site are the two important factors in the catalysis of the lipase family. Therefore, cutinase-type enzymes in the lipase superfamily are most probably type II-PLA depolymerase.

References

1. Williams, D. F. *Eng. Med.* **1981**, *10*, 5–7.
2. Fukuzaki, H.; Yoshida, M.; Asano, M.; Kumakura, M. *Eur. Polym. J.* **1989**, *25*, 1019–1025.
3. Akutsu, Y.; Nakajima-Kambe, T.; Nomura, N.; Nakahara, T. *Appl. Environ. Microbiol.* **1998**, *64*, 62–67.
4. Tokiwa, Y.; Jarerat, A. *Biotechnol. Lett.* **2004**, *26*, 771–777.
5. Akutsu-Shigeno, T.; Teeraphatpornchai, T.; Teamtisong, K.; Nomura, N.; Uchiyama, H.; Nakamura, T.; Nakajima-Kambe, T. *Appl. Environ. Microbiol.* **2003**, *69*, 2498–2504.
6. Oda, Y.; Yonetsu, A.; Urakami, T.; Tonomura, K. *J. Polym. Environ.* **2000**, *8*, 29–32.
7. Hoshino, A.; Isono, Y. *Biodegradation* **2002**, *13*, 141–147.
8. Lim, H. A.; Raku, T.; Tokiwa, Y. *Biotechnol. Lett.* **2005**, *27*, 459–464.
9. Pranamuda, H.; Tokiwa, Y.; Tanaka, H. *Appl. Environ. Microbiol.* **1997**, *63*, 1637–1640.
10. Tokiwa, Y.; Calbia, B. P. *Appl. Microbiol. Biotechnol.* **2006**, *72*, 244–251.
11. Pranamuda, H.; Tsuchii, A.; Tokiwa, T. *Macromol. Biosci.* **2001**, *1*, 25–29.
12. Nakamura, K.; Tomita, T.; Abe, N.; Kamio, Y. *Appl. Environ. Microbiol.* **2001**, *67*, 345–353.
13. Matsuda, E.; Abe, N.; Tamakawa, H.; Kaneko, J.; Kamio, Y. *J. Bacteriol.* **2005**, *187*, 7333–7340.
14. Tokiwa, Y.; Jarerat, A.; Tsuchiya, A. Japanese Patent 2003-61676, March 4, 2003.
15. Reeve, M. S.; McCarthy, S. P.; Downew, M. J.; Gross, R. A. *Macromolecules* **1994**, *27*, 825–831.
16. Tomita, K.; Tsuji, T.; Nakajima, H.; Kikuchi, Y.; Ikarashi, K.; Ikeda, N. *Polym. Degrad. Stab.* **2003**, *81*, 167–171.
17. Pranamuda, H.; Tokiwa, Y. *Biotechnol. Lett.* **1999**, *21*, 901–905.
18. Sakai, K.; Kawano, H.; Iwami, A.; Nakamura, M.; Moriguchi, M. *J. Biosci. Bioeng.* **2001**, *92*, 298–300.
19. Tomita, K.; Kuroki, Y.; Nagai, K. *J. Biosci. Bioeng.* **1999**, *87*, 752–755.
20. Tomita, K.; Nakajima, T.; Kikuchi, Y.; Miwa, N. *Polym. Degrad. Stab.* **2003**, *81*, 167–171.
21. Shigeno, Y. A.; Teeraphatpornchai, T.; Teamtisong, K.; Nomura, N.; Uchiyama, H.; Nakamura, T.; Kambe, T. N. *Appl. Environ. Microbiol.* **2003**, *69*, 2498–2504.
22. Mayumi, D.; Shigeno, Y.; Uchiyama, H.; Nomura, N.; Nakajima, K. T. *Appl. Microbiol. Biotechnol.* **2008**, *79*, 743–750.
23. Jarerat, A.; Tokiwa, Y. *Macromol. Biosci.* **2001**, *1*, 136–140.
24. Jarerat, A.; Tokiwa, Y.; Tanaka, H. *Appl. Microbiol. Biotechnol.* **2006**, *72*, 726–731.
25. Masaki, K.; Kamini, N. R.; Ikeda, H.; Iefuji, H. *Appl. Environ. Microbiol.* **2005**, *71*, 7548–7550.
26. Ikura, Y.; Kudo, T. *J. Gen. Appl. Microbiol.* **1999**, *45*, 247–251.
27. Tokiwa, Y.; Konno, M.; Nishida, H. *Chem. Lett.* **1999**, *1999*, 355–356.

28. Jarerat, A.; Tokiwa, Y. *Biotechnol. Lett.* **2003**, *25*, 401–404.
29. Jarerat, A.; Tokiwa, Y.; Tanaka, H. *Biotechnol. Lett.* **2003**, *25*, 2035–2038.
30. Maeda, H.; Yamagata, Y.; Abe, K.; Hasegawa, F.; Machida, M.; Ishioka, R.; Gomi, K.; Nakajima, T. *Appl. Microbiol. Biotechnol.* **2005**, *67*, 778–788.
31. Schmid, R. D.; Verger, R. *Angew. Chem., Int. Ed.* **1988**, *37*, 1608–1633.

Grafting and Functionalization Reactions

Chapter 28

Green Polymer Chemistry: Enzymatic Functionalization of Liquid Polymers in Bulk

Judit E. Puskas* and Mustafa Y. Sen

Department of Polymer Science, The University of Akron, 170 University Ave., Akron, OH 44325-3909, USA
*jpuskas@uakron.edu

The use of enzymes as catalysts for organic synthesis has become an increasingly attractive alternative to conventional chemical catalysis. Enzymes offer several advantages including high selectivity, the ability to operate under mild conditions, catalyst recyclability and biocompatilibity. Although there are many examples involving enzymes for the synthesis of polymers, only a few are in the area of polymer functionalization and most of the examples are characterized by low conversion. In this paper, we present examples of quantitative enzyme-catalyzed methacrylation of liquid polymers. Specifically, vinyl methacrylate was transesterified with liquid α,ω-dihydroxy polyisobutylenes, α,ω-dihydroxy polydimethylsiloxane (PDMS), PDMS-mono and –dicarbinol, and low molecular weight poly(ethylene glycol) in the presence of *Candida antarctica* lipase B (CALB; Novozym® 435) under solventless conditions. ^{1}H and ^{13}C NMR spectroscopy verified the structure of the functionalized polymers.

Introduction

Enzymes are nature's catalysts that accelerate specific metabolic reactions in living cells. However, many different types of enzymes are also known to catalyze the transformation of a wide range of "unnatural" substrates *in vitro* (*1*). In the area of polymer science this feature of enzymes has been well exploited for the synthesis of polymers (*2–4*), but the number of examples involving enzyme-catalyzed polymer modifications is rather limited (5) and generally characterized by low conversions (6, 7). As an environmentally friendly alternative to conventional chemical catalysts, enzymes offer several advantages including high selectivity, the ability to operate under mild conditions, catalyst recyclability and biocompatilibily (8). We have previously shown the quantitative chain end functionalization of poly(ethylene glycol)s (PEG)s and polyisobutylenes (PIB)s in solution by *Candida antarctica* lipase B (CALB)-catalyzed transesterification (9, 10). In the framework of on-going enzymatic polymer functionalization research in our laboratory, we recently reported the synthesis of α,ω-thymine-functionalized PEG via Michael addition using Amano lipase M from *Mucor javanicus* as the enzyme (11). Moreover, we utilized enzyme regioselectivity to methacrylate asymmetric α,ω-hydroxyl-functionalized PIBs exclusively at the ω-termini (11). In this paper, we present examples of CALB-catalyzed methacrylation of liquid hydroxyl-functionalized PIBs, α,ω-dihydroxy poly(dimethyl siloxane) (HO-PDMS-OH), hydroxyl-ethoxypropyl-terminated PDMSs (PDMS-mono- and dicarbinol) and poly(ethylene glycol) (PEG) under solventless conditions via transesterification of vinyl methacrylate (VMA).

Functionalization of Polyisobutylenes in Bulk

The quantitative enzymatic functionalization of hydroxyl-terminated polyisobutylenes (PIB-OHs) with various chain-end structures in hexane via CALB-catalyzed transesterification of VMA was previously achieved (10). Transesterification was the preferred synthetic pathway as some enzymes, i.e. lipases, are well-known to catalyze acyl transfer reactions of small molecules (8). Among several lipases used in transesterification, CALB was preferred due to its high stability and reactivity (12). VMA, which is an enolate ester, was chosen as the acyl donor. Enolate esters liberate unstable enols as by-products which instantly tautomerize to give the corresponding aldehydes or ketones and thus render the transesterification irreversible (13). Hexane solvent was used since the catalytically active conformation of the enzyme is best maintained in low polarity solvents (8). Following the same synthetic pathway, liquid hydroxyl-functionalized PIBs, i.e. PIB-CH$_2$-C(CH$_3$)-CH$_2$-OH (M$_n$=1500 g/mol, M$_w$/M$_n$=1.29) and Glissopal-OH (M$_n$=3600 g/mol, M$_w$/M$_n$=1.34), were enzymatically methacrylated in the absence of solvent. PIB-CH$_2$-C(CH$_3$)-CH$_2$-OH (M$_n$=1500 g/mol, M$_w$/M$_n$=1.29) was prepared by dehydrochlorination of the terminal –C(CH$_3$)$_2$Cl of a PIB-Cl (14), followed by hydroboration/oxidation of the resulting olefinic chain end (15). The PIB-Cl was synthesized by the 2-chloro-2,4,4-trimethylpentane (TMPCl)/BCl$_3$ initiated carbocationic polymerization of isobutylene (16, 17). Glissopal-OH, which also has the chain end structure -CH$_2$-C(CH$_3$)-CH$_2$-OH, was obtained by hydroboration/oxidation of Glissopal®2300 (BASF), a commercially available PIB with ~82% exo and ~18% endo terminal double bonds (18). Figure 1 shows the conditions we developed for the transesterification of VMA with these hydroxyl-functionalized PIBs in bulk.

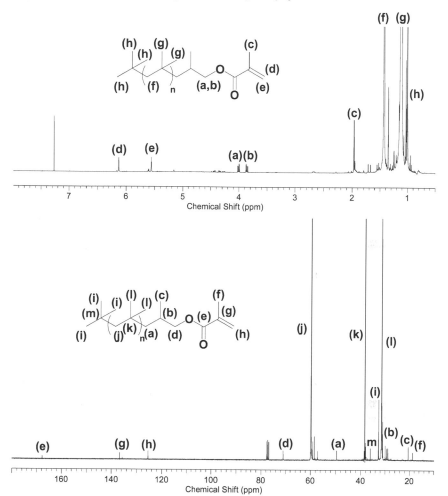

Figure 1. Transesterification of VMA with hydroxyl-functionalized PIBs in bulk.

Figure 2. ^1H NMR spectrum of the methacrylation product of a Glissopal-OH ($M_n=3600$ g/mol, $M_w/M_n=1.34$) in bulk (top) and its corresponding ^{13}C NMR (bottom) (NMR solvent: CDCl$_3$).

Both PIB-CH$_2$-C(CH$_3$)-CH$_2$-OHs derived from PIB-Cl and Glissopal®2300 reacted quantitatively. The ^1H and ^{13}C NMR spectra of the methacrylation product of Glissopal-OH are shown in Figure 2. The methylene protons adjacent to the hydroxyl group at δ=3.31-3.51 ppm shifted downfield to δ=3.84-4.02 ppm after methacrylation and the vinylidene [δ=5.56 (e) and 6.13 ppm (d)] and methyl [δ=1.97 ppm (c)] protons of the newly formed methacrylate-end appeared at the expected positions. The ^{13}C NMR also confirmed the structure of the

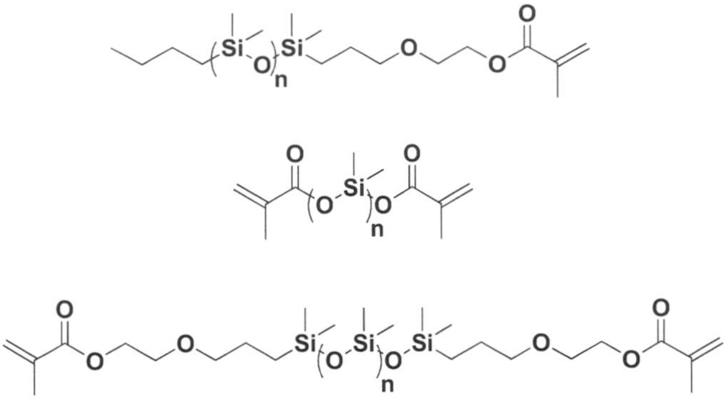

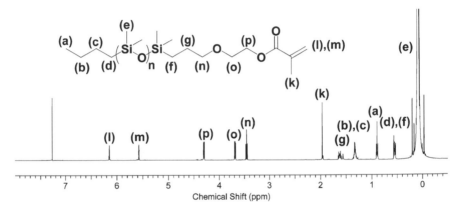

Figure 3. Methacrylated PDMS products.

Figure 4. ^1H NMR spectrum of the methacrylation product of a
PDMS-monocarbinol ($M_n=5000$ g/mol) in bulk (NMR solvent: $CDCl_3$).

methacrylated polymer where the carbonyl carbon was observed at $\delta=167.73$
ppm (e), and resonances of the vinyl ($\underline{C}H_2=C(CH_3)$-), methyl ($CH_2=C(\underline{C}H_3)$-),
and the alpha carbon ($CH_2=\underline{C}(CH_3)$-) of the methacrylate group were observed at
$\delta=125.38$ ppm (h), $\delta=18.60$ ppm (f) and $\delta=136.71$ ppm (g), respectively. The ^1H
NMR shows the presence of PIB with endo-olefin terminus ($\delta=5.16$ ppm) from
the starting material which was not converted into PIB-OH.

Functionalization of Polydimethylsiloxanes in Bulk

The enzymatic methacrylation of commercially available liquid PDMSs in
bulk was also quantitative. The list of methacrylated PDMS products is presented
in Figure 3.

Figure 4 shows the ^1H NMR spectrum of the methacrylation product of
PDMS-monocarbinol. The methylene protons adjacent to the hydroxyl group
in the starting material at $\delta=3.74$ ppm shifted downfield to $\delta=4.31$ ppm (p)
upon methacrylation; and vinylidene (l and m) and methyl protons (k) of the
methacrylate chain-end appeared at $\delta=6.15$ ppm (l), $\delta=5.58$ ppm (m) and $\delta=1.97$

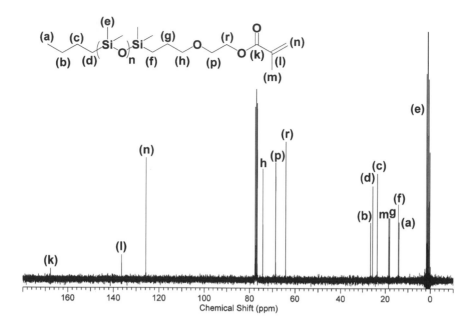

Figure 5. ^{13}C *NMR spectrum of the methacrylation product of a PDMS-monocarbinol (M_n=5000 g/mol) in bulk (NMR solvent: CDCl$_3$).*

ppm (k), respectively, with relative integral ratios of 2:1:1:3 demonstrating quantitative functionalization in 24 hours.

Figure 5 shows the ^{13}C NMR spectrum of the bulk methacrylation product of PDMS-monocarbinol. Compared to the ^{13}C NMR of the starting material, new peaks appeared at δ=18.54 ppm (m), δ=125.92 ppm (n), δ=136.49 ppm (l) and δ=167.60 ppm (k) corresponding to the carbons of the methacrylate chain-end. The peaks (p) and (r) shifted upfield and downfield after methacrylation and appeared at δ=68.70 ppm and δ=64.25 ppm, respectively.

The progress of the reactions was monitored using ^{1}H NMR. Figure 6 is given as an example where PDMS-dicarbinol was methacrylated. It was observed that the protons of the two methylene groups next to the hydroxyl group in the starting material at δ=3.74 ppm (b) and δ=3.55 ppm (c) shifted downfield to δ=4.31 (r) and 3.69 ppm (s), respectively, within only 2 hours when 1.5 eq. of VMA was used. The relative integration ratio of these protons to the methacrylate-end protons confirmed quantitative conversion.

Functionalization of Poly(ethylene glycol) in Bulk

Poly(ethylene glycol) (PEG)-dimethacrylate was prepared by the transesterification of VMA with HO-PEG-OH (M_n=1000 g/mol and M_w/M_n=1.08) in the presence of CALB (Figure 7). The low molecular weight PEG became liquid when warmed to 50 °C and was miscible with VMA. Monitoring the reaction with ^{1}H NMR revealed that the reaction was quantitative within 4 hours when 5 eq. of VMA per OH group in the HO-PEG-OH was used.

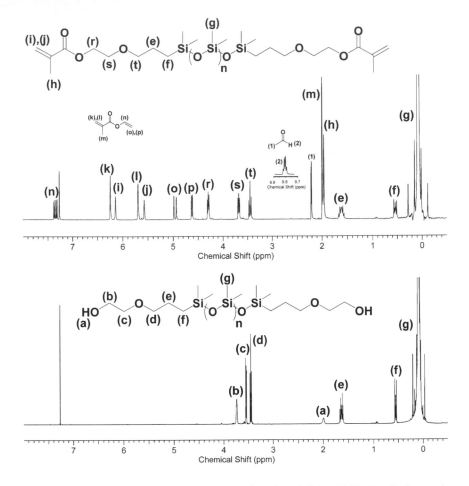

Figure 6. ^1H *NMR spectrum of a PDMS-dicarbinol ($M_n=4500$ g/mol) (bottom)*
and its methacrylation product in bulk after 2 hours of reaction time (top) (NMR
solvent: $CDCl_3$).

Figure 8 shows the ^1H NMR spectrum of the PEG-dimethacrylate. The hydroxyl protons at $\delta=4.55$ ppm from the HO-PEG-OH completely disappeared and the peak corresponding to the methylene protons adjacent to hydroxyl group shifted downfield from $\delta=3.50$ to $\delta=4.42$ ppm (c) after the reaction. The new peaks corresponding to the methyl [$\delta=1.73$ ppm (e)] and vinyl [$\delta=6.07$ ppm (f) and $\delta=5.81$ ppm (g)] protons of the methacrylate group were observed at the expected positions with integral values of 2:3:1:1 [(c):(e):(f):(g)] confirming successful functionalization.

The ^{13}C NMR spectrum of the methacrylation product also confirmed the structure of the polymer (Figure 9). The carbons connected to the hydroxyl group in the starting material at $\delta=60.13$ ppm shifted downfield to $\delta=63.89$ ppm (c) after the reaction and the carbon resonances of the methacrylate group appeared at $\delta=166.97$ ppm (d), $\delta=18.12$ ppm (e), $\delta=136.24$ ppm (f) and $\delta=126.11$ ppm

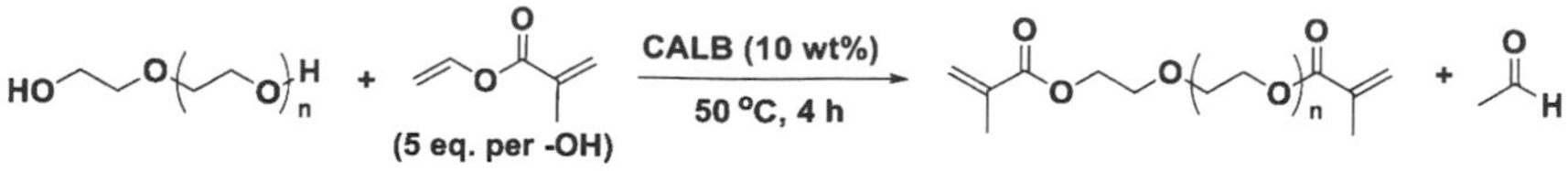

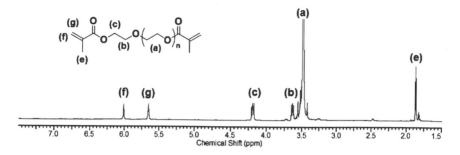

Figure 7. *Transesterification of VMA with HO-PEG-OH in bulk.*

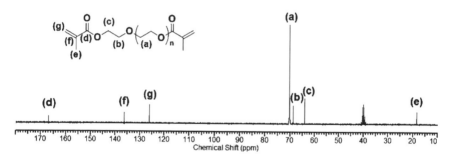

Figure 8. *¹H NMR spectrum of the methacrylation product of a HO-PEG-OH (Mₙ=1000 g/mol, Mw/Mₙ=1.08) in bulk after 4 hours of reaction time (NMR solvent: DMSO-d₆).*

Figure 9. *¹³C NMR spectrum of the methacrylation product of a HO-PEG-OH (Mₙ=1000 g/mol, Mw/Mₙ=1.08) in bulk after 4 hours of reaction time (NMR solvent: DMSO-d₆).*

(g) corresponding to carbonyl carbon, methyl carbon, alpha carbon and the vinyl carbon connected to the alpha carbon, respectively.

Conclusion

In conclusion, we were able to functionalize liquid PDMSs and HO-PIB-OH, and low molecular weight PEGs very effectively under solventless conditions via CALB-catalyzed transesterification of vinyl methacrylate. Both ¹H and ¹³C NMR spectroscopy verified the structure of the products. The absence of solvent in these transformations renders this approach attractive as it provides a cost-effective and environmentally benign way of producing telechelic polymers for biomedical applications.

Acknowledgments

This material is based upon work supported by The National Science Foundation under DMR-0509687 and #0804878. We wish to thank The Ohio Board of Regents and The National Science Foundation (CHE-0341701 and DMR-0414599) for funds used to purchase the NMR instrument used in this work. We are also grateful for the contribution of Kwang Su Seo and Dr. Serap Hayat-Soytas.

References

1. Koeller, K. M.; Wong, C. H. *Nature* **2001**, *409*, 232–240.
2. Varma, I. K.; Albertsson, A. C.; Rajkhowa, R.; Srivastava, R. K. *Prog. Polym. Sci.* **2005**, *30*, 949–981.
3. Kobayashi, S.; Uyama, H.; Kimura, S. *Chem. Rev.* **2001**, *101*, 3793–3818.
4. Gross, R. A.; Kumar, A.; Kalra, B. *Chem. Rev.* **2001**, *101*, 2097–2124.
5. Gubitz, G. M.; Paulo, A. C. *Curr. Opin. Biotechnol.* **2003**, *14*, 577–582.
6. Cheng, H. N.; Gu, Q.-M. In *Glycochemistry: Principles, Synthesis and Applications*; Wang, P. G., Bertozzi, C. R., Eds.; Marcel Dekker: New York, 2001; pp 567–580.
7. Jarvie, A. W. P.; Overton, N.; St Pourcain, C. B. *J. Chem. Soc., Perkin Trans. 1* **1999**, 2171–2176.
8. Faber, K. *Biotransformations in Organic Chemistry*, 5th ed.; Springer-Verlag: New York, 2004.
9. Puskas, J. E.; Sen, M. Y.; Kasper, J. R. *J. Polym. Sci., Part A: Polym. Chem.* **2008**, *46*, 3024–3028.
10. Sen, M. Y.; Puskas, J. E.; Ummadisetty, S.; Kennedy, J. P. *Macromol. Rapid Commun.* **2008**, *29*, 1598–1602.
11. Puskas, J. E.; Sen, M. Y.; Seo, K. S. *J. Polym. Sci., Part A: Polym. Chem.* **2009**, *47*, 2959–2976.
12. Yadav, G. D.; Trivedi, A. H. *Enzyme Microb. Technol.* **2003**, *32*, 783–789.
13. Faber, K.; Riva, S. *Synthesis* **1992**, 895–910.
14. Kennedy, J. P.; Chang, V. S. C.; Smith, R. A.; Ivan, B. *Polym. Bull.* **1979**, *1*, 575–580.
15. Ivan, B.; Kennedy, J. P.; Chang, V. S. C. *J. Polym. Sci., Polym. Chem. Ed.* **1980**, *18*, 3177–3191.
16. Hayat-Soytas, S. Ph.D. Dissertation, The University of Akron, 2009.
17. Kaszas, G.; Gyor, M.; Kennedy, J. P.; Tudos, F. *J. Macromol. Sci., Chem.* **1983**, *A18*, 1367–1382.
18. Technical information for Glissopal® 1000, 1300 and 2300. http://www2.basf.us/pib_derivatives/pdfs/Glissopal_1000_2300.pdf (accessed June 2, 2009).

Chapter 29

Bio-Based and Biodegradable Aliphatic Polyesters Modified by a Continuous Alcoholysis Reaction

James H. Wang* and Aimin He[+]

Corporate Research & Engineering, Kimberly-Clark Corporation, 2100 Winchester Road, Neenah, WI 54956, U.S.A.
*jhwang@kcc.com
[+]Present address: 9957 Autry Vue Lane, Alpharetta, GA 30022

A novel approach was developed to chemically tailor biodegradable aliphatic polyesters such as polylactic acid (PLA) and polybutylene succinate (PBS) for targeted applications. The reactive extrusion process utilized a catalyzed alcoholysis reaction to controllably cut the aliphatic polyester polymer chains to the desired lengths. During this continuous reaction, aliphatic polyesters reacted with a solution of a diol or functionalized alcohol and a catalyst in melt phase, resulting in modified aliphatic polyesters with hydroxyalkyl chain ends or other functional chain ends useful for further reactions or modifications. Titanium propoxide and dibutyltin diacetate were used as catalysts for alcoholysis reactions of PLA and PBS respectively. It was found that by selectively controlling the alcoholysis conditions, the molecular weights and melt rheology of modified aliphatic polyester were modified to make them suitable for meltblown nonwoven processing. Meltblown nonwovens were successfully spun from both modified PLA and modified PBS, fiber-to-fiber bonding and excellent mechanical properties were achieved from the modified bio-based and biodegradable meltblown nonwoven.

Introduction

Interest in bio-based and biodegradable polymers has increased significantly in recent years to deal with global climate change, fluctuating crude oil price, feedstock sustainability, etc. (*1*) Bio-based and biodegradable aliphatic polyesters were developed and commercialized including polylactic acid (PLA) (*2*, *3*), polyhydroxybutyrate (PHB) and polyhydroxybutyrate-co-valerate (PHBV) from polyhydroxyalkanoate (PHA) family (*4*, *5*), etc. PLA is chemically synthesized from lactide or lactic acid derived from the fermentation of renewable raw materials (e.g. starch) (*2*), while PHA is a microbial polyester accumulated inside cells of microorganisms using renewable carbon sources such as sugars, vegetable oils, organic acids, etc. (*3*) Synthetic biodegradable aliphatic polyesters such as polylactone (PCL), polybutylene succinate (PBS), polyethylene succinate (PES), polyesteramide, and biodegradable aliphatic-aromatic copolyesters (e.g. polybutylene adipate terephthalate, PBAT) were also developed and commercialized (*6–8*).

To be commercially viable, bio-based and biodegradable polymers need to meet a number of key criteria: 1) satisfactory performance meeting the requirements of intended applications, 2) acceptable processability in terms of line speed and cycle time, etc. 3) at market acceptable cost; 4) environmentally sustainable based on resource sustainability and life cycle assessment (LCA), etc.

Although the above bio-based or biodegradable polymers have been processed in a number of applications such as film extrusion, injection molding, thermoforming, sheeting, foaming, etc. a number of limitations of these polymers were identified. One of the limitations is the unavailability of bio-based and biodegradable polymers suitable for meltblown nonwoven applications. For a polymer to be useful for meltblown applications, the polymer must have the right melt rheology and molecular weight distribution required by the meltblown process. Since meltblown nonwoven represents a major type of meltspun nonwoven materials with broad applications in health, hygiene, and industrial applications (*9*), it is of great technological importance and value to develop the enabling chemistry and processes to modify bio-based and biodegradable aliphatic polyesters for meltblown nonwoven applications.

Experimental

Materials

Polylactic acid (PLA) L-9000 was purchased from Biomer (Krailing, Germany). Polybutylene succinate (PBS) G-4500 was purchased from Ire Chemical Inc. (Seoul, Korea). 1,4-Butandiol, titanium propoxide, ethylene glycol diacetate, and dibutyltin diacetate (DBTDAc) from Aldrich were used as received.

Alcoholysis of Polylactic Acid (PLA)

A polylactic acid resin was supplied by Biomer, Inc. under the designation BIOMER™ L-9000. A co-rotating, twin-screw extruder (ZSK-30) manufactured by Werner and Pfleiderer Corporation of Ramsey, N.J. was used to conduct alcoholysis reaction. The screw length was 1328 millimeters. The extruder had 14 barrels, numbered consecutively 1-14 from the feed hopper to the die. The first barrel (#1) received the BIOMER™ L-9000 resin *via* a gravimetric feeder at a throughput of 40 pounds per hour. The fifth barrel (#5) received varying percentages of a reactant solution *via* a pressurized injector connected with an Eldex pump. The reactant solution contained 1,4-butanediol (87.5 wt.%), ethanol (6.25 wt.%), and titanium propoxide (6.25 wt.%). The screw speed was 150 revolutions per minute (rpm). The die used to extrude the resin had 4 die openings (6 millimeters in diameter) that were separated by 4 millimeters. Upon formation, the extruded resin was cooled on a fan-cooled conveyor belt and cut into pellets by a Conair pelletizer. Reactive extrusion parameters were monitored during the reactive extrusion process. This process produced hydroxybutyl terminated PLA, which is chemically distinct from unmodified PLA.

Alcoholysis of Polybutylene Succinate (PBS)

PBS resin was extruded with a reactant solution to form a mixture containing varying percentages of the resin, alcohol ("reactant"), and dibutyltin diacetate catalyst in an extruder. The reactant mixture was fed by an Eldex pump to the Feed/Vent port of a co-rotating, twin-screw extruder (USALAB Prism H16, diameter: 16 mm, L/D of 40/1) manufactured by Thermo Electron Corporation. The screw length was 25 inches. The extruder had one die opening having a diameter of 3 millimeters. Upon formation, the extruded resin was cooled on a fan-cooled conveyor belt and formed into pellets by a Conair pelletizer. Reactive extrusion parameters were monitored on the USALAB Prism H16 extruder during the reactive extrusion process. In one typical experiment, PBS was fed to the extruder at a rate of 2 lbs/hr, 1,4-butanediol was fed at 4% of the rate of the PBS, and dibutyltin diacetate was fed at 0.08% of the PBS rate. The temperature profile on the extruder was 90, 125, 165, 125, and 110 °C respectively for zones 1, 2, 3-8, 9, and 10. The screw speed was 150 rpm.

Gel Permeation Chromatography (GPC)

GPC analysis of the polymer samples was performed using a Waters 600E gradient pump and controller, Waters 717 auto sampler, and a Waters 2414 differential refractometer detector at a sensitivity of 30 at 40 °C and a scale factor of 20. The columns were Styragel HR 1,2,3,4 & 5E at 41°C. The samples were prepared by making a 0.5% wt/v solution in chloroform. The dissolved sample was filtered through a 0.45-micrometer PTFE membrane. The injection volume was 50 microliters, the chloroform flow rate was 1 milliliter per minute. The average molecular weights and molecular weight distribution were determined using narrow molecular eight distribution polystyrene standards (Polysciences).

Melt Rheology

The rheological properties of polymers were determined using a Göttfert Rheograph 2003 capillary rheometer with WinRHEO version 2.31 analysis software. The setup included a 2000-bar pressure transducer and a 30/1 L/D round capillary die. Sample loading was done by alternating between sample addition and packing with a ramrod. A 2-minute melt time preceded each test to allow the polymer to completely melt at the test temperature (usually 150°C to 220 °C). The capillary rheometer determined the apparent viscosity (Pa·s) at various shear rates, such as 100, 200, 500, 1000, 2000, and 4000 s^{-1}. The resultant rheology curve of apparent shear rate versus apparent viscosity gave an indication of how the polymer would run at that temperature in an extrusion process.

Thermal Properties

The melting temperature (T_m), glass transition temperature (T_g), and latent heat of fusion (ΔH_f) were determined by differential scanning calorimetry (DSC). The differential scanning calorimeter was a THERMAL ANALYST 2910 Differential Scanning Calorimeter, which was outfitted with a liquid nitrogen cooling accessory and with a THERMAL ANALYST 2200 (version 8.10) analysis software program, both of which are available from T.A. Instruments Inc. of New Castle, Delaware. For resin pellet samples, the heating and cooling program was a 2-cycle test that began with an equilibration of the chamber to -50 °C, followed by a first heating period at a heating rate of 10 °C per minute to a temperature of 200 °C, followed by equilibration of the sample at 200 °C for 3 minutes, followed by a first cooling period at a cooling rate of 20 °C per minute to a temperature of -50 °C, followed by equilibration of the sample at -50 °C for 3 minutes, and then a second heating period at a heating rate of 10 °C per minute to a temperature of 200 °C. For fiber samples, the heating and cooling program was a 1-cycle test that began with an equilibration of the chamber to -50 °C, followed by a heating period at a heating rate of 20 °C per minute to a temperature of 200°C, followed by equilibration of the sample at 200 °C for 3 minutes, and then a cooling period at a cooling rate of 10 °C per minute to a temperature of -50 °C.

Meltblown Spinning of PLA Modified by Alcoholysis

A PLA sample modified by alcoholysis (M_n: 70,500 g/mol, M_w: 97,200 g/mol, M_w/M_n: 1.38, and a melt viscosity of 42 Pa.s at a shear rate of 1000 s^{-1} and 190 °C) was spun on a pilot meltblown line that included a Killion extruder (Verona, NY), a 10-feet hose from Dekoron/Unitherm (Riviera Beach, FL), and a 14-inch meltblown die with an 11.5-inch spray and an orifice size of 0.015 inch. The modified PLA resin was fed *via* gravity into the extruder and then transferred into the hose connected with the meltblown die.

The meltblown spinning was performed at a temperature profile of 190 to 196 °C, a primary air temperature of 204 °C, and a primary air pressure of 34 psi. The meltblown die was kept at a temperature of 193 °C. A meltblown PLA nonwoven was obtained.

Tensile Properties

The strip tensile strength values were determined in substantial accordance with ASTM Standard D-5034. Specifically, a nonwoven web sample was cut or otherwise provided with size dimensions that measured 25 millimeters (width) x 127 millimeters (length). A constant-rate-of-extension type of tensile tester was employed. The tensile testing system was a Sintech Tensile Tester, which is available from Sintech Corp. of Cary, North Carolina.

Results and Discussion

Alcoholysis Reaction for Modifying Bio-Based Polyesters

Alcoholysis reaction was reported to effectively reduce the molecular weight of an aliphatic polyester-polyhydroxyoctanoate (PHO) (10). The reaction of PHO with ethylene glycol was catalyzed by dibutyltin dilaurate in a solution of diglyme. It was found alcoholysis reaction yielded telechelic PHO of significantly lower molecular weights useful for as precursor for block polyester synthesis (11). However, the reaction was slow when performed in solution, taking up to 10 hours (10).

It is of scientific and technological interest to explore the use a fast reactive extrusion process for effectively modifying aliphatic polyesters. Reactive extrusion has a number of advantages over a traditional solution process for industrial production: 1) It does not use a solvent and thus avoid emission of volatile organic compounds (VOC) typically encountered during solution process, 2) reaction can be completed in melt phase in a very short time (typically ranging from about 30 seconds to several minutes), 3) it has less energy input due to the elimination of dissolution, solvent recovery, and post reaction work-up steps, 4) lower capital cost and operating cost (12). Reactive extrusion is very effective in making grafted polymers of bio-based PLA (13, 14), microbial polyhydroxybutyrate-co-valerate (PHBV) (15), biodegradable PBS (16), and grafted polyolefins (17–19). Reactive extrusion was also used to reduce molecular weights of polymers by free radical reactions (20–22).

To adapt the alcoholysis reaction for modifying bio-based polyesters, it requires the alcoholysis have a fast reaction rate to be compatible with reaction extrusion process, i.e. completing the reaction within the residence of a reactive extrusion process. It is the objective of this study to develop a solvent-free, fast reactive extrusion process to controllably tailor the molecular weights of bio-based and biodegradable polyesters via a continuous alcoholysis reaction.

Continuous Alcoholysis Reaction of Polylactic Acid

Several Group IVB and Group IVA metal compounds were found to be highly effective as catalysts for the alcoholysis of a biodegradable, aliphatic-aromatic copolyester (polybutylene adipate terephthalate) in a melt phase reactive extrusion process (23). The catalysts investigated include dibutyltin diacetate, titanium (IV) propoxide, titanium (IV) isopropoxide, and titanium (IV) butoxide. Various

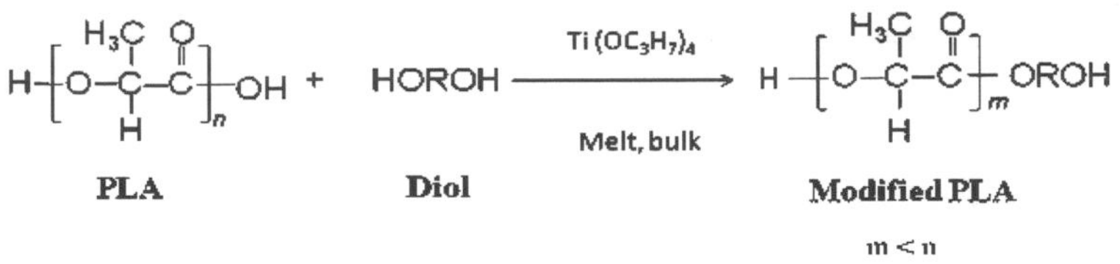

PLA   Diol   Modified PLA

$m < n$

Figure 1. Catalyzed alcoholysis reaction of polylactic acid.

concentrations of diol such as butanediol and monofunctional alcohols including 1-butanol, 2-propanol, and 2-ethoxyethanol were also studied. It was found that a diol is significantly more effective than a monofunctional alcohol for the alcoholysis reaction (23). As a result, 1,4-butanediol and titanium (IV) propoxide were selected as the alcohol and catalyst for alcoholysis reactions of PLA.

The alcoholysis reaction of polylactic acid (PLA) is shown in Figure 1. A PLA molecule has two different chain ends, a hydroxyl group at one chain end and a carboxyl group at the other chain end. Therefore, oligomeric PLA cannot be used as a prepolymer for synthesizing polyurethanes or other condensation polymers. Through the alcoholysis reaction of PLA with a diol like butanediol, the carboxyl chain end in PLA is transformed into a hydroxyl chain end. This reaction synthesizes a PLA-diol useful a wide range of condensation reactions. Furthermore, the new hydroxyl group is a primary hydroxyl instead of the secondary hydroxyl in the original PLA.

The continuous reactive extrusion process was conducted on a twin screw extruder with a diol reactant and a catalyst metered and injected into the melt of PLA (24). The PLA continuous alcoholysis reaction conditions are listed in Table 1. The PLA base resin rate was 40 lb/h for all the experiments. A PLA control was extruded without addition of any diol reactant and catalyst.

Experiment 1 was conducted at 0.7% butanediol in the absence of a catalyst. The Experiments 2 to 4 were performed at different levels of diol reactant and catalyst. Butanediol (0.7 to 1.2% by weight of PLA rate) and titanium propoxide (ranging from 235 to 820 ppm relative to PLA rate) were used as the reactants for the alcoholysis reaction.

The screw speed was 150 rpm. The reaction temperatures ranged from 130 to 194 °C for different reaction zones on the extruder. It was found that the melt temperature, melt pressure, and torque decreased upon the alcoholysis reaction in the presence of catalyst and alcohol. This indicated that the alcoholysis reaction had effectively reduced the melt viscosity of modified PLA which correspondingly decreased the melt temperature, torque, and melt pressure.

The average molecular weights and melt rheological properties of the alcoholysis modified PLA were determined; the results are summarized in Table 2. It was found that the alcoholysis modified PLA had significantly reduced molecular weights and melt viscosity. The PLA Control extruded without catalyst and diol reactant had essentially the same average molecular weights (M_w and M_n), polydispersity, and apparent viscosity as the PLA sample prior to melt extrusion.

430

Table 1. PLA Continuous Alcoholysis Reaction Conditions

Exp. No.	Butanediol (% of PLA)	Titanium Propoxide (ppm of PLA)	Melt Temperature (°C)	Melt Pressure (psi)	Torque (%)
PLA L-9000 Control	0	0	156	300	92-102
1	0.7	0	156	130	88-99
2	0.7	470	145	30	62-72
3	0.35	235	142	70	79-86
4	1.2	820	142	10	58-62

Table 2. Molecular weights and rheological properties of the modified PLA

Exp. No.	Average Mol. Wt. (g/mol)		Polydispersity (M_w/M_n)	Apparent Viscosity* (Pa.s)
	M_w	M_n		
PLA L-9000	143500	109300	1.31	257
Control	141900	107900	1.32	252
1	138900	105900	1.31	164
2	97200	70500	1.38	42
3	112900	83500	1.35	90
4	64900	41600	1.56	12

* The apparent viscosity was measured at a shear rate of 1000 s^{-1} and at 190 °C.

The results also showed that the PLA alcoholysis can be controlled by the levels of diol and catalyst. Butanediol alone (Experiment 1) resulted in very limited reduction of molecular weights and melt viscosity; this demonstrated that un-catalyzed alcoholysis of PLA with 1,4-butanediol proceeded only to a very limited extent during the short residence time (about 40 seconds) of reactive extrusion process. By adding titanium propoxide catalyst, the alcoholysis reaction is substantially accelerated and the resulting modified PLA had significantly reduced molecular weights and melt viscosities. It was found that the magnitude of the reduction corresponded very well with the increase of the addition levels of butanediol and catalyst.

It is theoretically possible that both esterification and transesterification could occur during the alcoholysis reaction of PLA with butanediol, since the titanate catalyst is also a catalyst for esterification reaction under appropriate reaction conditions. Esterification reaction at the carboxyl chain end of PLA leads to either no significant change in molecular weight of PLA or increased molecular weight as further esterification with the –COOH chain end of another PLA molecule. Typically, esterification reaction occurs at a higher temperature than used in

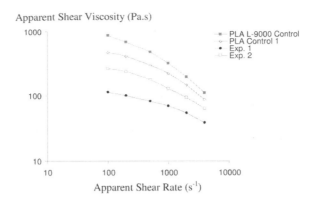

Apparent Shear Viscosity (Pa.s)

- PLA L-9000 Control
- PLA Control 1
- Exp. 1
- Exp. 2

Apparent Shear Rate (s⁻¹)

Figure 2. Melt rheology of alcoholysis PLA modified PLA versus control.

Figure 3. SEM image of a modified PLA meltblown nonwoven at a magnification of 100X.

the reactive extrusion (Table 1) and distillation of water is needed to shift the equilibrium. The transesterification reaction will result in chain scission of PLA. The results of significant molecular weight reduction in Table 2 suggested that the transesterification was the dominant reaction during alcoholysis of PLA.

The melt rheology of several modified PLA versus control samples were plotted in Figure 2. It showed that the melt viscosity of alcoholysis modified PLA had significantly reduced melt viscosity throughout the shear rate range. The modified PLA samples exhibited shear thinning behaviors. The sample from Exp. 2 conducted in the presence of both diol and catalyst had significantly lower capillary melt viscosity than the sample from Exp. 1 which was extruded with only diol but no catalyst. Reduction of melt viscosity of PLA in a controllable manner is one of key goals of this study to produce modified PLA with melt processability not previously available.

The thermal properties of the modified PLA and controls were analyzed. It was found that the alcoholysis modification had resulted in a slight reduction of both glass transition temperature (T_g) and melt peak temperature of PLA.

Figure 4. SEM Image of a modified PLA meltblown nonwoven at a magnification of 1000X.

$$H \!\!-\!\!\left[\!O(CH_2)_4O\overset{\overset{\displaystyle O}{\|}}{C}(CH_2)_2\overset{\overset{\displaystyle O}{\|}}{C}\!-\!\right]_{\!\!n}\!\!-OH \;+\; HOROH \quad\xrightarrow[\text{Melt, bulk}]{\substack{\text{Dibutyltin diacetate} \\ \text{or Ti (OC}_3\text{H}_7)_4}}\quad H\!\!-\!\!\left[\!O(CH_2)_4O\overset{\overset{\displaystyle O}{\|}}{C}(CH_2)_2\overset{\overset{\displaystyle O}{\|}}{C}\!-\!\right]_{\!\!m}\!\!-ROH$$

$$m < n$$

PBS   **Diol**   **Modified PBS**

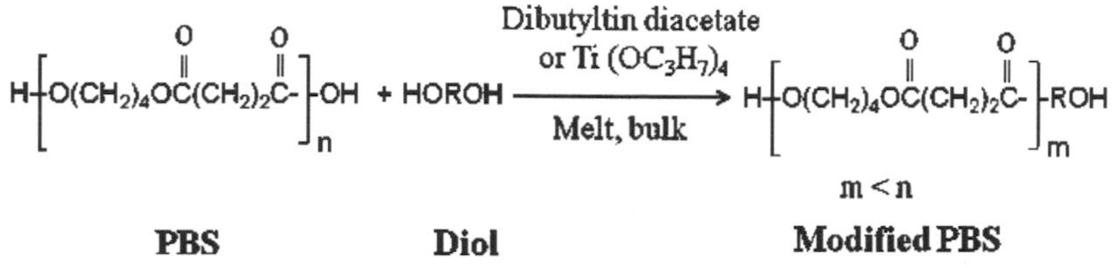

Figure 5. Catalyzed alcoholysis reaction of polybutylene succinate (PBS).

Meltblown Spinning of PLA Modified Alcoholysis Reaction

The alcoholysis modified PLA showed molecular weights and melt rheology compatible with nonwoven application requirements. A meltblown spinning experiment was conducted using the procedure described in the Experimental section. Meltblown from the modified PLA was obtained.

A Scanning Electron Microscopy (SEM) image of the meltblown PLA nonwoven at 100X is shown in Figure 3.

Figure 4 shows the SEM image of the modified PLA meltblown nonwoven at a magnification of 1000X, inter-fiber bonding of PLA meltblown fibers was observed. Bonding is important for making strong nonwovens, however, strong bonding of unmodified PLA fibers in spunbond nonwoven is difficult to achieve, and this was commonly attributed to the narrow thermal bonding window of PLA. Bonding of meltblown nonwoven showed the uniqueness of the meltblown process in which PLA meltblown fibers exiting the orifices of meltblown die were heated by the primary air at a temperature higher than the melting point of PLA. The hot primary air was able to keep PLA meltblown filaments in molten state to allow fiber-to-fiber bond formation prior to solidification during cooling.

Alcoholysis Reaction of Polybutylene Succinate (PBS)

Polybutylene succinate (PBS) and polyethylene succinate (PES) are synthetic biodegradable polymers made by condensation polymerization of succinic acid and butanediol or ethylene glycol. Potentially, the succinic acid and diols could be derived from biomass via fermentation, leading to bio-based PBS or PES. Both

Table 3. Alcoholysis Reaction Conditions of PBS under Different Reaction Conditions

Experiment No.	Temperature (^{o}C) Zone 1, 2, 3-8, 9, 10					Screw Speed (Rpm)	Resin Rate (Lb/h)	Reactant % of resin rate	Catalyst % of the resin
Control 1	90	125	165	125	110	150	1.9	0	0
Control 2	90	125	165	125	110	150	1.9	4	0
Control 3	90	125	165	125	110	150	1.9	4(EGDA)	0.08
1	90	125	165	125	110	150	2	3.3	0.08
2	90	125	165	125	110	150	2	1.7	0.04
3	90	125	165	125	110	150	2	5.2	0.12
4	90	125	165	125	110	150	2	1.7	0.02
5	90	125	165	125	110	150	2	3.3	0.04
6	90	125	165	125	110	150	2	5.2	0.06
7	90	125	165	125	110	150	2	1.7	0.08
8	90	125	165	125	110	150	2	3.3	0.16
9	90	125	165	125	110	150	2	5.2	0.24

Table 4. Molecular weights and rheological properties of the modified PBS

Exp. No.	Apparent Viscosity (Pa.s) at Apparent Shear Rate of 1000	Melt Flow Rate (g/10 min at 170 ^{o}C and 2.16 kg)	Average Mol. Wt (g/mol)		
			Mw	Mn	Mw/Mn
Control 1	155	8	128000	73900	1.73
Control 2	68	86	96900	58200	1.66
Control 3	154	-	-	-	-
1	28.5	290	77200	42000	1.84
2	85	56	101900	64700	1.58
3	9.8	852	65800	35200	1.87
4	163	50	97500	57500	1.69
5	37	185	86400	53600	1.61
6	11.4	840	61100	32400	1.87
7	65	83	99900	59500	1.68
8	14	600	67200	37000	1.82
9	4.9	1100	58600	31600	1.85

PBS and PES have been used for packaging films, mulching films, bags, injection molding, and flushable hygiene products applications (25–28). Since PBS and PES are significantly softer and more ductile than PLA, there is a scientific and technical interest to prepare soft and ductile biodegradable meltblown nonwoven from PBS.

However, PBS polymer for meltblown nonwoven processing is not available on the market. In order to assess the PBS nonwoven application potential, there is a technological need to modify the available high molecular weight PBS resins typically synthesized for film application into meltblown processable resins.

Following the success of controllably decreasing the molecular weights and melt viscosity of PLA as discussed previously, the alcoholysis reaction was explored in order to decrease the molecular weights of PBS. The reaction scheme is shown in Figure 5.

The continuous alcoholysis reaction conditions via a reactive extrusion process are listed in Table 3 (29). In order to determine the critical factors impacting the controlled degradation of PBS, a number of control experiments were conducted to investigate reactant variables during the reactive extrusion.

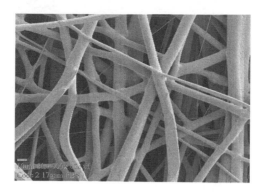

Figure 6. SEM Micrograph of meltblown nonwoven made from modified PBS at a magnification of 500X.

Control 1 was for the physical extrusion of PBS without an alcohol or a catalyst, it provided a baseline for PBS after the extrusion process to quantify the thermomechanical degradation of PBS. Control 2 was for the reactive extrusion PBS in the presence of 4% 1,4-butanediol, without a catalyst. Control 3 was performed in which PBS was extruded in the presence of 4% ethylene glycol diacetate and 0.08% dibutyltin diacetate catalyst; in this control, a diol capped with acetate groups was used to demonstrate the critical role of free hydroxyl functional groups for the reaction.

Experiments 1 though 9 were alcoholysis reactions in the presence of different levels of 1,4-butanediol and dibutyltin diacetate (catalyst).

The success of PBS alcoholysis reaction was demonstrated by the reduction of molecular weights and melt rheology of the resulting degraded PBS samples. Table 4 shows the apparent viscosity, melt flow rate, and average molecular weights of the modified PBS samples. Control 2 (4% diol, no catalyst) showed the alcoholysis without a catalyst was not very effectve. Control 3 showed that a diol ester (without any hydroxyl group) did not result in alcoholysis reaction, free hydroxyl groups are important for the alcoholysis reaction. Consequently, the apparent melt viscosity stayed about the same before and after the reactive extrusion reaction process.

For Experiments 1 through 9, the alcoholysis reactions occurred effectively in reducing the molecular weights and melt viscosity. The results showed that the relative ratios of alcohol and catalyst to PBS were critical to the success of alcoholysis reaction. The most significant reductions of molecular weights and melt viscosities were achieved at high alcohol and catalyst addition levels (Experiments 3, 6, and 9).

Meltblown nonwovens were successfully made from the modified PBS using the meltblown device used for making modified PLA meltblown. A representative SEM image of the modified PBS meltblown nonwoven is shown in Figure 6. There are two observations that can be made from the SEM micrograph. First, evidence of fiber-to-fiber bonding was shown on the micrograph. The bonding is important to achieve good mechanical properties of nonwovens. Secondly, there are fine fibers with diameter in submicron range, indicating the great extensibility of the PBS meltblown fibers allowing the fine fiber formation during meltblown process.

The mechanical properties of the biodegradable meltblown nonwoven were determined. At the same basis of 23 gsm (g/m²), the biodegradable PBS meltblown nonwoven had a 46% improvement in peak load and 67% improvement in elongation.

The reactive extrusion process using alcoholysis can be applied to broad classes of aliphatic polyesters including polycaprolactone, polyhydroxyalkanoate, and also copolyesters, etc. The modified aliphatic polyesters or copolyesters will have tailored melt rheology to make them suitable for various polymer processing methods such as fiber spinning, injection molding, etc.

Conclusions

A continuous alcoholysis reaction was demonstrated to be effective in significantly reducing the molecular weights and melt rheology of both polylactic acid (PLA) and polybutylene succinate (PBS). The alcoholysis modification was controlled by reactant stoichiometry. The most notable factors affecting the molecular weight reduction are the diol and catalysts levels relative to the base polymers. Multiple catalysts were found to be effective for modifying the biodegradable polymers for meltblown nonwoven applications. Modified PLA and modified PBS were successfully spun into meltblown nonwoven with good fiber-to-fiber bonding and mechanical properties.

Acknowledgments

The authors would like to thank Gregory Wideman of Kimberly-Clark Corporation for his assistance in the reactive extrusion process.

References

1. Stevens, C.; Verhe, R. *Renewable Bioresources*; John Wiley & Sons: 2004.
2. Lunt, J. *Polym. Degrad. Stab.* **1998**, *59*, 145.
3. Zhang, J. F.; Sun, X. In *Biodegradable Polymers for Industrial Applications*; Smith, R., Ed.; CRC Press: Boca Raton, 2005; p 251.
4. Doi, Y. *Microbial Polyesters*; Wiley-VCH: 1990.
5. Mobley, D. P., Ed. *Plastics from Microbes: Microbial synthesis of Polymers and Polymer Precursors*; Hanser: Munich, 2005.
6. Jérôme, R.; Lecomte, P. In *Biodegradable Polymers for Industrial Applications*; Smith, R., Ed.; CRC Press: Boca Raton, 2005; p 75.
7. Müller, R.-J. In *Handbook of Biodegradable Polymers*; Bastioli, C., Ed.; Rapra Technology: Shawbury, 2005; p 303.
8. Albertsson, A.-C.; Varma, I. K. In *Degradable Aliphatic Polyesters*; Albertsson, A.-C., Ed.; Springer: Berlin, 2002; p 1.
9. Russell, S., Ed. *Handbook of Nonwovens*; CRC: 2006.
10. Andrade, A. P.; Without, B.; Haney, R.; Eglin, T.; Li, T. *Macromolecules* **2002**, *35*, 684.

11. Andrade, A. P.; Witholt, B.; Chang, D.; Li, Z. *Macromolecules* **2003**, *36*, 9830.
12. Xantho, M., Ed. *Reactive Extrusion: Principles and Practice*; Hanser: Munich, 1992.
13. Wang, J. H.; Schertz, D. M. U.S. Patent 5,952,433, 1999.
14. Wang, J. H.; Schertz, D. M. U.S. Patent 5,945,490, 1999.
15. Wang, J. H.; Schertz, D. M. U.S. Patent 6,579,943 B, 2003.
16. Wang, J. H.; Schertz, D. M. U.S. Patent 6,500,897 B2, 2002.
17. Moad, G. *Prog. Polym. Sci.* **1999**, *24*, 81.
18. Wang, J. H.; Schertz, D. M. U.S. Patent 6,107,405, 2000.
19. Wang, J. H.; Schertz, D. M. U.S. Patent 6,297,326 B1, 2001.
20. Azizi, H.; Ghasemi, I. *Polym. Test.* **2004**, *23*, 107.
21. Wang, J. H.; Schertz, D. M. U.S. Patent 6,117,947, 2000.
22. Wang, J. H.; Schertz, D. M.; Soerens, D. A. U.S. Patent 6,172,177 B1, 2001.
23. Wang, J. H.; He, A. WO 2008/008068, 2008.
24. He, A.; Wang, J. H. WO 2008/008074, 2008.
25. Wang, J. H.; Schertz, D. M. U.S. Patent, 6,552,124 B2, 2003.
26. Wang, J. H.; Schertz, D. M. U.S. Patent, 6,579,934 B1, 2003.
27. Wang, J. H.; Schertz, D. M. U.S. Patent, 6,890,989 B2, 2003.
28. Wang, J. H.; Schertz, D. M. U.S. Patent, 7,053,151 B2, 2006.
29. He, A.; Wang, J. H. WO 2008/008067, 2008.

Chapter 30

Synthesis of Grafted Polylactic Acid and Polyhydroxyalkanoate by a Green Reactive Extrusion Process

James H. Wang* and David M. Schertz[†]

Corporate Research & Engineering, Kimberly-Clark Corporation,
2100 Winchester Road, Neenah, WI 54956, U.S.A.
*jhwang@kcc.com
[†]Present address: 820 Meadow Spring Ct., Alpharetta, GA 30004

Grafted polylactic acid (PLA) and polyhydroxyalkanoate (PHA) were synthesized by a reactive extrusion method. The grafted bio-based polymers had either polar functional groups such as hydroxyl (-OH) and polyethylene glycol (PEG) or non-polar functional groups. It was found that grafted biopolymers had significantly reduced melt viscosity, making them more suitable for certain polymer processing such injection molding and fiber spinning. The grafted biopolymers had improved compatibility in blending with other polar polymers such as polyvinyl alcohol and exhibited improved fiber spinning processability in polymer blends. Grafted PHA had a low crystallization rate making continuous reactive extrusion impossible. It was found that a novel co-grafting method, i.e. grafting PHA in the presence of PLA, was effective to overcome the process challenge of PHA. The reactive groups introduced to PLA or PHA can be used for further side chain reactions. The free radical initiated grafting reaction was a green reaction method. It eliminated the use and recovery of organic solvents, the reaction rate was also significantly increased over solution grafting reaction, taking seconds to complete rather than hours.

Introduction

In today's global economy, polymers and products made from various polymers play a major role by providing the society with high performance and cost-viable materials. The high volume applications of plastics include packaging (29%), construction (15%), consumer products (14%), and transportation (5%) (*1*). However, almost all of the polymers currently used are made from non-renewable mineral resources such as petroleum-based feedstock. From a sustainability point and life cycle assessment (LCA) perspective, the petroleum derived polymers have challenges of lacking long-term sustainability, especially when considering diminishing oil reserves, high energy input from production, and high level of emissions.

Renewable and bio-based polymers offer an attractive alternative to petroleum-derived polymers, since they are made from renewable resources such as agricultural, forest, or marine based products or wastes (*2*). The desirable attributes of bio-based polymers include: 1) produced from annually renewable feedstock; 2) significantly reduced carbon footprint; 3) reduced emission of green house gases; 4) some bio-based polymers are also biodegradable.

In spite of the advantages of bio-based polymers, there are limitations of the currently available bio-based polymers: 1) relatively narrow performance profile making them not possible for broad applications like common commodity polymers (*3*); 2) narrow and often challenging processability; 3) high cost compared to low-cost commodity polymers currently used; 4) limited volume and availability making it not possible to broadly replace current polymers, although the bioplastic industry has experienced rapid growth the last few years (*4, 5*).

Among the bio-based polymers, polylactic acid (PLA) (*6, 7*) and polyhydroxyalkanoate (PHA) are aliphatic polyesters (*8, 9*). PLA is synthesized from bio-derived lactide or lactic acid derived from the fermentation of renewable raw materials (e.g. starch or cellulose) (*6*), while PHA is a family of microbial polyesters accumulated inside cells of microorganisms using renewable carbon sources such as sugars, vegetable oils, organic acids. Both polymers have the backbone linked by carboxylic esters.

Both PLA and PHA do not have other reactive functional groups except terminal hydroxyl and carboxyl groups. However, the role of these chain end groups for reaction becomes practically diminished as molecular weights increase.

In this article, grafting functionalization of PLA and PHA was accomplished through a green, continuous extrusion reaction. This reaction was demonstrated by introducing reactive functional groups to fundamentally change the polarity of PLA and PHA and making grafted PLA and PHA with improved melt processability and compatibility with other polymers. PLA and PHA were found to have changed melt rheology and thermal properties after grafting. Furthermore, the reactive groups attached to PLA and PHA also make the polymers capable for additional modification reactions.

Experimental

Materials

The polylactic acid (PLA) was purchased from Aldrich. It had a number average molecular weight (M_n) of 60,000 g/mol, a weight average molecular weight (M_w) of 144,000 g/mol, and a glass transition temperature (T_g) of 60 °C. The polyvinyl alcohol (PVOH) used in the experiments was Ecomaty AX 10000 supplied by Nippon Gohsei (Osaka, Japan). It was a cold water-soluble, partially hydrolyzed PVOH with the melt flow rate of 100 g/10 minutes at 230 °C and 2.16 kg. The PHA used was a poly(3-hydroxybutyrate-co-3-hydroxyvalerate) (PHBV, BIOPOL® D600G) purchased from Zeneca Bio Products with a valerate content of 12% and a melt flow rate of 12 g/10 minutes at 2.16 kg and 190 °C. 2-Hydroxyethyl methacrylate (HEMA, 97%) and poly(ethylene glycol) ethyl ether methacrylate (PEGMA, molecular weight of 246 g/mol) were used as received from Aldrich. Lupersol® 101 (2,5-dimethyl-2,5-bis(tert-butylperoxy)hexane, CAS no. 78-63-7) was supplied by Elf Atochem North America.

Preparation of Grafted Polylactic Acid (PLA)

The grafting reaction was conducted in a co-rotating, twin-screw extruder (ZSK-30) manufactured by Werner and Pfleiderer Corporation of Ramsey, N.J. The screw length was 1328 millimeters, the screw diameter was 30 mm and the L/D was 44. The extruder had 14 barrels, numbered consecutively from 1 to 14 from the feed port to die. The first barrel (#1) received the PLA *via* a gravimetric feeder at a throughput of 20 pounds per hour (i.e. 9.08 kg/hr). HEMA was injected to the polymer melt in barrel #5 by a pressured injector at a rate of 1.8 lb/hr (792 g/hr). Lupersol 101 was injected via a pressurized injector to barrel #6 at a rate of 0.09 lb/hr (40.9 g/hr). Both liquids were delivered by reciprocating pumps made by Eldex. A vacuum port for devolatilization was located at barrel #11. The die for extruding the grafted PLA strands had 4 openings of 3 mm in diameter which were separated by 7 mm. The extruder had 7 heating zones. The temperatures for seven heating zones from the feed port to die were 180, 180, 180, 180, 180, 176, and 160 °C, respectively. The screw speed was 300 rpm. The resulting grafted PLA was cooled in a water bath and subsequently pelletized by a Conair pelletizer.

Preparation of Grafted Poly(3-Hydroxybutyrate-co-3-Valerate) (PHBV)

Grafting reaction was conducted using a counter-rotating, twin-screw extruder manufactured by HAAKE (now part of Thermo Electron Corporation). The extruder had two conical screws of 30 mm diameter at the feed port and 20 mm at the die. The screw was 300 mm in length. The PHBV resin was fed to the feed port by a calibrated volumetric feeder, HEMA, PEGMA, or butyl acrylate monomers and peroxide (Lupersol 101) were delivered by Eldex pump to the feed port directly. The extruder had one barrel with four heating zones respectively numbered Zone 1 to Zone 4 from feed port to die. In one experiment, PHBV was delivered into the extruder barrel *via* a volumetric feeder at a throughput of 5 pounds per hour (i.e. 2270 g/hr). Simultaneously, HEMA was fed to the feed port at a rate of 0.25 lb/hr (113.5 g/hr). Lupersol 101 was fed to the feed port at a rate of 0.02 lb/hr (9.08 g/hr). The screw speed was 150 rpm. The die had two openings of 3 mm in diameter which were separated by 5 mm. PHBV exhibited extremely slow solidification after initial cooling in water. In fact, the PHBV strands could not even be cooled sufficiently in ice water (~0 °C). The solidification of both unmodified PHBV and grafted PHBV took several minutes to complete, so the samples were collected in small chunks after solidification and then chipped into small pieces.

Preparation of Polymer Blends of Grafted PLA or PLA with Polyvinyl Alcohol

Polymer blends of various compositions were prepared using the same extruder used to graft PHBV. Polymer blends of grafted PLA or PLA/polyvinyl alcohol (PVOH) at 20/80, 30/70, and 40/60 weight ratios were fed to the extruder at a rate of 10 lbs/hr (4.54 kg/hr). The extruder was set at a temperature profile of 170, 180, 180, and 168 °C. The screw speed was 150 rpm. The resulting polymer blends were cooled on a fan-cooled conveyor belt and then pelletized.

Fiber Spinning of Grafted PLA/PVOH and PLA/PVOH Blends

Fiber spinning was conducted on a laboratory fiber spinning device consisting of a vertically mounted cylinder extruder heated by cartridge heaters. During the spinning of HEMA grafted PLA/PVOH 20/80 blend, the temperature of cylinder barrel was set at 360 °C. A vertically mounted Worm Gear Jacuator (Model: PKN-1801-3-1, manufactured by Duff-Northon Company, Charlotte, N.C.) was used to extrude the polymer blend into fibers. The fibers were spun from a spin plate with 3 openings of 0.356 mm. The fibers exiting the die were wound up on a drum having both reciprocating and rotary movements for fiber collecting.

Melt Rheology

The rheological properties of polymers were determined using a Göttfert Rheograph 2003 capillary rheometer with WinRHEO version 2.31 analysis software. The setup included a 2000-bar pressure transducer and a 30/1 L/D round capillary die. Sample loading was done by alternating between sample addition and packing with a ramrod. A 2-minute melt time preceded each test to allow the polymer to completely melt at the test temperature (usually 150°C to 220 °C). The capillary rheometer determined the apparent viscosity (Pa·s) at various shear rates, such as 100, 200, 500, 1000, 2000, and 4000 s^{-1}. The resultant rheology curve of apparent shear rate versus apparent viscosity gave an indication of how the polymer would run at that temperature in an extrusion process.

Thermal Properties

The melting temperature (T_m), glass transition temperature (T_g), and latent heat of fusion (ΔH_f) were determined by differential scanning calorimetry

(DSC). The differential scanning calorimeter was a THERMAL ANALYST 2910 Differential Scanning Calorimeter, which was outfitted with a liquid nitrogen cooling accessory and with a THERMAL ANALYST 2200 (version 8.10) analysis software program, both of which are available from T.A. Instruments Inc. of New Castle, Delaware. For resin pellet samples, the heating and cooling program was a 2-cycle test that began with an equilibration of the chamber to -50 °C, followed by a first heating period at a heating rate of 10 °C per minute to a temperature of 200 °C, followed by equilibration of the sample at 200 °C for 3 minutes, followed by a first cooling period at a cooling rate of 20 °C per minute to a temperature of -50 °C, followed by equilibration of the sample at -50 °C for 3 minutes, and then a second heating period at a heating rate of 10 °C per minute to a temperature of 200 °C. For fiber samples, the heating and cooling program was a 1-cycle test that began with an equilibration of the chamber to -50 °C, followed by a heating period at a heating rate of 20 °C per minute to a temperature of 200°C, followed by equilibration of the sample at 200 °C for 3 minutes, and then a cooling period at a cooling rate of 10 °C per minute to a temperature of -50 °C.

Results and Discussion

Grafted Polylactic Acid (PLA)

Grafting reactions have been used to introduce functional groups onto polymers. For example, maleic anhydride (MAH) grafted polyolefins were the subject of extensive scientific research and commercial development (10–13). Grafting of polyolefins led to improvements of the properties of polyolefins and new applications. Maleic anhydride grafted polyolefins had adhesive properties with polar substrates such as metal foil, polyesters and nylon and found applications as a tie-layers. Grafting of vinyl alkoxysilanes onto polyolefins is another well-known application for making crosslinked polyolefins for wires and cable industry (14, 15). The grafted polyolefins were also used as compatibilizer for polymer blends containing polyolefins and polar polymers.

There are a number of methods to prepare graft polymers. Grafting reaction in solution involves multiple steps: 1) dissolve a polymer in a solvent or mixture of solvents, 2) react the base polymer with a monomers such as maleic anhydride or acrylic acid and a free radical initiator (typically a peroxide or azo compound) in a reaction vessel for extended time, 3) precipitate the reaction mixture in a non-solvent to remove unreacted monomer and homopolymers; 4) filter to collect grafted polymer, 5) drying. Due to the low solubility of most high molecular eight polymers in organic solvents, only low concentration of base polymer is obtained. This typically led to a slow reaction and long reaction time.

There is a side reaction which is difficult to prevent during solution grafting reaction. Some monomers will tend to form homopolymers during the grafting reaction, resulting in waste of monomers and a need to separate homopolymers. Although grafting reactions can proceed via photochemical, radiation, sonic reactions, and others, the most preferred method is a melt phase reactive extrusion process.

Grafting reaction by reactive extrusion process has a number of advantages over a solution process for industrial production: 1) homopolymerization can be controlled by selecting the extrusion conditions and designing the reaction process; 2) fast reaction can be achieved; as the reaction is conducted at a high concentration of base polymer in the bulk (polymer melt) phase, the reaction typically takes seconds to complete versus hours for solution based process; 3) no solvent is used and it is an environmentally friendly green process, eliminating the emissions of volatile organic compounds (VOC) typically encountered during solution process, 4) it has less energy input due to the elimination of dissolution, solvent recovery, and post reaction work-up steps, 5) lower capital cost and operating cost (16). Reactive extrusion is very effective in making grafted polymers of bio-based PLA (17, 18), microbial polyhydroxybutyrate-co-valerate (PHBV) (19), biodegradable PBS (20), and grafted polyolefins (21–23). Reactive extrusion was also used to reduce molecular weights of polymers by free radical reactions (24, 25).

The grafting reaction of PLA is shown in Figure 1. The vinyl monomer can contain a variety of polar functional groups such as hydroxyl, carboxyl, polyethylene glycol, etc. The grafting reaction is initiated by a free radical initiator, typically a peroxide. Due to the short reaction time of the reactive extrusion process, the selection of peroxide has to be matched with the half life of the peroxide at the extrusion conditions. The PLA grafting was performed at 180 °C and a screw speed of 300 rpm (17, 18). 2,5-Dimethyl-2,5-bis(tert-butoxy)hexane (Lupersol®101) was selected and was found to be effective for the grafting reaction.

The grafting reaction was performed continuously at a PLA throughput of 20 lbs/hr (9.08 kg/hr), the monomer (2-hydroxyethyl methacrylate, HEMA) feeding rate was 1.6 lb/hr (792 g/hr), and peroxide rate was 0.09 lb/hr (40.9 g/hr). It is important to have the right stoichiometry of monomer to peroxide ratio during reactive extrusion. On one hand, if the Monomer/Initiator (M/I) ratio is too low, the peroxide initiator can cause crosslinking of PLA, one undesirable side reaction during grafting. On the other hand, if the M/I ratio is too high, it could lead to a lower grafting efficiency of the monomer as well as potential homopolymerization of the vinyl monomer should the monomer be easily homopolymerizable.

The reactive extrusion process is shown in Figure 2. Due to the reaction requirements, it is important that the extruder has sufficient length to diameter (L/D) ratio to allow the grafting reaction to complete. Essentially, the twin screw extruder incorporates multiple unit operations of a conventional batchwise process: 1) melting stage: accomplished in the first few barrels of the extruder, under heat and intensive shear force, the PLA pellets were melted into a PLA melt; 2) grafting reaction: monomer and peroxide were injected into the PLA melt, under intensive dispersive and distributive mixing the monomer and peroxide droplets were repeatedly divided into smaller sizes, creating large surface areas for the interfacial reaction to occur, the free radicals formed during the decomposition of the peroxide initiate grafting reactions. 3) devolatilization: unreacted monomer was removed from the reaction mixture at this stage by increasing the surface areas of polymer melt exposed to the vacuum vent port. The resulting PLA strands were cooled and cut into pellets. A long extruder of

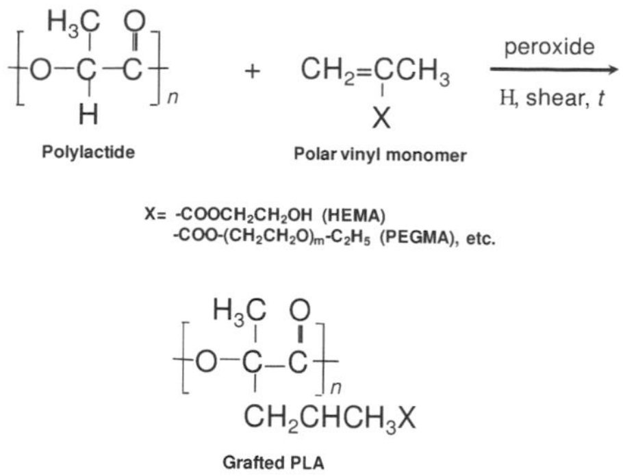
peroxide
H, shear, t
Polylactide
Polar vinyl monomer
X= -COOCH2CH2OH (HEMA)
-COO-(CH2CH2O)m-C2H5 (PEGMA), etc.
Grafted PLA

Figure 1. Grafting reaction of PLA

L/D of 44/1 was used for this reactive extrusion process. The residence time was about 40 seconds.

The melt rheology of the resulting PLA graft copolymer was characterized by capillary rheology. The rheological properties of the grafted PLA and PLA starting polymer is plotted in Figure 3. Both PLA and grafted PLA exhibited shear thinning behavior, the apparent melt viscosity decreased as apparent shear rate increased. It was also found that that grafted PLA has significantly reduced melt viscosity than the PLA starting material. For example, the melt viscosity of grafted PLA was 46.4 Pa.s versus 112.3 Pa.s of PLA starting polymer at a shear rate of 1000 s^{-1}. The reduction in melt viscosity could be attributed to free radical mediated degradation of PLA as observed during the grafting reaction of polar vinyl monomers onto polyethylene oxide (*24, 25*). The reduction of melt viscosity is favorable for polymer processing which requires a low melt viscosity. Such processes include injection molding, fiber spinning extrusion coatings.

Applications of Grafted Polylactic Acid (PLA)

By grafting polar functional monomers onto PLA, it is expected to have improved compatibility with polar polymers such as polyvinyl alcohol (PVOH), polyethylene oxide (PEO), or thermoplastic starch (TPS). PVOH is a broadly used water-soluble polymer, its water solubility can be controlled by the degree of hydrolysis. In certain applications, there is a need to control the water sensitivity of PVOH to match the application needs. One such way to reduce the water-solubility of PVOH is to create polymer blends with different levels of water-insoluble polymers such as biodegradable and bio-based PLA.

Polymer blends containing both grafted PLA/PVOH and PLA/PVOH were studied to determine the improvement of grafted PLA in polymer blends. Films were made from various ratios of PLA or grafted PLA and PVOH. Figure 4A shows the cross-sectional image of PLA/PVOH at 30/70 weight ratio by scanning electron microscopy (SEM). PLA existed as the dispersed phase shaped in nearly spherical droplets with diameter ranging from about 0.1 μm to 0.7 μm, there was a large variation in the size of PLA spherical entities, a large portion of the PLA droplets were large (~0.5 μm).

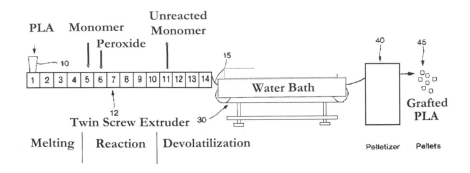

Figure 2. Reactive extrusion grafting process of PLA

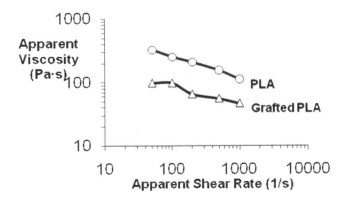

Figure 3. Melt rheology of Grafted PLA versus PLA starting polymer.

Figure 4B shows the SEM morphology of grafted PLA/PVOH 30/70 blend. It was found that the blend containing grafted PLA with PVOH had significantly reduced dispersed PLA phase size, majority of the grafted PLA droplets were small. Figure 4B shows improved compatibility of grafted PLA with PVOH than PLA in blends with PVOH (26). It was also found that it was possible to control the water sensitivity by changing polymer blend composition. Fibers of controlled water sensitivity are desirable for industrial or commercial applications.

Several polymers blends containing from 20, 30, and 40% of either grafted PLA or PLA with polyvinyl alcohol were prepared using twin screw extrusion (17). The fiber spinning experiments were performed for blends containing either PLA or grafted PLA. It was found that HEMA grafted PLA blends had significantly improved melt strength allowing them to spin into fine fibers. The improvement is important to overcome the deficiency of poor fiber spinning processability of PLA/PVOH blends. The improvement in melt processability of the blends resulted from the increased compatibility of the HEMA grafted PLA/PVOH blends, by increased hydrogen bonding of the hydroxyl group on the grafted HEMA with the hydroxyls of polyvinyl alcohol.

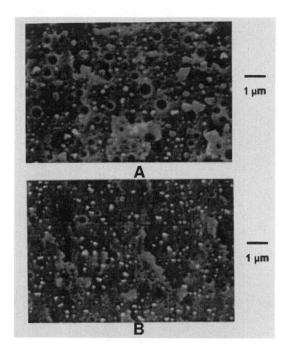

Figure 4. Cross sectional morphology of PLA/PVOH 30/70 film (A) and grafted PLA/PVOH (30/70) film (B).

Grafted Polyhydroxyalkanoate (PHA)

Polyhydroxyalkanoate (PHA) is a family of microbially produced aliphatic polyesters. There are many copolymers and terpolymers in the PHA family. Polyhydroxybutyrate (PHB) is the simplest PHA as a structurally equivalent homopolymer of 3-hydroxybutyrate, however, due to its high crystallinity PHB had high stiffness, low ductility, and narrow thermal processing window (*9*). Copolymers have been explored to improve both the mechanical properties (mainly to increase ductility and decrease stiffness) and processability. Other commonly investigated PHA's include poly(3-hydroxybutyrate-co-3-hydroxyvalerate) (PHBV), poly(3-hydroxybutyrate-co-4-hydroxybutyrate) (P3HB4HB), poly(3-hydroxybutyrate-co-3-hydroxyhexanoate) (PHBH), etc. (*8, 9*).

Similar to PLA, PHA does not have any reactive functional groups beyond the chain end groups. As the molecular weights of PHA increases, the role of the chain end groups diminishes. Grafting modification of PHA represents a versatile way to add reactive functionality onto PHA. The grafting reaction of PHA is shown Figure 5. The PHA grafting studied was also a melt phase reaction due to the same advantages of reactive grafting method as discussed for PLA case.

The grafting reaction of PHBV as an example of PHA was performed in a twin screw extruder (*20*). The reaction conditions are listed in Table 1. 2-Hydroxyethyl methacrylate (HEMA) was a hydroxyl functional monomer used to demonstrate the grafting reaction. Two grafting reactions were included in Table 1 at PHBV throughput of 5 lbs/hr (2270 g/hr). The first grafted PHBV sample was prepared

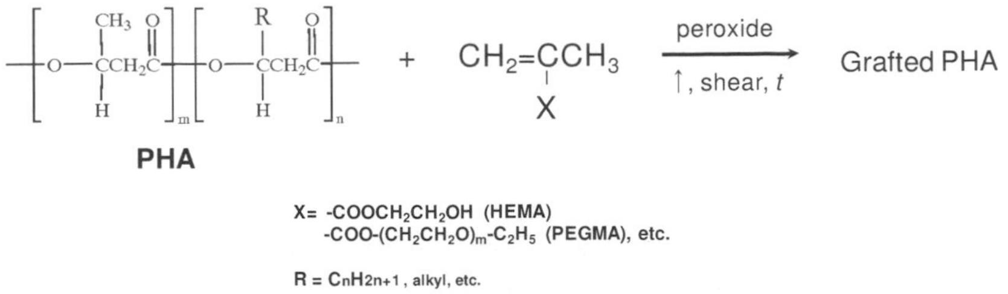

$$X= \text{-COOCH}_2\text{CH}_2\text{OH} \quad \text{(HEMA)}$$
$$\text{-COO-(CH}_2\text{CH}_2\text{O})_m\text{-C}_2\text{H}_5 \quad \text{(PEGMA), etc.}$$

$$R = C_nH_{2n+1}, \text{alkyl, etc.}$$

Figure 5. Free radical initiated melt phase grafting reaction of PHA.

at a temperature profile of 170, 180, 180, and 175 °C. HEMA monomer was added at 5% of PHBV base polymer, the peroxide was at 0.4% of PHBV. For the second grafting reaction, the monomer to PHBV base polymer ratio was doubled to 10%, while the peroxide initiator ratio to PHBV was slightly increased to 0.5% in order to promote better grafting but also to avoid crosslinking. To provide a baseline for comparison, a PHBV control sample was extruded at similar extrusion conditions (temperatures, screw speed, etc.) without adding any HEMA monomer or peroxide.

During the grafting reactions, the torque of the extruder was monitored. Figure 6 shows a chart of the torque versus time during the process. At the beginning of the reaction, when HEMA and peroxide were added to the extruder, the torque decreased significantly, about 40% from about 14 Nm to 10 Nm. This indicated the lubrication effect and/or plasticization effect of introduced monomer. As the grafting reaction reached a steady state, the torque tended to stabilize over a narrow range. As soon as HEMA and peroxide additions were stopped, the torque increased the level before grafting reaction. The reduction of the torque during the steady state grafting stage suggested that the grafted PHBV may have a reduced melt viscosity versus unreacted PHBV.

There are multiple sites on PHBV for hydrogen abstraction by free radicals and for subsequent grafting reactions. Figure 7 shows the proton NMR spectrum for HEMA grafted PHBV. The grafted PHBV showed a characteristic methyl peak at 2.0 ppm. The PHBV starting material did not exhibit this peak on its NMR spectrum. This confirmed that HEMA was grafted onto PHBV. However, the NMR spectroscopy was unable to differentiate the grafting sites on PHBV.

Polymer melt rheology is important for polymer processing. The capillary melt rheological measurements were performed on grafted PHBV and the extruded PHBV control. The melt rheology curves of grafted PHBV and unmodified PHBV at 180 °C are shown in Figure 8. The grafted PHBV sample was prepared at 10% HEMA monomer and 0.5% peroxide. At the same shear rate, it was found that the grafted PHBV had significantly reduced melt viscosity than extruded PHBV control which did not have grafting reaction. The results were in agreement with the melt rheology of grafted PLA versus un-grafted PLA as discussed previously.

The thermal properties of the grafted PHBV (5% HEMA, Table 1) and PHBV control were studied by Differential Scanning Calorimetry (DSC), the results are plotted in Figure 9. Figure 9A shows the DSC trace of PHBV with two overlapping

Table 1. Reactive Extrusion Conditions for Grafting HEMA onto PHBV

Process Conditions for Preparing HEMA Grafted PHBV

Sample I.D.	Temperature (° C.) Zones 1, 2, 3, 4	Screw Speed (rpm)	Resin Rate (lb/hr)	HEMA rate Lb/hr (% of resin rate)	Initiator rate (% of resin rate)
Control	170, 180, 180, 180	150	5.0	0	0
g-PHBV-1	170, 180, 180, 175	150	5.0	0.25 (5.0)	0.020 (0.40)
g-PHBV-2	170, 180, 180, 180	150	5.0	0.50 (10)	0.025 (0.50)

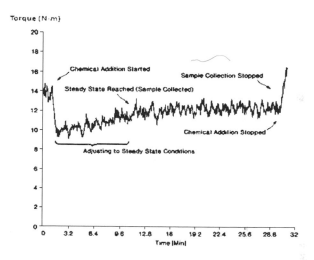

Figure 6. Torque changes during grafting reaction of PHBV.

peaks of similar intensity of 159 °C and 151 °C, i.e. separated by 8 °C. The enthalpy of the combined melting peaks was 84.0 J/g.

As shown in Figure 9B, the grafted PHBV also had also two melting peaks which are more separated apart than PHBV, at 159 °C and 144 °C respectively. Two peaks were separated by 15 °C. The enthalpy of melting was found to have reduced to 76.1 J/g. The reduction of both melting points and enthalpy could be accounted for by the presence of grafted HEMA side chains, which affected both the crystallization of grafted PHBV and the packing of grafted PHBV chains.

The DSC data are summarized in Table 2. The grafted PHBV-2 at 10% nominal HEMA grafting level also had reduced melting point and enthalpy of melting as compared to PHBV. The grafted PHBV had modified crystalline structure from the PHBV starting material, providing another evidence of the effect of grafting. In previous reactive extrusion grafting work on polyethylene

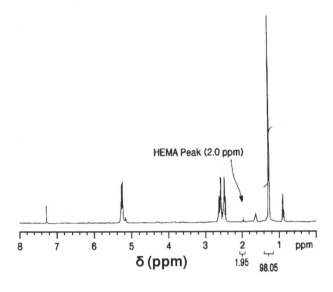

Figure 7. ¹H-NMR spectrum of HEMA grafted PHBV.

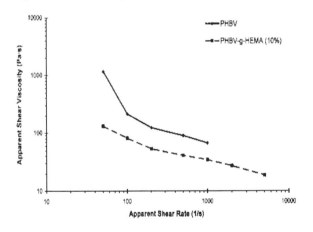

Figure 8. Melt rheology of grafted PHBV versus PHBV at 180 °C.

oxide (*24*), polybutylene succinate (PBS) (*20*), similar reduction in melting peak temperatures and enthalpy of melting were also observed.

Grafted of PHA in the Presence of PLA

Due to the fast crystallization rate of PLA, the grafting reaction of PLA can be performed continuously by cutting the grafted PLA produced during the process. However, PHBV and other PHA copolymers had significantly lower crystallization rates than PLA which made the continuous grafting reaction of PHBV impossible. As described in the experimental section, the resulting grafted PHBV strands had to be cooled for several minutes to allow them to solidify.

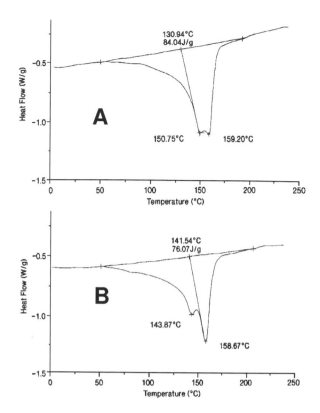

Figure 9. DSC traces of PHBV (A) and grafted PHBV (B).

Table 2. Thermal Properties of PHBV and Grafted PHBV

<u>DSC Analysis of PHBV and HEMA Grafted PHBV</u>

Sample I.D.	Melting peak-1 (° C.)	Melting peak-2 (° C.)	Enthalpy of melting (J/g)
Control	150.75	159.20	84.04
g-PHBV-1	143.87	158.67	76.06
g-PHBV-2	148.37	157.99	74.53

Even though PHBV was quite soft and elastic in melt state, the solidified PHBV strands were quite brittle making them difficult to cut into pellets.

To overcome this process challenge, experiments were conducted by grafting of PHBV in the presence of PLA, i.e. a blend of PHBV/PLA was used as a polymer substrate mixture. Co-grafting experiments of PHBV/PLA (50/50) were conducted on a HAAKE twin screw extruder, at a temperature profile of 170, 200,

190, and 190 °C respectively (*27*). The screw speed was 150 rpm, PLA feeding rate was 5.0 lb/hr (2270 g/hr). HEMA and peroxide rates were 0.5 lb/hr (227 g/hr) and 0.025 lb/hr (11.4 g/hr). The resulting grafted PHBV/PLA was found to solidify at much faster rate than pure grafted PHBV. As such, a continuous grafting reaction was achieved.

Besides polar functional monomer HEMA, a less polar butyl acrylate was also grafted onto PHBV/PLA (50/50) at a rate of 8.7 lb/hr (1950 g/hr) (*27*).

Conclusions

Bio-based polylactic acid (PLA) and a microbial polyhydroxyalkanoate (PHA)-polyhydroxybutyrate-co-valerate (PHBV) were grafted with a polar vinyl monomer or a non-polar vinyl monomer under a continuous reactive extrusion conditions. The reactive extrusion grafting reaction is a green reaction process which is conducted in the melt phase, under intensive shear, heat, and the action of a free radical initiator. Both the bio-based and biodegradable polymers had changed properties, resulting in reduced melt viscosity, reduced melt peak temperatures, and decreased enthalpy of melting. The grafted PLA was found to have improved compatibility with polar polymers such as polyvinyl alcohol, the increased polarity of the hydroxyl functionalized PLA had significantly improved dispersion in both the size of dispersed grafted PLA phase and the uniformity of dispersed phase sizes. Improved melt processability of grafted PLA/PVOH was found over PLA/PVOH, resulting in better fiber spinning processability as well as improved color of the resulting fibers from the polymer blends.

Acknowledgments

The authors would like to thank Gregory Wideman for his assistance in the reactive extrusion process.

References

1. Stevens, E. S. *Green Plastics*; Princeton University Press: Princeton, 2002; p 3.
2. Stevens, C.; Verhe, R. *Renewable Bioresources*; John Wiley & Sons: 2004.
3. Berins, M. L., Ed. *Plastics Engineering Handbook of the Society of the Plastics Indutsry, Inc.*; Van Nostrand Reinhold: New York, 1991.
4. Schut, J. H. *Plast. Technol.* **2008** (Feb), 62.
5. Mapleston, P. *Plast. Eng.* **2008** (Jan), 9.
6. Lunt, J. *Polym. Degrad. Stab.* **1998**, *59*, 145.
7. Zhang, J. F.; Sun, X. In *Biodegradable Polymers for Industrial Applications*; Smith, R., Ed.; CRC Press: Boca Raton, 2005; p 251.
8. Doi, Y. *Microbial Polyesters*; Wiley-VCH: 1990.
9. Mobley, D. P., Ed. *Plastics from Microbes: Microbial Synthesis of Polymers and Polymer Precursors*; Hanser: Munich, 2005.

10. Brown, S. B.; Orlando, C. M. Reactive Extrusion. *Encyclopedia of Polymer Science and Engineering* **1988**, *14*, 169.
11. Lambla, M. *Polym. Process Eng.* **1988**, *5*, 297.
12. Tzoganakis, C. *Adv. Polym. Technol.* **1989**, *9*, 321.
13. Moad, G. *Prog. Polym. Sci.* **1999**, *24*, 81.
14. Scott, H. G. U.S. Patent 3,646,155, 1972.
15. Ultsch, S.; Fritz, H. G. *Plast. Rubber Process. Appl.* **1990**, *13* (2), 81.
16. Xantho, M., Ed. *Reactive Extrusion: Principles and Practice*; Hanser: Munich, 1992.
17. Wang, J. H.; Schertz, D. M. U.S. Patent, 5,952,433, 1999.
18. Wang, J. H.; Schertz, D. M. U.S. Patent 5,945,480, 1999.
19. Wang, J. H.; Schertz, D. M. U.S. Patent 6,579,934 B1, 2003.
20. Wang, J. H.; Schertz, D. M. U.S. Patent 6,500,897 B2, 2002.
21. Wang, J. H.; Schertz, D. M. U.S. Patent 6,107,405, 2000.
22. Wang, J. H.; Schertz, D. M. U.S. Patent 6,297,326 B1, 2001.
23. Azizi, H.; Ghasemi, I. *Polym. Test.* **2004**, *23*, 107.
24. Wang, J. H.; Schertz, D. M. U.S. Patent 6,117,947, 2000.
25. Wang, J. H.; Schertz, D. M.; Soerens, D. A. U.S. Patent 6,172,177 B1, 2001.
26. Wang, J. H.; Schertz, D. M. U.S. Patent 6,664,333 B2, 2003.
27. Wang, J. H.; Schertz, D. M. U.S. Patent 7,053,151 B2, 2006.

Epilogue

What have we learnt about green chemistry?
   Cleaner and better products being sold,
   Improved environment being foretold,
And a new image for our industry.
It'll be a display of our artistry
   To fix up scars and smudges of the old,
   And to chart a new vista, brash and bold,
That stretches from healthcare to forestry.
Indeed the world is beautiful when green;
   Observe the trees and leaves in nature's store:
They dance and wave, looking lovely and clean.
   If they stay green, we would enjoy them more.
Since such a bright future can be foreseen,
   Let's work together with esprit de corps!

H. N. Cheng

March 2010

Indexes

Author Index

Subject Index

A

α,ω-Telechelic poly(ε-caprolactone) diols, 229*f*
Acetonide, 194*t*
Acinetobacter sp., 345*t*, 347
Acyl-enzyme complex, 273*f*
Alcohol and IA-Me, reactivity, 248*t*
Alcoholysis reaction
  PBS, 433*f*
  polylactic acid, 430*f*
Aliphatic polyesters, 425
Alkaline soluble polysaccharide
  compositon analysis, 79*t*
  Mark-Houwink plot, 81*f*
  molecular properties, 80*t*, 82*t*
  neutral sugar recovery, 80*t*
  weight percentage recovery, 79*t*
Amphiphilic conetwork, 23*f*
Amycolatopsis sp., 410*f*
Antigen-responsive hybrid hydrogel, 24*f*
AoC, 147*f*
Aptamer–hemin complexes, 121*t*
Aptazymes, 120*f*, 121*t*
ASP. *See* Alkaline soluble polysaccharide
ASP I, 83*f*
ASP II, 84*f*
Aspergillus oryzae cutinase. *See* AoC
Azidohomoalanine, 127*s*

B

Bacterial cells
  and 4-ketovaleric acid, 164*t*
  and valeric acid, 164*t*
Bacterial polysaccharides
  biosynthesis, 284
  remodeling, 291
Baeyer-Villiger monooxygenases. *See* BVMO
B antigen, 286*f*
BD
  lipase CA, 244*f*
  lipase PS-D, 239*s*
  polycondensation, 239*s*
  polymer yield, 241*f*, 244*f*
β-Hydroxy-2-ketones, 363*t*
Bicyclo[3.2.0]hept-2-en-6-one, 366*f*
Biobased elastomer, 237
Biobased materials

lipids, 5
polysaccharides, 5
triglycerides, 5
Biocatalysis, 1, 8*t*
  and polymers, 201
Biocatalysts, 2, 6
Biocatalytic redox polymerizations, 7
Biofabrication, 35
Biomaterials, 1, 6
Biopolymers, 35, 37
Biotransformations, cutinase, 141, 152*f*
β-Lactams
  activation, 273*f*
  Cal-B mediated polymerization, 272*f*
*Bo*PET, 389*f*
β-Propiolactam
  enzymatic ring-opening polymerization, 268*f*
Bulk urea crystals, 67*f*
Butane-1,4-diol. *See* BD
BVMO, 345*t*
  β-hydroxy-2-ketones, oxidation, 363*t*
  bicyclo[3.2.0]hept-2-en-6-one, oxidation, 366*f*
  crystal structures, 349*f*
  4-hydroxy-2-ketones, 360*f*
  oxidations, 363*t*
  type I, 349*f*
  Xanthobacter sp., 360*f*

C

C_{16}, C_{18} epoxy fatty acids, 143*f*
C_{16}, C_{18} ω-hydroxy fatty acids, 143*f*
CALB, 267*f*
  with AOT, 377*s*
  dimerization, 128*s*, 129*f*, 132*f*
  dimers, 128*s*
  hydrolytic activity, 129*f*
  methionine, 130*f*
  mutants, 130*t*
  and polymerization, 272*f*
CALB embedded PCL, 378*f*
  FITC tagged CALB, distribution, 382*f*
  vs. external addition, 379*t*
  PCL films, 380*f*
Candida antarctica lipase B. *See* CalB
Cationic polymerization, soybean oils, 88
$CDCl_3$, 246*f*
Chitosan

461

V

Valeric acid, 164*t*, 165*t*
Vegetable oils
  bioplastics and biocomposites, 88, 91, 93
  structure, 89*s*
  waterborne polyurethane dispersions, 95
Vinyl methacrylate. *See* VMA
VMA
  transesterification, 419*f*, 422*f*

W

Waterborne polyurethane dispersions, 97*s*
ω-Pentadecalactone. *See* PDl
Wzy, 292*f*

X

Xanthobacter sp., 360*f*

Printed in the USA/Agawam, MA
June 28, 2012

567076.036